Microorganisms for Sustainability

Volume 37

Sagar Chhabra • Ram Prasad •
Naga Raju Maddela • Narendra Tuteja
Editors

Plant Microbiome for Plant Productivity and Sustainable Agriculture

Springer

Editors
Sagar Chhabra
Department of Biosciences
School of Basic and Applied Sciences
Galgotias University
Greater Noida, Uttar Pradesh, India

Ram Prasad
Department of Botany
Mahatma Gandhi Central University
Motihari, Bihar, India

Naga Raju Maddela
Instituto de Investigación
Universidad Técnica de Manabí
Manabí, Ecuador

Narendra Tuteja
Plant Molecular Biology Group
International Centre for Genetic Engineering
New Delhi, India

ISSN 2512-1901 ISSN 2512-1898 (electronic)
Microorganisms for Sustainability
ISBN 978-981-19-5028-5 ISBN 978-981-19-5029-2 (eBook)
https://doi.org/10.1007/978-981-19-5029-2

This Springer imprint is published by the registered company Springer Nature Singapore Pte Ltd.
The registered company address is: 152 Beach Road, #21-01/04 Gateway East, Singapore 189721, Singapore

Preface

Plants are colonized by a multitude of microorganisms, collectively called "microbiome." In this new era of ecology knowing plants at par with microbiome is essential to address the growing demand and for sustainable agriculture. Microbiomes associate with plants as commensals, symbionts, or pathogens and are associated with multifarious functions such as nutrient acquisition and augmenting plant defense against biotic and abiotic stresses in many cases. These microbiomes occupy various niches in plants such as rhizosphere, root, seed, stem, and leaves. Global agriculture is facing urgent challenges to sustainable food production and to meet its growing demand. The quest to harness the potential of useful microorganisms from ecological niches has shown many interesting results.

The book on *Plant Microbiome for Plant Productivity and Sustainable Agriculture* features 18 chapters written by numerous renowned experts. The chapters provide a comprehensive knowledge and understanding on the topic. The diversity, distribution, functional potential of plant microbiota, and its associations with plants in agroecosystems are described. The various technological constraints, approaches, challenges to study microorganisms and prospects of using microbiomes for agriculture and sustainable production, and environment sustainability concepts are described in this book.

Chapter 1 describes fungal microbiomes and plant growth-promoting potential and opportunities for agriculture. Chapters 2 and 3 describe the rhizosphere and root microbiomes and functional traits that are essential for plants productivity. Chapter 4 covers nanotechnology and prospects of microbiomes for sustainable agriculture. Chapters 5 and 6 cover changes in plant microbiome and functional potential of microbiomes in response to abiotic stress. Chapters 7, 8, and 9 cover beneficial plant-microbe associations, and their diversity, distribution, and decoding using latest techniques. Chapters 10, 11, and 12 highlight the essential functional traits and function of Bacillus and Streptomyces for the management of biotic stresses in plants and the omic route to utilize endophytes and their functional potentials in plant growth promotion. Chapters 13, 14, 15, and 16 describe the functional traits of microbiomes in siderophore production in iron uptake and plant biofortification, plant microbiome diversity and potential for crops, endophytic microbiomes in plant hormone production, and endophyte microbiomes with potential in phosphate solubilization and phytoremediation of degraded lands. Chapters 17 and 18 cover the molecular mechanisms of plant probiotics for the improvement of soil health and plant microbiomes and the sustainability concepts for agriculture and in enhancing the productivity in agroecosystems.

Most of the chapters also contain technical advancements which will be very useful for microbiologists, agriculturalists, environmentalists, and Ph.D. students. The book will be of interest to industrial experts, policymakers, and those interested in environmental stewardship. The editors and contributing authors of this book make an excellent attempt to bring together a deep understanding from the primitive to the modern concepts about microorganism and their importance for crop improvement. This volume is expected to attract professionals from all over the world.

The editors are thankful to Akansha Tyagi, Senior Editor, and Jayesh Kalleri, Production Editor at Springer Nature, for their valuable help in formatting and incorporating editorial changes in the manuscripts.

Greater Noida, Uttar Pradesh, India — Sagar Chhabra
Motihari, Bihar, India — Ram Prasad
Manabí, Ecuador — Naga Raju Maddela
New Delhi, India — Narendra Tuteja

Contents

Editors and Contributors

About the Series Editor

Naveen Kumar Arora is Professor and Head, Department of Environmental Science at Babasaheb Bhimrao Ambedkar University (A Central University), Lucknow, Uttar Pradesh, India. He is a renowned researcher in the field of environmental microbiology and biotechnology. His specific area of research is plant-microbe interactions, particularly plant growth-promoting rhizobacteria. He has more than 75 research articles published in premium international journals and several articles published in magazines and dailies. He is an editor of 25 books, published by Springer. He is a member of several national and international societies, Secretary General of Society for Environmental Sustainability, on the editorial board of four journals, and a reviewer of several international journals. He is also the editor in chief of the journal *Environmental Sustainability* published by Springer Nature. He has delivered lectures in conferences and seminars around the globe. He has a long-standing interest in teaching at the PG level and is involved in taking courses in bacteriology, microbial physiology, environmental microbiology, agriculture microbiology, and industrial microbiology. He has been advisor to 134 postgraduate and 11 doctoral students. He has been awarded for excellence in research by several societies and national and international bodies/organizations. Although an academician and researcher by profession he has a huge obsession for the wildlife and its conservation and has authored a book, *Splendid Wilds*. He is the President of Society for Conservation of Wildlife and has a dedicated website www.naveenarora.co.in for the cause of wildlife and environment conservation.

About the Editors

Sagar Chhabra received his Ph.D. in Environmental Microbiology from the South East Technological University, Ireland, and has several years of postdoctoral experience working in Canada and Ireland in the field of agricultural and environmental microbiology and biotechnology. Dr. Chhabra is currently working as an Assistant Professor (Grade III) in the Department of Biosciences at Galgotias University, India, and previously worked as a Senior Researcher and Scientist in an industry in Ireland working on microbial product development for plant growth promotion and environmental clean-up. Dr. Chhabra has published several papers in the areas of plant and soil microbiomes and microbial ecology and has several national and international fellowships, travel grants, and other awards as a researcher.

Ram Prasad, Ph.D. is associated with the Department of Botany, Mahatma Gandhi Central University, Motihari, Bihar, India. His research interest includes applied and environmental microbiology, plant-microbe interactions, sustainable agriculture, and nanobiotechnology. Dr. Prasad has more than two hundred fifty publications to his credit, including research papers, review articles, and book chapters and seven patents issued or pending, and edited or authored several books. Dr. Prasad has thirteen years of teaching experience and has been awarded the Young Scientist Award and Prof. J.S. Datta Munshi Gold Medal by the International Society for Ecological Communications; FSAB fellowship by the Society for Applied Biotechnology; the American Cancer Society UICC International Fellowship for Beginning Investigators, USA; Outstanding Scientist Award in the field of Microbiology; BRICPL Science Investigator Award and Research Excellence Award, Fellow of The Biotech Research Society, India, etc. He has been serving as an editorial board member of *BMC Microbiology*, *BMC Biotechnology*, *Current Microbiology*, *Archives of Microbiology*, *Annals of Microbiology*, *IET Nanobiotechnology*, *Journal of Nanomaterials*, *Journal of Renewable Material*, *Archives of Phytopathology and Plant Protection*,

Journal of Agriculture and Food Research, *Environmental Science & Ecotechnology* including Series editor of Nanotechnology in the Life Sciences, Springer Nature, USA. Previously, Dr. Prasad served as Assistant Professor at Amity University Uttar Pradesh, India; Visiting Assistant Professor, Whiting School of Engineering, Department of Mechanical Engineering at Johns Hopkins University, Baltimore, United States; and Research Associate Professor at School of Environmental Science and Engineering, Sun Yat-sen University, Guangzhou, China. Citation: >11,500, h-index: >53, i10-index: >164.

Naga Raju Maddela received his M.Sc. (1996–1998) and Ph.D. (2012) in Microbiology from Sri Krishnadevaraya University, Anantapuramu, India. During his doctoral program in the area of Environmental Microbiology, he investigated the effects of industrial effluents/insecticides on soil microorganisms and their biological activities and worked as a Faculty in Microbiology for 17 years, teaching undergraduate and postgraduate students. He received "Prometeo Investigator Fellowship" (2013–2015) from Secretaría de Educación Superior, Ciencia, Tecnología e Innovación (SENESCYT), Ecuador, and "Postdoctoral Fellowship" (2016–2018) from Sun Yat-sen University, China. He also received external funding from "China Postdoctoral Science Foundation" in 2017, internal funding from "Universidad Técnica de Manabí" in 2020, worked in the area of Environmental Biotechnology, participated in 20 national/international conferences, and presented research data in China, Cuba, Ecuador, India, and Singapore. Currently, he is working as a full-time Professor at the Facultad de Ciencias de la Salud, Universidad Técnica de Manabí, Portoviejo, Ecuador. He has published 7 books (Springer), 15 chapters (InTech Open/Springer), and 50 research papers.

Narendra Tuteja An elected fellow of numerous national and international science academies, Dr. Narendra Tuteja is former Director and Professor at Amity Institute of Microbial Technology, NOIDA, India, and former Group leader and Senior Scientist at International Centre for Genetic Engineering & Biotechnology (ICGEB), New Delhi, India. He has made significant contributions to crop improvement under adverse conditions, reporting the first helicase from plant and human cells and demonstrating new roles of Ku autoantigen, nucleolin, and eIF4A as DNA helicases. Furthermore, he discovered novel functions of helicases, G proteins, CBL-CIPK and LecRLK in plant stress tolerance, and PLC and MAP kinase as effectors for Gα and Gβ G proteins. Narendra Tuteja also reported several high salinity stress tolerant genes from plants and fungi and developed salt/drought-tolerant plants. Recently Dr. Narendra Tuteja developed and published the marker-free and reporter-free PDH45 transgenic rice by salt screening. Citation: >35,000, h-index: >75, i10-index: >292.

Contributors

Angela T. Alleyne Department of Biological and Chemical Sciences, Faculty of Science and Technology, The University of the West Indies, Cave Hill Campus, Bridgetown, Barbados

R. Anandham Department of Agricultural Microbiology, Tamil Nadu Agricultural University (TNAU), Coimbatore, Tamil Nadu, India

Hira Aslam Department of Bioinformatics and Biotechnology, Government College University Faisalabad, Faisalabad, Pakistan

Mohd Fadzelly Abu Bakar Faculty of Applied Sciences and Technology, Universiti Tun Hussein Onn Malaysia (UTHM), Johor, Malaysia

Pooja Bhadrecha University Institute of Biotechnology, Chandigarh University, Mohali, Punjab, India

Priyanka Bhati Department of Biosciences, School of Basic and Applied Sciences, Galgotias University, Greater Noida, Uttar Pradesh, India

Rahul Chandnani Global Institute for Food Security, University of Saskatchewan, Saskatoon, SK, Canada

Shivani Chaudhary Institute of Environment and Sustainable Development, Banaras Hindu University, Varanasi, Uttar Pradesh, India

Sagar Chhabra Department of Biosciences, School of Basic and Applied Sciences, Galgotias University, Greater Noida, Uttar Pradesh, India

Nisha Choudhary Department of Botany, Central University of Jammu, Samba, Jammu and Kashmir, India

Farhana Tasnim Chowdhury Department of Biochemistry and Molecular Biology, Faculty of Biological Sciences, University of Dhaka, Dhaka, Bangladesh

Manoharan Devaprakash Department of Molecular Microbiology, School of Biotechnology, Madurai Kamaraj University, Madurai, Tamil Nadu, India

Nabin Kumar Dhal Environment Sustainability Department, CSIR-Institute of Minerals and Materials Technology, Bhubaneswar, Odisha, India

Jahanvi Ganotra Department of Botany, Central University of Jammu, Samba, Jammu and Kashmir, India

Dipita Ghosh Ecological Restoration Laboratory, Department of Environmental Science & Engineering, Indian Institute of Technology (ISM), Dhanbad, Jharkhand, India

N. O. Gopal Department of Agricultural Microbiology, Tamil Nadu Agricultural University (TNAU), Coimbatore, Tamil Nadu, India

Subramaniam Gopalakrishnan International Crops Research Institute for the Semi-Arid Tropics (ICRISAT), Patancheru, Telangana, India

Charu Gupta Amity Institute of Herbal Research and Studies, Amity University-Uttar Pradesh, Noida, Uttar Pradesh, India

Hitakshi Gupta Department of Botany, Central University of Jammu, Samba, Jammu and Kashmir, India

Mishaal Irfan Department of Bioinformatics and Biotechnology, Government College University Faisalabad, Faisalabad, Pakistan

Mohammad Riazul Islam Department of Biochemistry and Molecular Biology, Faculty of Biological Sciences, University of Dhaka, Dhaka, Bangladesh

Bhawna Jangra University Institute of Biotechnology, Chandigarh University, Mohali, Punjab, India

Chaitanya Kumar Jha Microbiology Department, Gujarat Arts and Science College, Ahmedabad, Gujarat, India

I. Johnson Department of Plant Pathology, Tamil Nadu Agricultural University (TNAU), Coimbatore, Tamil Nadu, India

Nishra Joshi Microbiology Department, Gujarat Arts and Science College, Ahmedabad, Gujarat, India

Haseena Khan Department of Biochemistry and Molecular Biology, Faculty of Biological Sciences, University of Dhaka, Dhaka, Bangladesh

Worapong Kitdamrongtham Faculty of Engineering and Technology, North-Chiang Mai University, Chiang Mai, Thailand

Leon V. Kochian Global Institute for Food Security, University of Saskatchewan, Saskatoon, SK, Canada

R. Krishnamoorthy Department of Crop Management, Vanavarayar Institute of Agriculture, Pollachi, Tamil Nadu, India

Deepak Kumar Centre for Advanced Study in Botany, Institute of Science, Banaras Hindu University, Varanasi, Uttar Pradesh, India

Jitendra Kumar Department of Biosciences, UIBT, Chandigarh University, Mohali, Punjab, India

Mahendra Kumar Institute of Environment and Sustainable Development, Banaras Hindu University, Varanasi, Uttar Pradesh, India

Manish Kumar Environment Sustainability Department, CSIR-Institute of Minerals and Materials Technology, Bhubaneswar, Odisha, India

Pankaj Kumar Department of Microbiology, Dolphin (PG) Institute of Biomedical and Natural Science, Dehradun, Uttarakhand, India

Rajaram Shyam Kumar Department of Biotechnology, Kamaraj College of Engineering and Technology, Virudhunagar, Tamil Nadu, India

Naga Raju Maddela Departamento de Ciencias Biológicas, Facultad de Ciencias de la Salud, Universidad Técnica de Manabí, Portoviejo, Ecuador

Faisal Mahmood Department of Environmental Science and Engineering, Government College University Faisalabad, Faisalabad, Pakistan

Subodh Kumar Maiti Ecological Restoration Laboratory, Department of Environmental Science & Engineering, Indian Institute of Technology (ISM), Dhanbad, Jharkhand, India

A. Manikandan Department of Agricultural Microbiology, Tamil Nadu Agricultural University (TNAU), Coimbatore, Tamil Nadu, India

B. S. Manisha Singh Environment Sustainability Department, CSIR-Institute of Minerals and Materials Technology, Bhubaneswar, Odisha, India
Academy of Scientific and Innovative Research (AcSIR), CSIR-Human Resource Development Centre, (CSIR-HRDC) Campus, Ghaziabad, Uttar Pradesh, India

Awais Maqsood Department of Bioinformatics and Biotechnology, Government College University Faisalabad, Faisalabad, Pakistan

Shweta Mehta Department of Biosciences, School of Basic and Applied Sciences, Galgotias University, Greater Noida, Uttar Pradesh, India

P. K. Mukherjee Nuclear Agriculture and Biotechnology Division, Bhabha Atomic Research Centre, Trombay, Mumbai, India

Vikas Pahal Department of Microbiology, Dolphin (PG) College of Science and Agriculture, Fatehgarh Sahib, Punjab, India

Ajay Kumar Pandey Department of Life Sciences and Biotechnology, Chhatrapati Shahu Ji Maharaj University (CSJMU), Kanpur, Uttar Pradesh, India

Shiv Shankar Pandey Department of Biotechnology, CSIR-IHBT, Palampur, Himachal Pradesh, India

Tirth Bhargavbhai Patel Biotechnology Department, Shri A.N. Patel PG Institute of Science and Research, Anand, Gujarat, India

Nikita Pathania Department of Botany, Central University of Jammu, Samba, Jammu and Kashmir, India

Laurent Penet INRAE, UR ASTRO, Petit-Bourg, Guadeloupe, France

Dhan Prakash Amity Institute of Herbal Research & Studies, Amity University-Uttar Pradesh, Noida, Uttar Pradesh, India

Ram Prasad Department of Botany, School of Life Sciences, Mahatma Gandhi Central University, Motihari, Bihar, India

Vishal Prasad Institute of Environment and Sustainable Development, Banaras Hindu University, Varanasi, Uttar Pradesh, India

R. Raghu Department of Agricultural Microbiology, Tamil Nadu Agricultural University (TNAU), Coimbatore, Tamil Nadu, India

Reia Rani Department of Botany, Central University of Jammu, Samba, Jammu and Kashmir, India

Janki A. Ruparelia Microbiology Department, Gujarat Arts and Science College, Ahmedabad, Gujarat, India

Pratyusha Sambangi International Crops Research Institute for the Semi-Arid Tropics (ICRISAT), Patancheru, Telangana, India

Meenu Saraf Department of Microbiology and Biotechnology, University School of Science, Gujarat University, Ahmedabad, Gujarat, India

M. Senthil Kumar Agricultural College and Research Institute, Tamil Nadu Agricultural University (TNAU), Eachangkottai, Tamil Nadu, India

Muhammad Shahid Department of Bioinformatics and Biotechnology, Government College University Faisalabad, Faisalabad, Pakistan

Ajay Shankar Institute of Environment and Sustainable Development, Banaras Hindu University, Varanasi, Uttar Pradesh, India

Aahuti Sharma Department of Biotechnology, CSIR-IHBT, Palampur, Himachal Pradesh, India

Divya Singh Plant Stress Biology Laboratory, Institute of Environment and Sustainable Development, Banaras Hindu University, Varanasi, Uttar Pradesh, India

Natesan Sivakumar Department of Molecular Microbiology, School of Biotechnology, Madurai Kamaraj University, Madurai, Tamil Nadu, India

Songita Sonowal Department of Botany, School of Life Sciences, Mahatma Gandhi Central University, Motihari, Bihar, India

Vadlamudi Srinivas International Crops Research Institute for the Semi-Arid Tropics (ICRISAT), Patancheru, Telangana, India

Neerja Srivastava Department of Life Sciences and Biotechnology, School of Sciences, CSJM University, Kanpur, India

Sudhakar Srivastava Plant Stress Biology Laboratory, Institute of Environment and Sustainable Development, Banaras Hindu University, Varanasi, Uttar Pradesh, India

Damini Supolia Department of Botany, Central University of Jammu, Samba, Jammu and Kashmir, India

Syeda Khola Tazeen Department of Bioinformatics and Biotechnology, Government College University Faisalabad, Faisalabad, Pakistan

Ramachandran Thirumalaivasan Department of Molecular Microbiology, School of Biotechnology, Madurai Kamaraj University, Madurai, Tamil Nadu, India

Narendra Tuteja International Centre for Genetic Engineering, Plant Molecular Biology Group, New Delhi, India

Nazia Rifat Zaman Department of Biochemsitry, Primeasia University, Dhaka, Bangladesh

Chapter 1
Fungal Microbiomes: The Functional Potential for Plant Growth Promotion and Opportunities for Agriculture

Angela T. Alleyne and Laurent Penet

Abstract Fungal microbial communities in plant hosts, or the mycobiome, are critical to key agriculture functions for increasing plant growth and improving disease management. In this chapter, the fungal microbiome is explored for its use as a tool in disease diagnosis, agroecology, and disease management in modern agriculture. Molecular markers and specifically the internal transcribed regions (ITS) are examined in their use as discriminants of community members in the mycobiome. While widely used, the ITS is sometimes incapable of resolving taxonomic complexities; thus, mycobiome analysis may require the use of several additional barcode genes in fungi. Further, the impact of abiotic and biotic stresses are examined and we discuss the application of microbiome therapy or using microbial communities as a tool in combatting plant diseases while also improving soil health.

Keywords Mycobiome · Fungi · ITS · Agroecology

1.1 Introduction

Living organisms are currently being re-examined in a wholistic manner as holobionts and hologenomes in relation to their microbiomes (Bordenstein and Theis 2015; Berg et al. 2020). Therefore, microbial communities or metagenomes have been recognized as a second genome of multicellular eukaryotes and part of their extended phenotype. Indeed, they may have effects beyond the organism that influence fitness. In plants, they play beneficial roles in plant growth and development, and stress responses in their hosts (Andreote et al. 2014; Patel et al. 2015; Zhang et al. 2021;

A. T. Alleyne (✉)
Department of Biological and Chemical Sciences, Faculty of Science and Technology, The University of the West Indies, Cave Hill Campus, Bridgetown, Barbados
e-mail: angela.alleyne@cavehill.uwi.edu

L. Penet
INRAE, UR ASTRO, Petit-Bourg, Guadeloupe, France

S. Chhabra et al. (eds.), *Plant Microbiome for Plant Productivity and Sustainable Agriculture*, Microorganisms for Sustainability 37,
https://doi.org/10.1007/978-981-19-5029-2_1

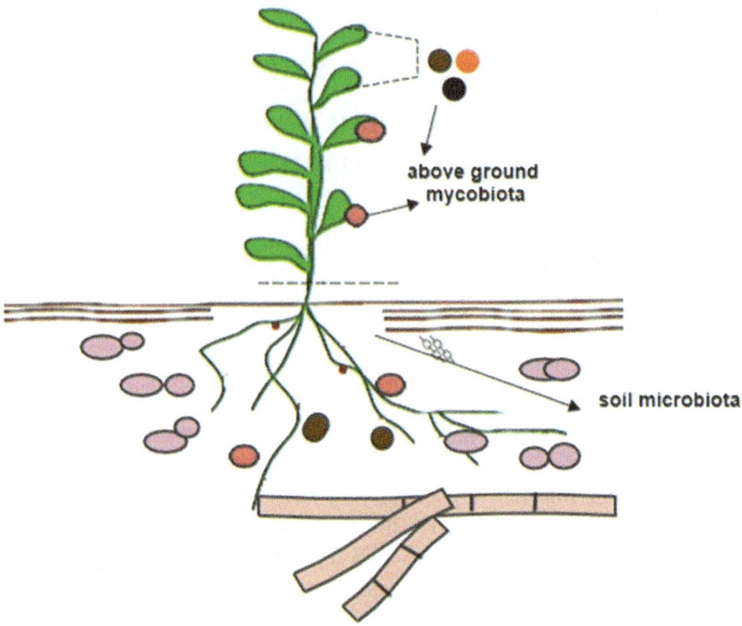

Fig. 1.1 The mycobiome: dynamic fungal community interactions in the phyllosphere and rhizosphere with its plant host

Prasad and Zhang 2022). Fungi perform several critical functions in their plant hosts in their roles as endosymbionts including biotrophic, necrotrophic, or saprophytic activities (Bonfante et al. 2019). Consequently, the fungal microbiome also known as the mycobiome (Fig. 1.1) and the genomes of endophytic fungal communities contribute to our understanding of major fields in agriculture such as agroecology and plant pathology (Gao et al. 2021; Meunier and Bayır 2021).

One of the major applications of metagenomics is in disease diagnosis where researchers can identify disease-causing agents that cannot be detected by traditional methods (Raja et al. 2017). Metabarcoding has also shown that plant diseases are sometimes more biologically complex than initially thought and disease symptoms can be caused by a complex of microbes (Raja et al. 2017; Nilsson et al. 2019). As such, the application of fungicides and other pest control agents can irreparably damage and remove beneficial organisms in the microbiome. These side effects can be reduced with a greater understanding of plant microbiomes, especially in fruits, vegetables, and staple cereal crops such as wheat, rice, maize, yam, and cassava (Clay and Schardl 2002; Edwards et al. 2015; Frediansyah 2021), in addition to understanding the impact of the fungal microbiome or mycobiome in soils (Dubey et al. 2019; Chen et al. 2021; Gao et al. 2021). Harnessing this major resource using biotechnology and bioengineering is therefore important for producing metabolites, improving crop yields, enhancing precision disease diagnosis, and providing alternatives to pesticide treatment in crops (Compant et al. 2019; Noman et al. 2021). In this chapter, we examine the impact of the mycobiome in plant disease and agro-ecological approaches to disease management, with applications for sustainable agriculture.

1.2 The Fungal Microbiome (Mycobiomes)

The plant mycobiome represents a dynamic network of fungal communities found throughout the plant holobiont (Fig. 1.1). Its importance is considered from the root systems and soil health, to the phyllosphere or plant leaf system and plant immunity. While the structure and composition of the mycobiome is important for its functional aspect in promoting plant growth, recruitment and organizing of the holobiontic plant system are also important characteristics left to be fully explored.

The vast majority of plant fungi in the plant holobiont are beneficial microorganisms (Pagano et al. 2017). The incredible speed and read length produced by modern NGS technologies are useful tools in probing the plant's mycobiome and have been used in the field of diagnostics, while aiding agronomic practices (Hartman et al. 2018). Additionally, the interaction of wild plants with cultivated varieties are being re-examined in terms of their microbiome characteristics and potential for contributing to local microbial ecology. As such, the communities of fungi among wild plants are more abundant than those of cultivated plants (Ma et al. 2021). Harnessing endosymbiotic microbiota for their benefits as an application in smart agriculture applications (Hartman et al. 2018) is therefore considered here.

1.2.1 Molecular Markers and Fungal Metagenomics in Agriculture

Sequence-based metagenomics provides information on species composition by either sequencing, annotating, and assembling all the genetic material obtained from mixed samples, or by sequencing targeted gene regions in fungi (Ahmed 2016; Noman et al. 2021). Generally, when used in disease diagnosis, molecular markers are based on a range of various hybridization techniques which are genome specific, sometimes are more precise than direct observations of symptoms, and less time-consuming and limiting than microbial cell culture (Schoch et al. 2012; Raja et al. 2017; Hariharan and Prasannath 2021). Polymerase chain reaction (PCR) and next-generation sequencing (NGS) primers that target the rapidly evolving and highly repetitive ribosomal DNA (rDNA) genes that transcribe rRNA and their internal transcribed regions (ITS 1 and 2) are widely used to identify fungi at both genus and species levels (Schoch et al. 2012; Crous et al. 2015; Raja et al. 2017; Tedersoo et al. 2018). The full ITS region is approximately 600–650 base pairs long and ITS1 and ITS2 are separated from each other by the 5.8 S rRNA (Schoch et al.

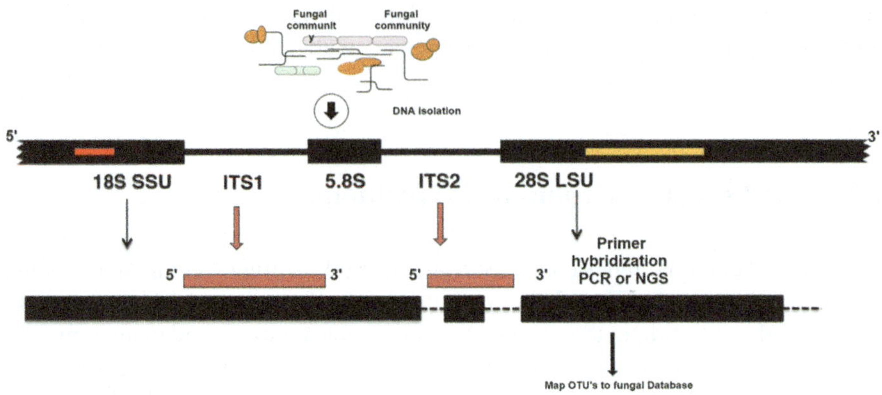

Fig. 1.2 Polymerase chain reaction amplication and next-generation sequencing of internal transcribed regions of fungal microbiome

2012; Edwards et al. 2017; Raja et al. 2017; Naranjo-Ortiz and Gabaldón 2019) (Fig. 1.2). Notwithstanding this, in some instances, hybridization to the ITS is insufficient in resolving taxonomic complexities in certain fungi, and therefore secondary barcode genes are necessary to reach accurate taxonomic allocation or identification. These include protein coding genes such as glyceraldehyde-3 phosphate dehydrogenase (GADPH), fungal-specific translation elongation factors (*TEF*), beta tubulin (*TUB*), actin (*ACT*), and RNA polymerase II (RP2) (Schoch et al. 2012; Stielow et al. 2015). Moreover, the border regions of the ITS designated as the small subunit (SSU) or the large subunit (LSU) also provide further taxonomic discrimination of fungi (Edwards et al. 2017) (Fig. 1.2). Additionally, in several NGS studies of the plant mycobiome, ITS1 and ITS2 do not yield significant differences in taxa assignment based on operational taxonomic units (OTUs) (Blaalid et al. 2013), but ITS2 may provide better characterization of Ascomycota fungi than the ITS1 region (Scibetta et al. 2018).

Notably, besides targeted amplicon sequencing of molecular markers, shotgun metagenomics can be used to discern fungal taxa. Although more expensive than amplicon-based NGS, shotgun metagenomics removes PCR bias and aids in the discovery of novel or new lineages while also performing a quantitative function in microbiome studies (Tedersoo et al. 2015; Donovan et al. 2018; Nilsson et al. 2019). This is especially important in agriculture where knowledge of the extent of fungal infection is usually desired but difficult to ascertain through real-time PCR (qPCR) or other means that may not be readily available in field studies. Significantly, metagenomic studies using molecular markers usually yield large numbers of unidentified taxa (Cuadros-Orellana et al. 2013) many of which belong to unidentified Ascomycota, which is challenging in well-established crops, and even more so in less studied but important tropical crops such as the roots and tubers including yam, cassava, and sweet potato (Gao et al. 2013; Chen et al. 2021; Xiong et al. 2021). In fact, the partial genomes of these crops have only recently been reported (Prochnik et al. 2012; Tamiru et al. 2017; Isobe et al. 2019). The incomplete genomic data for these important agricultural crops including wheat, rice, cassava, yam, and sweet potato remain an untapped resource for future development in agriculture research and plant health in general.

1.2.2 Core Mycobiomes and Plant Health

General plant health has been attributed to the plant microbiome and healthy or diseased status may be gleaned from an observation of specific organisms in the rhizosphere, phyllosphere, or endosphere of the plant (Toju et al. 2018b). Variation in the plant mycobiome in time and spatial compartments such as the phyllosphere, rhizosphere, endosphere, and seed have been described for several major agricultural crops such as wheat, maize, and rice (Clay and Schardl 2002; Kim and Lee 2020; Zheng et al. 2021), and tropical roots such as cassava (Frediansyah 2021; Zhang et al. 2021). Therefore, fungal communities are determined structurally by the developmental plant growth stage, soil types, as well as the plant species (Carbone et al. 2021; Zhang et al. 2021). However, fungal endophytes of tropical plant systems are less well known than those present in temperate grasses and shrubs. The latter are also less diverse, are transmitted horizontally, and contribute less to host defense than those found in woody angiosperms (Arnold et al. 2003).

From a developmental perspective, it has also been shown that root colonization by microorganisms rapidly spreads throughout the plant and may be predicted by functional roles assigned to plant–microbe interactions (de Souza et al. 2016; Toju et al. 2016, 2018b; Bonfante et al. 2019; Varma et al. 2019a, b). Generally, studies that examine commodity crops such as sugarcane and bananas have notable impacts on such processes as biofuel production, food security, human health, and development in countries heavily dependent on commodity exports (Yuan et al. 2021; Aslam et al. 2022). In a critical review of the core fungal microbiome in sugarcane, it was shown that yeast and yeast-like fungi dominate the various plant compartments and, in the rhizosphere (de Souza et al. 2016), species such as *Candida*, *Debaryomyces*, *Hanseniaspora*, *Meyerozyma*, *Wickerhamiella*, and *Zygosaccharomyces* made up approximately up to 12% of all fungal diversity in the sugarcane stalk (de Souza et al. 2016).

Suppressive soils rely heavily on studies of the plant microbiome. Disease suppression in soils occurs when resident microorganisms compete with and limit newer arrivals from occupying their niches, thereby coincidently reducing the appearance of disease in plant hosts (Hartman et al. 2018; Pascale et al. 2020). In a study on health status, it was found that the population of *Fusarium* decreases during reductive soil disinfection (RSD) and when exposed to aerobic conditions during replanting is more disease suppressive than in diseased soils (Huang et al. 2015).

Additionally, mycobiome studies have relied heavily on arbuscular mycorrhizae (AM) and their function in improving plant health. As such, the vast array of pathogenic and non-pathogenic fungi that make up the core mycobiomes of plants and soils remain to be further explored.

1.3 The Mycobiome and Sustainable Agriculture

1.3.1 Mycobiomes Boost Plant Growth

An overreliance on agricultural inputs such as fertilizers and pesticides has been detrimental to plant biodiversity and soil microbial diversity or soil health (Hartman et al. 2018). Sustainable agricultural practices such as tree-based intercropping and use of RSD make use of microbial communities to improve soil health and encourage improved plant growth (Compant et al. 2019; Tedersoo et al. 2020). Mycorrhizae act as a bridge in the rhizosphere to microbial communities, and with the roots of its plant hosts. Importantly, established networks of AM fungi mediate nutrient transfers between host and holobiont and function as significant nutrient trading partners in plant–plant interactions in agricultural systems (Mendes et al. 2013; Fitzpatrick et al. 2020; Tedersoo et al. 2020; Pozo et al. 2021; Varma et al. 2017a, b, c). It has also been shown through network analysis that usually a small group of organisms are key microbial players in the rhizosphere and phyllosphere and they control these extended symbiont–symbiont networks (Toju et al. 2016). Three processes are determined by these early colonizers and network organizers during establishment of the microbiome network: (1) functional species recruitment, (2) pathogen/pest blocking, and (3) core reinforcement (Toju et al. 2018a, b; Gao et al. 2021). Nonetheless, the mycobiome is also conditioned by several ecological factors that drive species composition including host plants, ecological drift, dispersal, and evolution (Fitzpatrick et al. 2020). Furthermore, it has been shown that abiotic factors such as drought stresses may effect changes in the root microbiome. As such, higher relative abundances in the fungal composition of the root endosphere were seen in response to black-foot disease (*Dactylonectria* and *Cylindrocarpon*) and the biocontrol agent *Trichoderma*, than in non-irrigated systems (Dubey et al. 2019; Carbone et al. 2021).

1.3.2 Plant Growth-Promoting Fungal Microbiomes in Disease Management

Plant microbiota do not only demonstrate growth effects on plants, but they also trigger plant resistance and increased protection against diseases. Resistance is probably increased via different paths. Some commensals induce natural plant immunity reactions despite not being pathogenic and causing no damage to plants, thus yielding baseline responses before pathogens arrive and threaten plant health.

This increase in resistance is thus passive and possibly incurs a defensive cost to the plant, especially if no further immunity-based reaction is required. Increasing induced immunity when no pathogen is eventually encountered is an energetic loss (Heil and Bostock 2002). When this loss involves only local collateral damage with casual benefits (pathogens eventually arrive), then the whole process is still advantageous. On the other hand, if immunity triggering commensals are numerous and co-exist on the whole individual, then the global cost for the plant might exceed its benefit, possibly at the cost of diverting ineffectual commitment to defense that could have been spent on growth or other functions (Martin 2001; Fitzpatrick et al. 2020). Little is known yet whether these effects translate into the global fitness balance for plants. Besides the plant's natural self-defense via antibiotic secretions, microbiota can extend plant protection whenever antibiosis is the result of commensal microbes themselves. This would lend to self-selecting communities (based on resistance genes to local compounds). Still, since it involves a classical arms race scheme for microbiota components, any microbe may become part of a local niche, including pathogens themselves whenever they evolve resistance to the local antibiosis conditions. Nevertheless, enhanced microbiota communities that would allow for pathogen control via local out-competition or even antibiosis are promising prospects in agriculture practice (Fitzpatrick et al. 2020).

As a matter of fact, application of useful microbiota will rely on careful choice, and precise and descriptive characterization of species in use, side effects, and potential pitfalls. Yet, sometimes benefits can be expected even from closely related co-existing organisms, and even when both behave as pathogens. We will illustrate here an example from *Colletotrichum* genus, a well-known group of fungi species-complexes often drastically pathogenic among many crops, worldwide. A study on weed hosting ability during yam intercropping revealed a tendency for two closely related species, namely *C. gloeosporioides* and *C. truncatum*, while co-existing at local scale on various weed species to segregate among different sub-communities within field vegetation. So, *C. gloeosporioides* was found more often on erect weeds, while *C. truncatum* was more frequent on vines and creeping species (Dentika et al. 2021). This segregation hinted at potential competition for the two fungi species. Yet, while *C. gloeosporioides* may be strongly pathogenic on yams (*Dioscorea alata*), *C. truncatum* is not considered a threat to this crop. Co-inoculation experiences comparing either admixing or prior inoculation by *C. truncatum* demonstrated that disease symptoms are aggravated in the first case, but were strongly reduced in the second (Dentika et al. 2021). Prior presence of *C. truncatum* was thus detrimental to the more pathogenic species on yams. Clearly, this means that competition between species can be used to decrease disease onset, and potentially attenuate epidemics on crops. Still, an overall benefit will depend on the agricultural context. In this case, *C. truncatum* is not pathogenic on yams, but the species complex is heavily damaging on other crops such as peppers (Than et al. 2008; Weir et al. 2012). Potential for control will therefore preclude its use where both yams and peppers are intercropped.

1.4 Agroecology, Sustainable Agriculture, and Fungal Microbiomes

Microbiota help in at least two major functions that are relevant to agriculture: helping plant growth via production of hormones and related compounds associated with nutrient uptake, and inhibiting pathogenic organisms, thus helping crops defend themselves against antagonists (Andreote et al. 2014; Bonfante et al. 2019; Compant et al. 2019). Both phylloplane microbiota (on aerial plant parts) and rhizosphere microbiota (soil microorganism communities around the roots) can thus significantly contribute to food production once we work out the best components and combinations from natural microbiota (Arif et al. 2020). Engineering optimal communities may also involve cycling components, as phyllosphere microbes can also further degrade plant litter and so improve cycling processes in the soil and humus (Mueller and Sachs 2015; Arif et al. 2020). Consequently, recent progress in important and widely cultivated crops such as rice has already identified microbe species of interest with high potential for increasing yield and most probably with culturing potential for future use in agriculture (Chen et al. 2021). Conversely, designing optimal microbiota combinations will require assessing their potential turnover both during cropping and intercropping, especially if rotation is involved and alternated crops do not respond optimally to the same communities, or if turnover is faster than crop duration. This issue may likely be less relevant to soil microbiota, which demonstrate more stability over time, given that some latency is expected in modification of the physicochemical properties of soil microbiota (Yuan et al. 2021). Additionally, aerial microbiota of plants are less resilient than their rhizosphere counterparts to meteorological conditions such as rain, temperature, humidity, and their daily fluctuations, and crop management as well (especially spraying). So, some level of turnover is expected, especially since dispersal from neighboring environments is facilitated by the lack of a physical barrier to the flow and arrival of natural airborne propagules. This drawback in the use of microbiota will also naturally be exacerbated by the nature of the microorganism: fungi are probably more resilient actors of the microbiota than bacteria are, both in soil and as endophytes in the phyllosphere, and therefore the very nature of their facilitation in microbiome networks will be impacted (Hartman et al. 2018; Compant et al. 2019).

Meanwhile, although microbiota will no doubt become a major tool in more sustainable agriculture systems, several challenges still need to be addressed before their use can be generalized at a large scale. The first and most obvious challenge is to delineate taxonomic contributions to ecosystemic services by microbial communities that would benefit cropping systems, and especially delineate potential differences in composition that will differ in diverse environments, especially temperate vs. tropical, but also potential asymmetries for sub-temperate/sub-tropical areas or humid vs. arid tropics. Of course, combining optimal components will be a major asset of the use of microbiota, but these might also prove different for the different plant organs, especially contrasting effects between leaves and flower structures, and even tubers or fruits (Müller et al. 2015). While the emergent and promising science of microbiome is growing exponentially in terms of studies and knowledgeable insights, and it highlights the many opportunities for deploying sustainable agriculture practices by potentially reducing an overreliance on chemistry, we still need to identify core microbial species, and core communities in combination with all the different conditions they will face, before the optimization process of communities may begin. The task is even more crucial because of climate change, as it intensifies and may already be contributing to altering potentially useful, but as yet unidentified microorganisms, or their current niches. In fact, climate change is suggested to significantly impact shifts in microbial communities (Dubey et al. 2019). Meanwhile, another dichotomy exists in the divide between monocrop-intensive agriculture and small-scale diversified agroecological farming systems, and therefore challenges in microbiota use will differ between these two facets of current agriculture systems. We will explore this dynamic in more details in the next section.

1.5 Opportunities for New Applications of Beneficial Fungal Communities to Improve Soils, Plant Growth, and Plant Health

Intensive agriculture, even within an ecologically intensive conceptual framework, will often consist of simplified cropping systems when compared to agroecological and agroforestry systems, and will thus more quickly benefit from applied microbiota tools when those become available for large-scale use. Nevertheless, issues highlighted previously will paradoxically also prove either more complex or easier to include within highly diversified small-scale farming, which is still currently in widespread use in agriculture systems worldwide. We will review here how enhanced microbiota approaches may translate within such systems, and whether a common framework would apply in both cases, or if alternative solutions should be worked out by the scientific community. There are grossly three levels for which both approaches to agriculture are so divergent that they might rely on different strategies for microbiota to be fully successful: the first deals with soil management and fertilization, the second is concerned with crop diversity on a local scale, and the last to the agronomic dark triad—weeds, pests, and diseases.

1.5.1 Soil Management and Fertilization

Currently, intensive agriculture mostly consists of broad-scale monocultures and has historically given modest attention to soil (and by extension soil health), which was viewed as an inert substrate and simple recipient for crops (Fedoroff 1987). Moreover, the concept of nutrient return in commercial agriculture is often optimized as a simple plant-specific formula to correctly balance and broadly compensate for agronomic inputs and to compensate for exports from harvested crops, and less frequently considers nutrient cycling via manure or other biotic factors. As a result, many cultivated soils are damaged and impoverished, suffer from high levels of run-off, and have become lifeless substrates on which efficient productivity will require widespread use of external inputs (Pankhurst and Lynch 1995). Meanwhile, the modern paradigm shift toward envisioning the soil as a living community of the microbiome, together with the older substrate or subtractive narrative, and our need to restock carbon lost from soils due to cultivated agriculture will provide us with new tools or approaches to make food production both sustainable and more environmentally friendly. Hence, provided specific regional mitigation, there are already steps toward soil stewardship involving regenerative soil microbial communities (Pagano et al. 2017; Carbone et al. 2021). These will involve increased cycling of organic matter, possibly through the local circular economy since soil resilience involves large quantities of fungi biomass and fungi often require high levels of organic matter (Pritsch and Garbaye 2011). While such fungi-based microbiota composition needs to be precisely worked out, researchers should focus on fungal taxa that are analogous in vegetative-stage evolution and create pioneer communities that lend itself to natural evolution toward locally resilient communities, helping both crops grow and resist diseases, while restoring prior soil health (Compant et al. 2019; Pozo et al. 2021). This is even more important than maintaining nutrient levels in healthy soils based solely on intake and release by microbiota.

Contrastingly, agroecological systems are already based on cycling biomass and enriching soils, be it in the form of compost use, green- and animal manure, mulching, or even in the more extreme forms of recycling trees (hugelkultur approach). Rich soils are known to be more resilient to drought and nutrient run-off. Compared to impoverished soils, the challenge becomes more that of trying to improve communities directly toward specifically optimized microbiota/plants interactions. While little is known about communal stability of artificial microbial assemblages, even less is known about whether replacement of constituent species is even possible. Unless recurrent propagule import is brought up to a specific location, we currently do not know if species can simply add up and build improved microbiota with beneficial impacts on crops, even though classical ecology taught us a lot about species displacement and competitive replacement, and specific combinations might simply prove too complex to engineer. Yet, systems that are naturally highly diverse often are resilient and self-sustained, so maybe the issue becomes not so much optimizing individual components within the community, but maintaining a functional collective entity with systematic efficiency. In this case, scientific investigations should focus on the impact of practices and how they can provide opportunities for improved functioning, stability, and benefit to crops. Clearly, while still poorly documented from an academic perspective, several practices sometimes fancied by farmers who have adopted agroecological or permaculture systems rely on the idea that microbiota are an important component of their system, specifically indigenous microorganism enrichment procedures. The common basis for such practices is often conducted by collecting humus from local forests and application of fermented solutions to seed fields with enriched solutions. There are probably numerous avenues for research in this regard, especially to study whether and at which rate beneficial organisms are transferred to cultivated plots, and whether recipes can be specifically designed to orientate the taxonomic composition for specific functions. One of the advantages of these approaches is that both local and low-tech solutions amenable to scaling up are included for small farmers worldwide.

1.5.2 Crop Diversity at Local Scale

Although intensive agriculture is most often characterized by monoculture, notwithstanding recent increased frequency in intercropping of a few crops, developing microbiota tools will still target a single or only a few crops. The core concern of microbiota as a productivity tool will thus develop around the question as to whether there exist simple communities that will be generalist enough to confer growth or resistance benefits to a broad and diverse array of crops, or whether optimal microbiota will be more specific and dependent on the crop species cultivated. Focus on core components of beneficial microbiota communities will probably first lead to rapid advances and early use, until more specialist organisms can be proposed for use whenever their benefits outperform those of core components. There is a serious question around communal identity of commensal microbes in the wild. Current literature highlights both the importance of space and local influence of species components. So, microbiota broadly shared between plants growing here and there, sometimes even on a large scale or region, and more specific interactions with microbiota species in close relationship with specific plants are critical. There is thus an important issue to resolve as to whether shared species components between local plants could demonstrate benefits in general, or if they behave more simply as casual commensals without strong interactions. Understanding both core components of microbiota and specific specialized components is thus an important goal for applied agriculture science.

In contrast, agroecological systems strive for diversity at every level of structural organization. Agroecological agrodiversity generally involves both infra-specific and inter-specific levels of diversity. In addition to that, spatial structuring of diversity is not necessarily homogenous or controlled, but often occurs with quite high degrees of admixing, especially in more traditional and small-scale agriculture contexts. In this case, the potential benefits of microbiota strongly rely on whether generalist beneficial communities can be proposed. When it comes to specific components, the natural diversity will preclude any artificial optimization of specific components. Therefore, as previously discussed, the focus should shift to integrated management of naturally occurring microbiota, with the hope to realize the full potential for improvement with targeted practices influencing microbiota evolution and equilibrium.

1.5.3 The Agronomic Dark Triad: Weeds, Pests, and Diseases

The agronomic dark triad (Fig. 1.3) affects any crop, independently of the nature of the cultivation system being used, be it under intensive or agroecological management.

Traditionally, intensive cropping systems relied on a suppressive approach via chemical control of the triad curatively or even preventively, at the cost of side effects extending way outside the field targets, and resulting in high levels of environmental damage. Agroecological methods have diverse strategies for each point of the triad, and no single generalization can be drawn, ranging from locally suppressive (e.g., control of weeds via mulching, control via bio-stimulants) to liberal/indifferent or even making use of that diversity (e.g., weeds used as pillars of push–pull strategies). Also, crop admixing levels are facilitated due to its barrier effect on plant diseases, which is a natural control of pathogens under agroecological management.

How would a microbiota approach translate in this case? First, we should note that historically, single microbial control was used very early in agricultural practice. For example, the fungus *Colletotrichum gloeosporioides* is regularly suggested as a control agent for weeds (or invasive plants), because it is a generalist, as a pathogen (Cai et al. 2009). Obviously, this approach is useful if the control agent only attacks its target, without collateral damage on crops. This was also the case with bacteria *Bacillus thuringiensis* for insect pest control, to the point its insecticidal protein was the gene used for developing transgenic crops very early in the history of agricultural biotechnology. As expected, single, but specific, solutions can flourish and solve precise issues in agriculture. However, the question regarding microbiota thus becomes once again a more general one: are there any communities that could be used in preventive management against the dark triad? While hard to predict with

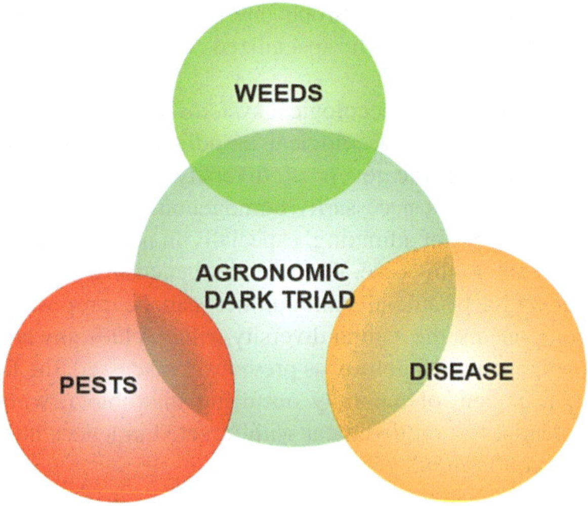

Fig. 1.3 Schematic of the three threats to sustainable agriculture

certainty, little is probably to be expected regarding weed control, because a microbiome approach would rely on component species behaving antagonistically toward weeds, without doing so on crops. Similarly, control of pests may prove very difficult as a communal strategy, and a single-species approach might be much easier to develop for efficient management. This concern is greater for intensive agriculture than it is for agroecological systems, as alternative ways beside microbiota are already employed to manage insect pests, especially natural control via naturally occurring pest antagonists (e.g., parasitoids) or push–pull strategies. Conversely, in the dark triad, the feature most amenable to control via microbiota are crop diseases. Current evidence seems to associate disease development with strong alteration and even decrease in foliar fungal diversity. Scientists need to understand basic processes involved in community changes, especially if these observations are simply reflecting competitive displacement, including secretion of chemical compounds by microbiota, predation, direct competition for local resources like nutrients, and whether other processes might intervene. Little is known about interaction levels. This is true of community or network composition, and in the case of the infection process and plant disease we need to resolve not only mutualistic or commensal interactions, but antagonisms as well, and how they could affect crop pathogens and suppress disease. Accordingly, a large factor in the efficiency of microbiomes as suppressive disease tools will depend on whether biological interactions are passive and purely demographic as a competition effect, or if they can be active based on aggressiveness within microbial communities and networks.

1.6 Conclusion

In summary, the next phase of agriculture is exciting and offers an opportunity for researchers to adapt their knowledge of the microbiome to farming in practice. Microbiota will certainly be an adequate tool to help agriculture evolve toward improved sustainability without impairing productivity, and even possibly resulting in increased yields. Nevertheless, there are still several challenges impeding its general implementation on farms, given our current knowledge and the research context. These include firstly, identification of core genera with relevant functions for plants, to propose efficient synthetic communities; secondly, identification of specific component species that match local (regional) or organ-specific characteristics for improving baseline synthetic communities; and thirdly, investigating the stability of synthetic communities over cropping cycles, to ensure benefits for food production.

In this chapter, we examined the structure and function of the fungal microbiome and provide specific recommendations for both intensive agriculture and small-scale agroecological agriculture. Microbiota engineering and its application in agriculture are the new frontiers in removing an unstainable dependency on chemical inputs for both increased yields and enhanced disease management. Moreover, adopting integrated pest management strategies which are key components in agroecology provides opportunities for a chemical independent, but microbiota-driven sustainable agriculture for increased plant growth and resistance to disease.

References

Ahmed A (2016) Analysis of metagenomics next generation sequence data for fungal ITS barcoding: do you need advance bioinformatics experience? Front Microbiol 7:1061. https://doi.org/10.3389/fmicb.2016.01061

Andreote FD, Gumiere T, Durrer A (2014) Exploring interactions of plant microbiomes. Sci Agric 71:528–539. https://doi.org/10.1590/0103-9016-2014-0195

Arif I, Batool M, Schenk PM (2020) Plant microbiome engineering: expected benefits for improved crop growth and resilience. Trends Biotechnol 38:1385–1396. https://doi.org/10.1016/j.tibtech.2020.04.015

Arnold AE, Mejía LC, Kyllo D et al (2003) Fungal endophytes limit pathogen damage in a tropical tree. Proc Natl Acad Sci 100:15649–15654. https://doi.org/10.1073/pnas.2533483100

Aslam MM, Okal EJ, Idris AL et al (2022) Rhizosphere microbiomes can regulate plant drought tolerance. Pedosphere 32:61–74. https://doi.org/10.1016/s1002-0160(21)60061-9

Berg G, Rybakova D, Fischer D et al (2020) Microbiome definition re-visited: old concepts and new challenges. Microbiome 8:103. https://doi.org/10.1186/s40168-020-00875-0

Blaalid R, Kumar S, Nilsson RH et al (2013) ITS1 versus ITS2 as DNA metabarcodes for fungi. Mol Ecol Resour 13:218–224. https://doi.org/10.1111/1755-0998.12065

Bonfante P, Venice F, Lanfranco L (2019) The mycobiota: fungi take their place between plants and bacteria. Curr Opin Microbiol 49:18–25. https://doi.org/10.1016/j.mib.2019.08.004

Bordenstein SR, Theis KR (2015) Host biology in light of the microbiome: ten principles of Holobionts and Hologenomes. PLoS Biol 13:e1002226. https://doi.org/10.1371/journal.pbio.1002226

Cai L, Hyde KD, Taylor PWJ et al (2009) A polyphasic approach for studying Colletotrichum. Fungal Divers 39:183–204

Carbone MJ, Alaniz S, Mondino P et al (2021) Drought influences fungal community dynamics in the grapevine rhizosphere and root microbiome. J Fungi 7:686. https://doi.org/10.3390/jof7090686

Chen Q-L, Hu H-W, He Z-Y et al (2021) Potential of indigenous crop microbiomes for sustainable agriculture. Nat Food 2:233–240. https://doi.org/10.1038/s43016-021-00253-5

Clay K, Schardl C (2002) Evolutionary origins and ecological consequences of endophyte symbiosis with grasses. Am Nat 160:S99–S127. https://doi.org/10.1086/342161

Compant S, Samad A, Faist H, Sessitsch A (2019) A review on the plant microbiome: ecology, functions, and emerging trends in microbial application. J Adv Res 19:29–37. https://doi.org/10.1016/j.jare.2019.03.004

Crous PW, Hawksworth DL, Wingfield MJ (2015) Identifying and naming plant-pathogenic fungi: past, present, and future. Annu Rev Phytopathol 53:1–21. https://doi.org/10.1146/annurev-phyto-080614-120245

Cuadros-Orellana S, Leite LR, Smith A et al (2013) Assessment of fungal diversity in the environment using metagenomics: a decade in review. Fungal Genom Biol 3:1–13. https://doi.org/10.4172/2165-8056.1000110

de Souza RSC, Okura VK, Armanhi JSL et al (2016) Unlocking the bacterial and fungal communities assemblages of sugarcane microbiome. Sci Rep 6:28774. https://doi.org/10.1038/srep28774

Dentika P, Ozier-Lafontaine H, Penet L (2021) Weeds as pathogen hosts and disease risk for crops in the wake of a reduced use of herbicides: evidence from yam (Dioscorea alata) fields and Colletotrichum pathogens in the tropics. J Fungi 7:283. https://doi.org/10.3390/jof7040283

Donovan PD, Gonzalez G, Higgins DG et al (2018) Identification of fungi in shotgun metagenomics datasets. PLoS One 13:e0192898. https://doi.org/10.1371/journal.pone.0192898
Dubey A, Malla MA, Khan F et al (2019) Soil microbiome: a key player for conservation of soil health under changing climate. Biodivers Conserv 28:2405–2429. https://doi.org/10.1007/s10531-019-01760-5
Edwards J, Johnson C, Santos-Medellín C et al (2015) Structure, variation, and assembly of the root-associated microbiomes of rice. Proc Natl Acad Sci 112:E911–E920. https://doi.org/10.1073/pnas.1414592112
Edwards JE, Forster RJ, Callaghan TM et al (2017) PCR and omics based techniques to study the diversity, ecology and biology of anaerobic fungi: insights, challenges and opportunities. Front Microbiol 8:1657. https://doi.org/10.3389/fmicb.2017.01657
Fedoroff N (1987) Scientific basis for soil protection in the European Community. Springer, New York, pp 65–85. https://doi.org/10.1007/978-94-009-3451-1_5
Fitzpatrick CR, Salas-González I, Conway JM et al (2020) The plant microbiome: from ecology to reductionism and beyond. Annu Rev Microbiol 74:1–20. https://doi.org/10.1146/annurev-micro-022620-014327
Frediansyah A (ed) (2021) Cassava: biology, production, and use. IntechOpen, London. https://doi.org/10.5772/intechopen.87488
Gao L, Tu Z, Millett BP, Bradeen JM (2013) Insights into organ-specific pathogen defense responses in plants: RNA-seq analysis of potato tuber-*Phytophthora infestans* interactions. BMC Genomics 14:340. https://doi.org/10.1186/1471-2164-14-340
Gao M, Xiong C, Gao C et al (2021) Disease-induced changes in plant microbiome assembly and functional adaptation. Microbiome 9:187. https://doi.org/10.1186/s40168-021-01138-2
Hariharan G, Prasannath K (2021) Recent advances in molecular diagnostics of fungal plant pathogens: a mini review. Front Cell Infect Microbiol 10:600234. https://doi.org/10.3389/fcimb.2020.600234
Hartman K, van der Heijden MGA, Wittwer RA et al (2018) Cropping practices manipulate abundance patterns of root and soil microbiome members paving the way to smart farming. Microbiome 6:14. https://doi.org/10.1186/s40168-017-0389-9
Heil M, Bostock RM (2002) Induced systemic resistance (ISR) against pathogens in the context of induced plant defences. Ann Bot 89:503–512. https://doi.org/10.1093/aob/mcf076
Huang X, Liu L, Wen T et al (2015) Illumina MiSeq investigations on the changes of microbial community in the *Fusarium oxysporum* f. sp. cubense infected soil during and after reductive soil disinfestation. Microbiol Res 181:33–42. https://doi.org/10.1016/j.micres.2015.08.004
Isobe S, Shirasawa K, Hirakawa H (2019) Current status in whole genome sequencing and analysis of ipomoea spp. Plant Cell Rep 38:1365–1371. https://doi.org/10.1007/s00299-019-02464-4
Kim H, Lee Y-H (2020) The rice microbiome: a model platform for crop Holobiome. Phytobiomes J 4:5–18. https://doi.org/10.1094/pbiomes-07-19-0035-rvw
Ma Y, Fort T, Marais A et al (2021) Leaf-associated fungal and viral communities of wild plant populations differ between cultivated and natural ecosystems. Plant Environ Interact 2:87–99. https://doi.org/10.1002/pei3.10043
Martin H (2001) The ecological concept of costs of induced systemic resistance (ISR). Eur J Plant Pathol 107:137–146. https://doi.org/10.1023/A:1008793009517
Mendes R, Garbeva P, Raaijmakers JM (2013) The rhizosphere microbiome: significance of plant beneficial, plant pathogenic, and human pathogenic microorganisms. FEMS Microbiol Rev 37:634–663. https://doi.org/10.1111/1574-6976.12028
Meunier R, Bayır S (2021) Metagenomics approaches in microbial ecology and research for sustainable agriculture. TATuP 30:24–29. https://doi.org/10.14512/tatup.30.2.24
Mueller UG, Sachs JL (2015) Engineering microbiomes to improve plant and animal health. Trends Microbiol 23:606–617. https://doi.org/10.1016/j.tim.2015.07.009
Müller DB, Vogel C, Bai Y, Vorholt JA (2015) The plant microbiota: systems-level insights and perspectives. Annu Rev Genet 50:211–234. https://doi.org/10.1146/annurev-genet-120215-034952

Naranjo-Ortiz MA, Gabaldón T (2019) Fungal evolution: diversity, taxonomy and phylogeny of the fungi. Biol Rev 94:2101–2137. https://doi.org/10.1111/brv.12550

Nilsson RH, Anslan S, Bahram M et al (2019) Mycobiome diversity: high-throughput sequencing and identification of fungi. Nat Rev Microbiol 17:95–109. https://doi.org/10.1038/s41579-018-0116-y

Noman M, Ahmed T, Ijaz U et al (2021) Plant–microbiome crosstalk: dawning from composition and assembly of microbial community to improvement of disease resilience in plants. Int J Mol Sci 22:6852. https://doi.org/10.3390/ijms22136852

Pagano MC, Correa EJA, Duarte NF et al (2017) Advances in eco-efficient agriculture: the plant-soil mycobiome. Agriculture 7:14. https://doi.org/10.3390/agriculture7020014

Pankhurst CE, Lynch JM (1995) 12 the role of soil microbiology in sustainable intensive agriculture. Adv Plant Pathol 11:229–247. https://doi.org/10.1016/s0736-4539(06)80014-0

Pascale A, Proietti S, Pantelides IS, Stringlis IA (2020) Modulation of the root microbiome by plant molecules: the basis for targeted disease suppression and plant growth promotion. Front Plant Sci 10:1741. https://doi.org/10.3389/fpls.2019.01741

Patel JS, Singh A, Singh HB, Sarma BK (2015) Plant genotype, microbial recruitment and nutritional security. Front Plant Sci 6:608. https://doi.org/10.3389/fpls.2015.00608

Pozo MJ, Zabalgogeazcoa I, de Aldana BRV, Martinez-Medina A (2021) Untapping the potential of plant mycobiomes for applications in agriculture. Curr Opin Plant Biol 60:102034. https://doi.org/10.1016/j.pbi.2021.102034

Prasad R, Zhang S-H (2022) Benefical microorganisms in agriculture. Springer, Singapore. https://link.springer.com/book/9789811907326

Pritsch K, Garbaye J (2011) Enzyme secretion by ECM fungi and exploitation of mineral nutrients from soil organic matter. Ann Forest Sci 68:25–32. https://doi.org/10.1007/s13595-010-0004-8

Prochnik S, Marri PR, Desany B et al (2012) The cassava genome: current progress, future directions. Trop Plant Biol 5:88–94. https://doi.org/10.1007/s12042-011-9088-z

Raja HA, Miller AN, Pearce CJ, Oberlies NH (2017) Fungal identification using molecular tools: a primer for the natural products research community. J Nat Prod 80:756–770. https://doi.org/10.1021/acs.jnatprod.6b01085

Schoch CL, Seifert KA, Huhndorf S et al (2012) Nuclear ribosomal internal transcribed spacer (ITS) region as a universal DNA barcode marker for Fungi. Proc Natl Acad Sci 109:6241–6246. https://doi.org/10.1073/pnas.1117018109

Scibetta S, Schena L, Abdelfattah A et al (2018) Selection and experimental evaluation of universal primers to study the fungal microbiome of higher plants. Phytobiomes J 2:225–236. https://doi.org/10.1094/pbiomes-02-18-0009-r

Stielow JB, Lévesque CA, Seifert KA et al (2015) One fungus, which genes? Development and assessment of universal primers for potential secondary fungal DNA barcodes. Persoonia Mol Phylogeny Evol Fungi 35:242–263. https://doi.org/10.3767/003158515x689135

Tamiru M, Natsume S, Takagi H et al (2017) Genome sequencing of the staple food crop white Guinea yam enables the development of a molecular marker for sex determination. BMC Biol 15:86. https://doi.org/10.1186/s12915-017-0419-x

Tedersoo L, Anslan S, Bahram M et al (2015) Shotgun metagenomes and multiple primer pair-barcode combinations of amplicons reveal biases in metabarcoding analyses of fungi. Mycokeys 10:1–43. https://doi.org/10.3897/mycokeys.10.4852

Tedersoo L, Sánchez-Ramírez S, Kõljalg U et al (2018) High-level classification of the Fungi and a tool for evolutionary ecological analyses. Fungal Divers 90:135–159. https://doi.org/10.1007/s13225-018-0401-0

Tedersoo L, Bahram M, Zobel M (2020) How mycorrhizal associations drive plant population and community biology. Science 367(6480):eaba1223. https://doi.org/10.1126/science.aba1223

Than PP, Jeewon R, Hyde KD et al (2008) Characterization and pathogenicity of Colletotrichum species associated with anthracnose on chilli (Capsicum spp.) in Thailand. Plant Pathol 57:562–572. https://doi.org/10.1111/j.1365-3059.2007.01782.x

Toju H, Yamamoto S, Tanabe AS et al (2016) Network modules and hubs in plant-root fungal biomes. J Roy Soc Interface 13:20151097. https://doi.org/10.1098/rsif.2015.1097

Toju H, Kurokawa H, Kenta T (2018a) Factors influencing leaf- and root-associated communities of bacteria and fungi across 33 plant orders in a grassland. Biorxiv. 439646. https://doi.org/10.1101/439646

Toju H, Peay KG, Yamamichi M et al (2018b) Core microbiomes for sustainable agroecosystems. Nat Plants 4:247–257. https://doi.org/10.1038/s41477-018-0139-4

Varma A, Prasad R, Tuteja N (2017a) Mycorrhiza: function, diversity and state-of-art. Springer International Publishing. ISBN: 978-3-319-53064-2. http://www.springer.com/us/book/9783319530635

Varma A, Prasad R, Tuteja N (2017b) Mycorrhiza: eco-physiology, secondary metabolites, nanomaterials. Springer International Publishing. ISBN: 978-3-319-57849-1. http://www.springer.com/us/book/9783319578484

Varma A, Prasad R, Tuteja N (2017c) Mycorrhiza: nutrient uptake, biocontrol, ecorestoration. Springer International Publishing. ISBN: 978-3-319-68867-1. http://www.springer.com/us/book/9783319688664

Varma A, Swati T, Prasad R (2019a) Plant biotic interactions: state of art. Springer International Publishing. ISBN 978-3-030-26657-8. https://www.springer.com/gp/book/9783030266561

Varma A, Swati T, Prasad R (2019b) Plant microbe interface. Springer International Publishing. ISBN 978-3-030-19831-2. https://www.springer.com/gp/book/9783030198305

Weir BS, Johnston PR, Damm U (2012) The Colletotrichum gloeosporioides species complex. Stud Mycol 73:115–180. https://doi.org/10.3114/sim0011

Xiong C, Zhu Y, Wang J et al (2021) Host selection shapes crop microbiome assembly and network complexity. New Phytol 229:1091–1104. https://doi.org/10.1111/nph.16890

Yuan X, Hong S, Xiong W et al (2021) Development of fungal-mediated soil suppressiveness against Fusarium wilt disease via plant residue manipulation. Microbiome 9:200. https://doi.org/10.1186/s40168-021-01133-7

Zhang L, Zhang J, Wei Y et al (2021) Microbiome-wide association studies reveal correlations between the structure and metabolism of the rhizosphere microbiome and disease resistance in cassava. Plant Biotechnol J 19:689–701. https://doi.org/10.1111/pbi.13495

Zheng Y, Li X, Cao H et al (2021) The assembly of wheat-associated fungal community differs across growth stages. Appl Microbiol Biot 105:7427–7438. https://doi.org/10.1007/s00253-021-11550-1

Chapter 2
Unearthing the Modern Trends and Concepts of Rhizosphere Microbiome in Relation to Plant Productivity

Hitakshi Gupta, Jahanvi Ganotra, Nikita Pathania, Tirth Bhargavbhai Patel, Nisha Choudhary, Reia Rani, Damini Supolia, and Deepak Kumar

Abstract The rhizosphere microbiome is crucial for plant growth and health, providing defense against plethora of surrounding potential pathogens, improving crops' nutrient acquisition, and aiding in withstanding series of abiotic stresses. Exchange of resources between plants and their associated soil environment is supported by a pivotal interface termed as rhizosphere. Plants being sessile are incessantly exposed to a diverse array of abiotic and biotic stresses under natural conditions, thereby acting as major bottlenecks to hinder their growth and production. However, it is possible to engineer plant rhizosphere microbiome as revealed by the recent advances in research in context to the plant–microbe interactions. Therefore, shaping of rhizosphere microbiome for developing promising strategies can be vital in relation to plant productivity. Furthermore, exploring the structure and dynamics of plant rhizosphere microbiome represents an exciting frontier of research to protect plants from potential phytopathogens in a sustainable manner. Advances in molecular tools are beneficial to unravel the concepts of plant–microbe association, which could act as a key driver in drafting the future "biofertilizers." Owing to the current challenges in crop production, there is an urgent need to understand plant–microbiome interactions in the rhizosphere to bring microbiome-based strategies for incorporating beneficial resident microbial communities into practice. The present chapter uncovers the concepts of plant rhizosphere microbiome, its diversity, abundance, composition, and interplay with the plants. Additionally, elucidation of

---

H. Gupta · J. Ganotra · N. Pathania · N. Choudhary · R. Rani · D. Supolia
Department of Botany, Central University of Jammu, Samba, Jammu and Kashmir, India

T. B. Patel
Biotechnology Department, Shri A.N. Patel PG Institute of Science and Research, Anand, Gujarat, India

D. Kumar (✉)
Centre for Advanced Study in Botany, Institute of Science, Banaras Hindu University, Varanasi, Uttar Pradesh, India
e-mail: deepakkumar@bhu.ac.in

S. Chhabra et al. (eds.), *Plant Microbiome for Plant Productivity and Sustainable Agriculture*, Microorganisms for Sustainability 37,
https://doi.org/10.1007/978-981-19-5029-2_2

plant rhizosphere microbiome engineering and various factors and techniques involved therein for enhancing plant productivity have been presented, thereby helping meet food requirements of exponentially growing global population.

Keywords Rhizosphere microbiome · Holobiont · Root exudation · Rhizosphere engineering · Bioinformatics

Abbreviations

CRISPR	Clustered regularly interspaced short palindrome repeats
Fish	Fluorescent in situ hybridization
MALDI-TOF	Matrix-assisted laser desorption/light ionization time
NGS	Next-generation sequencing
PCR	Polymerase chain reaction
PGPR	Plant growth-promoting rhizobacteria
RFLP	Restriction fragment length polymorphism

2.1 Introduction

Plants share their habitat with diversity of microbes such as bacteria, fungi, and viruses. The constitution of the plant microbiota is framed by complex interactions between the abiotic environment and its biotic natives. Depending on the outcome of an interaction for the host, microbes are considered as mutualistic, commensal, and pathogenic. The term microbiome was first used by Joshua Lederberg and it refers to the microorganisms inhabiting our body as commensals, symbionts, or pathogens (Lynch 1990), and plant microbiome is defined as the dynamic community of microorganisms associated with the plant. Regions of the plant which provide niche for microbial community are phyllosphere, rhizosphere, and endosphere (Berg et al. 2016).

The rhizosphere is represented as that zone of the soil which is in direct association with the plant roots (Shrivastava et al. 2014). In other words, it is an environment under the influence of plant. Rhizosphere is relatively stable and nutrient-rich environment and the rhizoplane encompasses the root surface and its adhering soil. Various organic compounds are released from plant roots that primarily participate in symbiotic functioning in the soil area which is under the influence of plant roots (rhizosphere) (Barea et al. 2005; Gupta et al. 2022). Moreover, rhizosphere provides a nutrient-rich environment for diazotrophic bacteria that fix atmospheric nitrogen, thereby making the nitrogen available to plants. Number of bacteria, fungi, and archaea is high in the rhizosphere due to the presence of nutrient-rich environment (Egamberdiyeva et al. 2008; Mendes et al. 2011). Therefore, profound knowledge in context to the diverse array of microorganisms and their respective functions in the rhizosphere acts as a requisite for enhancing the growth and productivity of crop plants. The present chapter embodies the information on rhizosphere microbiome, its diversity, abundance, composition, and communication with the plants. The targeted application of beneficial plant microbiome to counteract abiotic and biotic stresses is gaining importance and is considered as an exciting frontier of research. Advancement in next-generation sequencing (NGS) platform, gene editing technologies, and metagenomics and bioinformatics approaches allows us to untangle webs of interactions of holobionts and core microbiomes for efficiently deploying the microbiome to increase crops' nutrient acquisition as well as resistance in response to abiotic and biotic stresses. Henceforth, the present chapter also provides insights into the concept of plant rhizosphere microbiome engineering, and various factors and aforementioned techniques involved therein.

2.2 Composition, Abundance, and Diversity of Rhizosphere Microbiome

The interface between plant and soil is termed as rhizosphere, and it functions as the primary step of root microbiome recruitment and plant defense. It features a specialized microbial community, intensive microbe–plant and microbe–microbe interactions, and sophisticated signal communication. It has already been recognized that microbial life is present in quite many trifling areas of soil which are localized in hot spots like rhizosphere, where the microorganisms have uninterrupted ingress to the flow of number of plant-root-derived organic substrates (Nannipieri and Badalucco 2003). Flow of such nutrients along with biological factors and physicochemical factors can influence microbial community structure and performance of rhizosphere (Sorensen 1997; Brimecombe et al. 2001). Myriad of microorganisms inhabit the plant rhizosphere. Figure 2.1 depicts the composition, abundance, and diversity of microorganisms present in the rhizosphere zone.

2.3 Types of Interactions Between Microbes and Plants

Plants are non-motile but they constantly encounter the abiotic and biotic stresses. There is a constant war between the microbes and the host plant. The rhizosphere is a hot spot for potentially important microbes and copious organisms. Different types of unicellular and multicellular organisms such as bacteria, archaea, algae, fungi, protozoans, and arthropods together with plant roots form the most complex ecosystem on earth (Raaijmakers et al. 2009; Kushwaha et al. 2020). Plants release adequate amount of nutrients in the form of rhizodeposits which determine both the type and composition of rhizosphere microbiome. Various compounds are exuded from plant roots such as sugars, organic acids, nucleotides, peptides, enzymes, and

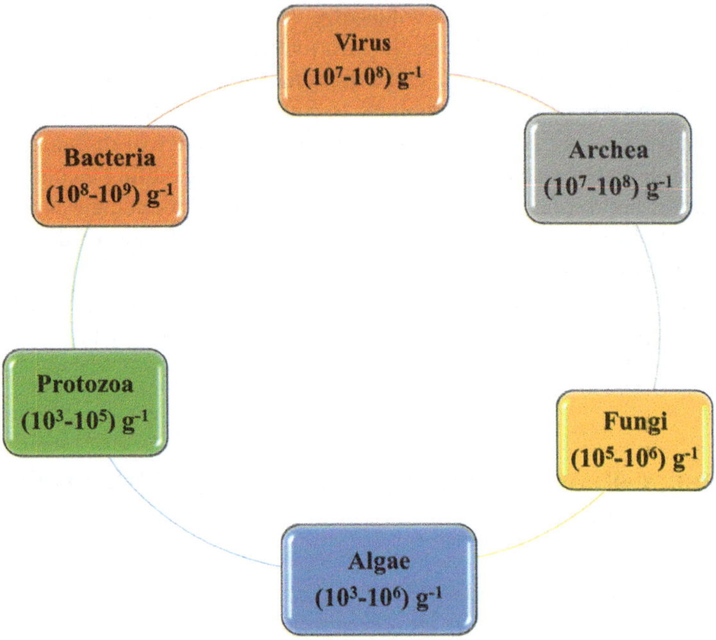

Fig. 2.1 Generalized diagram showing abundance of microorganisms present in the rhizosphere. The size of the circle is a measure of abundance of group of particular microbial community

other secondary metabolites which together regulate the microbial diversity and activity inside the rhizosphere. The plants sometimes also exert selective pressure by releasing unique rhizodeposits to stimulate the growth of beneficial microorganisms for their growth and development (Cook et al. 1995). Rhizospheric microorganisms can impart ecological fitness to their host plant and vice versa (Huang et al. 2014). Plant–microbiome communication is still fundamental and plant–microbiome interactions vary between crops species (and even cultivars), between individual members of the microbiome, and with environmental conditions. There is ample evidence suggesting the plant–microbe relationship is uncertain to health, productivity, and the overall condition of the plant. There are many different kinds of interactions between plants and microbes that traverse the whole variety from beneficial to pathogenic, and the outcome of the interaction between a plant and a microbe can vary among this range depending on plant species. Depending upon the type of microorganisms, host plants, as well as existing environmental conditions, both positive and negative plant–microbe interactions can persist in the rhizosphere.

2.3.1 Negative Interactions in the Rhizosphere

Beneficial microorganisms colonize plant roots in response to root exudates, but they can attract pathogenic population as well that can harm the plant. Plant diseases are directly involved in damaging crop plants and destructing agricultural production and thus agricultural economy. Soil-borne pathogens cause significant damage to the crops, whereas fungi are the most devastating. Their damaging effects include both mild and severe symptoms causing inconsiderate crop losses. Thus, they are major threat to food production and economic stability worldwide. The most common fungal pathogens include fungi of genus *Aspergillus*, *Fusarium*, *Pythium*, *Phytophthora*, *Mucor*, *Rhizopus*, and *Verticillium* (Tournas and Katsoudas 2005), as well as the common forest fungi, viz., *Armillaria* and *Poria* (Asiegbu et al. 2005). Pathogens belonging to the genus *Pseudomonas*, *Erwinia*, *Ralstonia*, and *Xanthomonas* are most common, as well as widely studied bacterial pathogens (Tournas and Katsoudas 2005).

2.3.2 Positive Interactions in the Rhizosphere

In the rhizosphere, plant–microbe interactions are involved in various crucial ecosystem functioning processes, such as nutrient mineralization and immobilization in biogeochemical cycles. To establish a symbiotic relation with plants, microorganisms form symbiotic associations with plants such as colonization of rhizosphere by plant growth-promoting rhizobacteria (PGPR), mycorrhizae, and legume–rhizobium association. These interactions are beneficial for plants and are of three types, namely biofertilizers, biocontrol agents, and biostimulants. Biofertilizers are the type of microorganisms that increases the availability of the nutrients to plants (Giri et al. 2019). Biocontrol agents are the group of microbes that protects the plants from pathogen attack and thus increases their growth indirectly. Biostimulants, on the other hand, stimulate the plant growth by secreting growth-promoting hormones and growth regulators such as auxins, cytokinins, and gibberellins (Sharma et al. 2022).

2.4 Evolution of Plant–Microbe Interaction

Co-evolution of microbes and plants is what has helped plants in growing and adapting under varied environmental conditions. Microbes help plants in nutrient acquisition and protection against various environmental stresses, thereby helping in stimulating plant growth and development. They are known to be associated with a specific group of microbes interacting with one another forming accumulation of individuals often known as a "holobiont" (Bordenstein and Theis 2015; Theis et al. 2016). To select plants that are associated with microbial community often requires a wide range of selective pressure which majorly acts upon different components of holobiont that put great impact on fitness of plant.

However, the varied variety of microbes found on various tissues of plant, together with more early origin of microorganism and their fast generation time as compared to their host, suggests that the microbe–microbe interactions are vital selective force that forms a composite assemblage of microbes in different compartments such as rhizosphere, phyllosphere, and endosphere. Understanding these microbial exchanges for shaping more convoluted plant-associated communities of microbes, along with their consequence for host health in a more natural environment, remains scarce. Plants secrete carbon-rich substrates with the help of their roots; those substrates are likely favored by microbes that could quickly assimilate them (Doornbos et al. 2012; Alqarawi et al. 2018). There are many success stories in context to rhizosphere microbiome engineering (Chaparro et al. 2014), wherein most of the antique lineages of plants depict a strong competence to alter the relative abundance of rhizospheric microbes (Valverde et al. 2016). The differences in the root exudate had resulted in the selection of contrasting microbiomes (Bais et al. 2006; Bell et al. 2015; Rasmann and Turlings 2016). The microbiome exerts profound impact upon plant health and similarly the plants can also influence the rhizosphere microbiome through a variety of mechanisms (Mohanram and Kumar 2019; Mendes et al. 2018). The reason behind this is phenotypic and genotypic variations in plant traits that guide the specific microbiome responsible for enhancing the plant growth in a variety of ways (Kumar and Dubey 2020). The advances in the plant–microbe interactions reveal that the plants are able to form their rhizosphere microbiome, as evidenced by the fact that different plant species which are host-specific microbial communities grow on the same soil. This complex plant-associated microbial community is also known as the second genome of the plant (Berendsen et al. 2012).

The plant–microbiome interactions are complex and depend on plant species, soil type, and environmental conditions such as abiotic stress, biotic stress, climatic conditions, and human intervention. Different textures of soils as well as different kinds of environmental stresses (e.g., nutrient deficiencies, metal toxicity, and pathogen attack) have been shown to prompt the plant-species-dependent physiological responses and consequently exudation patterns (Quiza et al. 2015).

2.5 Rhizosphere Microbiome Assembly

Notably, there is crucial participation of plant root exudation in recruitment and modulation of the rhizosphere microbiome. The root cap border cells serve as rhizodeposit to enhance plant growth and also facilitate recruitment of new microbes in the rhizosphere (Dennis et al. 2010; Hawes et al. 2000; Vermeer and McCully 1982). These rhizodeposits are a rich source of carbon and nitrogen along with additional primary and secondary metabolites. Thus, they influence species-specific growth in the rhizosphere microbiomes and function as vital components for the plant–microbe interface assembly. Along with factors such as plant species, genotypes, and developmental stages, diurnal cycling of the plant also affects the rhizosphere microassembly. Root exudates have been found to be affected and secreted within a diurnal secretion pattern. Some pioneer plant species having high levels of genetic diversity can naturally colonize sites which are important environmental concern such as abandoned unrestored mines and many other kindred sites that typically remain impoverished for decades, exposing to erosion, vast amounts of waste, and various other factors limiting the revegetation of these sites, including unfavorable abiotic and biotic conditions. This suggests that some plant genotypes are likely more fitted for acclimation to the conditions with toxic waste around. In some studies, it has been found that such plants improve soil health locally by modifying the physicochemical properties such as higher nutrient content and pH of the toxic waste and cause important shift in the microbial community composition, from lithotrophic communities that dominate toxic waste environments to heterotrophic communities involved in nutrient cycling. Plant genotype and the type of substrate in the rhizosphere are the main drivers of rhizosphere microbiome diversity and community structure, in cases assessing the effect of genotype-by-environment reciprocity by quantifying the physicochemical properties of the substrates and the swap in microbial community congregation. Notably, the plant genotype act as a selective pressure in modifying physicochemical properties of the substrate and structuring rhizosphere microbial communities, particularly bacterial taxa. Also genotype-by-environment interactions impact on the physicochemical properties of substrates and the composition of the rhizosphere microbiome. The cooperative harmonizing role of the soil type and host genotype bring forth the importance of homogenized consideration of condition of soil and genetic variability of plant for future development and synthetic microbiomes application. Besides, diagnosis of the attune role by specific plant genotype in rhizosphere microbiome assembly delivers us with an auspicious way for future breeding project to blend host traits engaged in assembly of beneficial microbiota. The rhizosphere microbiome make-up and interactions of microbe with microbe between soil types and plant genotypes forge ahead our keen understanding of regulating the role of both the factors in the plant rhizosphere microbiome assemblage.

2.5.1 Factors Affecting the Assembly of Microbial Community in the Rhizosphere

The phenomenon that the rhizosphere microbial community differs from the community in bulk soil, suggesting that plant roots recruit and accumulate specific microorganisms in the rhizosphere from bulk soil, is known as rhizosphere effect. There are different evidences that suggested the presence of species-specific microbial communities in the rhizosphere. A study of microbial communities in the rhizosphere of 19 herbaceous plant species using the 16S rRNA gene amplicon sequencing showed that each tested plant species had 18–111 special operational taxonomic units (OTUs) and that same species can also develop distinct rhizosphere microbial communities, although the effect of the plant genotypes on the microbiome assembly is quite weak (Bulgarelli et al. 2015; Walters et al. 2018). These findings demonstrate that host genetics contribute to plant microbiome assemblies. A number of studies clarify that root phenotypes (Massalha et al. 2017), soil type (Lundberg et al. 2012), and plant growth stages (Schlemper et al. 2017) also shape the rhizosphere microbial communities. The assembly of rhizosphere microbiomes was also being prone to different plant diseases. Quorum sensing (QS) is well known as a signaling mechanism that depends on the manner in which the rhizobacteria cell density regulates the gene expression of microbial physiological activity (Hartmann 2020). Many plant-associated bacteria are enriched in plant-associated environments by QS and require QS to regulate a series of important processes, such as rhizosphere competition, conjugation, and biofilm maturation (Newton and Fray 2004; Frederix and Downie 2011). The important factors affecting the assemblage of microbial community in rhizosphere are mentioned below.

2.5.1.1 Plant Growth Changes Root Metabolite and Assembly of the Rhizosphere Microbiome

Different factors influence the rhizosphere microbiome assemblage including plant growth, aging, and propagation. For instance, plant roots release root cap border cells into the rhizosphere as a kind of rhizodeposition to enhance the rhizospheric effect and to recruit specific microorganisms (Dennis et al. 2010; Hawes et al. 2000; Vermeer and McCully 1982). It has been reported that *A. thaliana* root caps release border-like cells to promote *Rhizobium* sp. YAS34 accumulation (Vicré et al. 2005). About ~10% of photosynthetically fixed carbon and ~15% of total plant nitrogen are observed in root exudates, including primary metabolites and secondary metabolites, like sugars, organic acids, amino acids, mucilage, and so on (Jones et al. 2009; Gargallo-Garriga et al. 2018). Plant roots release these metabolites with complex transmembrane system (Canarini et al. 2019). Different plant species release different root exudates. For example, cucumber roots secrete citric acid to assemble *B. amyloliquefaciens* SQR9 and to shape a special biofilm (Zhang et al. 2014). *A. thaliana* releases several amino acids in root exudate that aid in colonizing *B. subtilis* in specific root segment. Antifungal *Pseudomonas* colonizes in the root exudates of tomato root containing organic acids and sugar. Genotype and physiological status of plant determine the root exudation. By different metabolomics methods, it was observed that there were 19 *A. thaliana* genotypes that were present with different exudation patterns (Mönchgesang et al. 2016). The root exudates in *A. thaliana* were present at different development stages like seedling, vegetative, bolting, and flowering stages; the abundances of four phyla, namely Acidobacteria, Actinobacteria, Bacteroidetes, and Cyanobacteria, were significantly different at varied developmental time points (Chaparro et al. 2014). The root exudate compounds mentioned such as amino acids, phenolics, sugars, or sugar alcohols released by these four phyla show significant correlation. In addition to plant species, genotypes, developmental stages, and diurnal cycling of the plant also affect the rhizosphere microbiome, which breaks our orthodox notion that the rhizosphere microbiome does not change within the short time. The circadian clock plays an important role and is a key regulator of the plant diurnal physiological processes. An impairment in circadian clock strongly influences the *A. thaliana* rhizosphere microbiome, particularly the rare taxa; however, the important factors changing the rhizosphere microbiota were not revealed (Hubbard et al. 2018). Thus, traits including plant species, genotypes, developmental stages, as well as other characteristics are strongly interlinked with the rhizosphere microbiome assembly by modulating the physiological states of plants eventually resulting in the influence on anabolism and root exudation. Plant root exudation is a key mechanism in recruiting and modulation of the rhizosphere microbiome.

2.5.1.2 Abiotic and Biotic Stresses Modulate Root Exudation and Recruit the Rhizosphere Microbiome

Different abiotic and biotic stresses affect the assemblage of rhizosphere microbiome. In nature, plants are exposed to several abiotic and biotic stresses in their lifetime (Sharma et al. 2021). The effect of these environmental stresses on the microbiome assembly is of key concern while studying the plant–root interface. Different environmental stresses shaping the rhizosphere microbiome, including nutritional stress, such as phosphorus limitation, can activate phosphate starvation responses (PSRs) to regulate important genes to mediate the synthesis of primary and secondary metabolites and modify the root microbiome composition to alleviate phosphate stress (Castrillo et al. 2017). Environmental contaminants like pesticide residues modify the rhizosphere communities. A direct impact was detected as inoculation of soil with atrazine significantly recruited three OTUs: *Halobacillus*, *Bacillus decolorationis*, and *Cesiribacter* sp. JJ02 (Xu et al. 2018). Additionally, maize can transfer glyphosate sprayed on the surface of leaves to soils and thus significantly increase the amount of *Fusarium* on maize roots (Kremer and Means 2009). It was noticed that pesticides can also alter the rhizosphere microbiome through modulating plant exudation. For example, in rice seedling after treatment with diclofop methyl the root exudation of amino acids, fatty acids, and organic acids increased. This increased the relative abundance of *Massilia* and *Anderseniella* genera and changed the richness and diversity of rhizosphere microorganisms (Chen et al. 2017).

Biotic stress such as plant pathogen infection modifies the rhizosphere community. Infection of *A. thaliana* leaves with *Pseudomonas syringae* pv. tomato modifies plant root exudation patterns and significantly enriches the *Roseiflexus* genus in the rhizosphere by increasing the exudation of amino acids, nucleotides, and long-chain organic acids, and by decreasing the exudation of sugars, alcohols, and short-chain organic acids. This change in exudation patterns recruits more beneficial rhizosphere microbes which help *A. thaliana* to resist above-ground pathogens (Yuan et al. 2018). Likewise, in accordance to a study, *Carex arenaria* root releases a set of volatile organic compounds (VOCs) into the rhizosphere post *Fusarium culmorum* infection, and recruits a special rhizosphere community that varies from the rhizosphere of healthy plants (Schulz-Bohm et al. 2018). During pathogen attack, the salicylic acid (SA) *A. thaliana* mutants establish specific bacterial taxa in the root microbiome, possibly due to SA mutants changing the root exudate profiles (Lebeis et al. 2015). Jasmonic acid (JA) mutants of *A. thaliana* (myc2, med25) possess a different root microbiome composition and root exudates with lower levels of ornithine, tryptophan, and asparagine, but harbor a higher abundance of Enterobacteriaceae in the myc2 rhizosphere and *Bacillus*, *Lysinibacillus*, and *Streptomyces* in the med25 rhizosphere (Carvalhais et al. 2015). From these studies it can be concluded that the plant immune system and their rhizosphere microbiome are closely interlinked. However, the mechanism behind this relationship is not known and intense future research is required in this field.

2.6 Impact of Rhizosphere Communities on Plant Growth and Diseases Resistance

It is well known that the resident rhizosphere communities of plant root microbiota influence plant growth in more than one way. There are certain molecular mechanisms that are at play, which influence this pattern. Selecting inoculants from a well-growing microbiome community and applying it for growth purposes has been successfully demonstrated in certain plant species. Examples have surfaced where *Rhizobia* inoculants have been used for increasing crop yield and as an alternative to chemical and synthetic fertilizers in agricultural communities. In addition to the facilitation of biological atmospheric nitrogen fixation for plants by certain bacterial communities, microbes also showcase involvement in overcoming various nutrient limitations such as phosphate by making it accessible for plants. Additionally, microorganisms have also been involved in providing siderophores to plants under iron-limiting conditions.

Plant growth-promoting microorganisms (PGPM) impact plant yield by determining rhizosphere microbiome diversity. Different microbial communities exert synergistic effects to promote growth activities. Additionally, increased diversity also confers resistance against plethora of plant pathogens. Certain microbes belonging to rhizosphere community release phytohormonal substances. In drought stress, root growth and lateral root formation in wheat were stimulated by indole acetic acid (IAA) derived from *Azospirillum* generated, which is beneficial to increase water and nutrient absorption (Arzanesh et al. 2011). In addition, abscisic acid (ABA) was also produced and released in response to water deficit stress. ABA helps in maintaining plant transpiration rate and regulating stomatal behavior. According to a study, inoculation of *Arabidopsis thaliana* with *Phyllobacterium brassicacearum*

STM196 decreased leaf transpiration to elevate drought stress tolerance by secreting ABA. *Pseudomonas chlororaphis* subsp. *aureofaciens* strain M71 synthesizes ABA to modulate a higher extent of stomatal closure to enhance water use efficiency and tomato growth (Brilli et al. 2019).

The term suppression deals with the resilient attitude of certain rhizospheric communities against soil-borne pathogens. This property of microbes can be successfully transferred between soils, thus conferring the "non-resistant soils" with a form of induced resilience. Certain microbes synthesize antimicrobial compounds imparting an inhibitory effect to host plant pathogens. Antibiotics and related compounds are represented as major classes functioning to inhibit pathogen growth (Mhlongo et al. 2018). *Bacillus subtilis*, a model Gram-positive bacterium, can shape the biofilm on plant roots and has been utilized as a biopesticide. The biological control activity of *B. subtilis* depends on the secretion of antibacterial compounds. For example, *B. subtilis* strain 330-2 produce lytic enzymes (laminarase, cellulase, and protease) responsible for degrading the pathogenic fungi cell wall and also inhibit *Rhizoctonia solani*. In addition to disease suppressing of soils and production of antimicrobial compounds, plant disease-resisting microorganisms (PDRM) are capable of resisting diseases in the aerial parts of plants by long-term signaling mechanisms involving induced systemic resistance.

Thus, understanding the role of microorganisms in promoting the growth and disease resistance of crops is key to future development of bioproducts. However, environmental factors and microbiome colonization are still crucial factors that influence the benefit of microbial products. Determining the precise mechanisms of rhizosphere microbiome colonization and assembly is still difficult. Further exploration of the mechanisms will allow for a breakthrough in the application of beneficial microbiomes in practical agricultural development.

2.6.1 Rhizosphere Engineering

Rhizosphere is central to microbial and nutrient dynamics and describes the zone of soil surrounding roots of plant species which release organic substances (Dommergues 1978). It is the main zone where the plant roots interact with its environment and major activities like nutrient uptake and water absorption occur. This region is highly susceptible to abiotic and biotic stresses (Giri et al. 2018). The physical and chemical properties of the rhizosphere comprise the integration of many competing processes that depend on the soil type, water content, composition, biological activities of root-associated microbial communities, and the physiology of the plant itself (Pinton et al. 2007). Plants can be engineered to modify the rhizosphere pH or to release compounds to improve nutrient availability, protect against abiotic and biotic stresses, or encourage the proliferation of beneficial microorganisms (Bowen and Rovira 1999). Microorganisms form a vital component of the rhizosphere where the total biomass and composition of rhizosphere microbial populations distinctly aid in interactions between plants and the soil environment

(Arshad et al. 1993). Microorganisms engineered in the rhizosphere exude exogenous compounds that improve plant nutrition, suppress pathogenic microbes, and minimize the consequences of abiotic or biotic stresses (Ryan et al. 2009). The rhizosphere engineering results in the release of inorganic and organic substances from the plant roots and this process is termed as rhizodeposition. The exudates released by the roots are rhizodeposits. These exudates enhance nutrient accession to avoid mineral stresses and also favor the growth of salutatory microorganisms. A systemic approach is required for successful engineering of the rhizosphere microbiome. By understanding the underlying mechanism behind shaping of the associated rhizosphere, the sustainability and efficiency of crop production can be enhanced just by imitating the beneficial symbiotic associations existing between the soils, microbes, and plants (Kaushal and Prasad 2021).

2.6.2 Plant-Mediated Engineering

The beneficial plant traits are manipulated via two different approaches: genetic engineering and plant breeding. Using plant breeding techniques for selecting a specific microbial community is an interesting approach. Aim of this technique is to increase crop yield, by providing plant resistance toward a variety of stresses (Ryan et al. 2009). The main process of modification of rhizosphere through plants is by the release of root exudates and this process is known as rhizodeposition. These exudates can enrich nutrient accession, aid to elude mineral stresses, or cultivate the growth of favorable microorganisms. Generation of membrane potential difference and an electrochemical gradient for H^+ helps in many transport processes across the plasma membrane in plants like nutrient uptake. In addition to generating a driving force for membrane transport, H^+ efflux can contribute to nutrient acquisition by acidifying the rhizosphere (Hinsinger et al. 2003). The release of organic anions such as citrate and malate, as well as phytases and phosphatases, helps some species to access poorly soluble organic and inorganic phosphorus in a similar way (Dinkelaker et al. 1995; Richardson et al. 2001; Vance et al. 2003). Henceforth, the discharge of organic anions from roots exhibits an immense influence on plant growth and nutrition. Through genetic engineering, rhizosphere can be modified by manipulating the expression of genes controlling the exudates. Neal et al. (1973) in their study used the substitution of chromosome between two wheat lines for improving tolerance toward root rot disease and thereby preserving the group of beneficial bacterial populations present in rhizosphere. In accordance to a study, transgenic plants have greater ability to secrete citrate from the roots, which grow better on phosphate-limited soil as compared to the wild type. In addition, this study also suggested that crop plants with an enhanced ability to use aluminum phosphate developed an enhanced ability to grow in acidic soils and tolerance toward aluminum (Koyama et al. 2000). Therefore, different studies are being conducted to manipulate the genes controlling the useful exudates which help in developing plant–microbiome interactions.

2.6.3 Microbiome-Mediated Engineering

The microbiome is often called as secondary genome of the plant as plant and the microbiome are interdependent. So, microbiome is also called as meta-organism or holobiont (Sharma et al. 2022). This brings the "opine concept" that combines the harmony of the host plants to secrete particular root exudates simultaneously with the inoculation of microbes that are engineered to degrade this substrate. This often results in the colonization of the rhizosphere by a specific type of microbial community. Hence, it was also noticed that the opines produced by transgenic plants lead toward the selection of the host-specific microbial community that can maintain themselves at very high concentrations, even after the transgenic plant is removed (Savka et al. 2002). These approaches are highly peculiar as specific metabolites are being utilized.

Bioengineering of synthetic microbial communities presents several opportunities for plant/crop growth promotion, disease resistance, and stress tolerance/regulation. It has been identified that hundreds of bacterial strains possess many beneficial effects; it is a challenge in engineering a sustainable synthetic microbial community. For example, in a simple two-strain co-culture, six ecological interaction factors must be taken into account (Grosskopf and Soyer 2014), including: (1) Commensalism, in which one strain benefits from the other without affecting it, for instance products from one strain serves as substrates for the other; (2) Competition, in which two strains compete for the same product, for instance substrate competition; (3) Predation, in which predator benefits while the prey is harmed; (4) No interaction, in which two strains have no or net zero effect on each other with no shared substrates, no predator–prey relations, and no competition; (5) Cooperation, in which both strains are profited from each other, and (6) Amensalism, in which one strain is negatively affected while the other strains has no effect. The complexity of these possible ecological interactions will scale linearly with the addition of extra strains (Grosskopf and Soyer 2014). During rhizosphere engineering, the main challenges lie in minimizing the harmful or negative interactions like parasitism and competition while maximizing beneficial effects and cooperation. As even in two-strain co-cultures, competition tends to dominate temperature, nutrient availability, and host plant exudates, which affect growth rates, seeding rate, stabilization, susceptibility to pathogens, and sustainability of the synthetic microbial community once applied. So, minimizing competition is quite challenging.

Many microbial genera are known that colonize the rhizosphere and can be manipulated in genetic engineering efforts. Many plant growth-promoting rhizobacteria (PGPR), free-living strains, are able to colonize roots and stimulate plant growth (Prasad et al. 2015). Growth stimulation can be mediated directly via enhanced nutrient acquisition or modulation of phytohormone synthesis, and indirectly via induction of the plant's own defense responses or antagonism of soil-borne pathogens. Moreover, a number of other mechanisms of promotion can function simultaneously in a single strain. These genera include *Bacillus* (Dong and Zhang 2014), *Paenibacillus* (Kim and Timmusk 2013), *Streptomyces* (Medema et al. 2011), and *Rhizobium* (Patel and Sinha 2011). While *Streptomyces* spp. offer great samples of PGPRs with tractable genetic systems and much of obtainable complete genome sequences, *Streptomyces* spp. have some drawbacks in that they need very large genomes (on average ~ 7 MB) (Köberl et al. 2015) and have many mobile elements, which make them difficult to engineer and/or grow in cooperative synthetic microbial communities. *Bacillus* is the base of the synthetic microbial community, because it is comparatively easy to genetically engineer (Dong and Zhang 2014), contains many isolates that have plant growth proprieties (Köberl et al. 2013; Köberl et al. 2015), and/or is currently utilized in biocontrol applications. For example, *Bacillus* could be engineered to contain a nitrogen-fixation machinery (e.g., NifH from *Paenibacillus*) (Kim and Timmusk 2013), produce high concentrations of plant hormones (Arkhipova et al. 2005), or add pathways from other *Bacillus* sp. to control pathogens (Köberl et al. 2013). For increasing nitrogen fixation, *Pseudomonas*, *Rhizobium*, and/or *Bradyrhizobium* genera might be added. A simple three-strain member consortium, including an engineered *Bacillus* with two natural or engineered nitrogen fixers like *Pseudomonas*, *Rhizobium*, and *Bradyrhizobium*, could provide many of the benefits of a more complex natural rhizosphere community. The potential ecological functional interactions increase with the number of strains added; a three-strain consortium would potentially contain approximately 729 predicted interactions, and a four-member consortium 531,441 predicted interactions (Grosskopf and Soyer 2014). Efforts should be taken to limit the number of strains within a synthetic microbial community to three strains in order to exert control of potential interactions rather quickly (Foster and Bell 2012). Thus, by modifying the microbial community the crop productivity can be increased. Table 2.1 enlists some more instances of crop productivity enhancement resulting due to modification of microbial community.

2.6.4 Engineering the Interactions Between Plants and Microbes

Engineering the interactions between plants and microbes is an exciting approach. Several recent studies in this context have dealt with plant endophytes and phytoremediation. For example, an endophytic *Burkholderia cepacia* strain was transformed with a plasmid encoding toluene degradation, and was re-introduced to yellow lupine (*Lupinus luteus* L.). Plants inoculated with the transformed strain showed no sign of phytotoxicity at a high toluene concentration (1000 mg/L) and sustained growth, while the control plants experienced phytotoxicity at levels above 100 mg/L. It has been observed that 50–70% reduction in toluene evapotranspiration through the leaves occurred after plant–microbe association engineering. A similar experiment using another strain of *B. cepacia* and another plasmid encoding toluene degradation was conducted. Inoculation of poplar hybrids (*P. trichocarpa* × *P. deltoides*) with the *B. cepacia* strain harboring the toluene-degrading plasmid

Table 2.1 Crop productivity enhancement in response to modification of microbial community

Crop	Microbial community	Remarks	References
Wheat (var. H1105 and PBW660) and cowpea (var. PL-1 and PL-2)	*Variovorax paradoxus* RAA3 and M11	Enhancement in plant growth of wheat and cowpea by seed bacterization; significant increase in shoot/root length (root parameter was not observed in wheat), shoot and root fresh/dry weight, chlorophyll content, and also increase in the NPK contents	Kumar et al. (2021)
Glycine max	Co-inoculation of *Bradyrhizobium* strain along with *Streptomyces griseoflavus*	Enhanced nodulation, nitrogen fixation, and nutrients uptake	Htwe et al. (2018)
Phaseolus vulgaris	Co-inoculation of *Bacillus megaterium* and *Paenibacillus polymyxa* along with *Rhizobium*	Enhancement in plant biomass as compared to *Rhizobium* inoculation alone	Korir et al. (2017)
Maize, wheat, sunflower, and lettuce	*Pseudomonas*, *Azospirillum*, *Azotobacter*, and *Bacillus*	19–40% enhancement in yield	Rubin et al. (2017)
Cicer arietinum	Application of *Serratia* strain 5D as an inoculum	25.55% and 30.85% increase in the grain yield of crops grown on fertile soil in irrigated areas and nutrient-deficient soil in rainfed areas, respectively	Zaheer et al. (2016)

had a positive effect on plant growth in the presence of toluene, and reduced the amount of toluene released via evapotranspiration (Taghavi et al. 2005). Plant inoculation with PGPR endophytes has also been reported. Antimicrobial metabolites produced by a number of these bacteria, such as 2,4-diacetylphloroglucinol (DAPG), can indeed enhance disease suppression in plants. This has been observed, for instance, in eggplants (*Solanum melongena*) inoculated with DAPG producing *Pseudomonas*, *Enterobacter*, or *Bacillus* isolates. Inoculated plants became partly resistant to *R. solanacearum* as judged from the wilt incidence that was reduced by over 70% compared to non-inoculated plants (Ramesh et al. 2009). The potential value of plant–endophyte interaction engineering and the associated difficulties have been extensively reviewed in the cases of phytoremediation (Weyens et al. 2009) and plant growth promotion (Gaiero et al. 2013).

Opines are low-molecular-weight molecules that are typically synthesized in crown gall tumors or hairy roots inoculated by *Agrobacterium* sp. The synthesis is directed by the products of genes transferred by the bacteria to the host plant cells. Opines play a major role in the *Agrobacterium* plant interaction and are used as growth substrates by the inciting bacteria and as signals to induce the conjugal transfer of the Ti-plasmid, the genetic element that bears most of the pathogenicity determinants (Dessaux et al. 1998). The opine concept stipulates that these molecules help in the expansion of the pathogen and therefore dissemination of pathogenicity. Interestingly, this approach can lead to the selection of indigenous strains and is independent of plant species and soil type. Root exudation can also be manipulated indirectly. The application of the flagellar peptideflg22 or that of the bacterial toxin coronatine to the foliar system of *Arabidopsis* induced the expression of the malic acid transporter ALMT1, which led to an increased malic acid concentration in the rhizosphere of the plant *Arabidopsis*. Induced systemic resistance response in plants against *P. syringae* pv. tomato was observed by increase in concentration of malic acid, which led to increase in *Bacillus subtilis* strain FB1A7. From all the above discussion, we can conclude that rhizosphere engineering is clearly more than a promising way to reduce the usage of agrochemicals, and to improve soil and crop quality and productivity.

2.7 Techniques Associated with Rhizosphere Microbiome in Relation to Plant Productivity

Biological agents are attracting worldwide attention for sustainable plant production, yet investigating their potential for a multilayer environment such as the rhizosphere is difficult. Therefore, knowledge of genomics as an independent cell tool to understand the diversity and importance of the functioning of the rhizosphere microbiome in sustainable agriculture is essential, although recent research in the rhizosphere is carried out using evolving techniques such as metagenomics, metaproteomics, and metatranscriptomics, as well as taking into account their challenges, internal issues, impacts, and possible solutions.

In understanding soil compaction, the role of functional diversity and the taxonomy of soil microbiota and the role played by microbial metabolites in this process have been analyzed and discussed in the context of a method known as "omics." "Omics" techniques are used to reveal important information about the diversity of viruses, their response to various abiotic and biotic diseases, and the physiology of stress for various diseases. The undoubted role of the rhizospheric microbiome in plant production has been demonstrated by conventional methods, but the approach to understand the non-invasive species and their ecology requires a combination of multilayer technologies (Mendes et al. 2013). Thus, the molecular techniques and "omics" studies have helped researchers to link the functional roles of the rhizosphere microbiota to their ecosystem (Singh et al. 2017).

Metagenomics emerges as a genomics tool by combining a set of mechanisms, in which the microbial genome is involved as a model for functional genetic identification and diversity in conjunction with the help of bioinformatics to obtain information from acquired data. Opportunities for rhizospheric microbiome studies as well as metatranscriptomics and metaproteomics studies are introduced as a combination of molecular techniques and bioinformatics tools to present complex microbial interactions in a rhizospheric active metagenomic environment.

2.7.1 Genomics

Genomics refers to the gene branch responsible for genome research. Genome refers to a haploid set of genes or chromosomes present in living organisms. Genomics fall under the genetic branch that can be described as structural and functional genomics (Wang et al. 2020). Structural genomics includes location, physical characterization, and genetic sequence present in the genome. The genomics approach is extended to the functional part of the whole genome, including transcriptomics (RNA research), proteomics (protein research), and metabolomics (metabolite research) (Soni et al. 2015; Suyal et al. 2019). In addition, the combination of many "meta-" and "omics" technologies seems to benefit mankind, especially in the agricultural, industrial, and medical fields (Rawat and Rangarajan 2019; Suyal et al. 2019). Several genomics tools and techniques that appear on a daily basis are discussed here.

2.7.1.1 Polymerase Chain Reaction

The polymerase chain reaction (PCR) method was originally developed by Kary Banks Mullis in 1983, who won the Nobel Prize in 1993 for the same that transformed the entire field of molecular biology and still works today (Mullis 1990). This process involves the development of targeted DNA fragments extracted from any source. Quantitative PCR (qPCR) technique involves measuring the dynamic changes and relative abundance of fungal pathogens associated with black-foot disease (Berlanas et al. 2019). In addition, PCR offers many benefits to researchers in combination with gel electrophoresis approaches, i.e., agarose gel electrophoresis, and temperature gradient gel electrophoresis, and enhances our understanding in microbial community analysis.

2.7.2 Restriction Fragment Length Polymorphism

Inhibitory enzymes are part of endonucleases that can separate DNA only in certain areas. They are also considered to be cellular scissors. These enzymes are widely used to make genome mapping (O'Donnell et al. 2020). Some of the methods that test the endonuclease block system include restriction fragment length polymorphism (RFLP), amplified fragment length polymorphism, and plasmid fingerprinting.

2.7.3 DNA Sequencing

DNA sequencing is involved in identifying nucleotide sequences in a particular genome. Nowadays, it is easy to track and analyze the entire genome by using high-quality, automated, efficient, and reliable next-generation sequencing techniques (Kumar et al. 2014). Several methods such as microfluidics and fluorescent activated cell sorting (FACS) are popular in single cell tracking. This process involves a variety of mechanisms such as DNA marking, classification, and fluorescent cell sequence (O'Donnell et al. 2020).

2.7.4 Rhizospheric Microbiome Characterization by Next-Generation Sequencing

Next-generation sequencing (NGS) is one of the most expensive and time-saving tools used in the sequence of metagenomes. Prior to the discovery of NGS, studies were mainly aimed at analyzing genetically modified genes consisting of consistent sequences of genes with medical significance. This process involves a sequence of high output because millions of DNA fragments from a single sample are in sequence.

2.7.5 DNA Cloning

DNA cloning involves the transfer of a portion of DNA from one cell to another in order to produce duplicate copies in vivo (O'Donnell et al. 2020). Various cloning vectors have been developed in recent years, which can accommodate different types and sizes of DNA fragments, including plasmids, hybrid vectors, and synthetic chromosomes.

2.7.6 Blending Strategies

Hybridization method measures the level of genetic similarity between two different nucleic acid molecules. The basic methods used by this system include: DNA analysis/kinetic recombination and fluorescent in situ hybridization (FISH). Also, DNA microarray is an advanced mechanism based on it. Microarray involves the mixing of a piece of DNA and a probe into a chip called a DNA chip. In most cases, a DNA chip involves a single genome, but many genomes can also be explored. The method known as the "representational difference analysis" (RDA) analyzes the differences between the types of variables based on a previously tracked representative. This process involves a combination of techniques such as PCR and DNA sequencing. It is a very popular method of analyzing prokaryotic genomes as they can vary greatly in their genetic size (Barcellos et al. 2009).

However, a combination of genomics and other omics technologies is commonly used in rhizospheric microbiome research (Goel et al. 2018). Also, the merging of genetic approach with bioremediation is important to evaluate relationship among microbial and plant communities and ecosystem, which possess the aim of improvement of phytoremediation of contaminated regions (Agarwal et al. 2020). The advent of bioinformatics tools with these technologies has opened up new avenues for research and development in the field of microbial ecology.

2.8 Metagenomics

It is accepted that only about 0.1–1% (depending on local sample) of germs can be grown in artificial growth sources and more than 99% of microbe variants remain unused (Suyal et al. 2019). Germs to enter under active but indestructible conditions are forced by various environmental pressures which also reduce their access through the use of genomics. Biodiversity can therefore be underestimated in identifying plant-based pathogens. In order to overcome the limitations and complexities associated with farming technology, the metagenomics approach has already emerged as a potential tool (Soni and Goel 2011; Soni et al. 2017; Joshi et al. 2017; Kumar et al. 2019).

Metagenomics is concerned with the study of a genome collection of microbes (metagenome) from any surrounding environment to provide information on ecology and the diversity of small forms in a particular area. Due to the continued reduction in costs of nucleotide sequencing and the development of high-throughput sequences, it is now possible to sequence large amounts of DNA from biological mix (Metzker 2010) and thus provide a deeper understanding of the rhizosphere. The knowledge of certain members of the microbiome living in the rhizosphere is provided by a high-resolution analysis of the taxonomic structure of the rhizosphere soil (Lagos et al. 2015). In addition, metagenomics not only is effective in determining bacterial diversity but also helps assess the number of fungi in the rhizosphere (LeBlanc et al. 2015). Explanation of soil profiles with a culturally independent mold community and the rhizosphere of sugarcane planted in the field reveals that the concentration of nitrogen fertilizer significantly alters the formation but not the taxon richness of fungal communities in the rhizosphere and soil (Paungfoo-Lonhienne et al. 2015).

2.8.1 Integrated Metagenomics Methods

Various cellular techniques are successfully applied to microbial diversity analysis, including polymerase chain reactions, cloning and ribosomal gene sequence, denaturing gradient gel electrophoresis, borderline polymorphism, polymorphism-length terminal-restriction, and fluorescent hybridization. Additionally, the 16S rRNA gene is used as a phylogenetic marker to analyze genetic diversity, as this gene has been remarkably well preserved in the first few years of evolution.

There are different types of next-generation sequencing technology (NGS) that have recently been used in microbial studies. These include internal transcribed spacer (ITS) and amplicon gene sequence (target enlargement) of 16S rDNA stored to investigate viral and fungal variants. Metagenome sequences obtain information about the life force and genetic diversity of all viral communities present in a particular area. Metatranscriptomics involves the sequence of cDNA (a modified mRNA of active genes) to measure genetic expression in relation to genetic and metagenome reference and thus to identify potential functional functions and active microbes. Metaproteomics sequence of proteins is used to evaluate the expressed proteins and their richness to provide information on the active function between plants and living organisms. Metabolomics profile of metabolites is extracted using a state-of-the-art mass spectrometry (MS) method to detect and measure molecular growth and its possible involvement in the metabolic response of plants and bacterial communities.

Recently, a combination of advanced chemical analysis techniques and molecular biology such as gas chromatography-mass spectrometry (GC-MS), capillary electrophoresis-mass spectrometry (CE-MS), and liquid chromatography-mass spectrometry (LC-MS) has significantly improved the qualitative and quantitative analyses of the chemical and or metabolites of any plant or tissue component, the rhizosphere, and the natural niche (Zhang et al. 2012; Zhang et al. 2015).

2.9 Bioinformatics Tools

2.9.1 Metagenome Analysis Software

Metagenome sequence generates a large amount of nucleotide sequence data that need to be re-analyzed to obtain accurate results. Bioinformatics software is required to process DNA sequences generated by Sanger sequence and a separate next-generation sequencing (NGS) platform, i.e., 454 pyrosequencing and Illumina producing long and short readings, respectively. There are various software available for amplicon analysis and used for Sanger sequencing having 454 ribosomal pyrotag sequence. Quantitative Insights into Microbial Ecology (QIIME), MEGAN, and CARMA are among the most important and widely used software for metagenomic analysis (Caporaso et al. 2014; Gerlach and Stoye 2011; Huson and Weber 2013).

Recently, software such as the Illumina study and PacBio study have been developed for the purpose of metagenomic analysis of short-term readings and very long sequences, respectively. There are a few metagenomic forums available that provide information about the analysis of microbial diversity. One of the major challenges in analyzing ecological sequences is data integration and the question of how to analyze different types of data in an integrated manner that can provide information in both taxonomic and operational analyses. In order to cope with these challenges, community enabling cloud compatibility platform is available, which includes IMG/M (Markowitz et al. 2006), CAMERA (Seshadri et al. 2007), and WebCARMA (Gerlach et al. 2009). The Joint Genome Institute (JGI) supports IMG (Integrated Microbial Genomes), which provides tools for analyzing microbial genomes, genes, and functions. A community database for metagenomic data deposition CAMERA (Cyberinfrastructure for Advanced Marine Microbial Ecology Research and Analysis) is an important first step in developing methods for monitoring microbial communities. CARMA is a new software which characterizes the species composition and the genetic potential of microbial samples using short, unassembled reads. The Galaxy framework supports Cloud Virtual Resource (CloVR) (Angiuoli et al. 2011), basic metagenomic analysis (Pond et al. 2009), and Metagenomics Rapid Annotation using Subsystem Technology (MG-RAST) (Wilke et al. 2011).

2.9.2 Transcriptomics

Rhizosphere biology has revolutionized recent advances in DNA sequencing technologies by recording microbial formation by deep integration with metagenomics. However, functional understanding has not been provided by metagenomics, which thus provides a functional role for the obscure functional rhizospheric microbiome. Transcriptomics and metatranscriptomics both are desirable as they are able to determine both the structure and function of the active rhizospheric microbiome and thus complete metagenomics data. Transcriptomics is about the study of the corresponding RNA content that is produced under a certain natural environment. On the other hand, metatranscriptomics refers to the high sequence of complete RNA isolated from a natural sample. And two of the most popular metatranscriptomics tools used to study the rhizosphere include RNA sequence and gene expression microarray.

RNA sequencing is a method used to sequence and measure RNA molecules in a sample using next-generation sequencing (NGS) technology. RNA-seq reveals a complete transcriptome with quantitative and qualitative properties of mRNA, rRNA, and tRNA and is currently considered the gold standard for genetic analysis. The first step in this RNA-seq process involves the breakdown of high-quality RNA in rhizospheric soils, which is followed by conversion to cDNA fragments (cDNA library) followed by NGS. Urich et al. (2008) for the first time used the "Double-RNA approach" to determine the formation and function of bacterial communities in the soil by sequencing both rRNA and mRNA in a single transcriptome.

Microarray is a technique that involves the collection of microscope probes attached to a solid surface used to analyze high-resolution expression and comparative studies of genomics hybridization (Martínez et al. 2015). This process is also used to monitor genetic expression and to detect viruses present in different natural samples. Mendes et al. (2011) used a microarray-based approach to characterize the rhizosphere microbiome and identified 33,000 species of bacteria and fossils. *Rhizobium leguminosarum* biovar *viciae* inoculation effect on gene expression of pea, alfalfa, and sugar beet rhizosphere was previously studied with the help of a microarray method that revealed the presence of preserved plant colonies (Ramachandran et al. 2011).

2.9.3 Proteomics Methods

Proteomics has provided new opportunities to test soil biodiversity and functions. It is one of the most relevant and alternative metagenomics methods that provide useful information for key biological factors that perform key metabolic functions to solve the mystery of soil-acquisition skills in a particular ecology (Ploetze et al. 2015).

Proteomics is a systematic biological approach and is considered a sensible option to investigate plant and bacterial interactions. The investigation here was based on a two-dimensional electrophoresis (2-DE) gel. The identification of classified proteins is determined by the sequential library available on the website. Initially, the protein site is derived from agarose gel, which is then extracted and subjected to electrolytic cleavage in which peptide fragments are formed. Strategies such as matrix-assisted laser desorption/light ionization time (MALDI-TOF) mass spectrometry (MS) analysis will be performed to analyze broken peptide fragments. MALDI-TOF analysis produces a peptide list of broken fragments. A special feature of the protein is the predictable fraction size from the gene sequence that is also compared to the site of the peptide enumerated/sent for each open reading frame (ORF) in the genome. If no similarity is found in the sequential database, then the proteins can be analyzed in peptide sequence.

Also, proteomics has a missing list of applications including the diagnosis of plant diseases and their management and the analysis of the diversity of microbiomes by contact with soil, water, and other organisms. In addition, studying plant diseases, resistance, or infection is highly beneficial for proteomics. Microarray technology will be used to develop proteomics in such a way that the immune system analyzes changes in protein levels with a large amount of protein in a chip in a similar way to how mRNA mutations are currently measured. Proteomics is widely used to classify intercellular proteins that provide insight into microbial activity in the soil of the rhizosphere. Biological control of brown rice disease, caused by the deadly virus *Helminthosporium oryzae*, was investigated and tripartite interactions between the pathogen-biocontrol agent (*Bacillus*) and rice were investigated using a proteomics method. About nine proteins including ribulose 1,5-bisphosphate carboxylase, ATP synthase, serine/threonine protein kinases, 2-cys-peroxiredoxin, trehalose-phosphatase, and 50S ribosomal proteins were obtained using 2-D polyacrylamide gel electrophoresis (PAGE) analysis followed by a different expression using strategies such as MALDI-TOF mass spectrometry (Prabhukarthikeyan et al. 2019). These proteins can however help in plant utilization and produce a protective response against the brown spot pathogen. The proteomics method to study the expression of protein in the soil of the rhizosphere during interactions between plant organisms and soil was used by Wang et al. (2011). Therefore, proteomics is an appropriate and alternative method and one of the most effective ways to solve complex plant–bacterial interactions.

2.9.4 Metaproteomics Methods

Within the umbrella of the "omics," metaproteomics is among the recent and most innovative trends. This method examines the expression pattern of proteins within a complex biological system and thus provides direct evidence of the physiological and metabolic functions of the microbiome. Therefore, metaproteome expression will help to improve the understanding of the microbial world and thus favor microbial communities in natural functions (Wang et al. 2014). This process incorporates high-performance mass spectrometry to identify a suite of proteins that can regulate metabolic functions in microbial communities (Hettich et al. 2013). Also, the discovery of extensive metagenomic sequences from different microbial communities has extended the genomics era to a new exit area of research in recent years.

However, metaproteomics is one of the best ways to analyze the bacterial community in the soil. This process is basically done in four main steps, which includes: (1) rhizosphere soil collection sample, (2) protein extraction, (3) purification and separation, and analysis of MS, and finally (4) defining a protein and analyzing it using bioinformatics (Wang et al. 2014). Also, two major workflows have been developed for metaproteomics analysis including sodium dodecyl sulfate polyacrylamide gel electrophoresis (SDS-PAGE) combined with matrix-assisted laser desorption/ionization time-of-flight (MALDI-TOF-TOF) mass spectrometry analysis (MS) or electrospray ionization source tandem MS analysis (ESI-MS/MS) and liquid chromatography combined with electrospray ionization source tandem MS (LC-ESI-MS/MS).

To solve the mystery of interactions between plants and bacteria in the soil ecosystem, the metaproteomics analysis of the rhizosphere soil is a very powerful and useful scientific method. A standardized method was developed by Wang et al. (2011) for protein extraction from different soil samples with the identification of 1000 different sites with high density and contaminated 2-DE gels. It was noted that 189 spots represent 122 proteins in 2-DE gel, which are rice samples identified by MALDI-TOF/TOF-MS successfully. And the proteins identified mainly from rice and bacteria were involved in various metabolic activities including nucleotide, energy, and secondary metabolism, as well as signal transmission and anti-stress-related condition. In the sugarcane field, a physiological profile analysis (CLPP) of rhizosphere soil involving metaproteomics analysis was performed to determine the cause of the decline in sugarcane yields. It was noted that the amount of sugarcane made significant changes in the activity of the enzyme in the soil, the catabolic microbial community, and changes in the level of protein production in the soil. In addition, they were also found to influence biochemical processes in the rhizosphere ecosystem and thus contribute to intermediate sugarcane interactions and bacterial interactions (Lin et al. 2013). On the basis of comparative metaproteomics analysis, it was found that 38 proteins have been shown to vary in number in the soil of sugarcane, which was responsible for the decline in yield. Also, metaproteomics analysis of bacterial communities (bacteria and archaea) was performed in the soil of the phyllosphere and the rhizosphere and the phyllosphere of rice by Knief et al. (2012). Also, in the metaproteomics information analysis, a total of 4600 detected proteins were found, and they reflected the process of single carbon conversion in the phyllosphere and rhizosphere. Rhizosphere, however, was dominated by proteins involved in methanogenesis and methanotrophy and phyllosphere was dominated by methylobacterium. Also, the enzyme namely dinitrogenase reductase was found entirely in the rhizosphere despite the presence of nifH genes (Knief et al. 2012).

2.9.5 Metabolomics

Metabolomics is a method of qualitative and quantitative study of low-molecular-weight metabolites (<1 kDa). It serves as an important tool for the detection, measurement, and clarification of cell interactions within the rhizosphere. Most plant-to-microbial contact and microbe-to-microbe communication in the rhizospheric niche are associated with group of secondary metabolites. Thus, exploring these metabolites in the rhizosphere enhances the interaction of the various cells operating in the plant–microbe interface that further reveals a number of important signaling pathways involved in promoting plant growth and stress defense, and creates systemic resistance against plant diseases (Nath et al. 2017). Metabolomics therefore strengthens our understanding of the mechanisms of cells and cells operating in the rhizosphere.

There are usually three major steps in the normal course of mass spectrometry (MS)-based metabolomics. The first step is to prepare a sample that involves the release of metabolites using organic solvents or by using a solid-phase extraction method. The second step is concerned with differentiation and extraction, in which metabolites are categorized by different chromatographic techniques based on the nature of the metabolites and then "obtained through the use of quantitative analysis." For flexible and stable thermal compresses, gas chromatography-mass spectrometry (GC-MS) is popular, which can separate the metabolites by gas chromatography and obtain them by quadrupole, quadrupole time-of-flight (qTOF), or triple-quadrupole (QqQ) mass analysis (van Dam and Bouwmeester 2016). However, on the other hand, liquid chromatography-mass spectrometry (LC-MS) usually uses standard phase (NP) or regression phase (RP) chromatography to classify metabolites on the basis of their variability. Finally, data is tested with free software such as MarVis1, MAVEN, Mzine, Metaboanalyst, and MetAlign, or available commercial software such as Markerlynx, Profile Solutions, and Mass Profiler Pro. Rhizosphere metabolomics was previously used to study various plant growth compounds such as ACC deaminase, auxins, cytokinins, abscisic acid, gibberellins, jasmonic acid, salicylic acid, and bacterial-derived siderophores (Mhlongo et al. 2018). Also, metabolomics is one of the best tools to study molecules that show root and nodule symbiosis, i.e., flavonoids. In addition, the role of acyl homoserine lactones (AHLs) and its degrading products in the rhizospheric microbial response was investigated by Rothballer et al. (2018). However, metabolomics also documented changes in the microbial community in the grass (*Avena barbata*) rhizosphere during the period of development and biological sequence in relation to substrate preferences in altering root exudates (Zhalnina et al. 2018). However, due to the limited public index site, the cost of equipment and the lack of appropriate technology make metabolomics much more complex than DNA-based sequencing methods.

2.9.6 Phenomics

Phenomics is a systematic study of phenotypes on a wide range of genome or a set of multiple ways to study how the genome of living organism translates a complete set of phenotypic traits. Because of large number of genes interacting with each other and the nature to produce phenotype, the prediction of phenotype from genotype does not get reflected. However, the metagenomics approach provided access to a complete genotype of rhizospheric microbes up to genus, species, and subspecies levels.

In addition to the traditional techniques for making phenotypic characters, methods such as transcriptomics, proteomics, and metabolomics are widely used tools that provide large-scale phenomics data, and thus explain the phenomics of rhizospheric microbes (Houle et al. 2010). Rhizobia phenome (phenotypic traits) has been studied to differentiate and thus classify into different candidate groups. Phenomics has also been very useful for the study and management of pathogen co-evolution and the interaction of pathogens at the cellular level. Also, the complexity of biological processes at various stages of development needs to be addressed with high-quality digital phenotypic data. Recently, the work on the global *Escherichia coli* promoter was infiltrated through PFI boxes to obtain high-level genetic data amid antibiotic stress (French et al. 2018). To achieve microbial compatibility in any ecosystem, growth measurement is a key phenotype. However, there are limited tools available to study the phenomics of rhizospheric bacteria. However, there is a need for the development of advanced phenotyping tools and

Table 2.2 Advanced molecular techniques involving metagenomics for the characterization of rhizosphere microbial communities

Techniques	Objective of the study	Plant community of microbes	References
Metagenome sequencing	Grassland plant community richness and soil edaphic 454 pyrosequencing to analyze rhizosphere fungal communities during soybean growth	Soybean (*Glycine max*) rhizosphere	LeBlanc et al. (2015), Sugiyama et al. (2014)
Amplicon gene sequencing of conserved marker genes, 16S rRNA	Bacterial and fungal rhizosphere communities in hydrocarbon-contaminated soils	Rhizobacterial population of *Arachis hypogea* Rhizosphere of apple (*Malus pumila*) nurseries	Bell et al. (2014), Haldar and Sengupta (2015), Sun et al. (2014)
Metatranscriptome sequencing	Rhizosphere microbiome assemblage affected by plant development Root surface microbiome	*Arabidopsis* rhizosphere	Chaparro et al. (2014), Ofek-Lalzar et al. (2014)
Metaproteomics profiling	Phyllosphere and rhizosphere study	Sugarcane (*Saccharum officinarum*) rhizosphere	Lin et al. (2013)
Metabolomics profiling	Mycorrhizal study of roots	Tomato (*Solanum lycopersicum*) rhizosphere	Rivero et al. (2015)

advanced adaptation to deal with detailed phenomics of rhizospheric microorganisms. Table 2.2 enlists various molecular techniques employed for the characterization of rhizosphere microbial community.

2.10 The Role of CRISPR for Plant Development

CRISPR—Clustered Regularly Interspaced Short Palindrome Repeats—and CRISPR-compliant nine protein pathways have been developed in more than 20 plants to date (Sedeek et al. 2019). Genome engineering for targeted crop development for improvement of crop yield and stress management, including management of biotic and abiotic conditions, is a promising tool for future improvement of plant sciences. Biotic stress caused by various phytopathogenic bacteria possesses serious challenges in crop loss around the globe, which can be overcome by using CRISPR technology by which we can develop disease-resisting tolerant cultivars that reduce the crop losses in future.

A study by Wang et al. (2016) strongly reported an increase in plant resistance to *Magnaporthe* outbreak using a targeted CRISPR/Cas9 transformation process in ethylene responsive factor (ERF), OsERF922 in rice. Similarly, successful CRISPR/Cas9-mediated genome engineering in soybeans (*Glycine max*)—using a different transgene sgRNA and six sgRNAs that targeted multiple sites of two endogenous soybeans (GmSHR and GmFEI2) and testing the efficacy of sgRNAs in the hairy root system—was the first report by Cai et al. (2015). Several other studies employed CRISPR to release Avr4/6, a genetic pathogen virulence in *Phytophthora sojae* (Fang and Tyler 2016). Also, the Avolog4/6 mutation (NPT II), which is an induced gene for the CRISPR/Cas9 gene, focuses on the contribution made by the gene virulence to pathogen-induced pathogen production in soybeans, namely Rps4 and Rps6. In addition, by identifying the CRISPR/Cas9 tools, two OsSWEET13s are formed, which are flexible in directing its promoter, which thus led to improved bacterial tolerance for rice. There are various important factors such as crop yield and abiotic resistance that are controlled by influence of more than one gene. However, in a variety of crop development programs, a few studies are trying to map out these multifactor QTL, which plays a role in managing important aspects of agriculture. There are a number of identified quantitative regions introduced to promote advanced species in selected lines. Therefore, the CRISPR/Cas9 process can be a useful tool for introducing and learning unusual mutations in crop plants. Therefore, the use of the CRISPR/Cas9 genetics and genetically modified approach accelerates research and increases researchers' ability to produce genetic models (Borrelli et al. 2018).

2.11 Basics of CRISPR-Mediated Plant–Microbial Interactions in Agriculture

Plant–bacterial interactions depend on genetics in both the microbiome and host (Levy et al. 2018). Because of the important role of microorganisms in plant stress resilience, identification of candidate gene responsible for plant–microbe interaction is important in regulating the agronomic factor, which will help to improve desired plant characteristics for agricultural and industrial purposes. CRISPR-based tools have given novel insights into learning the genetic functions by initiating genetic mutations in plants or viruses or bacteria. CRISPR-based tools have the unique advantage of completely eliminating the target gene compared to genetic mutation as part of the RNA interference (RNAi) method that produces a phenotype component (Bisht et al. 2019; Sarma et al. 2021).

Therefore, with the help of CRISPR-based tools, one can gain accurate genetic information, especially genetic function in plant–microbe interactions at the cellular level. More recently, the CRISPR/Cas and ssDNA recombineering method was developed in the rhizospheric bacterium, i.e., *Pseudomonas putida* KT2440, to modify various genes, including gene removal, replacement, and insertion (Sun et al. 2018). However, mechanical research is important to assess the genetic link for harmful or pathogenic plant–microbe interactions in the case of non-model microbial isolation. Thus, genome engineering of non-model microbes with powerful CRISPR/Cas tools enables studies to make links between particular genes and their respective functions. In addition, the latest method of using biomaterials either in DNA, mRNA, or protein precedes the unique solution for the delivery of CRISPR/Cas into biological implants in normal ways (Eoh and Gu 2019).

2.12 Conclusion

The purpose of rhizosphere microbiome to implement and maintain the ecosystem of plant is well established. However, various conventional techniques used for understanding the function are still in infancy stage for vast majority of microorganisms present in the rhizosphere. Unraveling new plant signal molecules and exudates from roots in particular environment make available biochemical and microbial markers for study about how beneficial microbes are recruited and stimulated by plants in rhizosphere. The association present between crops and rhizospheric microbes has catalyzed the view of plants as a holobiont. For the management of rhizosphere, all the present components can be manipulated to favor the plant growth and productivity. Global climate change has one of the impacts on structure and function of various microbial species in the rhizosphere microbiome. To decrypt the extent of the effect of climate change on rhizosphere is still in need to be explored.

Acknowledgments D.K. acknowledges Institution of Eminence Seed Grant [Dev. No. 6031 (B)], Banaras Hindu University, Varanasi, India, for providing the financial support to the laboratory. D.K. also acknowledges the Centre of Advanced Study in Botany, Institute of Science, Banaras Hindu University, Varanasi, India, for providing necessary facilities and infrastructural support.

References

Agarwal P, Giri BS, Rani R (2020) Unravelling the role of rhizospheric plant-microbe synergy in phytoremediation: a genomic perspective. Curr Genomics 21:334–342

Alqarawi AA, Hashem A, Kumar A, Al-Arjani ABF, Abd Allah EF, Dar BA, Wirth S, Davranov K, Egamberdieva D (2018) Allelopathic effects of the aqueous extract of *Rhazya stricta* on growth and metabolism of *Salsola villosa*. Plant Biosyst 152:1263–1273

Angiuoli SV, Matalka M, Gussman A, Galens K, Vangala M, Riley DR, Arze C, White JR, White O, Fricke WF (2011) CloVR: a virtual machine for automated and portable sequence analysis from the desktop using cloud computing. BMC Bioinformatics 12(1):1–5

Arkhipova TN, Veselov SU, Melentiev AI, Martynenko EV, Kudoyarova GR (2005) Ability of bacterium *Bacillus subtilis* to produce cytokinin and to influence the growth and endogenous hormone content of lettuce plants. Plant Soil 272:201–209

Arshad M, Hussain A, Javed M, Frankenberger WT (1993) Effect of soil applied L-methionine on growth, nodulation and chemical composition of *Albizia lebbeck* L. Plant Soil 148:129–135

Arzanesh MH, Alikhani HA, Khavazi K, Rahimian HA, Miransari M (2011) Wheat (*Triticum aestivum* L.) growth enhancement by *Azospirillum* sp. under drought stress. World J Microbiol Biotechnol 27:197–205

Asiegbu FO, Nahalkova J, Li G (2005) Pathogen-inducible cDNAs from the interaction of the root rot fungus *Heterobasidion annosum* with Scots pine (*Pinus sylvestris* L). Plant Sci 168:365–372

Bais HP, Weir TL, Perry LG, Gilroy S, Vivanco JM (2006) The role of root exudates in rhizosphere interactions with plants and other organisms. Annu Rev Plant Biol 57:233–266

Barcellos FG, Batista JSS, Menna P et al (2009) Genetic differences between Bradyrhizobium japonicum variant strains contrasting in N2-fixation efficiency revealed by representational difference analysis. Arch Microbiol 191:113–122

Barea JM, Pozo MJ, Azcon R, Aguilar CA (2005) Microbial co-operation in the rhizosphere. J Exp Bot 56:1761–1778

Bell TH, Hassan SE, Lauron-Moreau A, Al-Otaibi F, Hijri M, Yergeau E, St-Arnaud M (2014) Linkage between bacterial and fungal rhizosphere communities in hydrocarbon-contaminated soils is related to plant phylogeny. ISME J 8:331–343

Bell CW, Asao S, Calderon F, Wolk B, Wallenstein MD (2015) Plant nitrogen uptake drives rhizosphere bacterial community assembly during plant growth. Soil Biol Biochem 85:170–182

Berendsen RL, Pieterse CM, Bakker PA (2012) The rhizosphere microbiome and plant health. Trends Plant Sci 17:478–486

Berg G, Rybakova D, Grube M, Koberl M (2016) The plant microbiome explored: implications for experimental botany. J Exp Bot 67:995–1002

Berlanas C, Berbegal M, Elena G, Laidani M, Cibriain JF, Sagües A, Gramaje D (2019) The fungal and bacterial rhizosphere microbiome associated with grapevine rootstock genotypes in mature and young vineyards. Front Microbiol 10:1142

Bisht DS, Bhatia V, Bhattacharya R (2019) Improving plant-resistance to insect-pests and pathogens: the new opportunities through targeted genome editing. Semin Cell Dev Biol 96:65–76

Bordenstein SR, Theis KR (2015) Host biology in light of the microbiome: ten principles of Holobionts and Hologenomes. PLoS Biol 13:e1002226

Borrelli VM, Brambilla V, Rogowsky P, Marocco A, Lanubile A (2018) The enhancement of plant disease resistance using CRISPR/Cas9 technology. Front Plant Sci 9:1245

Bowen GD, Rovira AD (1999) The rhizosphere and its management to improve plant growth. Adv Agron 66:1–102

Brilli F, Pollastri S, Raio A, Baraldi R, Neri L, Bartolini P, Podda A, Loreto F, Maserti BE, Balestrini R (2019) Root colonization by *Pseudomonas chlororaphis* primes tomato (*Lycopersicum esculentum*) plants for enhanced tolerance to water stress. J Plant Physiol 232: 82–93

Brimecombe MJ, De Leij FAAM, Lynch JM (2001) Effect of introduced *Pseudomonas fluorescens* strains on soil nematode and protozoan populations in the rhizosphere of wheat and pea. Microb Ecol 38:387–397

Bulgarelli D, Garrido-Oter R, Münch PC, Weiman A, Dröge J, Pan Y, McHardy AC, Schulze-Lefert P (2015) Structure and function of the bacterial root microbiota in wild and domesticated barley. Cell Host Microbe 17:392–403

Cai Y, Chen L, Liu X, Sun S, Wu C, Jiang B, Han T, Hou W (2015) CRISPR/CAS9-mediated genome editing in soybean hairy roots. PLoS One 10(8):e0136064

Canarini A, Kaiser C, Merchant A, Richter A, Wanek W (2019) Root exudation of primary metabolites: mechanisms and their roles in plant responses to environmental stimuli. Front Plant Sci 10:157

Caporaso JG, Kuczynski J, Stombaugh J, Bittinger K, Bushman FD, Costello EK, Fierer N, Peña AG, Goodrich JK, Gordon JI, Huttley GA, Kelley ST, Knights D, Koenig JE, Ley RE, Chaparro JM, Badri DV, Vivanco JM (2014) Rhizosphere microbiome assemblage is affected by plant development. ISME J 8:790–803

Carvalhais LC, Dennis PG, Badri DV, Kidd BN, Vivanco JM, Schenk PM (2015) Linking jasmonic acid signaling, root exudates, and rhizosphere microbiomes. Mol Plant-Microbe Interact 28: 1049–1058

Castrillo G, Teixeira PJ, Paredes SH, Law TF, de Lorenzo L, Feltcher ME, Finkel OM, Breakfield NW, Mieczkowski P, Jones CD, Paz-Ares J (2017) Root microbiota drive direct integration of phosphate stress and immunity. Nature 543:513–518

Chaparro JM, Badri DV, Vivanco JM (2014) Rhizosphere microbiome assemblage is affected by plant development. ISME J 8:790–803

Chen S, Li X, Lavoie M, Jin Y, Xu J, Fu Z, Qian H (2017) Diclofop-methyl affects microbial rhizosphere community and induces systemic acquired resistance in rice. J Environ Sci 51:352–360

Cook RJ, Thomashow LS, Weller DM, Fujimoto D, Mazzola M, Bangera G, Kim DS (1995) Molecular mechanisms of defense by rhizobacteria against root disease. Proc Natl Acad Sci U S A 92:4197

Dennis PG, Miller AJ, Hirsch PR (2010) Are root exudates more important than other sources of rhizodeposits in structuring rhizosphere bacterial communities? FEMS Microbiol Ecol 72:313–327

Dessaux Y, Petit A, Farrand SK, Murphy PJ (1998) Opines and opine-like molecules involved in plant-Rhizobiaceae interactions. In: The rhizobiaceae. Springer, Dordrecht, pp 173–197

Dinkelaker B, Hengeler C, Marschner HJBA (1995) Distribution and function of proteoid roots and other root clusters. Bot Acta 108:183–200

Dommergues YR (1978) The plant–microorganism system. In: Developments in agricultural and managed forest ecology, vol 4. Elsevier, pp 1–37

Dong H, Zhang D (2014) Current development in genetic engineering strategies of *Bacillus* species. Microb Cell Factories 13:1–11

Doornbos RF, Van Loon LC, Bakker PA (2012) Impact of root exudates and plant defense signaling on bacterial communities in the rhizosphere: a review. Agron Sustain Dev 32:227–243

Egamberdiyeva D, Kamiliva F, Validov S, Gafurova L, Kucharova Z, Lugentenberg B (2008) High incidence of plant growth stimulating bacteria associated with the rhizosphere of wheat grown in salinated soil in Uzbekistan. Environ Microbiol 10:1–9

Eoh J, Gu L (2019) Biomaterials as vectors for the delivery of CRISPR–Cas9. Biomater Sci 7(4): 1240–1261

Fang Y, Tyler BM (2016) Efficient disruption and replacement of an effector gene in the oomycete *Phytophthora sojae* using CRISPR/Cas9. Mol Plant Pathol 17:127–139

Foster KR, Bell T (2012) Competition, not cooperation, dominates interactions among culturable microbial species. Curr Biol 22:1845–1850

Frederix M, Downie JA (2011) Quorum sensing: regulating the regulators. Adv Microb Physiol 58: 23–80

French S, Coutts BE, Brown ED (2018) Open-source high-throughput phenomics of bacterial promoter-reporter strains. Cell Syst 7:339–346

Gaiero JR, McCall CA, Thompson KA, Day NJ, Best AS, Dunfield KE (2013) Inside the root microbiome: bacterial root endophytes and plant growth promotion. Am J Bot 100:1738–1750

Gargallo-Garriga A, Preece C, Sardans J, Oravec M, Urban O, Peñuelas J (2018) Root exudate metabolomes change under drought and show limited capacity for recovery. Sci Rep 8(1):1–15

Gerlach W, Stoye J (2011) Taxonomic classification of metagenomic shotgun sequences with CARMA3. Nucleic Acids Res 39:e91

Gerlach W, Jünemann S, Tille F, Goesmann A, Stoye J (2009) Web CARMA: a web application for the functional and taxonomic classification of unassembled metagenomic reads. BMC Bioinformatics 10:430

Giri B, Prasad R, Varma A (2018) Root biology. Springer International Publishing. ISBN 978-3-319-75910-4. https://www.springer.com/us/book/9783319759098

Giri B, Prasad R, Wu Q-S, Varma A (2019) Biofertilizers for sustainable agriculture and environment. Springer International Publishing. ISBN 978-3-030-18932-7. https://www.springer.com/gp/book/9783030189327

Goel R, Kumar V, Suyal DC, Soni R (2018) Toward the unculturable microbes for sustainable agricultural production. In: Meena VS, Mishra PK, Bisht JK, Pattanayak A (eds) Role of rhizospheric microbes in soil. Springer, Singapore, pp 107–123

Grosskopf T, Soyer OS (2014) Synthetic microbial communities. Curropin Microbiol 18:72–77

Gupta A, Raina M, Kumar D (2022) Endophytic microorganisms: utilization as a tool in present and future challenges in agriculture. In: Radhakrishnan EK, Kumar A, Aswani R (eds) Biocontrol mechanisms of endophytic microorganisms. Academic, New York, pp 285–301. ISBN: 978-0-323-88478-5.

Haldar S, Sengupta S (2015) Impact of plant development on the rhizobacterial population of *Arachis hypogaea*: a multifactorial analysis. J Basic Microbiol 55:922–928

Hartmann A (2020) Quorum sensing N-acyl-homoserine lactone signal molecules of plant beneficial gram-negative rhizobacteria support plant growth and resistance to pathogens. Rhizosphere 16:100258

Hawes MC, Gunawardena U, Miyasaka S, Zhao X (2000) The role of root border cells in plant defense. Trends Plant Sci 5:128–133

Hettich RL, Pan C, Chourey K, Giannone RJ (2013) Metaproteomics: harnessing the power of highperformance mass spectrometry to identify the suite of proteins that control metabolic activities in microbial communities. Anal Chem 85:4203–4214

Hinsinger P, Plassard C, Tang C, Jaillard B (2003) Origins of root-mediated pH changes in the rhizosphere and their responses to environmental constraints: a review. Plant Soil 248:43–59

Houle D, Govindaraju DR, Omholt S (2010) Phenomics: the next challenge. Nat Rev Genet 11: 855–866

Htwe AZ, Moh SM, Moe K, Yamakawa T (2018) Effects of co-inoculation of *Bradyrhizobium japonicum* SAY3-7 and *Streptomyces griseoflavus* P4 on plant growth, nodulation, nitrogen fixation, nutrient uptake, and yield of soybean in a field condition. Soil Sci Plant Nutr 64:222–229

Huang XF, Chaparro JM, Reardon KF, Zhang R, Shen Q, Vivanco JM (2014) Rhizosphere interactions: root exudates, microbes, and microbial communities. Botany 92:267–275

Hubbard CJ, Brock MT, van Diepen LT, Maignien L, Ewers BE, Weinig C (2018) The plant circadian clock influences rhizosphere community structure and function. ISME J 12:400–410

Huson DH, Weber N (2013) Microbial community analysis using MEGAN. Methods Enzymol 531:465–485

Jones DL, Nguyen C, Finlay RD (2009) Carbon flow in the rhizosphere: carbon trading at the soil–root interface. Plant Soil 321:5–33

Joshi D, Kumar S, Suyal DC, Goel R (2017) The microbiome of the Himalayan ecosystem. In: Mining of microbial wealth and metagenomics. Springer, New York, pp 101–116

Kaushal M, Prasad R (2021) Microbial biotechnology in crop protection. Springer, Singapore. ISBN: 978-981-16-0049-4. https://www.springer.com/gp/book/9789811600487

Kim SB, Timmusk S (2013) A simplified method for gene knockout and direct screening of recombinant clones for application in *Paenibacilluspolymyxa*. PLoS One 8:e68092

Knief C, Delmotte N, Chaffron S, Stark M, Innerebner G, Wassmann R, von Mering C, Vorholt JA (2012) Metaproteogenomic analysis of microbial communities in the phyllosphere and rhizosphere of rice. ISME J 6:1378–1390

Köberl M, Ramadan EM, Adam M, Cardinale M, Hallmann J, Heuer H, Smalla K, Berg G (2013) Bacillus and Streptomyces were selected as broad-spectrum antagonists against soilborne pathogens from arid areas in Egypt. FEMS Microbiol Lett 342:168–178

Köberl M, White RA III, Erschen S, El-Arabi TF, Jansson JK, Berg G (2015) Draft genome sequence of Streptomyces sp. strain Wb2n-11, a desert isolate with broad-spectrum antagonism against soilborne phytopathogens. Genome Announc 3:e00860–e00815

Korir H, Mungai NW, Thuita M, Hamba Y, Masso C (2017) Co-inoculation effect of rhizobia and plant growth promoting rhizobacteria on common bean growth in a low phosphorus soil. Front Plant Sci 8:141

Koyama H, Kawamura A, Kihara T, Hara T, Takita E, Shibata D (2000) Overexpression of mitochondrial citrate synthase in *Arabidopsis thaliana* improved growth on a phosphorus-limited soil. Plant Cell Physiol 41:1030–1037

Kremer RJ, Means NE (2009) Glyphosate and glyphosate-resistant crop interactions with rhizosphere microorganisms. Eur J Agron 31:153–161

Kumar A, Dubey A (2020) Rhizosphere microbiome: engineering bacterial competitiveness for enhancing crop production. J Adv Res 24:337–352

Kumar R, Eipers P, Little RB, Crowley M, Crossman DK, Lefkowitz EJ, Morrow CD (2014) Getting started with microbiome analysis: sample acquisition to bioinformatics. Curr Protoc Hum Genet 82(1):18–18

Kumar J, Sharma N, Kaushal G, Samurailatpam S, Sahoo D, Rai AK, Singh SP (2019) Metagenomic insights into the taxonomic and functional features of kinema, a traditional fermented soybean product of Sikkim Himalaya. Front Microbiol 1744

Kumar A, Chandra D, Sharma AK (2021) Impact of seed applied Rhizobacterial inoculants on growth of wheat (*Triticum aestivum*) and cowpea (*Vigna unguiculata*) and their influence on Rhizospheric microbial diversity. Agric Res 11:1–14

Kushwaha RK, Rodrigues V, Kumar V, Patel H, Raina M, Kumar D (2020) Soil microbes-medicinal plants interactions: ecological diversity and future prospect. In: Varma A, Tripathi S, Prasad R (eds) Plant microbe symbiosis. Springer, Cham, pp 263–286

Lagos L, Maruyama F, Nannipieri P, Mora ML, Ogram A, Jorquera MA (2015) Current overview on the study of bacteria in the rhizosphere by modern molecular techniques: a mini–review. J Soil Sci Plant Nutr 15:504–523

Lebeis SL, Paredes SH, Lundberg DS, Breakfield N, Gehring J, McDonald M, Malfatti S, Del Rio TG, Jones CD, Tringe SG, Dangl JL (2015) Salicylic acid modulates colonization of the root microbiome by specific bacterial taxa. Science 349:860–864

LeBlanc N, Kinkel LL, Kistler HC (2015) Soil fungal communities respond to grassland plant community richness and soil edaphics. Microb Ecol 70:188–195

Levy A, Conway JM, Dangl JL, Woyke T (2018) Elucidating bacterial gene functions in the plant microbiome. Cell Host Microbe 24:475–485

Lin W, Wu L, Lin S et al (2013) Metaproteomic analysis of ratoon sugarcane rhizospheric soil. BMC Microbiol 13:135

Lundberg DS, Lebeis SL, Paredes SH, Yourstone S, Gehring J, Malfatti S et al (2012) Defining the core Arabidopsis thaliana root microbiome. Nature 488:86–90

Lynch JM (1990) Soil biotechnology: beneficial interactions between microorganisms and roots. Biotechnol Adv 8:335–346

Markowitz VM, Korzeniewski F, Palaniappan K, Szeto E, Werner G, Padki A, Zhao X, Dubchak I, Hugenholtz P, Anderson I et al (2006) The integrated microbial genomes (IMG) system. Nucleic Acids Res 34:D344–D348

Martínez MA, Soto-Del Río Mde L, Gutiérrez RM et al (2015) DNA microarray for detection of gastrointestinal viruses. J Clin Microbiol 53:136–145

Massalha H, Korenblum E, Tholl D, Aharoni A (2017) Small molecules below-ground: the role of specialized metabolites in the rhizosphere. Plant J 90:788–807

Medema MH, Blin K, Cimermancic P, De Jager V, Zakrzewski P, Fischbach MA, Weber T, Takano E, Breitling R (2011) Anti SMASH: rapid identification, annotation and analysis of secondary metabolite biosynthesis gene clusters in bacterial and fungal genome sequences. Nucleic Acids Res 39:W339–W346

Mendes R, Kruijt M, de Bruijn I, Dekker E, der Voort V, Schneider M (2011) Deciphering the rhizosphere microbiome for disease-suppressive bacteria. Science 332:1097–1100

Mendes R, Garbeva P, Raaijmakers JM (2013) The rhizosphere microbiome: significance of plant beneficial, plant pathogenic, and human pathogenic microorganisms. FEMS Microbiol Rev 37: 634–663

Mendes LW, Raaijmakers JM, De Hollander M, Mendes R, Tsai SM (2018) Influence of resistance breeding in common bean on rhizosphere microbiome composition and function. ISME J 12: 212–224

Metzker ML (2010) Sequencing technologies—the next generation. Nat Rev Genet 11:31–46

Mhlongo MI, Piater LA, Madala NE, Labuschagne N, Dubery IA (2018) The chemistry of plant–microbe interactions in the rhizosphere and the potential for metabolomics to reveal signaling related to defense priming and induced systemic resistance. Front Plant Sci 9:112

Mohanram S, Kumar P (2019) Rhizosphere microbiome: revisiting the synergy of plant-microbe interactions. Ann Microbiol 69:307–320

Mönchgesang S, Strehmel N, Schmidt S, Westphal L, Taruttis F, Müller E, Herklotz S, Neumann S, Scheel D (2016) Natural variation of root exudates in *Arabidopsis thaliana* linking metabolomic and genomic data. Sci Rep 6:1–1

Mullis KB (1990) The unusual origin of the polymerase chain reaction. Sci Am 262(4):56–65

Nannipieri P, Badalucco L (2003) Biological processes. In: Bembi DK, Nieder R (eds) Handbook of processes in the soil–plant system: modelling concepts and applications. The Haworth Press, Binghamton, NY, pp 57–76

Nath M, Bhatt D, Prasad R, Tuteja N (2017) Reactive oxygen species (ROS) metabolism and signaling in plant-mycorrhizal association under biotic and abiotic stress conditions. In: Varma A, Prasad R, Tuteja N (eds) Mycorrhiza. Springer International Publishing AG, pp 223–232

Neal JL, Larson RI, Atkinson TG (1973) Changes in rhizosphere populations of selected physiological groups of bacteria related to substitution of specific pairs of chromosomes in spring wheat. Plant Soil 39:209–212

Newton JA, Fray RG (2004) Integration of environmental and host-derived signals with quorum sensing during plant–microbe interactions. Cell Microbiol 6:213–224

O'Donnell ST, Ross RP, Stanton C (2020) The progress of multi-omics technologies: determining function in lactic acid bacteria using a systems level approach. Front Microbiol 10:3084

Ofek-Lalzar M, Sela N, Goldman-Voronov M, Green SJ, Hadar Y, Minz D (2014) Niche and host-associated functional signatures of the root surface microbiome. Nat Commun 5:1–9

Patel U, Sinha S (2011) Rhizobia species: a boon for "plant genetic engineering". Indian J Microbiol 51:521–527

Paungfoo-Lonhienne C, Yeoh YK, Kasinadhuni NR, Lonhienne TG, Robinson N, Hugenholtz P, Ragan MA, Schmidt S (2015) Nitrogen fertilizer dose alters fungal communities in sugarcane soil and rhizosphere. Sci Rep 5:1–6

Pinton R, Varanini Z, Nannipieri P (2007) The rhizosphere: biochemistry and organic substances at the soil-plant interface. CRC Press, West Palm Beach, FL

Ploetze FA, Bertin PN, Carapito C (2015) Proteomic tools to decipher microbial community structure and functioning. Environ Sci Pollut Res 22:13566–13612

Pond SK, Wadhawan S, Chiaromonte F, Ananda G, Chung WY, Taylor J, Nekrutenko A, Team G (2009) Windshield splatter analysis with the Galaxy metagenomic pipeline. Genome Res 19 (11):2144–2153

Prabhukarthikeyan SR, Yadav MK, Anandan A, Aravindan S, Keerthana U, Raghu S, Baite MS, Parameswaran C, Panneerselvam P, Rath PC (2019) Bio-protection of brown spot disease of rice and insight into the molecular basis of interaction between *Oryza sativa*, *Bipolaris oryzae* and *Bacillus amyloliquefaciens*. Biol Control 137:104018

Prasad R, Kumar M, Varma A (2015) Role of PGPR in soil fertility and plant health. In: Egamberdieva D, Shrivastava S, Varma A (eds) Plant growth-promoting rhizobacteria (PGPR) and medicinal plants. Springer International Publishing, Cham, pp 247–260

Quiza L, St-Arnaud M, Yergeau E (2015) Harnessing phytomicrobiome signaling for rhizosphere microbiome engineering. Front Plant Sci 6:507

Raaijmakers JM, Paulitz TC, Steinberg C, Alabouvette C, Moenne-Loccoz Y (2009) The rhizosphere: a playground and battlefield for soilborne pathogens and beneficial microorganisms. Plant Soil 321:341–361

Ramachandran VK, East AK, Karunakaran R, Downie JA, Poole PS (2011) Adaptation of *Rhizobium leguminosarum* to pea, alfalfa and sugar beet rhizospheres investigated by comparative transcriptomics. Genome Biol 12:R106

Ramesh R, Ghanekar A, Joshi M (2009) Pseudomonads: major antagonistic endophytes to suppress bacterial wilt pathogen, *Ralstonia solanacearum* in eggplant (*Solanum melanogena* L.). World J Microbiol Biotechnol 25:47–55

Rasmann S, Turlings TC (2016) Root signals that mediate mutualistic interactions in the rhizosphere. Curr Opin Plant Biol 32:62–68

Rawat M, Rangarajan S (2019) Omics approaches for elucidating molecular mechanisms of microbial bioremediation. In: Smart bioremediation technologies. Academic, New York, pp 191–203

Richardson AE, Hadobas PA, Hayes JE (2001) Extracellular secretion of *Aspergillus* phytase from *Arabidopsis* roots enables plants to obtain phosphorus from phytate. Plant J 25:641–649

Rivero J, Gamir J, Aroca R, Pozo MJ, Flors V (2015) Metabolic transition in mycorrhizal tomato roots. Front Microbiol 6:598

Rothballer M, Uhl J, Kunze J, Schmitt-Kopplin P, Hartmann A (2018) Detection of the bacterial quorum-sensing signaling molecules N-acyl-homoserine lactones (HSL) and N-acyl-homoserine (HS) with an enzyme-linked immunosorbent assay (ELISA) and via ultrahigh-performance liquid chromatography coupled to mass spectrometry (UHPLC-MS). Methods Mol Biol 1673:61–72

Rubin RL, van Groenigen KJ, Hungate BA (2017) Plant growth promoting rhizobacteria are more effective under drought: a meta-analysis. Plant Soil 416:309–323

Ryan PR, Dessaux Y, Thomashow LS, Weller DM (2009) Rhizosphere engineering and management for sustainable agriculture. Plant Soil 321:363–383

Sarma H, Forid N, Prasad R, Prasad MNV, Ma LQ, Rinklebe J (2021) Enhancing phytoremediation of hazardous metal(loid)s using genome engineering CRISPR–Cas9 technology. J Hazard Mater. https://doi.org/10.1016/j.jhazmat.2021.125493

Savka MA, Dessaux Y, Oger P, Rossbach S (2002) Engineering bacterial competitiveness and persistence in the phytosphere. Mol Plant-Microbe Interact 15:866–874

Schlemper TR, Leite MF, Lucheta AR, Shimels M, Bouwmeester HJ, van Veen JA, Kuramae EE (2017) Rhizobacterial community structure differences among *Sorghum* cultivars in different growth stages and soils. FEMS Microbiol Ecol 93

Schulz-Bohm K, Gerards S, Hundscheid M, Melenhorst J, de Boer W, Garbeva P (2018) Calling from distance: attraction of soil bacteria by plant root volatiles. ISME J 12:1252–1262

Sedeek KEM, Mahas A, Mahfouz M (2019) Plant genome engineering for targeted improvement of crop traits. Front Plant Sci 10:114

Seshadri R, Kravitz SA, Smarr L, Gilna P, Frazier M (2007) CAMERA: a community resource for metagenomics. PLoS Biol 5:e75

Sharma A, Singh A, Raina M, Kumar D (2021) Plant-fungal association: an ideal contrivance for combating plant stress tolerance. In: Prasad R, Nayak SC, Kharwar RN, Dubey NK (eds) Mycoremediation and environmental sustainability. Springer Nature, Cham. https://doi.org/10.1007/978-3-030-54422-5_13

Sharma A, Raina M, Kumar D, Singh A, Chugh S, Jain S, Kumar M, Rustagi A (2022) Harnessing phytomicrobiome signals for phytopathogenic stress management. J Biosci 47:6

Shrivastava S, Prasad R, Varma A (2014) Anatomy of root from eyes of a microbiologist. In: Morte A, Varma A (eds) Root engineering, vol 40. Springer-Verlag, Berlin, pp 3–22

Singh RP, Kothari R, Koringa PG, Singh SP (2017) Plant-pathogen interactions: a proteomic approach. In: Kaur A, Kumar A, Reddy S (eds) Understanding host-microbiome interactions—an omics approach. Springer, New York, pp 207–226

Soni R, Goel R (2011) nif-H homologs from soil metagenome. Ekologija 57:87–95

Soni P, Nutan KK, Soda N, Nongpiur RC, Roy S, Singla-Pareek SL, Pareek A (2015) Towards understanding abiotic stress signaling in plants: convergence of genomic, transcriptomic, proteomic, and metabolomic approaches. In: Elucidation of abiotic stress signaling in plants. Springer, pp 3–40

Soni R, Kumar V, Suyal DC, Jain L, Goel R (2017) Metagenomics of plant rhizosphere microbiome. In: Understanding host-microbiome interactions—an omics approach. Springer, pp 193–205

Sorensen J (1997) The rhizosphere as a habitat for soil microorganisms. In: Elsas JDV, Trevors JT, EMH W (eds) Modern microbiology. Marcel Dekker, New York, pp 21–45

Sugiyama A, Ueda Y, Zushi T, Takase H, Yazaki K (2014) Changes in the bacterial community of soybean rhizospheres during growth in the field. PLoS One 9:e100709

Sun J, Zhang Q, Zhou J, Wei Q (2014) Illumina amplicon sequencing of 16S rRNA tag reveals bacterial community development in the rhizosphere of apple nurseries at a replant disease site and a new planting site. PLoS One 9:e111744

Sun J, Wang Q, Jiang Y et al (2018) Genome editing and transcriptional repression in pseudomonas putida KT2440 via the type II CRISPR system. Microb Cell Factories 17:41

Suyal DC, Joshi D, Debbarma P, Soni R, Das B, Goel R (2019) Soil metagenomics: unculturable microbial diversity and its function. In: Mycorrhizosphere and Pedogenesis. Springer, New York, pp 355–362

Taghavi S, Barac T, Greenberg B, Borremans B, Vangronsveld J, van der Lelie D (2005) Horizontal gene transfer to endogenous endophytic bacteria from poplar improves phytoremediation of toluene. Appl Environ Microbiol 71:8500–8505

Theis KR, Dheilly NM, Klassen JL, Brucker RM, Baines JF, Bosch TC, Cryan JF, Gilbert SF, Goodnight CJ, Lloyd EA, Sapp J (2016) Getting the hologenome concept right: an eco-evolutionary framework for hosts and their microbiomes. Msystems 1:e00028–e00016

Tournas VH, Katsoudas E (2005) Mould and yeast flora in fresh berries, grapes and citrus fruits. Int J Food Microbiol 105:11–17

Urich T, Lanzén A, Qi J, Huson DH, Schleper C, Schuster SC (2008) Simultaneous assessment of soil microbial community structure and function through analysis of the meta-transcriptome. PLoS One 3:e2527

Valverde A, De Maayer P, Oberholster T, Henschel J, Louw MK, Cowan D (2016) Specific microbial communities associate with the rhizosphere of Welwitschia mirabilis, a living fossil. PLoS One 11:e0153353

van Dam NM, Bouwmeester HJ (2016) Metabolomics in the rhizosphere: tapping into belowground chemical communication. Trends Plant Sci 21:256–265

Vance CP, Uhde-Stone C, Allan DL (2003) Phosphorus acquisition and use: critical adaptations by plants for securing a nonrenewable resource. New Phytol 157:423–447

Vermeer J, McCully ME (1982) The rhizosphere in *Zea*: new insight into its structure and development. Planta 156:45–61

Vicré M, Santaella C, Blanchet S, Gateau A, Driouich A (2005) Root border-like cells of Arabidopsis. Microscopical characterization and role in the interaction with rhizobacteria. Plant Physiol 138:998–1008

Walters WA, Jin Z, Youngblut N, Wallace JG, Sutter J, Zhang W, González-Peña A, Peiffer J, Koren O, Shi Q, Knight R (2018) Large-scale replicated field study of maize rhizosphere identifies heritable microbes. PNAS 115:7368–7373

Wang HB, Zhang ZX, Li H et al (2011) Characterization of meta-proteomics in crop rhizospheric soil. J Proteome Res 10:932–940

Wang DZ, Xie ZX, Zhang SF (2014) Marine meta-proteomics: current status and future directions. J Proteome 97:27–35

Wang F, Wang C, Liu P, Lei C, Hao W, Gao Y, Liu YG, Zhao K (2016) Enhanced Rice blast resistance by CRISPR/Cas9-targeted mutagenesis of the ERF transcription factor gene *OsERF922*. PLoS One 11(4):e0154027

Wang D, Li F, Cao S, Zhang K (2020) Genomic and functional genomics analyses of gluten proteins and prospect for simultaneous improvement of end-use and health-related traits in wheat. Theor Appl Genet 133(5):1521–1539

Weyens N, van der Lelie D, Taghavi S, Vangronsveld J (2009) Phytoremediation: plant–endophyte partnerships take the challenge. Curr Opin Biotechnol 20:248–254

Wilke A, Wilkening J, Glass EM, Desai NL, Meyer F (2011) An experience report: porting the MG-RAST rapid metagenomics analysis pipeline to the cloud. Concurr Comput Pract Exp 23: 2250–2257

Xu J, Zhang Y, Zhang P, Trivedi P, Riera N, Wang Y, Liu X, Fan G, Tang J, Coletta-Filho HD, Cubero J (2018) The structure and function of the global citrus rhizosphere microbiome. Nat Commun 9:4894

Yuan J, Zhao J, Wen T, Zhao M, Li R, Goossens P, Huang Q, Bai Y, Vivanco JM, Kowalchuk GA, Berendsen RL (2018) Root exudates drive the soil-borne legacy of aboveground pathogen infection. Microbiome 6:1–2

Zaheer A, Mirza BS, Mclean JE, Yasmin S, Shah TM, Malik KA, Mirza MS (2016) Association of plant growth-promoting *Serratia* spp. with the root nodules of chickpea. Res Microbiol 167: 510–520

Zhalnina K, Louie KB, Hao Z, Mansoori N, da Rocha UN, Shi S, Cho H, Karaoz U, Loqué D, Bowen BP et al (2018) Dynamic root exudates chemistry and microbial substrate preferences drive patterns in rhizosphere microbial community assembly. Nat Microbiol 3:470–480

Zhang A, Sun H, Wang P, Han Y, Wang X (2012) Modern analytical techniques in metabolomics analysis. Analyst 137(2):293–300

Zhang N, Wang D, Liu Y, Li S, Shen Q, Zhang R (2014) Effects of different plant root exudates and their organic acid components on chemotaxis, biofilm formation and colonization by beneficial rhizosphere-associated bacterial strains. Plant Soil 374:689–700

Zhang X, Qian Y, Cao C (2015) Effects of straw mulching on maize photosynthetic characteristics 1378 and rhizosphere soil micro-ecological environment. Chil J Agric Res 75:481–487

Chapter 3
The Role of the Root Microbiome in the Utilization of Functional Traits for Increasing Plant Productivity

Rahul Chandnani and Leon V. Kochian

Abstract It is now clear that the root microbiome, which consists of bacteria, archaea, and fungi that colonize both the rhizosphere and the internal space of the root, is one of the most complex ecosystems in nature and is very important for root and plant health and function.

In this chapter we have focused on the role of the root microbiome functional traits in improvement of nutrient acquisition and abiotic stress tolerance, with a focus on drought stress, the biocontrol of root and shoot plant diseases, and the role of root-associated microbes in both producing plant growth-promoting hormones and impacting the plant hormone metabolism and signaling pathways to alter root growth. Additionally, we have also endeavored to give the readers an introduction into the rapid advances in this field, from the metagenomic analyses that now have become relatively routine for the study of "what is there" in the root microbiome, regarding microbial composition, diversity, and abundance, to nascent studies beginning to study the plant and microbial molecular and physiological mechanisms and processes that underlie how the microbiome is assembled, and how the microbiome confers improved functional crop traits. Furthermore, given the incredible complexity of this ecosystem, we discuss the recent research involving systems biology analysis of the root microbiome, which will be critical in deciphering the trait–function links and interactions between roots and soil microbes. Finally, we also discuss the agricultural and genetic interventions that are being employed to modify the root microbiome via inoculation of the seed and plant with potentially beneficial soil microbes, as well as the studies looking at the role of plant genetic and molecular variation in impacting the composition and function of the microbiome.

Rahul Chandnani and Leon V. Kochian contributed equally with all other contributors.

R. Chandnani (✉) · L. V. Kochian (✉)
Global Institute for Food Security, University of Saskatchewan, Saskatoon, SK, Canada
e-mail: rahul.chandnani@gifs.ca; leon.kochian@gifs.ca

S. Chhabra et al. (eds.), *Plant Microbiome for Plant Productivity and Sustainable Agriculture*, Microorganisms for Sustainability 37,
https://doi.org/10.1007/978-981-19-5029-2_3

Keywords Root microbiome · Rhizobacteria · Functional traits · Metagenomics · Plant growth

3.1 Introduction

Diverse microbial communities are an integral part of plant and animal life. A root–microbial ecosystem that consists of soil microbes that colonize and inhabit the soil at the root surface, the rhizosphere, and live and function in the root, comprise what plant biologists commonly refer to as the root microbiome.

The composition, diversity, and microbial species structure of the root microbiome can be plant species-specific and contain both beneficial and harmful microorganisms (Bressan et al. 2009; Lugtenberg and Kamilova 2009; Takeuchi et al. 1996). Root microbiome composition is shaped by the host plant and, often to a greater degree, the soil microflora, and many of the microorganisms in this biome are beneficial for host plant and root health and function. Research findings have shown that beneficial microbes in the root microbiome can increase the solubility and uptake of soil macro- and micronutrients such as phosphorus, nitrogen, and iron (Yadav et al. 2021).

Several soil-based abiotic stresses such as drought and salinity can impact the composition of the root microbiome, resulting in microbiome shifts that may confer increased tolerance to these stresses (Chen et al. 2017; Marasco et al. 2012; Vurukonda et al. 2016; Giri et al. 2018). Beneficial microorganisms in the root microbiome can also enhance plant growth and development by both synthesizing growth phytohormones and altering plant hormone metabolism required for plant growth and thereby increase root system growth and crop productivity (Arkhipova et al. 2005; Duca et al. 2018; Kudoyarova et al. 2014; Ping and Boland 2004; Prasad and Zhang 2022). On the other hand, there are microbial species in the root microbiome that are harmful and can negatively affect plant and human health such as disease-causing bacteria and fungi (Takeuchi et al. 1996). A saying that "nothing is free" also stands true for root and soil microbial interactions as plant roots and root exudates serve as a source of carbon and sugars that are required for the survival, growth, and replication of the microorganisms in the root microbiome (Bais et al. 2006; Foster 1986). Furthermore, specific root exudates play roles as signaling molecules between the root and soil microbes, influencing the microbial composition of the root microbiome. The root microbiome is one of the largest and most complex biomes in nature and plays significant roles in the maintenance and growth of plants.

The exponential growth of the human population and diminishing agricultural lands and input resources (e.g., water, fertilizer) required for efficient agricultural output, warrants novel and sustainable ways to achieve food security. Employment of genetically narrow crop germplasm (seeds or tissue specific to a species, genotype, or population maintained for plant breeding purposes) and intensive selection techniques in current breeding programs have enhanced the vulnerability to agricultural pests and diseases (Hammons 1976). Significant recent advances in molecular-based plant breeding include tools such as genomic selection, which has been shown to improve genetic gain via prediction of crop performance without phenotypic analysis of novel germplasm. Further improvements to genomic selection is enabling plant breeders to improve genetic gain while not losing genetic diversity (see, for example, Daetwyler et al. 2015). However powerful these advances in plant breeding are, they have yet to be intensively applied to the "hidden half" of the plant—the root, and especially the root microbiome.

It is important now to focus on the role of the root microbiome in functional plant traits to continue to improve crop resiliency and sustainability. For example, application of inorganic fertilizers and pesticides has significantly improved crop yields since the dawn of the Green Revolution. However, increasing pesticide and pathogen resistance, the increasing cost of fertilizers, and especially the finite availability of phosphorous fertilizer, as well as environmental degradation due to leaching of a significant portion of applied fertilizer into surface and ground waters, make these approaches unsustainable in the future (Denholm et al. 1998; Savci 2012). To address these complex problems, a systems approach integrating advances in precision fertilizer and water management with fundamental research innovations resulting in more nutrient and water-efficient crop varieties is needed (Macintosh et al. 2019). A critical component of the research aimed at improving crop nutrient and water acquisition will be the investigations that enhance our understanding of the development and function of the root microbiome. This will provide new strategies for improving crop yields and agricultural sustainability through modification of the root microbiome in part via use of biofertilizers that enhance the availability and acquisition of essential mineral nutrients and also through microbial-based biocontrol approaches to enhance disease and pest resistance.

It has been known for well over 100 years that certain root–microbe interactions, specifically N_2-fixing bacteria in legume root nodules as well as arbuscular mycorrhizae (AM), play key roles in root N and P acquisition from the soil (Beijerinck 1901; Frank 1885; Hellriegel and Wilfarth 1888). In recent years with the realization that microbiomes associated with eukaryotic organisms play key roles in that organism's health, well-being, and function, research on the root microbiome is clearly demonstrating that in addition to N_2-fixing bacteria and AM, many other free-living or more intimately root-associated bacteria and fungi play roles in improving mineral nutrient availability in the rhizosphere and enhance biotic and abiotic stress tolerance. One example of this is microbial-mediated solubilization in the rhizosphere of sparingly soluble essential minerals such as P, Fe, and Zn (Fabiańska et al. 2019; Gururani et al. 2013; Harbort et al. 2020; Hiruma et al. 2016; Weiß et al. 2016). However, the molecular basis and physiological mechanisms underlying these improvements in plant function are still quite poorly understood. To effect changes in the root microbiome as a strategy to improve crop resiliency, improve yields using less inputs, and enhance agricultural sustainability, it is of great importance to understand the genetic and functional (molecular and physiological) mechanisms and regulation underlying the structure and function of root microbiomes. Quite a few studies have been successful in the identification and the use of plant growth-promoting bacteria from the soil and plant tissues to improve crop production (Ji et al. 2014; Kloepper et al. 1980; Li et al. 2021).

Certainly, the most significant advance in microbiome research over the past decade has been the metagenomic analysis of microbiomes involving isolation of microbial genomic DNA from bulk soil, the rhizosphere, and the root endosphere, followed by amplification of specific highly variable regions of bacterial and fungal genomes (16S rRNA for bacteria and the internal transcribed spacer [ITS]) region for fungi. For a review, see Sczyrba et al. 2017. Next-generation sequencing of these amplified bacterial and fungal gDNA regions and subsequent computational analysis of the sequence is used to identify microbial operational taxonomic units, microbial structure, and diversity. Furthermore, these types of studies have made it possible to begin to probe possible functions of the root microbiome community and to postulate about gene-function links that may be useful in identifying or designing specific root-associated microbes or communities of microbes that can be used in crop improvement (Chen et al. 2017; Naylor and Coleman-Derr 2018; Shulse et al. 2019; Xu et al. 2018, 2021). Comparatively recent advances in genome-resolved metagenomic and holo-omics approaches are finally enabling the researchers to identify the changes in host plant metabolomes and possibly important causal interactions between the root and microbes that could enhance nutrient uptake, microbial phytohormone production resulting in root growth promotion, and greater tolerance to abiotic and biotic stresses (Xu et al. 2021). In this chapter, we focus on possible advantages the microbiome could confer to the root systems in terms of acquiring resources for better plant growth, improving resistance to pathogens, and microbiome gene-functional links that have been and can be exploited further for these purposes.

3.2 Overview of the Root Microbiome

Dynamic and diverse groups of root-associated microorganisms, which include plant-beneficial microbes, are recruited by plant roots via signaling between the root and soil microbes (Hartman and Tringe 2019), to comprise the extremely complex microhabitat, that is, the root microbiome. This ecosystem can have impacts on plant health, growth, and function via direct or indirect pathways. The root microbiome consists of bacteria and archaea, algae, fungi, and protozoa, with bacteria being the most abundant component. The root microbiome can be quite diverse, and can consist of microbiota with as many as 33,000 different bacterial and archaea species (Mendes et al. 2011). Despite this amazing diversity, it has been reported that there are primarily two bacterial phyla, Actinobacteria and Proteobacteria, that dominate the global soil microbiome and root microbiome, for example, in disease suppressive soils (Delgado-Baquerizo et al. 2018; Mendes et al. 2011).

The root microbiome is actively recruited by plant roots and specific microbial communities are certainly influenced by the unique chemical composition of root exudates (Berendsen et al. 2012; Doornbos et al. 2012). Plants acquire the majority of their nutrients and water from the roots and, in return, low-molecular-weight organic compounds such as organic acids, sugars, phenolics, and amino acids are secreted as root exudates (Antoun 2013). Published research findings have shown that there can be an enrichment, for example, of unique organic acids in the root exudates in different plant species that can play an important role in shaping the root microbiome (Cotton et al. 2019; Hu et al. 2018; Huang et al. 2019; Tan et al. 2013; Wang et al. 2021). Kamilova et al. (2006) demonstrated that tomato root-tip-specific colonizing microbes had better growth and enrichment than other rhizobial microbes, when tomatoes were grown on minimal media with citrate, a major tomato root-tip organic exudate, as the primary carbon source. Furthermore, by introducing the stable carbon isotope, $^{13}CO_2$, into the rhizosphere of wheat, maize, rape, and barrel clover plants, it was shown that there were differences in the sources of carbon released from roots and used by different groups of microbes in the rhizosphere (el Zahar Haichar et al. 2008).

Root exudate metabolites often play a beneficial role by modifying the rhizobiome, which in turn can alter plant hormonal content and function. For example, Hu et al. (2018) identified the secondary metabolite, benzoxazinoids, which is released by maize roots and can shape the rhizobiome community. In turn, this can favor plant defenses by increasing the production of the plant defense hormone, jasmonic acid, conferring protection against herbivores. Differences in the genotype of a plant species, which can significantly impact the composition of root exudates, are another factor that not surprisingly influences the microbial composition of the root microbiome. It has been reported that transgenic *Arabidopsis thaliana* plants expressing the sorghum *CYP79A1* gene, resulted in the accumulation of high levels of a derivative of the sulfur secondary metabolite, glucosinolate, p-hydroxybenzylglucosinolate. This transgenic line exhibited significant alterations in the profile of the root exudation of glucosinolate compounds, which altered the microbial composition of root microbiome. This study showed that even small modifications in root metabolism can have significant effects on root exudates and the microbial composition of the root microbiome (Bressan et al. 2009).

There is a specific nomenclature used to define the microbiome on versus inside the root. Root microbes residing inside the root tissue are known as endophytes whereas the rhizomicrobiome is defined as microbes inhabiting the rhizosphere, the thin layer of soil intimately associated with the root surface (Bulgarelli et al. 2012; Edwards et al. 2015; Lundberg et al. 2012; Schlaeppi et al. 2014). Bulgarelli et al. (2012) demonstrated that the endosphere microbiome is distinct and does not have the same variation in microbial composition as does the rhizosphere, which is certainly more strongly influenced by the microbial composition of the bulk soil surrounding the root. Later, it was shown that the microbes residing on the surface of roots, sometimes termed the rhizoplane microbiome, are also distinct from the other two root microbiomes (the endomicrobiome and the rhizomicrobiome; Edwards et al. 2015).

3.3 Functional Traits to Enhance Plant Productivity

3.3.1 Biofertilizers that Impact Mineral Nutrient Availability and Acquisition by Roots

3.3.1.1 Nitrogen Fixation

Nitrogen is one of the most important mineral nutrients required for plant growth and can be a significant limiting factor to crop yields. Of course, nitrogen is essential for synthesis of amino acids, proteins, and enzymes that are prerequisite for many plant physiological processes (Novoa and Loomis 1981). Furthermore, nitrogen is the core component of chlorophyll that is required for photosynthesis, which provides the fixed carbon that underpins plant growth. Due to absence of large amounts of bioavailable nitrogen in the soil, most agricultural crop production relies upon the application of nitrogen fertilizers, usually as ammonia or urea. However, the increasing costs of nitrogen fertilizers and the significant environmental costs associated with the leaching of as much as 50–60% of N fertilizer before the plant roots can absorb it, often lead to nitrate pollution of ground waters and pose a greater risk to the environment (McCasland et al. 1985). Biological nitrogen fixation (BNF) is the process by which N_2 gas in the atmosphere is converted to NH_3 by nitrogenase enzyme activity in microbial diazotrophs, which are prokaryotes that have the ability to fix N_2 gas to ammonia (Kim and Rees 1994). BNF has the potential to be a sustainable alternative to fulfill the nitrogen requirement of plants. However, only a few species of plants have the ability to be colonized by nitrogen-fixing bacteria as symbiotic microbes. There are two types of diazotrophs defined based on their habitat: (1) symbiotic N_2 fixers which live within root nodules of primarily leguminous crop species, which include bacteria in the genera *Rhizobium* and *Frankia*; and (2) non-nodular diazotrophic bacteria that can establish either associative or free-living relationships with roots where the bacteria reside on or near the root surface (epiphytes). Other species of non-nodular N_2-fixing bacteria form endophytic relationships with a wide range of non-legumes, where the bacteria colonize inner plant tissues and reside within root and even shoot tissues. These free-living diazotrophic bacteria include species within the genera Azospirillum, Azoarcus, and Herbaspirillum (Santi et al. 2013).

Most current root-nodulated crops obtain fixed nitrogen by the activity of a molybdenum (Mo)-dependent nitrogenase complex (Boyd et al. 2011; Rubio and Ludden 2008). It has been shown that only three genes—*nifH*, *nifD*, and *nifK*—are required to encode the structural subunits of nitrogenase enzyme (Seefeldt et al. 2009; Yang et al. 2018). Mo-dependent nitrogenase (*Nif*) complexes are a two-component enzyme system. The dinitrogenase reductase component is a homodimeric iron (Fe) protein encoded by the *NifH* gene that donates electrons, and a dinitrogenase or heterotetrameric Mo–Fe protein component encoded by *NifDK* that contains the Fe–Mo cofactor that serves as the substrate reduction site, accepting electrons from the Fe–S electron transfer protein, ferredoxin (Bulen and

LeComte 1966). Apart from these two components, maturation of the nitrogenase enzyme complex for its activity also involves some regulatory proteins encoded by *nifE*, *nifN*, and *nifB* genes. Due to high similarity in 16S rRNA phylogeny and conserved nature of *nif* sequences, this similarity has been quite useful in identifying nitrogen-fixing bacteria from a soil sample. Quite a few studies have employed *nifH* as a phylogenetic marker to identify bacterial strains that make functional contributions to N_2 fixation (Boyd et al. 2011; Bürgmann et al. 2004; Coelho et al. 2009; Seefeldt et al. 2009).

Our understanding of the mechanisms and regulation of nitrogenase N_2 fixation and the nature of how and why these diazotrophs form these symbiotic relationships with the host plants open potential opportunities to enable non-leguminous crops to benefit from BNF. As cereals are the most widely grown food crops and are the source of the largest proportion of calories consumed by the human population, there has been considerable engineering biology research investigating and developing the tools to transfer nodule-based symbiotic nitrogen fixation to cereal crops (Burén et al. 2017; Ryu et al. 2020). This approach has been taken in the Burén et al. (2017) publication from the Voight Lab at MIT. The focus is to express the genes in the bacterial nitrogenase-dependent nitrogen-fixing pathway genes in the mitochondria or chloroplasts of eukaryotes, as these organelles have an ancient bacterial origin and thus are better suited for expression of bacterial genes than the nucleus of plant cells. In the Burén et al. (2017) publication, they have reengineered the 16 gene nitrogenase pathway from an N_2-fixing bacterium to remove its native regulation and replaced it with well-understood synthetic genetic parts. They have ultimately been able to express an important and functional section of the bacterial N_2-fixing pathway in the mitochondria eukaryotic model system, *Saccharomyces cerevisiae*, which is an important first step in generating a functional nitrogenase enzyme in plant cells.

A second approach to enhancing crop N nutrition via N_2-fixing soil microbes involves research focusing on free-living N_2-fixing bacteria. There are different approaches that are being explored for this and one of them is to transfer nitrogen-fixation ability from non-native diazotrophs to plant-host-colonizing rhizobacteria by introducing genomic islands that can encode the nitrogenase activity into free-living bacteria that readily colonize plant roots. Fox et al. (2016) demonstrated that transfer of X940 genomic island from *Pseudomonas* A1501 to the aerobic root-associated beneficial bacterium, *Pseudomonas protegens* Pf-5, followed by the inoculation of maize and wheat plants with this genetically modified bacterium, enabled the host plant's root surface (rhizoplane) and rhizosphere to be colonized by Pf-5, providing enough radiolabeled fixed nitrogen to the roots to confer higher grain and biomass yields.

3.3.1.2 Phosphorus Bioavailability and Uptake

Phosphorus is the second most important mineral nutrient for the plant (after N), and can be a major limiting factor in plant growth as P deficiency is important to a wide range of plant processes, including cell division, root and shoot development, biomass production, photosynthesis, and reproduction (Hu and Schmidhalter 2005; López-Arredondo et al. 2014; Razaq et al. 2017). Even if soils are not deficient in total P, inorganic phosphate, which is the primary inorganic form of P absorbed by plant roots, will readily bind to the Fe and Al oxides/hydroxides on the surface of soil clay minerals, rendering them unavailable for root P absorption (López-Arredondo et al. 2014). Root P acquisition is relatively inefficient, and only approximately 20–25% of the applied P fertilizer is taken up by the plant during the first season after fertilizer application (Roberts and Johnston 2015). Much of the remaining phosphate fertilizer is either fixed to soil minerals, or is lost as runoff into surface waters, resulting in environmental damage including algal blooms in lakes and streams, which are costly to remediate.

The availability of inorganic phosphate to plant roots also is highly dependent on soil pH (Marschner 1995). In acidic soils (pH < 5.5), as mentioned above, the phosphate anion binds to aluminum and iron in clay minerals, whereas at higher soil pH values, insoluble calcium phosphate is formed that cannot be absorbed by roots (Hinsinger 2001). Currently, more than half of the arable lands worldwide consist of either acidic or alkaline soil and are deficient in inorganic phosphate (López-Arredondo et al. 2014). Soil microbes can solubilize the phosphate bound to Fe and Al in the soil, as they can release organic acids such as citric and malic acids which are strong chelators of Fe and Al, which then will release the bound phosphate into the soil solution for absorption by plant roots. Phosphorus also accumulates in soil in relatively unavailable organic forms, thus being tied up in soil organisms and plant litter. For example, phytic acid or phytate, an organic storage form of P, accumulates in soil and plants and some rhizobacteria can release the enzyme, phytase, which hydrolyzes phytate, releasing inorganic phosphate into the soil solution where it also can be absorbed by roots (Ke et al. 2021; López-Arredondo et al. 2014; Shulse et al. 2019). These P-solubilizing soil microbes open additional avenues of research that may enable the development of sustainable microbial-based strategies to improve phosphate availability, enhancing root P acquisition efficiency in agricultural crops. Biofertilization is likely not sufficient to meet the complete phosphorus requirement of crops. However, it clearly could be a strategy to significantly improve P bioavailability from applied phosphate fertilizer, increasing farmer's yields per unit of P fertilizer applied, and reduce the environmental costs associated with remediating P pollution of waterways and ground water.

Research aimed at engineering rhizobacteria to increase P bioavailability was recently conducted by Shulse et al. (2019). In this study, they demonstrated that multiple rhizobacteria species (*Pseudomonas* and *Ralstonia* sp.), when transformed with phytase genes, significantly increased the release of soluble phosphate from soil phytate. In this study, a total of 6674 metagenomes were screened and 82 phylogenetically diverse phytase genes were selected and their sequences optimized for high gene expression in three species of rhizobacteria. The researchers identified 12 strains across three bacterial species that generated a significant increase in growth using *Arabidopsis thaliana* grown on phytate as the sole phosphate source. However, to take advantage of research findings such as the findings from Shulse et al. (2019)

using genetically engineered soil microbes, it will be necessary to develop a more comprehensive regulatory framework to foster the more facile movement of inventions such as genetically modified soil microbes from the laboratory to the marketplace. This will likely require the development of real partnerships between academia, government, and industry to produce a science-based regulatory system which is predictable and straightforward to navigate.

In another study, to investigate the microbial community structure of the root microbiome, metagenomic analysis of the root microbiome of *Lotus japonicus*, a wild legume that has a history of reasonable growth and yield without phosphate fertilizer, was done. It was found that specific bacterial phyla, including Bacteroidetes, Betaproteobacteria, Chlorobi, Dehalococcoidetes, and Methanobacteria, were specifically abundant in the root microbiome (Unno and Shinano 2013). These rhizobacterial genera include bacterial strains that could enhance phytic acid utilization, promoting plant growth using phytate as a P source. Furthermore, several gene clusters possibly involved in phytic acid utilization, including alkaline phosphatase and citrate synthase, were investigated by Chhabra et al. (2013) who characterized the mineral phytate as well as phosphate solubilization trait employing functional metagenomics of the barley rhizosphere. They discovered that mineral phosphate solubilization screening of fosmid clones in *E. coli* identified genes/operons related to phosphorus mineralization and uptake. Further, Yadav et al. (2021) discovered that genes involved in gluconic acid synthesis are involved in phosphate solubilization by creating transposon insertion mutant libraries of *Mesorhizobium ciceri* Ca181, which is a symbiotic rhizobacteria colonized with chickpea roots. Clearly, recent research in this area has identified several rhizobacterial species that could play a role in root microbiome-induced enhanced solubilization of P for root uptake from either fixed inorganic phosphate in the soil or from organic P released from phytate.

3.3.1.3 Increasing Soil Iron Bioavailability via Bacterial Siderophores

Iron is an essential micronutrient required for several significant cellular processes including chlorophyll synthesis, photosynthesis, respiration, nitrogen fixation, and hormone production in plants (Vert et al. 2002). Iron is available in the soil as ferric (Fe^{3+}) and ferrous (Fe^{2+}) ions depending on soil pH, and soil aeration and redox potential. In aerobic soils, ferric iron predominates, and it is quite insoluble across most soil pH values, except in highly acidic soils. Hence, most of the soluble ferrous iron exists as ligands chelated by compounds such as organic acids, phenolics, humic acids, and bacterial siderophores (Guerinot and Yi 1994). Root microbes play an important role in soil iron solubility and uptake by releasing siderophores, which are low-molecular-weight organic Fe^{3+} chelators that form strong ligands via high-affinity binding with Fe^{3+} ions (Das et al. 2007). These Fe^{3+} siderophore complexes are taken up into bacterial cells by specific bacterial membrane transporters under iron-limiting conditions (Neilands 1984). The major classes of compounds that function as bacterial siderophores include catecholates (also known as phenolates), hydroxamates, and carboxylates (derivatives of citric acid).

Plants have also evolved their own mechanisms to acquire Fe^{3+} from the soil. In dicots and non-gramineous monocots, they employ a root cell plasma membrane ferric reductase to reduce extracellular chelated Fe(III) iron and the resulting Fe^{2+} ions are released by the ferric chelate and rapidly transported into the root cell by iron-regulated transporter 1 (IRT1), a root cell plasma membrane Fe^{2+} transporter (Guerinot and Yi 1994; Marschner 1995). Grass species use their own phytosiderophore process, releasing non-protein amino acids in response to Fe deficiency, which chelate ferric (Fe^{3+}) ions in the soil and the Fe(III)-phytosiderophore complex is transported in toto across the root cell plasma membrane. Both types of plant Fe acquisition systems can absorb Fe chelated by bacterial siderophores. Using the reductase-based system, dicots and non-grass monocots can reduce the iron in the Fe(III)-siderophore complex at the root cell plasma membrane and then the released Fe^{2+} ion is transported into root cells via the IRT1 transporter. In grasses, it has been suggested that the uptake of Fe(III) complexes by bacterial siderophores can occur by Fe exchange between siderophores and phytosiderophores, or that the Fe-siderophore complex may be transported by the plant root plasma membrane transporter specifically functioning to absorb Fe(III)-phytosiderophore complexes (Bar-Ness et al. 1992; Crowley et al. 1988; Wang et al. 1993).

In addition to using bacterial siderophore release to enhance crop Fe acquisition and nutrition, the agronomic use of siderophore-producing plant-beneficial microbes can also be part of a strategy for biocontrol of plant pathogenic microbes. For example, the fluorescent Pseudomonads contain a number of plant-beneficial bacterial species. These fluorescent Pseudomonads release very high affinity Fe-binding siderophores which cannot generally be used by pathogenic bacteria. Hence, by decreasing the iron availability to phytopathogenic bacteria by chelating most of iron, these Pseudomonads promote plant growth by acting as a biocontrol agent against plant pathogens (Smarrelli and Castignetti 1986). Furthermore, fluorescent *Pseudomonas* sp. is one of initial bacterial strains explored that produced siderophores that have resulted in enhanced iron uptake in oat and mungbean (Crowley et al. 1988; Sharma et al. 2003). Sharma et al. (2003) showed that mungbean plants inoculated with a siderophore-producing *Pseudomonas* strain GRP_3, exhibited reduced chlorosis and increased chlorophyll content under low Fe growth conditions. The rhizobacterial species, *Rhizobium* and *Bradyrhizobuim*, have also been reported to produce siderophores under iron-deficient conditions (Nambiar and Sivaramakrishnan 1987). Molecular microbiologists are dissecting the gene pathways involved in regulating siderophore synthesis and release in response to Fe-limiting conditions. One key regulator of these processes is the ferric uptake regulator, or Fur protein, that regulates siderophore production in variable iron conditions (Hassan and Troxell 2013). As the understanding of the regulation of these processes advances, more molecular tools are becoming available that will better facilitate the modification of rhizobacterial genomes, to enable agricultural researchers to improve the root microbiome to enhance crop Fe nutrition in

Fe-limiting soils, and to provide microbial-based biocontrol against certain species of plant pathogenic bacteria.

3.3.2 Drought Tolerance

Drought stress is the most important abiotic stress limiting the yield of agricultural crops worldwide (Daryanto et al. 2016; Zipper et al. 2016). Additionally, climate change associated with global warming has exacerbated both the severity and frequency of droughts with more dire impacts on food crops, and increases the need for the identification of more sustainable approaches to improve crop performance under drought, mitigating drought-induced yield losses. Applications of beneficial rhizobacteria to both soils and via plant inoculation have been studied in various crop species to investigate the role of soil microbes and the root microbiome in enhanced resistance to drought stress, and have demonstrated some promising prospects for coping with drought (Chen et al. 2017; Marasco et al. 2012; Vurukonda et al. 2016). Beneficial rhizobacterial inoculants have been suggested to enhance performance under drought in different ways. First is production of enzymes that can catabolize stress-responsive phytohormones such as ACC deaminase enzyme that can degrade the ethylene precursor, ACC (1-aminocyclopropane-1-carboxylic acid), and thereby reduce ethylene production which at high levels inhibits root growth (Mayak et al. 2004). Another possible drought-resistance strategy involving beneficial rhizobacteria involves alterations in the levels of exopolysaccharides and the drought-associated amino acid, proline (Vardharajula et al. 2011). Proline acts as an osmolyte and when its concentration increases in the cell symplasm under drought, this creates a water potential gradient directed into the cell and allows the water to flow into the cell even though the water potential outside the cell has dropped due to drought. This confers the ability to increase the host plant's drought tolerance by increasing cellular osmotic and thus water potential, thus maintaining cell water content under drought stress. Also, proline can play a role in protecting cell membranes and other cell components against damage caused by free radical production during drought. Exopolysaccharides produced by rhizobacteria during drought stress can enhance soil aggregation and structure and thereby retain soil water near the roots during drought stress (Khan et al. 2017; Naseem and Bano 2014; Yoshiba et al. 1997).

Additionally, inoculation of plants with certain rhizobacteria can alter hormone-mediated drought responses including changes in root system architecture via modulation of phytohormones to increase lateral root growth and biomass, thereby enhancing the host plant's ability to acquire water from soils under drought (Arzanesh et al. 2011; Shakir et al. 2012; Zahir et al. 2008). Another example of this type of rhizobacterial drought response was a report about rhizobacterial-induced drought-resistance mechanisms that involved the production of volatile metabolites that led to a systemic response inducing stomatal closure and reduced water loss during drought stress. This systemic response appears to involve the volatile compound inducing interplay between ABA, jasmonic acid, and ethylene (Cho et al. 2008).

Although a few published studies have shown that colonization of roots with certain types of rhizobacteria resulted in improved plant performance under drought, the functional physiological and molecular mechanisms that drive the recruitment of a drought-responsive microbiome and the host plant responses remain poorly understood. Research based on 16S rRNA sequencing and metagenomic analysis has shed light on dynamic changes in the composition and function of the root microbiome under drought, by enrichment of bacterial phyla such as Actinobacteria, primarily in the root endosphere, but also in the rhizosphere (Edwards et al. 2015; Naylor and Coleman-Derr 2018; Naylor et al. 2017; Xu et al. 2018). In the Xu et al. (2018) study, a research approach integrating genome-resolved metagenomics, transcriptomics, and plant root metabolomics was conducted in sorghum during response to drought. The authors found that drought increased the abundance and activity of monoderm bacteria (Actinobacteria, Firmicutes, and Chloroflexi), which only have a single outer cell membrane and have thick cell walls. They also found that drought increased the production of certain metabolites in the root, including certain carbohydrates (especially glycerol-3-P) and amino acids, and that the monoderm bacteria showed a concomitant increase in the expression of transporters for these metabolites. Furthermore, inoculation of sorghum roots with monoderm isolates suggested that increased abundance of monoderm bacteria in the root microbiome increased plant growth during drought. These findings suggest that production of specific root metabolites by drought and their transfer to monoderm bacteria may play a role in reshaping the microbiome bacterial composition, to enhance plant growth under drought. These findings indicate that it may be possible to develop agricultural and/or genetic interventions that reshape the root microbiome to enhance crop performance under drought in agricultural crops.

3.3.3 Biocontrol of Plant Diseases

Crop diseases caused by pathogens (bacteria and fungi) can result in up to 30% yield loss in economically important crops (Soko et al. 2018; Tirnaz and Batley 2019). Hence, the impact of plant pathogens on crop production is a major challenge to global food security. For decades, chemical control measures have been applied to control plant diseases; however, the lack of rapid rates of breakdown of these chemicals into relatively safer constituents makes this approach detrimental to the environment (Gilden et al. 2010). Additionally, rapid trends in the evolution of pesticide resistance in plant pathogens have warranted the search for alternative and environmentally sustainable approaches, such as biocontrol (Lucas et al. 2015), which could be part of a broader strategy integrating biocontrol with plant genetics and breeding, better agronomic practices, and new classes of pesticides. Therefore, as the emphasis on environmentally friendly biocontrol strategies increases, over the past 20 years there have been an increasing number of publications on the use of beneficial bacteria to help control plant pathogens (Akgül and Mirik 2008; Girish and Umesha 2005; Mishra and Arora 2018; Sang et al. 2008). Even more recently, root and soil microbiome metagenomic studies have been published and these types of studies should ultimately increase our understanding of how the root microbiome operates as a system to protect especially against root diseases (Wille et al. 2019). For example, Lee et al. (2008) demonstrated the biocontrol of root diseases employing *Bacillus* subtilis extracted from rhizosphere soil to control *Phytophthora* blight of pepper caused by *Phytophthora capsici*, when seeds of pepper plants were inoculated with *Bacillus subtilis* before planting. There have been reports of several types of antibiotics produced by rhizobacteria that have been used to control plant root pathogens such as: pyrrolnitrin, phloroglucinols, phenazines, cyclic lipopeptides, and hydrogen cyanide (Haas and Défago 2005). Some of these antibiotics have been characterized functionally, including bacterial cyanide production and soil antifungal activity against root fungal pathogens that are both mediated by the *GacA/GacS* quorum sensing system in *Pseudomonas* sp. (Heeb and Haas 2001). The *GacA/GacS* two-component system is involved in the synthesis and release of secondary metabolites that have antifungal activity. The same *GacA/GacS* system is used by the beneficial rhizobacterium *Pseudomonas fluorescens* F113, and is responsible for HCN synthesis that enables *P. fluorescens* F113 to control the pathogenic fungal species, *Pythium ultimum*, which causes damping off disease in important agricultural crops (Aarons et al. 2000).

Another example of quorum sensing involved in production and release of antifungal compounds was shown in a study on the *PhzR/PhzI* system in the plant-beneficial bacterium, *Pseudomonas chlororaphis 30-84*. The *PhzR/PhzI* quorum sensing system regulates the synthesis and release of phenazine derivatives that are antibiotics with antifungal activity. The synthesis of phenazine is mediated by *PhzR*, whereas *PhzI* encodes for acyl-homoserine-lactone (AHL) synthase. AHL is the signal that activates *PhzR* to synthesize phenazine, which is then released into the soil. Using this two-component system, *Pseudomonas chlororaphis 30-84* was effective at controlling the serious fungal disease, Take-all, in wheat (Chin-A-Woeng et al. 2003; Pierson and Pierson 1996; Zhang and Pierson 2001).

In other publications, *Bacillus amyloliquefaciens* SQR9 has been reported to control *Fusarium* wilt disease in cucumber, and *Fusarium* head blight in wheat has been shown to be controlled by use of *Bacillus amyloliquefaciens* AS 43.3 (Cao et al. 2011; Dunlap et al. 2013). It has been shown that regulation of bacterial chemotaxis can impact biocontrol effectiveness. For example, Weng et al. (2013) reported on a mutation in the *AbrB* gene in *Bacillus amyloliquefaciens* SQR9 that regulates chemotaxis and biofilm formation around root, which resulted in the enhancement of *Bacillus amyloliquefaciens* SQR9 chemotaxis and biofilm formation, thereby improving root colonization and biocontrol efficacy. Dunlap et al. (2013) showed that whole-genome sequencing and analysis of *Bacillus amyloliquefaciens* AS 43.3 identified a number of biosynthetic gene clusters (ribosomal and non-ribosomal) for synthesis of secondary metabolites that act as antibiotics such as surfactin, difficidin, and plantazolicin. Finally, recent advances in genome-resolved metagenomics have enabled researchers to identify novel soil bacteria that can be resources for genes involved in the synthesis of important secondary metabolites (Crits-Christoph et al. 2018). In Crits-Christoph et al. (2018), the authors reconstructed hundreds of near-complete genomes of bacteria isolated from a natural grassland soil ecosystem and identified members of understudied bacterial phyla such as Acidobacteria, Rokubacteria, Gemmatimonadetes, and Verrucomicrobia that contain novel gene clusters encoding important plant-beneficial bacterial biosynthetic pathways. In this study, they found members of these phyla contained genes that encode diverse polyketides and non-ribosomal peptides, and these classes of compounds include many antibiotics, antifungals, and siderophores that thus could be used as novel beneficial rhizobacteria.

3.3.4 Plant Hormone-Producing Bacteria

Optimal situation-dependent levels of plant hormones such as indole-3-acetic acid (IAA), gibberellic acid (GA), the direct precursor of the hormone ethylene, 1-aminocyclopropane-1-carboxylic acid (ACC) deaminase, ethylene, and cytokinin are required for adequate plant growth due to their significant roles in all the stages of plant growth development from embryogenesis to seed development (Chen et al. 2014; Davies 2010). Certain taxa of plant growth and performance promoting bacteria enhance root and/or shoot growth by inducing alterations in plant hormonal homeostasis that can lead to improved overall plant growth, root growth, abiotic stress tolerance, and, ultimately, improved yield (Duca et al. 2018; Miransari 2014; Vessey 2003). Although quite a few species of rhizobacteria have been reported to produce phytohormones, *Pseudomonas* and *Bacillus* spp. have been studied more extensively than other bacterial species with regard to microbial biosynthesis of plant hormones (Duca et al. 2018; Patten and Glick 2002b; Shilev 2013; Vessey 2003).

3.3.4.1 Indole-3-Acetic Acid (IAA)

Plant root exudates often contain tryptophan, a precursor in the biosynthesis of the predominant class of auxin compounds, indole-3-acetic acid (IAA) via the indolepyruvic acid pathway (Patten and Glick 2002a). Root soil colonization in tea (*Camellia sinensis*) with the IAA-producing rhizobacteria, *Bacillus megaterium DE BARY TRS-4*, resulted in significantly improved tea plant growth and decreased level of the fungal disease, brown root rot (Chakraborty et al. 2006). Gene sequences and genetic pathways for IAA biosynthesis have been identified in *Pseudomonas* sp. UW4 in Duan et al. (2013), and later this information was used by Duca et al. (2018) to transform *Pseudomonas* sp. UW4, overexpressing four native IAA biosynthesis genes: *ami*, *nit*, *nthAB*, and *phe*. Overexpression of all four genes individually in *Pseudomonas* sp. UW4, resulted in significant increases in bacterial IAA concentrations. Canola seed inoculation with one of the four bacterial overexpression lines and subsequent canola growth demonstrated that all four overexpression lines had greater root growth in plants 10 days post inoculation. The *Pseudomonas* sp. UW4 *nit* overexpression line had the greatest stimulation in root growth (Duca et al. 2018). The authors also measured the activity of ACC (1-aminocyclopropane-1-carboxylic acid) deaminase in the transgenic and wildtype bacterial lines and found significant decreases (30–70%) in enzyme activity in the overexpression lines. As this enzyme decreases the amount of ACC, the direct precursor to ethylene, they speculated that the root growth increase could be due to a direct effect of IAA or one of the other auxin compounds released from the transgenic overexpression lines colonizing canola roots, or reduction in root ACC and ethylene by the increased levels of ACC deaminase could also increase root growth.

Low levels of auxin are often associated with lower plant growth; however, excessively high auxin levels can affect shoot and root growth adversely (Thimann 1939). Therefore, there have been a number of studies on bacterial genes and genetic pathways that enhance degradation of IAA and their impact on plant growth (Costacurta and Vanderleyden 1995; Patten and Glick 1996; Spaepen et al. 2007). Leveau and Gerards (2008) identified and characterized a putative IAA degrading *iac* gene cluster by using the *Pseudomonas putida 1290* strain originally isolated from pear tree foliage (Leveau and Lindow 2005). The *iac* gene cluster was shown to introduce IAA degradability in the *P. putida* KT2440 sp., which does not have the ability to degrade IAA, demonstrating the likeliness that some of the genes in this operon are involved in IAA catabolism (Leveau and Gerards 2008). Subsequently, Scott et al. (2013) conducted insertional inactivation of each of the genes in the *iac* cluster, and expression of the altered gene cluster in *E. coli* were combined with MS-based auxin metabolite analysis to demonstrate that *iac*-based degradation of IAA involved the first gene in the cluster, *iacA*, and transcript profiling of a knockout of another gene in the cluster, *iacR*, which encodes for a repressor of *iacA* expression, and this repression is overcome by exposure to IAA. As high levels of IAA also inhibit *Pseudomonas putida* 1290 growth as they can with regard to plant growth, the presence of this IAA degradation pathway in *Pseudomonas putida* might allow it to better colonize root tips where higher levels of root-synthesized IAA might occur.

3.3.4.2 Cytokinin

Cytokinin is an essential phytohormone of paramount importance for plant growth regulatory processes and cell division (Skoog and Armstrong 1970). Cytokinin production has been reported by quite a few rhizobacteria such as members of the genera *Azospirillum*, *Rhizobium*, *Pseudomonas*, and *Bacillus* (Cacciari et al. 1989; García de Salamone et al. 2001; Grover et al. 2021; Timmusk et al. 1999). It has been shown that inoculating plants with cytokinin-producing rhizobacteria can enhance plant growth and yield (Arkhipova et al. 2005; Kudoyarova et al. 2014; Ping and Boland 2004; Wang et al. 2018). Wang et al. (2018) showed that inoculation of

Arabidopsis cytokinin receptor knockout mutants with cytokinin-producing *Bacillus* sp. LZR216 alters shoot growth and root system architecture, significantly stimulates lateral root number and length, and inhibits tap root length, while increases leaf surface area and shoot biomass. *Bacillus* sp. LZR216 treatment upregulated the gene expression of the *Arabidopsis* cytokinin signaling genes *AHK3/AHK4*, *AHP1/AHP3*, and *ARR4/5/7/10/12/15* in root tips, suggesting that root-cytokinin-regulated genes may play a role in interactions and signaling between the root and cytokinin-producing bacteria such as *Bacillus* sp. LZR216.

3.3.4.3 ACC Deaminase Activity and Ethylene

Ethylene and its precursor 1-aminocyclopropane-1-carboxylate (ACC) play important roles not only in plant developmental but also in plant defense and symbiotic programs. Thus, it is likely that ethylene and ACC play a central role in the regulation of bacterial colonization and formation of root and shoot microbiomes. A number of rhizobacteria can produce significant amounts of the enzyme, 1-aminocyclopropane-1-carboxylate (ACC) deaminase, enabling rhizobacteria to degrade the ACC produced and exuded by plant roots and thereby reduce ethylene levels in roots (Glick et al. 1998; Jacobson et al. 1994). ACC is an immediate precursor to ethylene biosynthesis from methionine in plants (Adams and Yang 1979), and triggering of a surge in ethylene production can lead to root growth inhibition (Jackson 1991). Although the catalyzing function of ACC deaminase was first studied in free-living *Pseudomonas* spp., subsequently its activity was identified in many other bacteria including rhizobacteria from the genera *Rhizobium*, *Bacillus pumilus*, and *Rhodococcus*, and *Burkholderia phytofirmans* sp. Nov. (Glick 2005). The ACC deaminase gene, *acdS*, is well characterized in the beneficial rhizobacteria, *Pseudomonas putida* UW4, and it has been shown to be transcriptionally regulated by an *acdR* gene that encodes the leucine-responsive regulatory protein (Lrp). Lrp proteins have been found immediately upstream of many bacteria ACC deaminase genes, suggesting transcriptional regulation of these deaminase genes by Lrp's is a key feature in the regulation of bacterial ACC deaminases. On the other hand, the *nifA* gene, which is a master regulator of nitrogen fixation, has also been identified as a regulator of an *acdS* gene in *Mesorhizobium loti* MAFF303099 (Kaneko et al. 2000). These results suggest evolution of parallel pathways of regulation of *acdS* gene in different classes of proteobacteria that form symbiotic relationships with plant roots.

ACC deaminases in rhizobacteria, which apparently enable the bacteria to modulate the levels of root ethylene in the host organ and increase or decrease root growth where the bacteria reside, likely indicate this enzyme plays an important role in root–rhizobacterial interactions. The ability to enzymatically modify root ethylene levels is also a potential tool for agricultural researchers attempting to modulate the composition and function of the root microbiome. But in order to effectively use this information to facilitate root-microbiome-mediated enhancement of crop functional traits, it will be necessary to more deeply and completely understand the relationship between ACC deaminase containing rhizobacteria and the plant root.

3.4 Conclusions

3.4.1 Genome-Level Investigations of the Root Microbiome and Holo-Omics Are Required to Fully Exploit Microbiome Functional Traits

Plant growth and performance promoting microbes (PGPPM) in the root microbiome have the potential to be used as important alternative strategies to plant breeding or agronomic methods to enhance crop productivity, by enhancing nutrient acquisition, producing plant growth hormones, and enhancing disease and pest resistance. The genome of the root microbiome can be thought of as a second plant genome that needs to be decoded to provide the knowledge necessary to use natural and genetically modified root microbes to improve crop yields by enhancing root mineral nutrient and water acquisition, abiotic and biotic stress tolerance, and root system growth and vigor via altered plant hormone homeostasis (Fig. 3.1). Despite having great potential for the utilization of microbial functional traits, efficient root colonization with beneficial microbes is certainly one of the major challenges in enabling modification of root microbiomes to enhance crop functional traits. There have been major advances in this research area that ranges from the use of microbe-specific structural and functional markers to multi-omics approaches that combine two or more omics technologies such as metagenomics, transcriptomics, metabolomics, and proteomics to study the root microbiome at the genome level. However, functional studies of the root and its associated microbiome are still in their infancy, as scientists are developing and improving techniques to study the microbiome. As these types of studies advance and mature, they will enable the next steps, which will be to study the root microbiome as a functional ecosystem, using approaches such as deep-sequencing-based genome-resolved metagenomics, holo-omics, and microbiome/crop pan genome approaches. Using more systems-based approaches will be necessary to gain a deeper understanding of the detailed genetic, biochemical, and physiological networks underlying complex PGPPM-mediated processes and interactions of the root microbiome with the host plant roots and exudates, required for the microbial recruitment and root colonization resulting in enhancement of important crop functional traits. It is imperative that more of our efforts now focus on understanding the functional aspects of root–microbe interactions, which will be necessary to more effectively and efficiently use root microbiome modifications as a sustainable tool in our crop improvement toolbox to enhance plant resiliency and productivity, to help ensure global food security.

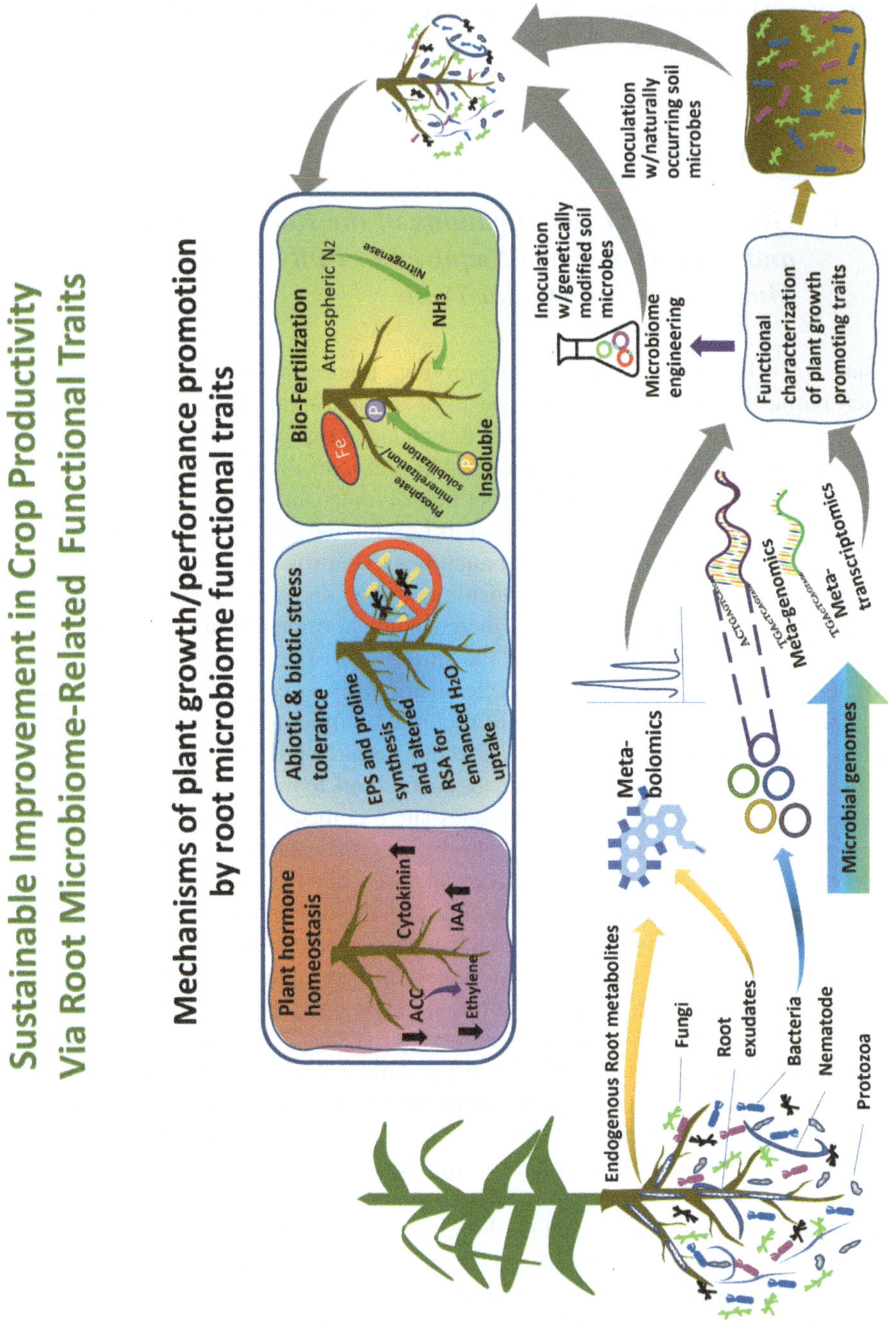

Fig. 3.1 Sustainable improvement in crop productivity via root microbiome-related functional traits. Genomic DNA or RNA is extracted from the root microbiome for meta-genomic and meta-transcriptomic analyses. Plant metabolites are extracted from the root and/or collected as root exudates (very difficult to currently do for roots in soil) for metabolomic analysis. The holo-omics approaches involves combining and analyzing the different meta-omics data from root

References

Aarons S, Abbas A, Adams C, Fenton A, O'Gara F (2000) A regulatory RNA (PrrB RNA) modulates expression of secondary metabolite genes in *Pseudomonas fluorescens* F113. J Bacteriol 182(14):3913–3919

Adams D, Yang SF (1979) Ethylene biosynthesis: identification of 1-aminocyclopropane-1-carboxylic acid as an intermediate in the conversion of methionine to ethylene. Proc Natl Acad Sci 76(1):170–174

Akgül D, Mirik M (2008) Biocontrol of Phytophthora capsici on pepper plants by Bacillus megaterium strains. J Plant Pathol 90:29–34

Antoun H (2013) Plant-growth-promoting rhizobacteria. Brenner's encyclopedia of genetics, vol 5, 2nd edn. Elsevier, Amsterdam

Arkhipova T, Veselov S, Melentiev A, Martynenko E, Kudoyarova G (2005) Ability of bacterium Bacillus subtilis to produce cytokinins and to influence the growth and endogenous hormone content of lettuce plants. Plant Soil 272(1):201–209

Arzanesh MH, Alikhani H, Khavazi K, Rahimian H, Miransari M (2011) Wheat (*Triticum aestivum* L.) growth enhancement by Azospirillum sp. under drought stress. World J Microbiol Biotechnol 27(2):197–205

Bais HP, Weir TL, Perry LG, Gilroy S, Vivanco JM (2006) The role of root exudates in rhizosphere interactions with plants and other organisms. Annu Rev Plant Biol 57:233–266

Bar-Ness E, Hadar Y, Chen Y, Shanzer A, Libman J (1992) Iron uptake by plants from microbial siderophores: a study with 7-nitrobenz-2 oxa-1, 3-diazole-desferrioxamine as fluorescent ferrioxamine B analog. Plant Physiol 99(4):1329–1335

Fig. 3.1 (continued) microbes, root tissue, and root exudates, which will be necessary for a more informative functional characterization of microbial plant performance promoting traits and to identify gene-functional links and networks. Functional information will be important to identify root beneficial microbes and to genetically manipulate naturally occurring bacteria that can be used as plant growth and performance promoting microbes (PGPPM). Inoculation with naturally occurring beneficial or genetically engineered rhizobacteria can possibly alter the root microbiome composition (if the introduced microbial species can compete and be established in this ecosystem), to increase plant productivity by different mechanisms such as biofertilization, abiotic and biotic stress tolerance, and plant hormone homeostasis. PGPPM can provide the major macronutrient, nitrogen, in the form of NH_3 by fixing atmospheric N_2 via nitrogenase in nodular and free-living root-associated N_2 fixing microbes. Inorganic phosphates that tend to be fixed in the soil, reducing their availability to roots, can be solubilized by PGPPM to provide bioavailable phosphate to the roots. Micronutrient iron can be provided to plant roots by siderophores released by bacteria in the root microbiome. Exopolysaccharides produced by PGPPM and microbe-induced increases in proline in plants can act as soil aggregants (exopolysaccharides) and root and shoot osmolytes (proline) to contribute to drought stress resistance. Additionally, microbiome-induced changes in plant hormone metabolism can alter root system architecture via increases in the number and length of lateral roots to promote increases in water and mineral nutrient acquisition efficiency. Antibiotics secreted by the root microbiome can act as biocontrol agents for fungal and bacterial soil-borne diseases, and competition by PGPPM with pathogenic bacteria for carbon, Fe, and other nutrients essential for microbial growth can also reduce the abundance of harmful microbes via competition within the root microbiome. Various plant hormones such as cytokinin and IAA required for adequate plant growth can be synthesized by specific members of the root microbiome and enhance plant growth, whereas reduction in concentrations of root growth inhibiting hormones such as ethylene by rhizobacterial-produced ACC deaminase can also promote root growth and plant productivity

Beijerinck M (1901) Uber oligonitrophile mikroben. Zentralbl Bakterol Parasitenkd Infektionskr Hyg Abt II 7:561–582

Berendsen RL, Pieterse CM, Bakker PA (2012) The rhizosphere microbiome and plant health. Trends Plant Sci 17(8):478–486

Boyd E, Anbar A, Miller S, Hamilton T, Lavin M, Peters J (2011) A late methanogen origin for molybdenum-dependent nitrogenase. Geobiology 9(3):221–232

Bressan M, Roncato M-A, Bellvert F, Comte G, el Zahar Haichar F, Achouak W, Berge O (2009) Exogenous glucosinolate produced by Arabidopsis thaliana has an impact on microbes in the rhizosphere and plant roots. ISME J 3(11):1243–1257

Bulen W, LeComte J (1966) The nitrogenase system from Azotobacter: two-enzyme requirement for N2 reduction, ATP-dependent H2 evolution, and ATP hydrolysis. Proc Natl Acad Sci U S A 56(3):979

Bulgarelli D, Rott M, Schlaeppi K, van Themaat EVL, Ahmadinejad N, Assenza F, Rauf P, Huettel B, Reinhardt R, Schmelzer E (2012) Revealing structure and assembly cues for Arabidopsis root-inhabiting bacterial microbiota. Nature 488(7409):91–95

Burén S, Jiang X, López-Torrejón G, Echavarri-Erasun C, Rubio LM (2017) Purification and in vitro activity of mitochondria targeted nitrogenase cofactor maturase NifB. Front Plant Sci 8:1567

Bürgmann H, Widmer F, Von Sigler W, Zeyer J (2004) New molecular screening tools for analysis of free-living diazotrophs in soil. Appl Environ Microbiol 70(1):240–247

Cacciari I, Lippi D, Pietrosanti T, Pietrosanti W (1989) Phytohormone-like substances produced by single and mixed diazotrophic cultures of Azospirillum and Arthrobacter. Plant Soil 115(1): 151–153

Cao Y, Zhang Z, Ling N, Yuan Y, Zheng X, Shen B, Shen Q (2011) Bacillus subtilis SQR 9 can control Fusarium wilt in cucumber by colonizing plant roots. Biol Fertil Soils 47(5):495–506

Chakraborty U, Chakraborty B, Basnet M (2006) Plant growth promotion and induction of resistance in Camellia sinensis by Bacillus megaterium. J Basic Microbiol 46(3):186–195

Chen J, Lausser A, Dresselhaus T (2014) Hormonal responses during early embryogenesis in maize. Portland Press, London

Chen C, Xin K, Liu H, Cheng J, Shen X, Wang Y, Zhang L (2017) Pantoea alhagi, a novel endophytic bacterium with ability to improve growth and drought tolerance in wheat. Sci Rep 7(1):1–14

Chhabra S, Brazil D, Morrissey J, Burke JI, O'Gara F, Dowling DN (2013) Characterization of mineral phosphate solubilization traits from a barley rhizosphere soil functional metagenome. Microbiology 2(5):717–724

Chin-A-Woeng TF, Bloemberg GV, Lugtenberg BJ (2003) Phenazines and their role in biocontrol by Pseudomonas bacteria. New Phytol 157(3):503–523

Cho SM, Kang BR, Han SH, Anderson AJ, Park J-Y, Lee Y-H, Cho BH, Yang K-Y, Ryu C-M, Kim YC (2008) 2R, 3R-butanediol, a bacterial volatile produced by Pseudomonas chlororaphis O6, is involved in induction of systemic tolerance to drought in Arabidopsis thaliana. Mol Plant-Microbe Interact 21(8):1067–1075

Coelho MR, Marriel IE, Jenkins SN, Lanyon CV, Seldin L, O'Donnell AG (2009) Molecular detection and quantification of nifH gene sequences in the rhizosphere of sorghum (Sorghum bicolor) sown with two levels of nitrogen fertilizer. Appl Soil Ecol 42(1):48–53

Costacurta A, Vanderleyden J (1995) Synthesis of phytohormones by plant-associated bacteria. Crit Rev Microbiol 21(1):1–18

Cotton TA, Pétriacq P, Cameron DD, Al Meselmani M, Schwarzenbacher R, Rolfe SA, Ton J (2019) Metabolic regulation of the maize rhizobiome by benzoxazinoids. ISME J 13(7): 1647–1658

Crits-Christoph A, Diamond S, Butterfield CN, Thomas BC, Banfield JF (2018) Novel soil bacteria possess diverse genes for secondary metabolite biosynthesis. Nature 558(7710):440–444

Crowley DE, Reid CP, Szaniszlo PJ (1988) Utilization of microbial siderophores in iron acquisition by oat. Plant Physiol 87(3):680–685

Daetwyler HD, Hayden MJ, Spangenberg GC, Hayes BJ (2015) Selection on optimal haploid value increases genetic gain and preserves more genetic diversity relative to genomic selection. Genetics 200(4):1341–1348

Daryanto S, Wang L, Jacinthe P-A (2016) Global synthesis of drought effects on maize and wheat production. PLoS One 11(5):e0156362

Das A, Prasad R, Srivastava A, Giang PH, Bhatnagar K, Varma A (2007) Fungal siderophores: structure, functions and regulations. In: Varma A, Chincholkar SB (eds) Microbial siderophores. Springer-Verlag, Berlin, pp 1–42

Davies PJ (2010) The plant hormones: their nature, occurrence, and functions. In: Plant hormones. Springer, New York, pp 1–15

Delgado-Baquerizo M, Oliverio AM, Brewer TE, Benavent-González A, Eldridge DJ, Bardgett RD, Maestre FT, Singh BK, Fierer N (2018) A global atlas of the dominant bacteria found in soil. Science 359(6373):320–325

Denholm I, Cahill M, Dennehy T, Horowitz A (1998) Challenges with managing insecticide resistance in agricultural pests, exemplisfied by the whitefly Bemisia tabaci. Philos Trans R Soc Lond B Biol Sci 353(1376):1757–1767

Doornbos RF, van Loon LC, Bakker PA (2012) Impact of root exudates and plant defense signaling on bacterial communities in the rhizosphere. A review. Agron Sustain Dev 32(1):227–243

Duan J, Jiang W, Cheng Z, Heikkila JJ, Glick BR (2013) The complete genome sequence of the plant growth-promoting bacterium Pseudomonas sp. UW4. PLoS One 8(3):e58640

Duca DR, Rose DR, Glick BR (2018) Indole acetic acid overproduction transformants of the rhizobacterium Pseudomonas sp. UW4. Antonie Van Leeuwenhoek 111(9):1645–1660

Dunlap CA, Bowman MJ, Schisler DA (2013) Genomic analysis and secondary metabolite production in Bacillus amyloliquefaciens AS 43.3: a biocontrol antagonist of Fusarium head blight. Biol Control 64(2):166–175

Edwards J, Johnson C, Santos-Medellín C, Lurie E, Podishetty NK, Bhatnagar S, Eisen JA, Sundaresan V (2015) Structure, variation, and assembly of the root-associated microbiomes of rice. Proc Natl Acad Sci 112(8):E911–E920

el Zahar Haichar F, Marol C, Berge O, Rangel-Castro JI, Prosser JI, Balesdent J, Heulin T, Achouak W (2008) Plant host habitat and root exudates shape soil bacterial community structure. ISME J 2(12):1221–1230

Fabiańska I, Gerlach N, Almario J, Bucher M (2019) Plant-mediated effects of soil phosphorus on the root-associated fungal microbiota in Arabidopsis thaliana. New Phytol 221(4):2123–2137

Foster R (1986) The ultrastructure of the rhizoplane and rhizosphere. Annu Rev Phytopathol 24(1): 211–234

Fox AR, Soto G, Valverde C, Russo D, Lagares A Jr, Zorreguieta Á, Alleva K, Pascuan C, Frare R, Mercado-Blanco J (2016) Major cereal crops benefit from biological nitrogen fixation when inoculated with the nitrogen-fixing bacterium Pseudomonas protegens Pf-5 X940. Environ Microbiol 18(10):3522–3534

Frank B (1885) Über die auf Wurzelsymbiose beruhende Ernährung gewisser Bäume durch unterirdische Pilze. Plant Biol 3(4):128–145

García de Salamone IE, Hynes RK, Nelson LM (2001) Cytokinin production by plant growth promoting rhizobacteria and selected mutants. Can J Microbiol 47(5):404–411

Gilden RC, Huffling K, Sattler B (2010) Pesticides and health risks. J Obstet Gynecol Neonatal Nurs 39(1):103–110

Giri B, Prasad R, Varma A (2018) Root biology. Springer International Publishing. ISBN 978-3-319-75910-4. https://www.springer.com/us/book/9783319759098

Girish N, Umesha S (2005) Effect of plant growth promoting rhizobacteria on bacterial canker of tomato. Arch Phytopathol Plant Protect 38(3):235–243

Glick BR (2005) Modulation of plant ethylene levels by the bacterial enzyme ACC deaminase. FEMS Microbiol Lett 251(1):1–7

Glick BR, Penrose DM, Li J (1998) A model for the lowering of plant ethylene concentrations by plant growth-promoting bacteria. J Theor Biol 190(1):63–68

Grover M, Bodhankar S, Sharma A, Sharma P, Singh J, Nain L (2021) PGPR mediated alterations in root traits: way toward sustainable crop production. Front Sustain Food Syst 4:287

Guerinot ML, Yi Y (1994) Iron: nutritious, noxious, and not readily available. Plant Physiol 104(3): 815

Gururani MA, Upadhyaya CP, Baskar V, Venkatesh J, Nookaraju A, Park SW (2013) Plant growth-promoting rhizobacteria enhance abiotic stress tolerance in *Solanum tuberosum* through inducing changes in the expression of ROS-scavenging enzymes and improved photosynthetic performance. J Plant Growth Regul 32(2):245–258

Haas D, Défago G (2005) Biological control of soil-borne pathogens by fluorescent pseudomonads. Nat Rev Microbiol 3(4):307–319

Hammons RO (1976) Peanuts: genetic vulnerability and breeding strategy 1. Crop Sci 16(4): 527–530

Harbort CJ, Hashimoto M, Inoue H, Niu Y, Guan R, Rombolà AD, Kopriva S, Voges MJ, Sattely ES, Garrido-Oter R (2020) Root-secreted coumarins and the microbiota interact to improve iron nutrition in Arabidopsis. Cell Host Microbe 28(6):825–837.e826

Hartman K, Tringe SG (2019) Interactions between plants and soil shaping the root microbiome under abiotic stress. Biochem J 476(19):2705–2724

Hassan HM, Troxell B (2013) Transcriptional regulation by ferric uptake regulator (FUR) in pathogenic bacteria. Front Cell Infect Microbiol 3:59

Heeb S, Haas D (2001) Regulatory roles of the GacS/GacA two-component system in plant-associated and other gram-negative bacteria. Mol Plant-Microbe Interact 14(12):1351–1363

Hellriegel H, Wilfarth H (1888) Untersuchungen uber die Stickstoffnahrung der Gramineen und Leguminosen. Buchdruckerei der "Post" Kayssler, Berlin

Hinsinger P (2001) Bioavailability of soil inorganic P in the rhizosphere as affected by root-induced chemical changes: a review. Plant Soil 237(2):173–195

Hiruma K, Gerlach N, Sacristán S, Nakano RT, Hacquard S, Kracher B, Neumann U, Ramírez D, Bucher M, O'Connell RJ (2016) Root endophyte Colletotrichum tofieldiae confers plant fitness benefits that are phosphate status dependent. Cell 165(2):464–474

Hu Y, Schmidhalter U (2005) Drought and salinity: a comparison of their effects on mineral nutrition of plants. J Plant Nutr Soil Sci 168(4):541–549

Hu L, Robert CA, Cadot S, Zhang X, Ye M, Li B, Manzo D, Chervet N, Steinger T, Van Der Heijden MG (2018) Root exudate metabolites drive plant-soil feedbacks on growth and defense by shaping the rhizosphere microbiota. Nat Commun 9(1):1–13

Huang AC, Jiang T, Liu Y-X, Bai Y-C, Reed J, Qu B, Goossens A, Nützmann H-W, Bai Y, Osbourn A (2019) A specialized metabolic network selectively modulates Arabidopsis root microbiota. Science 364(6440):eaau6389

Jackson M (1991) Ethylene in root growth and development. In: Matoo AK, Suttle JC (eds) The plant hormone ethylene. CRC Press, Boca Raton, FL

Jacobson C, Pasternak J, Glick B (1994) Partial purification and characterization of ACC deaminase from the plant growth-promoting rhizobacterium *Pseudomonas putida* GR12-2. Can J Microbiol 40(1):19–1025

Ji SH, Gururani MA, Chun S-C (2014) Isolation and characterization of plant growth promoting endophytic diazotrophic bacteria from Korean rice cultivars. Microbiol Res 169(1):83–98

Kamilova F, Kravchenko LV, Shaposhnikov AI, Azarova T, Makarova N, Lugtenberg B (2006) Organic acids, sugars, and L-tryptophane in exudates of vegetables growing on stonewool and their effects on activities of rhizosphere bacteria. Mol Plant-Microbe Interact 19(3):250–256

Kaneko T, Nakamura Y, Sato S, Asamizu E, Kato T, Sasamoto S, Watanabe A, Idesawa K, Ishikawa A, Kawashima K (2000) Complete genome structure of the nitrogen-fixing symbiotic bacterium Mesorhizobium loti. DNA Res 7(6):331–338

Ke J, Wang B, Yoshikuni Y (2021) Microbiome engineering: synthetic biology of plant-associated microbiomes in sustainable agriculture. Trends Biotechnol 39(3):244–261

Khan N, Bano A, Babar M (2017) The root growth of wheat plants, the water conservation and fertility status of sandy soils influenced by plant growth promoting rhizobacteria. Symbiosis 72(3):195–205

Kim J, Rees DC (1994) Nitrogenase and biological nitrogen fixation. Biochemistry 33(2):389–397

Kloepper JW, Leong J, Teintze M, Schroth MN (1980) Enhanced plant growth by siderophores produced by plant growth-promoting rhizobacteria. Nature 286(5776):885–886

Kudoyarova GR, Melentiev AI, Martynenko EV, Timergalina LN, Arkhipova TN, Shendel GV, Kuz'mina LY, Dodd IC, Veselov SY (2014) Cytokinin producing bacteria stimulate amino acid deposition by wheat roots. Plant Physiol Biochem 83:285–291

Lee KJ, Kamala-Kannan S, Sub HS, Seong CK, Lee GW (2008) Biological control of Phytophthora blight in red pepper (Capsicum annuum L.) using *Bacillus subtilis*. World J Microbiol Biotechnol 24(7):1139–1145

Leveau JH, Gerards S (2008) Discovery of a bacterial gene cluster for catabolism of the plant hormone indole 3-acetic acid. FEMS Microbiol Ecol 65(2):238–250

Leveau JH, Lindow SE (2005) Utilization of the plant hormone indole-3-acetic acid for growth by Pseudomonas putida strain 1290. Appl Environ Microbiol 71(5):2365–2371

Li M, Wang J, Yao T, Wang Z, Zhang H, Li C (2021) Isolation and characterization of cold-adapted PGPB and their effect on plant growth promotion. J Microbiol Biotechnol 31(9):1218–1230

López-Arredondo DL, Leyva-González MA, González-Morales SI, López-Bucio J, Herrera-Estrella L (2014) Phosphate nutrition: improving low-phosphate tolerance in crops. Annu Rev Plant Biol 65:95–123

Lucas JA, Hawkins NJ, Fraaije BA (2015) The evolution of fungicide resistance. Adv Appl Microbiol 90:29–92

Lugtenberg B, Kamilova F (2009) Plant-growth-promoting rhizobacteria. Annu Rev Microbiol 63: 541–556

Lundberg DS, Lebeis SL, Paredes SH, Yourstone S, Gehring J, Malfatti S, Tremblay J, Engelbrektson A, Kunin V, Del Rio TG (2012) Defining the core *Arabidopsis thaliana* root microbiome. Nature 488(7409):86–90

Macintosh KA, Doody DG, Withers PJ, McDowell RW, Smith DR, Johnson LT, Bruulsema TW, O'Flaherty V, McGrath JW (2019) Transforming soil phosphorus fertility management strategies to support the delivery of multiple ecosystem services from agricultural systems. Sci Total Environ 649:90–98

Marasco R, Rolli E, Ettoumi B, Vigani G, Mapelli F, Borin S, Abou-Hadid AF, El-Behairy UA, Sorlini C, Cherif A (2012) A drought resistance-promoting microbiome is selected by root system under desert farming. PLoS One 7(10):e48479

Marschner H (1995) Adaptation of plants to adverse chemical soil conditions. In: Mineral nutrition of higher plants. Academic, New York

Mayak S, Tirosh T, Glick BR (2004) Plant growth-promoting bacteria that confer resistance to water stress in tomatoes and peppers. Plant Sci 166(2):525–530

McCasland M, Trautmann NM, Wagenet RJ (1985) Nitrate: health effects in drinking water. Cornell Cooperative Extension, Ithaca, NY

Mendes R, Kruijt M, De Bruijn I, Dekkers E, van der Voort M, Schneider JH, Piceno YM, DeSantis TZ, Andersen GL, Bakker PA (2011) Deciphering the rhizosphere microbiome for disease-suppressive bacteria. Science 332(6033):1097–1100

Miransari M (2014) Plant growth promoting rhizobacteria. J Plant Nutr 37(14):2227–2235

Mishra J, Arora NK (2018) Secondary metabolites of fluorescent pseudomonads in biocontrol of phytopathogens for sustainable agriculture. Appl Soil Ecol 125:35–45

Nambiar P, Sivaramakrishnan S (1987) Detection and assay of siderophores in cowpea rhizobia (Bradyrhizobium) using radioactive Fe (59Fe). Lett Appl Microbiol 4(2):37–40

Naseem H, Bano A (2014) Role of plant growth-promoting rhizobacteria and their exopolysaccharide in drought tolerance of maize. J Plant Interact 9(1):689–701

Naylor D, Coleman-Derr D (2018) Drought stress and root-associated bacterial communities. Front Plant Sci 8:2223

Naylor D, DeGraaf S, Purdom E, Coleman-Derr D (2017) Drought and host selection influence bacterial community dynamics in the grass root microbiome. ISME J 11(12):2691–2704
Neilands JB (1984) Methodology of siderophores. In: Siderophores from microorganisms and plants. structure and bonding, vol 58. Springer, Berlin, Heidelberg. https://doi.org/10.1007/BFb0111309
Novoa R, Loomis R (1981) Nitrogen and plant production. Plant Soil 58(1):177–204
Patten CL, Glick BR (1996) Bacterial biosynthesis of indole-3-acetic acid. Can J Microbiol 42(3): 207–220
Patten CL, Glick BR (2002a) Regulation of indoleacetic acid production in *Pseudomonas putida* GR12-2 by tryptophan and the stationary-phase sigma factor RpoS. Can J Microbiol 48(7): 635–642
Patten CL, Glick BR (2002b) Role of Pseudomonas putida indoleacetic acid in development of the host plant root system. Appl Environ Microbiol 68(8):3795–3801
Pierson LS, Pierson EA (1996) Phenazine antibiotic production in *Pseudomonas aureofaciens*: role in rhizosphere ecology and pathogen suppression. FEMS Microbiol Lett 136(2):101–108
Ping L, Boland W (2004) Signals from the underground: bacterial volatiles promote growth in Arabidopsis. Trends Plant Sci 9(6):263–266
Prasad R, Zhang S-H (2022) Benefical microorganisms in agriculture. Springer, Singapore. https://link.springer.com/book/9789811907326
Razaq M, Zhang P, Shen H-l (2017) Influence of nitrogen and phosphorous on the growth and root morphology of Acer mono. PLoS One 12(2):e0171321
Roberts TL, Johnston AE (2015) Phosphorus use efficiency and management in agriculture. Resour Conserv Recycl 105:275–281
Rubio LM, Ludden PW (2008) Biosynthesis of the iron-molybdenum cofactor of nitrogenase. Annu Rev Microbiol 62:93–111. https://doi.org/10.1146/annurev.micro.62.081307.162737
Ryu M-H, Zhang J, Toth T, Khokhani D, Geddes BA, Mus F, Garcia-Costas A, Peters JW, Poole PS, Ané J-M (2020) Control of nitrogen fixation in bacteria that associate with cereals. Nat Microbiol 5(2):314–330
Sang MK, Chun S-C, Kim KD (2008) Biological control of Phytophthora blight of pepper by antagonistic rhizobacteria selected from a sequential screening procedure. Biol Control 46(3): 424–433
Santi C, Bogusz D, Franche C (2013) Biological nitrogen fixation in non-legume plants. Ann Bot 111(5):743–767
Savci S (2012) An agricultural pollutant: chemical fertilizer. Int J Environ Sci Dev 3(1):73
Schlaeppi K, Dombrowski N, Oter RG, van Themaat EVL, Schulze-Lefert P (2014) Quantitative divergence of the bacterial root microbiota in Arabidopsis thaliana relatives. Proc Natl Acad Sci 111(2):585–592
Scott JC, Greenhut IV, Leveau JH (2013) Functional characterization of the bacterial iac genes for degradation of the plant hormone indole-3-acetic acid. J Chem Ecol 39(7):942–951
Sczyrba A, Hofmann P, Belmann P, Koslicki D, Janssen S, Dröge J, Gregor I, Majda S, Fiedler J, Dahms E (2017) Critical assessment of metagenome interpretation—a benchmark of metagenomics software. Nat Methods 14(11):1063–1071
Seefeldt LC, Hoffman BM, Dean DR (2009) Mechanism of Mo-dependent nitrogenase. Annu Rev Biochem 78:701–722
Shakir MA, Asghari B, Muhammad A (2012) Rhizosphere bacteria containing ACC-deaminase conferred drought tolerance in wheat grown under semi-arid climate. Soil Environ 31(1): 108–112
Sharma A, Johri B, Sharma A, Glick B (2003) Plant growth-promoting bacterium Pseudomonas sp. strain GRP3 influences iron acquisition in mung bean (Vigna radiata L. Wilzeck). Soil Biol Biochem 35(7):887–894
Shilev S (2013) Soil rhizobacteria regulating the uptake of nutrients and undesirable elements by plants. In: Plant microbe symbiosis: fundamentals and advances. Springer, New York, pp 147–167

Shulse CN, Chovatia M, Agosto C, Wang G, Hamilton M, Deutsch S, Yoshikuni Y, Blow MJ (2019) Engineered root bacteria release plant-available phosphate from phytate. Appl Environ Microbiol 85(18):e01210–e01219

Skoog F, Armstrong DJ (1970) Cytokinins. Ann Rev Plant Physiol 21(1):359–384

Smarrelli J Jr, Castignetti D (1986) Iron acquisition by plants: the reduction of ferrisiderophores by higher plant NADH: nitrate reductase. Biochim Biophys Acta 882(3):337–342

Soko T, Bender CM, Prins R, Pretorius ZA (2018) Yield loss associated with different levels of stem rust resistance in bread wheat. Plant Dis 102(12):2531–2538

Spaepen S, Vanderleyden J, Remans R (2007) Indole-3-acetic acid in microbial and microorganism-plant signaling. FEMS Microbiol Rev 31(4):425–448

Takeuchi T, Sawada H, Tanaka F, Matsuda I (1996) Phylogenetic analysis of Streptomyces spp. causing potato scab based on 16S rRNA sequences. Int J Syst Evol Microbiol 46(2):476–479

Tan S, Yang C, Mei X, Shen S, Raza W, Shen Q, Xu Y (2013) The effect of organic acids from tomato root exudates on rhizosphere colonization of *Bacillus amyloliquefaciens* T-5. Appl Soil Ecol 64:15–22

Thimann KV (1939) Auxins and the inhibition of plant growth. Biol Rev 14(3):314–337

Timmusk S, Nicander B, Granhall U, Tillberg E (1999) Cytokinin production by Paenibacillus polymyxa. Soil Biol Biochem 31(13):1847–1852

Tirnaz S, Batley J (2019) DNA methylation: toward crop disease resistance improvement. Trends Plant Sci 24(12):1137–1150

Unno Y, Shinano T (2013) Metagenomic analysis of the rhizosphere soil microbial community. Mol Micro Ecol Rhizosp 1:1099–1103

Vardharajula S, Zulfikar Ali S, Grover M, Reddy G, Bandi V (2011) Drought-tolerant plant growth promoting Bacillus spp.: effect on growth, osmolytes, and antioxidant status of maize under drought stress. J Plant Interact 6(1):1–14

Vert G, Grotz N, Dédaldéchamp F, Gaymard F, Guerinot ML, Briat J-F, Curie C (2002) IRT1, an Arabidopsis transporter essential for iron uptake from the soil and for plant growth. Plant Cell 14(6):1223–1233

Vessey JK (2003) Plant growth promoting rhizobacteria as biofertilizers. Plant Soil 255(2):571–586

Vurukonda SSKP, Vardharajula S, Shrivastava M, SkZ A (2016) Enhancement of drought stress tolerance in crops by plant growth promoting rhizobacteria. Microbiol Res 184:13–24

Wang Y, Brown H, Crowley D, Szaniszlo P (1993) Evidence for direct utilization of a siderophore, ferrioxamine B, in axenically grown cucumber. Plant Cell Environ 16(5):579–585

Wang J, Zhang Y, Jin J, Li Q, Zhao C, Nan W, Wang X, Ma R, Bi Y (2018) An intact cytokinin-signaling pathway is required for Bacillus sp. LZR216-promoted plant growth and root system architecture altereation in Arabidopsis thaliana seedlings. Plant Growth Regul 84(3):507–518

Wang P, Chai YN, Roston R, Dayan FE, Schachtman DP (2021) The *Sorghum bicolor* root exudate Sorgoleone shapes bacterial communities and delays network formation. mSystems 6(2): e00749-20

Weiß M, Waller F, Zuccaro A, Selosse MA (2016) Sebacinales–one thousand and one interactions with land plants. New Phytol 211(1):20–40

Weng J, Wang Y, Li J, Shen Q, Zhang R (2013) Enhanced root colonization and biocontrol activity of *Bacillus amyloliquefaciens* SQR9 by abrB gene disruption. Appl Microbiol Biotechnol 97(19):8823–8830

Wille L, Messmer MM, Studer B, Hohmann P (2019) Insights to plant–microbe interactions provide opportunities to improve resistance breeding against root diseases in grain legumes. Plant Cell Environ 42(1):20–40

Xu L, Naylor D, Dong Z, Simmons T, Pierroz G, Hixson KK, Kim Y-M, Zink EM, Engbrecht KM, Wang Y (2018) Drought delays development of the sorghum root microbiome and enriches for monoderm bacteria. Proc Natl Acad Sci 115(18):E4284–E4293

Xu L, Pierroz G, Wipf HM-L, Gao C, Taylor JW, Lemaux PG, Coleman-Derr D (2021) Holo-omics for deciphering plant-microbiome interactions. Microbiome 9(1):1–11

Yadav A, Singh RP, Singh AL, Singh M (2021) Identification of genes involved in phosphate solubilization and drought stress tolerance in chickpea symbiont *Mesorhizobium ciceri* Ca181. Arch Microbiol 203(3):1167–1174

Yang J, Xie X, Xiang N, Tian Z-X, Dixon R, Wang Y-P (2018) Polyprotein strategy for stoichiometric assembly of nitrogen fixation components for synthetic biology. Proc Natl Acad Sci 115(36):E8509–E8517

Yoshiba Y, Kiyosue T, Nakashima K, Yamaguchi-Shinozaki K, Shinozaki K (1997) Regulation of levels of proline as an osmolyte in plants under water stress. Plant Cell Physiol 38(10): 1095–1102

Zahir Z, Munir A, Asghar H, Shaharoona B, Arshad M (2008) Effectiveness of rhizobacteria containing ACC deaminase for growth promotion of peas (Pisum sativum) under drought conditions. J Microbiol Biotechnol 18(5):958–963

Zhang Z, Pierson LS (2001) A second quorum-sensing system regulates cell surface properties but not phenazine antibiotic production in Pseudomonas aureofaciens. Appl Environ Microbiol 67(9):4305–4315

Zipper SC, Qiu J, Kucharik CJ (2016) Drought effects on US maize and soybean production: spatiotemporal patterns and historical changes. Environ Res Lett 11(9):094021

Chapter 4
Crop Microbiome for Sustainable Agriculture in Special Reference to Nanobiology

Pratyusha Sambangi, Vadlamudi Srinivas, and Subramaniam Gopalakrishnan

Abstract The microbial population surrounding plant is regarded as the crop microbiome and they are essential in crop development and sustenance. Specifically, the plant growth-promoting microbes play a key role as growth enhancers and disease suppressors. These plants associated microbiomes are often considered as one of the essential agricultural components toward an enhanced crop yield. This has led to a wide range of nanotechnological applications on the agricultural systems to boost crop output. In this scenario, the nanoparticles are largely seen interacting with the crop microbiomes and the plant systems. Hence, in the present chapter, the role and response of these crop microbiomes and nanoparticles will be discussed in detail. From the point of application of these nanoparticles, their effect on the plant growth-promoting bacterial systems and to their biosynthesis utilizing the crop microbiomes will be explored. Nanomaterials interaction with root colonizing microbes often promised enhanced plant health during both abiotic and biotic stress conditions through rhizobacterial metabolite changes. Nanotechnology, being a new frontier in the modern agriculture practices, the challenges of nanoparticle dosage, cost efficiency, and their footprint in the agricultural soils over a long time was one of the vital areas to observe. On the other hand, nanoparticles are widely reported to enhance the plant growth-promoting and defense traits of the crop microbiomes and play a key role in global food production. The present chapter discusses the key features of crop microbiomes, their response to nanoparticles, and together how they could influence the crop yield and biotic stress resistance.

Keywords Crop microbiome · PGPR · Symbiosis · Nanoparticles · Agriculture

---

P. Sambangi · V. Srinivas · S. Gopalakrishnan (✉)
International Crops Research Institute for the Semi-Arid Tropics (ICRISAT), Patancheru, Telangana, India
e-mail: s.pratyusha@cgiar.org; s.vadlamudi@cgiar.org; s.gopalakrishnan@cgiar.org

S. Chhabra et al. (eds.), *Plant Microbiome for Plant Productivity and Sustainable Agriculture*, Microorganisms for Sustainability 37,
https://doi.org/10.1007/978-981-19-5029-2_4

4.1 Introduction

The crops are surrounded by a whole system of environment and diverse microorganisms. These microbial communities associated with the crops are regarded as the crop microbiome, and they are essential for plant sustenance and growth (Khatoon et al. 2020). Crop microbiota consists of different types of living communities such as archaea, bacteria, fungi, protozoans, and viruses that enhance the soil diversity and fertility (Pedros-Ali and Manrubia 2016). They often support plant growth and development through some significant processes, namely hormone synthesis, nitrogen fixation, and nutrient recycling (Tilak et al. 2005). Apart from that, this soil microbiome also assists in plant protection against certain invading biotic and abiotic factors (Müller et al. 2016). Crops with healthy microbiome will often facilitate rich nutrients and metals into the agricultural cultivable lands (Fageria 2016). This results in fertile and healthy soils that enhance agricultural crop production, which is an important necessity at times of ever-increasing population and greater food demand. Hence, agriculturalists across the world are adopting various technological developments such as varied microbial inoculants, bio-fertilizers, and microbiome nano-engineering, further incorporating them into the traditional agricultural practices (Chen and Yada 2011; Compant et al. 2019). These will surely flourish the crop microbial ecosystem and sequentially enhance the crop resilience and yield.

Crop microbiomes are mostly influenced by the host plant and the environment surrounding it. These microbes colonize in different parts of the plant, such as the rhizosphere, endosphere, and phyllosphere. Among them, rhizosphere, i.e., plant root surface and its associated microbes, is of paramount importance for undertaking various metabolic and nutrient cycles (Lakshmanan et al. 2014). These root colonizing beneficial bacteria, which are known as Plant Growth Promoting Rhizobacteria (PGPR) and they, directly and indirectly, facilitate both plant growth and protection (Table 4.1). The advantages of PGPR are attributed to their characteristics features of nitrogen fixation, phosphate/potassium solubilization, phytohormone/siderophore production, antibiotic/hydrolytic enzyme production, and induced systemic resistance (Alekhya and Gopalakrishnan 2017; Vijayabharathi et al. 2018b; Anusha et al. 2019; Gopalakrishnan and Vadlamudi 2019; Sambangi et al. 2021; Prasad et al. 2015). Hence, these PGPR are used as an alternative to agrochemicals in the form of bio-fertilizers/inoculants toward sustainable agriculture management. Further, with the advanced technologies and multidisciplinary research, technologies such as nano-encapsulation and nanomaterials are being utilized for the better performance and productivity of these beneficial microbes (Dixshit et al. 2013; Rana et al. 2021; Harish et al. 2022). Hence in the present chapter, the applications of nanotechnology in the agriculture sector will be thoroughly analyzed with particular emphasis on the positive and negative impacts of these nanoparticles on the essential soil crop microbiomes.

Table 4.1 Role of some plant growth promoting rhizobacteria (PGPR)

PGPR	Beneficial effect	Crops	References
Azospirillum	Nitrogen fixation, phosphate solubilization, improved growth and nodulation	Bean, wheat, chickpea, rice, maize, sugarcane	Hamaoui et al. (2001), Lucy et al. (2004), Tejera et al. (2005)
Bacillus	Biofertilization, bioprotection, nitrogen fixation, improved plant resistance, growth and yield	Chickpea, pigeonpea, rice, groundnut, maize, tomato, cucumber	Ongena and Jacques (2008), Gopalakrishnan et al. (2011, 2012b), Beneduzi et al. (2012), Gopalakrishnan et al. (2016), Sharma et al. (2017), Vaikundamoorthy et al. (2018), Hashami et al. (2019)
Pseudomonas	Increase in plant yield, stress resistance, nitrogen fixation, P and Fe uptake, effective symbiosis with other beneficial bacteria	Rice, maize, chickpea, pigeonpea, wheat, cotton	Shaharoona et al. (2008), Yao et al. (2010), Gopalakrishnan et al. (2011, 2012b, 2016), Paramanandham et al. (2017), Cheng et al. (2019), Lawrance et al. (2019)
Rhizobia	Nitrogen fixation, bioremediation, bioprotection	Legumes, rice, tubers	Yanni et al. (2001), Garcia-Fraile et al. (2012), Flores-Felix et al. (2013), Gopalakrishnan et al. (2018), Vijayabharathi et al. (2018b)
Streptomyces	Enhanced nitrogen fixation, increased plant growth, yield, and disease resistance	Legumes, rice, tomato, chili	Gopalakrishnan et al. (2011, 2012a, b, 2016), Alekhya and Gopalakrishnan (2017), Vijayabharathi et al. (2018a), Ankati et al. (2021), Srinivas et al. (2020), Gopalakrishnan et al. (2020, 2021), Pratyusha et al. (2021)

4.2 Nanotechnology in Sustainable Agriculture

Nanotechnology, being one of the significant technological, scientific advancement in recent times, has shown a tremendous impact on many fields, including agriculture. It plays multiple roles in crop production/improvement, immunity against abiotic and biotic factors, and environmental remediation (Mukhopadhyay 2014; Moulick et al. 2020; Prasad 2014; Prasad et al. 2014, 2017a). These nanoparticles being characteristically smaller in size, have unique features such as a large surface area to volume, physiochemical stability, aggregation ability, high reactivity, and targeted delivery (Prasad et al. 2016; Srivastava et al. 2021). Thus, many nanoparticles, namely Ag, Au, Carbon nanotubes, Cu, Fe, Si, TiO_2, Zn, and ZnO, have exhibited some significant

Table 4.2 List of nano-products utilized on various agriculturally important crops

Types	Impact		References
	Positive	Negative	
Nanofertilizers (e.g., *NPs of urea, ZnO, chitosan, NPK*)	Quick and specific release, improved nutrient efficiency, anti-microbial water holding capacity	Overdose toxicity, indirect effect on PGPR and environment	Gogos et al. (2012), Chhipa (2017), Pitambara and Shukla (2019)
Nanopesticides (e.g., *Nano-Ag, Nano-FeO, Nano-silica, Nano-ZnO*)	Low dose, enhance bio-availability of active ingredient, targeted delivery	Unknown fate of the carrier, slow degradation, possible risk to humans and microbiome	John et al. (2017), Kah et al. (2018), Bundschuh et al. (2018)
Nano-biopesticides (e.g., *Azadirachtin-chitosan, Aloin-AgNPs, pheromones-nanofibers; garlic essential oil-polyethylene glycol*)	Crop protection, increase efficacy of biologically active compound, low dosage	Possible hazardous effect on environment and soil	Yang et al. (2009), Feng and Peng (2012), Shah et al. (2016), Bipin et al. (2017), Vijayabharathi et al. (2018a)
Nano-biosensors (e.g., *Aptasensors, nanofibers, nanocapsules*)	Monitor of plant and soil health, smart delivery	Risk of toxicity and untargeted dispersion into soil and water	Rai et al. (2012), Srivastava et al. (2018)

effects on plant development and protection (Pokropivny et al. 2007). Many researchers employed nanotechnology to promote quick and low-cost agriculture systems toward enhanced yield and crop diagnosis (Sekhon 2014; Kim et al. 2017; Abd-Elsalam and Prasad 2018, 2019; Abd-Elsalam et al. 2019). Also, these nano-engineered technologies have shown a promising increase in the agricultural yield through sustainable manner and a better alternative to agrochemicals (Deepti et al. 2020; Prasad et al. 2017b, c).

With the necessity to feed the ever-increasing population, research and innovation in agri-nanotechnology have led to varied applications leading to the development of nano-biosensors, nano-fertilizers, nano-pesticides, and nano-encapsulated products (Bhattacharyya et al. 2016). These different forms of nanomaterials have strongly exhibited their plant growth and protection properties along with agricultural soil remediation (Kah et al. 2019; Borah et al. 2022). Following (Table 4.2) are some of the significant developments of nanotechnology, utilized in the agriculture sector with strong impacts on agricultural crops, crop microbiomes, and soil status (Sangeetha et al. 2017a, b, c).

4.2.1 Nano-Agrochemicals

Among all nano-products that are widely in the application, nano-pesticides and nano-fertilizers together are considered as nano-agrochemicals and have gained

Table 4.3 List of marketing nano-agrochemicals produced by different agri-companies

Nano-agrochemical	Company	Mechanism
Nano-Gro	Agro Nanotechnology Corp, USA	Increases crop yield (FAO 2011)
Nano Green	United Phosphorus Limited, South Africa	Attacks pest respiratory system (FAO 2011)
PRIMo Maxx	Syngenta, Switzerland	Attacks pest neural system (FAO 2011)
Karate Zeon	Syngenta, Switzerland	Attacks pest neural system (FAO 2011)
Fruity Fresh	Tamil Nadu Agricultural University, India	Enhances shelf life of fruits and vegetables (TNAU 2019)
Nano-fert	Geolife Agritech India Pvt. Ltd., India	Nutrition (Geolife group, www.geolifegroup.com)
Fib-sol	Fib-sol Pvt. Ltd., India	PGPR benefits (www.fibsol.com)

much popularity due to their potential effectiveness. Nano-scale technology is applied to these pesticides and fertilizers to enable their precise management and better targeted-delivery than the conventional methods. Different types of nano-formulations are utilized for these agrochemical applications, namely nano-encapsulations, nano-emulsions, nano-tubes, nano-gels, and nano-suspensions (Liu and Lal 2015; Prasad et al. 2019). Agriculture being the backbone to many developing countries, globally, many countries accepted the significance of this nanotechnology in agribusiness. By 2030, agribusiness expected to have a 3 trillion USD market value globally. In that, nano-agrochemicals will contribute a significant share with their vibrant developments in the sectors of food yield and plant protection. For example, the leading agrochemical companies, namely Dow AgroSciences, Syngenta AG, Bayer crop Sciences AG, and Baden Aniline and Soda Factory (BASF), have already opened their nano-research divisions. In fact, Syngenta already marketed nano-pesticides such as Banner MAXX fungicide, Cruise MAXX Beans, and Primo MAXX Plant Growth Regulator (Bhan et al. 2018). Likewise, agrochemical companies in Asia and America also are actively participating in the manufacturing of these nano-agrochemicals (Table 4.3).

Along with this extensive application of nano-agrochemicals, the question of the risk assessment and the final fate of these products in the environment also been raised over longer periods. There is uncertainty regarding their usage and safety toward non-targeted living systems such as crop microbiomes and soil diversity. Some studies reported adverse effects of these nano-agrochemicals on smaller living organisms (Ostiguy et al. 2010). Apart from that, the biotransformation of these nano-agrochemicals could accumulate in the plants, rhizobacteria, and food products (Melanie and Rai 2020). Intrusion of plant-microbial symbiosis by these nano-agrochemicals must be thoroughly analyzed before their fullest utilization.

4.3 Nanoparticles and Plant Microbiomes

It is evident that from many research reports that in the last few years, nanotechnology has been playing a significant role in the agriculture sector in the forms of different nano-products. The behavior and the impact of these widely introduced nano-products on the environment are yet to be fully understood (Table 4.4). These nanoparticles might have direct and indirect interactions on the plants and their most closely associated soil bacterial communities (Shweta et al. 2018; Singh et al. 2019). Many researchers have investigated the efficacy and effect of these NPs on the soil rhizobacteria, and they include both positive and negative implications (Fig. 4.1).

Table 4.4 Nanoparticles effect on the soil microbiomes

Type of nanoparticles	Effect/impact	References
Ag	Effects on enzymes	Shin et al. (2012)
Ag	Effects on β-glucosidase, acid phosphatase, dehydrogenase, urease	Hänsch and Emmerling (2010)
Ag	Reduces enzymatic activities	Peyrot et al. (2014)
Ag	Changes/loss in the bacterial communities	Colman et al. (2013)
Ag, Al_2O_3	Reduces the bacterial and archaeal amoA gene abundance in soil	McGee et al. (2017)
Au	Effects the nutrients, enzymes, and microbial communities	Asadishad et al. (2017)
Cu	Reduces the C and N content and alters the microbial structures in the soil	Kumar et al. (2012)
Cu & Fe	Changes in soil humic contents	Ben-Moshe et al. (2013)
Ti	Alters soil bacterial structures	Ge et al. (2012)

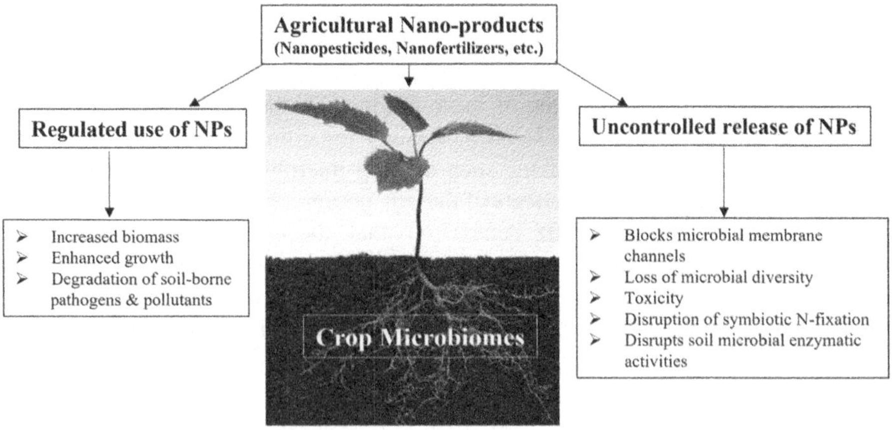

Fig. 4.1 Impact of nanoparticles on the crop microbiomes

There were also reports of constant percolations of NPs into the soil for more extended periods of time, leading to toxicity (Yadav et al. 2014). These NPs could also alter the microscopic properties of soil, i.e., reduced humic acid contents, resulting in disruption of the residing soil microbial diversity (Maurer-Jones et al. 2013). The introduction of these NPs into the natural environment along with unintentional releases in the form of water and sewage pose a great threat to the beneficial crop microbiomes (Coll et al. 2016).

4.3.1 Positive Impact

There are significant positive attributes of these NPs on the soil bacteria communities. For example, Au, Cu, and SiO_2 NPs enhanced the number of microbial communities in the soil and increased their metabolic rate (Shah and Belozerova 2009). In many reports, AgNPs in lower doses have increased the microbial density, the nitrifying ability of PGPR, and exhibited anti-microbial activity against pathogenic microbes (Pietrzak and Gutarowska 2015; Nawaz and Bano 2019). In addition, the TiO_2 and ZnO NPs have assisted in the degradation of organic pollutants in agricultural soils (Ge et al. 2012). The natural soil nanoparticles, namely TiO_2 have enhanced the performance of the PGPR and, in turn, increased plant growth and physiology (Timmusk et al. 2018). This implies that the NPs under proper monitor and dose restrictions could enhance the PGPR abilities and pave the way to an ecological solution of increased agricultural crop yields. It is also reported that some NPs, namely Cu, Ag, and Zn, could act against the soil-borne pathogens and play a vital role in protecting the plant and crop microbiomes against various biotic stress factors (Malandrakis et al. 2019).

Application of these NPs in the form of nano-fertilizers, nano-pesticides, and nano-formulations can replace chemical fertilizers to provide site-specific crop nutrition (Kah et al. 2019), suppress weeds (Amna et al. 2019), and insect-disease-specific resistance (Camara et al. 2019) with better efficiency for sustainable crop production. For instance, the silver nanoparticles biosynthesized from *Streptomyces griseoplanus* SAI-25 showed antifungal activity against the fungal pathogen *Macrophomina phaseolina* (Vijayabharathi et al. 2018a). In another study, the MgO nanoparticles had reported anti-fungal activity against the soil-borne pathogens, namely *Phytophthora nicotianae* and *Thielaviopsis basicola* (Juanni et al. 2020). Silver and copper NPs biosynthesized from *Pseudomonas aeruginosa* exhibited anti-fungal activity against three pathogenic fungi *Botrytis cinerea*, *Pilidium concavum*, and *Pestalotia* sp. (Bayat et al. 2021). These reports show us the positive impacts of NPs in the management of soil-borne pathogens. Thus, developing eco-friendly nano-formulations, establishing proper delivery system, and validation of these nano-formulations under field conditions to manage pests and weeds along with potential benefits such as biofortification in crops is the need of the hour for sustainability in productivity and conservation of biodiversity in agricultural production systems (Gupta et al. 2018; Ismail et al. 2017).

4.3.2 Negative Impact

The potential benefits of nanotechnology in the agriculture sector and its enormous usage and applications have led to detecting NPs in the soil above the threshold levels. This could significantly alter the biological and physiochemical properties of the residing soil crop microbiomes (Vittori Antisari et al. 2013). These metallic NPs were accumulated in the soil through different routes, and they could exhibit toxic effects on PGPR, which might influence the plant health and the soil structures. Apart from that, the accumulated NPs could also alter the associated microbial carbon and nitrogen ecological cycles in the environment (Simonin et al. 2018).

The composition and type of the metallic NPs play a significant role in the effects on the soil crop microbiomes. For example, the CuO NPs reduced the oxidative potential of the soil (Schlich and Hund-Rinke 2015), and ZnO NPs decreased the enzymatic activity of the residing rhizobacteria (Shen et al. 2015). Also, overdose and continuous percolation of these NPs could accumulate and block the ion channels of the bacteria. It was reported that the AgNPs impact the ion release structure of the bacteria (Zhai et al. 2016) and hinder the bacterial enzymatic activity (Lowry et al. 2012). The Fe_3O_4 NPs have shown increased toxicity toward the crop microbiomes (Frenk et al. 2013). Gold NPs at higher concentrations in the soil led to the loss of rhizobacterial diversity (Asadishad et al. 2017). In the end, these NPs being present in abnormal quantities in the soil could possibly alter the natural soil ecosystem and negatively effect on the plant-microbial symbiosis through membrane disruption and cell wall damage.

4.3.3 Nanomaterial's Role in Crop Abiotic Stress

Crops constantly expose to various abiotic stress factors such as salinity, drought, extreme temperatures, and heavy metal stress (Zhang et al. 2022). Many researchers are constantly working to utilize the applications of nanotechnology in minimizing the abiotic stresses in the crops. Nanomaterials are being investigated to overcome both the abiotic and biotic stresses in the plants through their promising and potential nano-sizes. It was observed that the application of nano-silicon dioxide had lowered the salinity stress in the strawberry crops (Avestan et al. 2019). Similarly, nano-titanium dioxide had reduced the NaCl levels by increasing the antioxidative enzymes and compatible solutes in tomato crops (Khan 2016). Nanoscale zero-valent iron (nZVI) nanomaterials are known to attract the heavy metals and immobilize them with their distinct structures (Fajardo et al. 2012). The application of Cu-Zn nanoparticles showed an effective reduction in the drought levels in the plants, especially in the wheat (Taran et al. 2017). In another study it was observed that the application of SiO_2 nanoparticles could increase the plant resistance to drought conditions (Ashkavand et al. 2015).

4.4 Future Trends and Challenges

With a wide range of potential applications, agri-nanotechnology has been actively researched at both institutional and industrial levels. The development of novel nano-tools could be encouraged in the future for a sustainable agriculture. At present, nanotechnology is the promising solution to meet the two most demanding issues, i.e., efficient utilization of resources and reducing the output waste. Hence, innovative nanomaterials and systems will be developed to improve the quality of the soil, stimulate plant/rhizobacterial growth, and intensify sustainable agricultural output (Yadav et al. 2020). The validation and safety limits of these nanoparticles must be studied before proceeding to a wide range of utilization and commercialization. A proper regulatory guidelines and legislative paperwork must be enforced, with risk management factors of nanoparticles. In this scenario, at both national and international level the regulatory bodies namely Department of Biotechnology, India, International Standard Organization, Organization for Economic Cooperation and Development, and US Food and Drug Administration have laid down guidelines and practices, to ensure the proper utilization of these nanomaterials in various sectors of agriculture (FAO/WHO 2013; DBT 2019). Strictly following these guidelines will minimize the adverse effects of these NPs on the natural environmental factors such as vegetation, microbes, and soil over a long period and exhibit their substantial positive impacts.

4.5 Conclusion

With the growing population and global warming application of nanotechnology undoubtedly promise sustainable agriculture and enhanced food yield. But, in view of the above-addressed reports of both positive and negative impacts of NPs on the crop microbiomes, it is necessary to perform a proper risk-benefit assessment before their application in the agricultural fields. It must explore a clear overview of the plant-microbe-soil trophic chain transfer of these NPs. Indeed, the green synthesis of nanoparticles is a safer approach and promising with less toxicity. In future, a thorough physio-chemical assessment of these beneficial NPs in natural experimental designs such as agricultural fields could be advantageous to evaluate their actual footprint in the ecological nature and on soil microbiota.

Acknowledgments We thank Mr. P.V.S. Prasad for his significant contribution in collecting the literatures.

Conflict of Interest None.

References

Abd-Elsalam KA, Prasad R (2018) Nanobiotechnology applications in plant protection. Springer International Publishing. ISBN 978-3-319-91161-8. https://www.springer.com/us/book/9783319911601

Abd-Elsalam K, Prasad R (2019) Nanobiotechnology applications in plant protection, vol 2. Springer International Publishing. ISBN 978-3-030-13295-8. https://www.springer.com/gp/book/9783030132958

Abd-Elsalam K, Mohamed AA, Prasad R (2019) Magnetic nanostructures: environmental and agricultural applications. Springer International Publishing. ISBN 978-3-030-16438-6. https://www.springer.com/gp/book/9783030164386

Alekhya G, Gopalakrishnan S (2017) Biological control and plant growth-promotion traits of *Streptomyces* species under greenhouse and field conditions in chickpea. Agribiol Res 6:410–420. https://doi.org/10.1007/s40003-017-0278-2

Amna AHF, Hakeem KR, Qureshi MI (2019) Weed control through herbicide-loaded nanoparticles. In: Husen A, Iqbal M (eds) Nanomaterials and plant potential. Springer, Cham. https://doi.org/10.1007/978-3-030-05569-1_20

Ankati S, Srinivas V, Pratyusha S, Gopalakrishnan S (2021) Streptomyces consortia-mediated plant defense against *Fusarium* wilt and plant growth-promotion in chickpea. Microbial Pathog 157: 1–8. https://doi.org/10.1016/j.micpath.2021.104961. ISSN 0882-4010

Anusha B, Gopalakrishnan S, Naik MK, Sharma M (2019) Evaluation of Streptomyces spp. and Bacillus spp. for biocontrol of *Fusarium* wilt in chickpea (*Cicer arietinum* L.). Arch Phytopathol Plant Prot 52:1–26. https://doi.org/10.1080/03235408.2019.1635302

Asadishad B, Chaha S, Cianciarelli V, Zhou K, Tufenkji N (2017) Effect of gold NPs on extracellular nutrient-cycling enzyme activity and bacterial community in soil slurries: role of nanoparticle size and surface coating. Environ Sci Nano 4:907–918. https://doi.org/10.1039/C6EN00567E

Ashkavand P, Tabari M, Zarafshar M, Tomášková I, Struve D (2015) Effect of SiO_2 nanoparticles on drought resistance in hawthorn seedlings. Forest Res Pap 76:350–359. https://doi.org/10.1515/frp-2015-0034

Avestan S, Ghasemnezhad M, Esfahani M, Byrt CS (2019) Application of nano-silicon dioxide improves salt stress tolerance in strawberry plants. Agronomy 9:246. https://doi.org/10.3390/agronomy9050246

Bayat M, Zargar M, Chudinova E, Astarkhanova T, Pakina E (2021) In vitro evaluation of antibacterial and antifungal activity of biogenic silver and copper nanoparticles: the first report of applying biogenic nanoparticles against *Pilidium concavum* and *Pestalotia* sp. fungi. Molecules 26:5402. https://doi.org/10.3390/molecules26175402

Beneduzi A, Ambrosini A, Passaglia LMP (2012) Plant growth-promoting rhizobacteria (PGPR): their potential as antagonists and biocontrol agents. Genet Mol Biol 35:1044–1051. https://doi.org/10.1590/S1415-47572012000600020

Ben-Moshe T, Frenk S, Dror I, Minz D, Berkowitz B (2013) Effects of metal oxide nanoparticles on soil properties. Chemosphere 90:640–646. https://doi.org/10.1016/j.chemosphere.2012.09.018

Bhattacharyya A, Duraisamy P, Govindarajan M, Buhroo AA, Prasad R (2016) Nano-biofungicides: emerging trend in insect pest control. In: Prasad R (ed) Advances and applications through fungal nanobiotechnology. Springer International Publishing Switzerland, pp 307–319

Bhan S, Mohan L, Srivastava CN (2018) Nanopesticides: a recent novel ecofriendly approach in insect pest management. J Entomol Res 42:263. https://doi.org/10.5958/0974-4576.2018.00044.0
Bipin L, Gogle DP, Nandeshwar SB (2017) Nano Bio pesticide to constraint plant destructive pests. J Nanomed Res 6:1–9. https://doi.org/10.15406/jnmr.2017.06.00158
Borah SN, Koch N, Sen S, Prasad R, Sarma H (2022) Novel nanomaterials for nanobioremediation of polyaromatic hydrocarbons. In: Sharma H, Dominguez DC, Lee W-Y (eds) Emerging contaminants in the environment. Elsevier, pp 643–667
Bundschuh M, Filser J, Lüderwald S, Mckee MS, Metriveli G, Schaumann GE, Schulz R, Wagner S (2018) Nanoparticles in the environment: where do we come from, where do we go to? Environ Sci Eur 30:6. https://doi.org/10.1186/s12302-018-0132-6
Camara MC, Campos EVR, Monteiro RA, Pereira AES, Proenca PLDF, Fraceto LF (2019) Development of stimuli-responsive nano-based pesticides: emerging opportunities for agriculture. J Nanobiotechnol 17:100. https://doi.org/10.1186/s12951-019-0533-8
Chen H, Yada R (2011) Nanotechnologies in agriculture: new tools for sustainable development. Trends Food Sci Technol 22:585–594. https://doi.org/10.1016/j.tifs.2011.09.004
Cheng KY, Karthikeyan R, Wong JW (2019) Microbial electrochemical remediation of organic contaminants: possibilities and perspective. Microb Electrochem Technol 25:613–640. https://doi.org/10.1016/B978-0-444-64052-9.00025-X
Chhipa H (2017) Nanofertilizers and nanopesticides for agriculture. Environ Chem Lett 15:15–22. https://doi.org/10.1007/s10311-016-0600-4
Coll C, Notter D, Gottschalk F, Sun T, Som C, Nowack B (2016) Probabilistic environmental risk assessment of five nanomaterials (nano-TiO2, nano-Ag, nano-ZnO, CNT, and fullerenes). Nanotoxicology 10:436–444. https://doi.org/10.3109/17435390.2015.1073812
Colman BP, Arnaout CL, Anciaux S, Gunsch CK, Hochella MF Jr, Kim B, Lowry GV, McGill BM, Reinsch BC, Richardson CJ, Unrine JM, Wright JP, Yin L, Bernhardt ES (2013) Low concentrations of silver nanoparticles in sewage sludge cause adverse ecosystem responses under realistic field scenario. PLoS One 8:57189. https://doi.org/10.1371/journal.pone.0057189
Compant S, Samad A, Faist H, Sessitsch A (2019) A review on the plant microbiome: ecology, functions, and emerging trends in microbial application. J Adv Res 19:29–37. https://doi.org/10.1016/j.jare.2019.03.004
DBT (2019) Guidelines for evaluation of nano-a. http://dbtindia.gov.in/sites/default/files/DBT_Draft1-Nano-Agri_Input_nd_Nano-Agri_Products.pdf. Agri input products and nano-agriproducts in India. pp 1–26
Deepti M, Gurjeet K, Parul S, Karmveer Y, Azmal AS (2020) Nanoparticle-based sustainable agriculture and food science: recent advances and future outlook. Front Nanotech 2:10. https://doi.org/10.3389/fnano.2020.579954
Dixshit A, Shukla SK, Mishra RK (2013) Exploring nanomaterials with PGPR in current agriculture scenario PGPR with special reference to nanomaterials lab Lambert. Academic, Berlin, p 51
Fageria NK (2016) In: Fageria NK (ed) The use of nutrients in crop plants. CRC Press, New York, NY
Fajardo C, Ortiz LT, Rodriguez-Membibre ML, Nande M, Lobo MC, Martin M (2012) Assessing the impact of zero-valent iron (ZVI) nanotechnology on soil microbial structure and functionality: a molecular approach. Chemosphere 86:802–808. https://doi.org/10.1016/j.chemosphere.2011.11.041
FAO (Food and Agriculture Organization of the United Nations) (2011) The International Plant Protection Convention at 60 years. http://www.fao.org/agriculture/crops/news-events-bulletins/detail/en/item/116461/icode/?no_cache=1
FAO/WHO (2013) State of the art on the initiatives and activities relevant to risk assessment and risk Management of Nanotechnologies in the food and agriculture sectors. FAO/WHO, Rome
Feng B, Peng L (2012) Synthesis and characterization of carboxymethyl chitosan carrying ricinoleic functions as an emulsifier for azadirachtin. Carbohdr Polym 88:576–582. https://doi.org/10.1016/j.carbpol.2012.01.002

fibsol (n.d.). https://www.fibsol.com/product-category/gels/

Flores-Felix JD, Menendez E, Rivera LP (2013) Use of *Rhizobium leguminosarum* as a potential biofertilizer for *Lactuca sativa* and *Daucus carota* crops. J Plant Nutr Soil Sci 176:876–882. https://doi.org/10.1002/jpln.201300116

Frenk S, Ben-Moshe T, Dror I, Berkowitz B, Minz D (2013) Effect of metal oxide nanoparticles on microbial community structure and function in two different soil types. PLoS One 8:84441. https://doi.org/10.1371/journal.pone.0084441

Garcia-Fraile P, Carro L, Robledo M (2012) Rhizobium promotes non-legumes growth and quality in several production steps: towards a biofertilization of edible raw vegetables healthy for humans. PLoS One 7:e38122. https://doi.org/10.1371/journal.pone.0038122

Ge Y, Schimel JP, Holdena PA (2012) Identification of soil bacteria susceptible to TiO2 and ZnO NPs. Appl Environ Microbiol 78:6749–6758. https://doi.org/10.1128/AEM.00941-12

geolifegroup (n.d.). https://www.geolifegroup.com/region/domestic-products/item/44-nano-fert

Gogos A, Knauer K, Bucheli TD (2012) Nanomaterials in plant protection and fertilization: current state, foreseen applications, and research priorities. J Agric Food Chem 60:9781–9792. https://doi.org/10.1021/jf302154y

Gopalakrishnan S, Vadlamudi S (2019) Management of soil-borne diseases of grain legumes through broad-spectrum actinomycetes having plant growth-promoting and biocontrol traits. In: Plant microbe interface. Springer, New York. https://doi.org/10.1007/978-3-030-19831-2_5

Gopalakrishnan S, Humayun P, Kiran BK, Kannan GK, Sree Vidya M, Deepthi K, Rupela O (2011) Evaluation of bacteria isolated from rice rhizosphere for biological control of charcoal rot of sorghum caused by Macrophomina phaseolina (Tassi) Goid. World J Microbiol Biotech 27: 1313–1321. https://doi.org/10.1007/s11274-010-0579-0. ISSN: 09593993

Gopalakrishnan S, Pagidi H, Vadlamudi S, Vijayabharathi R, Bhimineni R, Rupela O (2012a) Plant growth-promoting traits of *Streptomyces* with biocontrol potential isolated from herbal vermicompost. Biocont Sci Technol 22:1199–1210. https://doi.org/10.1080/09583157.2012.719151

Gopalakrishnan S, Upadhyaya HD, Vadlamudi S, Humayun P, Vidya MS, Alekhya G, Singh A, Vijayabharathi R, Bhimineni RK, Seema M, Rathore A, Rupela O (2012b) Plant growth-promoting traits of biocontrol potential bacteria isolated from rice rhizosphere. Springerplus 102:1–13. https://doi.org/10.1186/2193-1801-1-71

Gopalakrishnan S, Vadlamudi S, Samineni S, Sameer KCV (2016) Plant growth-promotion and biofortification of chickpea and pigeonpea through inoculation of biocontrol potential bacteria, isolated from organic soils. Springerplus 5:01–11. https://doi.org/10.1186/s40064-016-3590-6

Gopalakrishnan S, Srinivas V, Anil Kumar V, Samineni S, Rathore A (2018) Influence of diazotrophic bacteria on nodulation, nitrogen fixation, growth promotion and yield traits in five cultivars of chickpea. Biocatal Agri Biotech 15:35–42. https://doi.org/10.1016/j.bcab.2018.05.006

Gopalakrishnan S, Sharma R, Srinivas V, Naresh N, Mishra SP, Ankati S, Pratyusha S, Govindaraj M, Gonzalez SV, Nervik S, Simic N (2020) Identification and characterization of a *Streptomyces albus* strain and its secondary metabolite organophosphate against charcoal rot of sorghum. Plan Theory 9:1–14. https://doi.org/10.3390/plants9121727

Gopalakrishnan S, Srinivas V, Naresh N, Pratyusha S, Ankati S, Madhuprakash J, Govindaraj M, Sharma R (2021) Deciphering the antagonistic effect of *Streptomyces* spp. and host-plant resistance induction against charcoal rot of sorghum. Planta 253:1–12. https://doi.org/10.1007/s00425-021-03577-5

Gupta N, Upadhyaya CP, Singh A, Abd-Elsalam KA, Prasad R (2018) Applications of silver nanoparticles in plant protection. In: Abd-Elsalam K, Prasad R (eds) Nanobiotechnology applications in plant protection. Springer International Publishing AG, pp 247–266

Hamaoui B, Abbadi J, Burdman S, Rashid A, Sarig S, Okon Y (2001) Effects of inoculation with *Azospirillum brasilense* on chickpeas (*Cicer arietinum*) and faba beans (*Vicia faba*) under different growth conditions. Agronomy 21:553–560. https://doi.org/10.1051/agro:2001144

Hänsch M, Emmerling C (2010) Effects of silver nanoparticles on the microbiota and enzyme activity in soil. J Plant Nut Soil Sci 173:554–558

Harish V, Tewari D, Gaur M, Yadav AB, Swaroop S, Bechelany M, Barhoum A (2022) Review on nanoparticles and nanostructured materials: bioimaging, biosensing, drug delivery, tissue engineering, antimicrobial, and agro-food applications. Nano 12:457. https://doi.org/10.3390/nano12030457

Hashami SZ, Nakamura H, Ohkama-Ohtsu N, Kojima K, Djedidi S, Fukuhara I et al (2019) Evaluation of immune responses induced by simultaneous inoculations of soybean (*Glycine max* [L.] Merr.) with soil bacteria and rhizobia. Microbes Environ 34(1):64–75. https://doi.org/10.1264/jsme2.ME18110

Ismail M, Prasad R, Ibrahim AIM, Ahmed ISA (2017) Modern prospects of nanotechnology in plant pathology. In: Prasad R, Kumar M, Kumar V (eds) Nanotechnology. Springer Nature Singapore Pte Ltd, pp 305–317

John H, Lucas J, Clare W, Dusan L (2017) Nanopesticides: a review of current research and perspectives. In: Grumezescu AM (ed) New pesticides and soil sensors. Academic, New York, pp 193–225

Juanni C, Lintong W, Mei L, Shasha L, Ziyan L, Wei D (2020) Comparative study on the fungicidal activity of metallic MgO nanoparticles and macroscale MgO against soilborne fungal phytopathogens. Front Microbiol 11:365. https://doi.org/10.3389/fmicb.2020.00365

Kah M, Kookana RS, Gogos A, Bucheli TD (2018) A critical evaluation of nanopesticides and nanofertilizers against their conventional analogues. Nat Nanotechnol 13:677–684. https://doi.org/10.1038/s41565-018-0131-1

Kah M, Tufenkji N, White JC (2019) Nano-enabled strategies to enhance crop nutrition and protection. Nat Nanotechnol 14:532–540. https://doi.org/10.1038/s41565-019-0439-5

Khan MN (2016) Nano-titanium dioxide (Nano-TiO2) mitigates NaCl stress by enhancing antioxidative enzymes and accumulation of compatible solutes in tomato (*Lycopersicon esculentum* Mill.). J Plant Sci 11:1–11

Khatoon Z, Huang S, Rafique M, Fakhar A, Kamran MA, Santoyo G (2020) Unlocking the potential of plant growth-promoting rhizobacteria on soil health and the sustainability of agricultural systems. J Environ Manag 273:111118. https://doi.org/10.1016/j.jenvman.2020.111118

Kim DY, Kadam A, Shinde S, Saratale RG, Patra J, Ghodake GJ (2017) Recent developments in nanotechnology transforming the agricultural sector: a transition replete with opportunities. Sci Food Agric 98:849–864. https://doi.org/10.1002/jsfa.8749

Kumar N, Shah V, Walker VK (2012) Influence of a nanoparticle mixture on an arctic soil community. Environ Toxicol Chem 31:131–135. https://doi.org/10.1002/etc.721

Lakshmanan V, Selvaraj G, Bais HP (2014) Functional soil microbiome: below ground solutions to an above ground problem. Plant Physiol 166:689–700. https://doi.org/10.1104/pp.114.245811

Lawrance S, Varghese S, Varghese EM, Asok AK (2019) Quinoline derivatives producing *Pseudomonas aeruginosa* H6 as an efficient bioherbicide for weed management. Biocat Agric Biotechnol 18:101096. https://doi.org/10.1016/j.bcab.2019.101096

Liu R, Lal R (2015) Potentials of engineered nanoparticles as fertilizers for increasing agronomic productions. Sci Total Environ 514:131–139. https://doi.org/10.1016/j.scitotenv.2015.01.104

Lowry GV, Espinasse BP, Badireddy AR, Richardson CJ, Reinsch BC, Bryant LD, Colman BP (2012) Long-term transformation and fate of manufactured Ag nanoparticles in a simulated large scale freshwater emergent wetland. Environ Sci Technol 46:7027–7036. https://doi.org/10.1021/es204608d

Lucy M, Reed E, Glick BR (2004) Applications of free-living plant growth-promoting rhizobacteria. Antonie Van Leeuwenhoek 86:1–25. https://doi.org/10.1023/B:ANTO.0000024903.10757.6e

Malandrakis AA, Kavroulakis N, Chrysikopoulos CV (2019) Use of copper, silver and zinc nanoparticles against foliar and soil-borne plant pathogens. Sci Total Environ 670:292–299. https://doi.org/10.1016/j.scitotenv.2019.03.210

Maurer-Jones MA, Ian L, Gunsolus IL, Murphy CJ, Haynes CL (2013) Toxicity of engineered nanoparticles in the environment. Anal Chem 85:3036–3049. https://doi.org/10.1021/ac303636s

McGee CF, Storey S, Clipson N, Doyle E (2017) Soil microbial community responses to contamination with silver, aluminium oxide and silicon dioxide nanoparticles. Ecotoxicology 26:1–10. https://doi.org/10.1007/s10646-017-1776-5

Melanie K, Rai K (2020) Emerging investigator series: nanotechnology to develop novel agrochemicals: critical issues to consider in the global agricultural context. Environ Sci Nano 7: 1867–1873. https://doi.org/10.1039/D0EN00271B

Moulick RG, Das S, Debnath N, Bandyopadhyay K (2020) Potential use of nanotechnology in sustainable and 'smart' agriculture: advancements made in the last decade. Plant Biotechnol Rep 14:505. https://doi.org/10.1007/s11816-020-00636-3

Mukhopadhyay SS (2014) Nanotechnology in agriculture prospects and constraints. Nanotechnol Sci Appl 7:63–71. https://doi.org/10.2147/NSA.S39409

Müller DB, Vogel C, Bai Y, Vorholt JA (2016) The plant microbiota: systems-level insights and perspectives. Annu Rev Genet 50:211–234. https://doi.org/10.1146/annurev-genet-120215-034952

Nawaz S, Bano A (2019) Effects of PGPR (*Pseudomonas* sp.) and Ag-nanoparticles on enzymatic activity and physiology of cucumber. Recent Pat Food Nutr Agric 11:124–136. https://doi.org/10.2174/2212798410666190716162340

Ongena M, Jacques P (2008) Bacillus lipopeptides: versatile weapons for plant disease biocontrol. Trends Microbiol 16:115–125. https://doi.org/10.1016/j.tim.2007.12.009

Ostiguy C, Roberge B, Woods C, Soucy B (2010) Engineered nanoparticles: current knowledge about OHS risks and prevention measures. Studies and research projects, report R-656, 2nd edn. Institut Robert Sauvé de Recherche en Santé et Sécurité au Travail, Montréal

Paramanandham P, Rajkumari J, Pattnaik S, Busi S (2017) Biocontrol potential against *Fusarium oxysporum* sp. lycopersici and Alternaria solani and tomato plant growth due to Plant Growth–Promoting Rhizobacteria. Int J Veg Sci 23:294–303. https://doi.org/10.1080/19315260.2016.1271850

Pedros-Ali C, Manrubia S (2016) The vast unknown microbial biosphere. Proc Natl Acad Sci U S A 113:6585–6587. https://doi.org/10.1073/pnas.1606105113

Peyrot C, Wilkinson KJ, Desrosiers M, Sauvé S (2014) Effects of silver nanoparticles on soil enzyme activities with and without added organic matter. Environ Toxicol Chem 33:115–125. https://doi.org/10.1002/etc.2398

Pietrzak K, Gutarowska B (2015) Influence of the silver nanoparticles on microbial community indifferent environments. Acta Biochim Pol 62:72–724. https://doi.org/10.18388/abp.2015_1118

Pitambara A, Shukla YM (2019) Nanofertilizers: a recent approach in crop production. In: Panpatte D, Jhala Y (eds) Nanotechnology for agriculture: crop production & protection. Springer, Singapore. https://doi.org/10.1007/978-981-32-9374-8_2

Pokropivny V, Lohmus R, Hussainova I, Pokropivny A, Vlassov S (2007) Introduction to nanomaterials and nanotechnology. University of Tartu, Tartu, p 225

Prasad R (2014) Synthesis of silver nanoparticles in photosynthetic plants. J Nanopart 2014:963961. https://doi.org/10.1155/2014/963961

Prasad R, Kumar V, Prasad KS (2014) Nanotechnology in sustainable agriculture: present concerns and future aspects. Afr J Biotechnol 13(6):705–713

Prasad R, Kumar M, Varma A (2015) Role of PGPR in soil fertility and plant health. In: Egamberdieva D, Shrivastava S, Varma A (eds) Plant growth-promoting rhizobacteria (PGPR) and medicinal plants. Springer International Publishing, Cham, pp 247–260

Prasad R, Pandey R, Barman I (2016) Engineering tailored nanoparticles with microbes: quo vadis. WIREs Nanomed Nanobiotechnol 8:316–330. https://doi.org/10.1002/wnan.1363

Prasad R, Bhattacharyya A, Nguyen QD (2017a) Nanotechnology in sustainable agriculture: recent developments, challenges, and perspectives. Front Microbiol 8:1014. https://doi.org/10.3389/fmicb.2017.01014

Prasad R, Kumar M, Kumar V (2017b) Nanotechnology: an agriculture paradigm. Springer Nature Singapore Pte Ltd. ISBN: 978-981-10-4573-8

Prasad R, Kumar V, Kumar M (2017c) Nanotechnology: food and environmental paradigm. Springer Nature Singapore Pte Ltd. ISBN 978-981-10-4678-0

Prasad R, Kumar V, Kumar M, Choudhary D (2019) Nanobiotechnology in bioformulations. Springer International Publishing. ISBN 978-3-030-17061-5. https://www.springer.com/gp/book/9783030170608

Pratyusha S, Vadlamudi S, Gopalakrishnan S (2021) Understanding the evolution of plant growth-promoting rhizobacteria. In: Shrivastava N et al (eds) Symbiotic soil microorganisms, soil biology. Springer, New York, p 60. https://doi.org/10.1007/978-3-030-51916-2_12

Rai V, Acharya S, Dey N (2012) Implications of nanobiosensors in agriculture. J Biomat Nanobiotech 3:1–11. https://doi.org/10.4236/jbnb.2012.322039

Rana RA, Siddiqui MN, Skalicky M, Brestic M, Hossain A, Kayesh E, Popov M, Hejnak V, Gupta DR, Mahmud NU et al (2021) Prospects of nanotechnology in improving the productivity and quality of horticultural crops. Horticulturae 7:332. https://doi.org/10.3390/horticulturae7100332

Sambangi P, Srinivas V, Gopalakrishnan S (2021) Understanding the evolution of plant growth-promoting Rhizobacteria. In: Shrivastava N, Mahajan S, Varma A (eds) Symbiotic soil microorganisms. Soil biology, vol 60. Springer, Cham. https://doi.org/10.1007/978-3-030-51916-2_12

Sangeetha J, Thangadurai D, Hospet R, Harish ER, Purushotham P, Mujeeb MA, Shrinivas J, David M, Mundaragi AC, Thimmappa AC, Arakera SB, Prasad R (2017a) Nanoagrotechnology for soil quality, crop performance and environmental management. In: Prasad R, Kumar M, Kumar V (eds) Nanotechnology. Springer Nature Singapore Pte Ltd, pp 73–97

Sangeetha J, Thangadurai D, Hospet R, Purushotham P, Karekalammanavar G, Mundaragi AC, David M, Shinge MR, Thimmappa SC, Prasad R, Harish ER (2017b) Agricultural nanotechnology: concepts, benefits, and risks. In: Prasad R, Kumar M, Kumar V (eds) Nanotechnology. Springer Nature Singapore Pte Ltd, pp 1–17

Sangeetha J, Thangadurai D, Hospet R, Purushotham P, Manowade KR, Mujeeb MA, Mundaragi AC, Jogaiah S, David M, Thimmappa SC, Prasad R, Harish ER (2017c) Production of bionanomaterials from agricultural wastes. In: Prasad R, Kumar M, Kumar V (eds) Nanotechnology. Springer Nature Singapore Pte Ltd, pp 33–58

Schlich K, Hund-Rinke K (2015) Influence of soil properties on the effect of silver nanomaterials on microbial activity in five soils. Environ Pollut 196:321–330. https://doi.org/10.1016/j.envpol.2014.10.021

Sekhon BS (2014) Nanotechnology in agri-food production: an overview. Nanotechnol Sci Appl 7: 31–53. https://doi.org/10.2147/NSA.S39406

Shah V, Belozerova I (2009) Influence of metal nanoparticles on the soil microbial community and germination of lettuce seeds. Water Air Soil Pollut 197:143–148. https://doi.org/10.1007/s11270-008-9797-6

Shah M, Wani S, Khan A (2016) Nanotechnology and insecticidal formulations. J Food Bioeng Nanoproc 1:285–310

Shaharoona B, Naveed M, Arshad M (2008) Fertilizer-dependent efficiency of *Pseudomonas* for improving growth, yield, and nutrient use efficiency of wheat (Triticum aestivum L.). Appl Microbiol Biotechnol 79:147–155. https://doi.org/10.1007/s00253-008-1419-0

Sharma R, Paliwal JS, Chopra P, Dogra D, Pooniya V, Bisaria VS, Swarnalakshmi K, Sharma S (2017) Survival, efficacy and rhizospheric effects of bacterial inoculants on *Cajanus cajan*. Agric Ecosyst Environ 240:244–252. https://doi.org/10.1016/j.agee.2017.02.018

Shen Z, Chen Z, Hou Z, Li T, Lu X (2015) Ecotoxicological effect of zinc oxide NPs on soil microorganisms. Front Environ Sci Eng 9:912–918. https://doi.org/10.1007/s11783-015-0789-7

Shin YJ, Kwak JI, An YJ (2012) Evidence for the inhibitory effects of silver nanoparticles on the activities of soil exoenzymes. Chemosphere 88:524–529. https://doi.org/10.1016/j.chemosphere.2012.03.010

Shweta, Tripathi DK, Chauhan DK, Peralta-Videa JR (2018) Availability and risk assessment of nanoparticles in living systems: a virtue or a peril? In: Nanomaterials in plants, algae, and microorganisms. Academic, London, pp 1–31. https://doi.org/10.1007/978-3-030-34544-0

Simonin M, Cantarel AA, Crouzet A, Gervaix J, Martins JM, Richaume A (2018) Negative effects of copper oxide nanoparticles on carbon and nitrogen cycle microbial activities in contrasting agricultural soils and in presence of plants. Front Microbiol 9:3102. https://doi.org/10.3389/fmicb.2018.03102

Singh J, Vishwakarma K, Ramawat N, Rai P, Singh VK, Mishra RK, Kumar V, Tripathi DK, Sharma S (2019) Nanomaterials and microbes' interactions: a contemporary overview. 3 Biotech 9:68. https://doi.org/10.1007/s13205-019-1576-0

Srinivas V, Gopalakrishnan S, Kamidi JP, Chander G (2020) Effect of plant growth-promoting *Streptomyces* sp. on plant growth and yield of tomato and chilli. Andhra Pradesh J Agril Sci 6: 65–70. http://oar.icrisat.org/id/eprint/11668

Srivastava AK, Dev A, Karmakar S (2018) Nanosensors and nanobiosensors in food and agriculture. Environ Chem Lett 16:161–182. https://doi.org/10.1007/s10311-017-0674-7

Srivastava S, Usmani Z, Atanasov AG, Singh VK, Singh NP, Abdel-Azeem AM, Prasad R, Gupta G, Sharma M, Bhargava A (2021) Biological nanofactories: using living forms for metal nanoparticle synthesis. Mini-Rev Med Chem 21(2):245–265

Taran N, Storozhenko V, Svietlova N et al (2017) Effect of zinc and copper nanoparticles on drought resistance of wheat seedlings. Nanoscale Res Lett 12:60. https://doi.org/10.1186/s11671-017-1839-9

Tejera N, Lluch C, Martínez-Toledo MV (2005) Isolation and characterization of *Azotobacter* and *Azospirillum* strains from the sugarcane rhizosphere. Plant Soil 270:223–232. https://doi.org/10.1007/s11104-004-1522-7

Tilak KVBR, Ranganayaki N, Pal KK, De R, Saxena AK, Nautiyal CS, Mittal S, Tripathi AK, Johri BN (2005) Diversity of plant growth and soil health supporting bacteria. Curr Sci 89:136–150

Timmusk S, Seisenbaeva G, Behers L (2018) Titania (TiO_2) nanoparticles enhance the performance of growth-promoting rhizobacteria. Sci Rep 8:617. https://doi.org/10.1038/s41598-017-18939-x

TNAU (2019) Crop protection guide horticulture crops. http://agritech.tnau.ac.in/pdf/HORTICULTURE.pdf

Vaikundamoorthy R, Rajendran R, Selvaraju A, Moorthy K, Perumal S (2018) Development of thermostable amylase enzyme from *Bacillus cereus* for potential antibiofilm activity. Bioorg Chem 77:494–506. https://doi.org/10.1016/j.bioorg.2018.02.014

Vijayabharathi R, Sathya A, Gopalakrishnan S (2018a) Extracellular biosynthesis of silver nanoparticles using *Streptomyces griseoplanus* SAI-25 and its antifungal activity against *Macrophomina phaseolina*, the charcoal rot pathogen of sorghum. Biocatal Agri Biotech 14: 166–171. https://doi.org/10.1016/j.bcab.2018.03.006

Vijayabharathi R, Gopalakrishnan S, Arumugam S, Vadlamudi S, Sharma M (2018b) Deciphering the tri-dimensional effect of endophytic *Streptomyces* sp. on chickpea for plant growth promotion, helper effect with *Mesorhizobium ciceri* and host-plant resistance induction against *Botrytis cinerea*. Microbl Pathog 122:98–107. https://doi.org/10.1016/j.micpath.2018.06.019

Vittori Antisari L, Carbone S, Gatti A, Vianello G, Nannipieri P (2013) Toxicity of metal oxide (CeO2, Fe3O4, SnO2) engineered NPs on soil microbial biomass and their distribution in soil. Soil Biol Biochem 60:87–94. https://doi.org/10.1016/j.soilbio.2013.01.016

Yadav T, Mungray AA, Mungray AK (2014) Fabricated nanoparticles: current status and potential phytotoxic threats. Rev Environ Contam Toxicol 230:83–110. https://doi.org/10.1007/978-3-319-04411-8_4

Yadav RV, Singh NB, Singh A, Yadav V, Bano C, Khare S, Niharika (2020) Expanding the horizons of nanotechnology in agriculture: recent advances, challenges and future perspectives. Vegetos 33:203–221. https://doi.org/10.3389/fnano.2020.579954

Yang FL, Li XG, Zhu F, Lei CL (2009) Structural characterization of nanoparticles loaded with garlic essential oil and their insecticidal activity against *Tribolium castaneum* (Herbst) (Coleoptera: Tenebrionidae). J Agric Food Chem 57:10156–10162. https://doi.org/10.1021/jf9023118

Yanni Y, Rizk R, Abd-El Fattah F (2001) The beneficial plant growth-promoting association of *Rhizobium leguminosarum* bv. trifolii with rice roots. Aust J Plant Physiol 28:845–870. https://doi.org/10.1071/PP01069

Yao L, Wu Z, Zheng Y (2010) Growth promotion and protection against salt stress by *Pseudomonas putida* Rs-198 on cotton. Eur J Soil Biol 46:49–54. https://doi.org/10.1016/j.ejsobi.2009.11.002

Zhai Y, Hunting RE, Wouters M, Willie Peijnenburg WJGM, Vijver GM (2016) Silver nanoparticles, ions, and shape governing soil microbial functional diversity: nano shapes micro. Front Microbiol 7:1. https://doi.org/10.3389/fmicb.2016.01123

Zhang H, Zhu J, Gong Z et al (2022) Abiotic stress responses in plants. Nat Rev Genet 23:104–119. https://doi.org/10.1038/s41576-021-00413-0

Chapter 5
Changes in Plant Microbiome in Response to Abiotic Stress

Mishaal Irfan, Hira Aslam, Awais Maqsood, Syeda Khola Tazeen, Faisal Mahmood, and Muhammad Shahid

Abstract Heavy metal, drought, salinity, temperature, and nutrient shortage are among abiotic factors that influence crop plants and jeopardize agricultural output. The association of a variety of microbes with plants is very significant for the growth and health of the plants for nutrient uptake, protection against phytopathogens, hormonal signaling for homeostasis, and acquired tolerance against various abiotic stresses. A community of these microbes that associates with plant is termed as plant microbiome. Recent research reveals that phenotypic and genotypic traits including length, water storage content, water storage capacity, nitrogen content, the phosphorus content of leaf, transpiration rate, and net photosynthetic rate are correlated with the composition of bacterial, archaeal, and fungal communities. Hence, a shift in the population density of these microbes in response to the abiotic stresses could significantly affect plant morphology and physiology. For example, drought could alter the microbial compositions in the rhizosphere and endosphere. The experimental studies revealed the depletion of several *Acidobacteria*, *Verrucomicrobia*, and *Deltaproteobacteria* followed by the enrichment of *Actinobacteria* and *Chloroflexi* during drought stress. Halotolerant microbes survive in high salt concentration expressing such traits that help plants to live in elevated salinity environments. All microorganisms respond to extreme temperatures by making specific polypeptides called heat shock proteins (HSPs). Certain bacterial strains produce exopolysaccharides (EPSs) that possess distinctive characteristics of holding water and also has cementing characteristics. The higher temperature increased *Bacteroidetes* and *Verrucomicrobia* while causing a decrease in the abundance of *Proteobacteria*. Engineering the rhizosphere of plants with specific abiotic

M. Irfan · H. Aslam · A. Maqsood · S. K. Tazeen · M. Shahid (✉)
Department of Bioinformatics and Biotechnology, Government College University Faisalabad, Faisalabad, Pakistan
e-mail: mshahid@gcuf.edu.pk

F. Mahmood
Department of Environmental Science and Engineering, Government College University Faisalabad, Faisalabad, Pakistan

S. Chhabra et al. (eds.), *Plant Microbiome for Plant Productivity and Sustainable Agriculture*, Microorganisms for Sustainability 37,
https://doi.org/10.1007/978-981-19-5029-2_5

stress-tolerant microbes makes a specific environment for the plant to nourish. This chapter will focus on different abiotic factors that affect the plant microbiome.

Keywords Abiotic stress · Microbiome · Rhizosphere · PGPM

5.1 Introduction

Microbial communities are essential for the survival of all living forms on the planet, as they are involved in practically every biogeochemical cycle that occurs (van der Heijden and Hartmann 2016). Microbiomes are made up of a diverse group of microorganisms that live in close proximity to higher creatures. Microorganisms can be found in all higher organisms, including animals, plants, and fish, as well as insects, mice, apes, and people. The plant microbiome is a term used to describe the collective groups of plant-associated bacteria (Barea 2015). Plant microbiome is the key factor in maintaining plant's fitness. Almost every part of the plant harbor microorganisms, our focus is on the rhizosphere, phyllosphere (plant aerial parts), and endosphere (internal tissues) (Tosi et al. 2020). The rhizosphere is a rich region of soul-derived microbes colonizing root exudates. In contrast, phyllosphere is a nutrient-poor region and is subjected to extreme temperature, radiation, and moisture conditions. Epiphytes are microbial communities found in the rhizosphere and endosphere. Microbes residing within plant parts comprise in endosphere. Microbes are of immense importance as it is recognized that they have enormous potential to increase host health. Microbes associated with plants are classified into three groups, based on their effects on plants: beneficial, deleterious, and neutral (Spaepen et al. 2009). Some of the beneficial microbes can help plants maintain their productivity under stressful conditions. The interaction of plants with microbes encompasses multipart benefits in a plant's cellular system.

Plants are subjected to a variety of biotic and abiotic stresses throughout their lives, all of which have a favorable or negative impact on their growth. Changes in transcriptomics and metabolomics can result in changes in tree, root, and leaf exudates, affecting the plant-associated microbial population in the process. Plants with microbiomes, particularly those that can withstand various abiotic challenges, are not only beneficial to plants but also operate as a legacy and boost offspring fitness, according to a growing body of evidence and research (Liu et al. 2020). Microorganisms react to a wide range of stimuli, and their diverse stress responsive systems interact with one another and have a role in pathogenicity.

Abiotic stressors caused by bad climatic conditions are one of the most important limiting factors for agricultural output loss. According to a 2007 FAO assessment, only 3.5% of the total land area is undisturbed by environmental limitations. Dryness, low/high temperature, salt, and acidic environments, light intensity, submergence, nutritional inadequacies, drought, and heavy metal stress are all examples of abiotic stressors. Drought has affected 64% of the land area, whereas flood (anoxia) has affected 13%, salinity has affected 6%, mineral deficiency has affected 9%, acidic soils have affected 15%, and cold has affected 57%. Erosion, soil

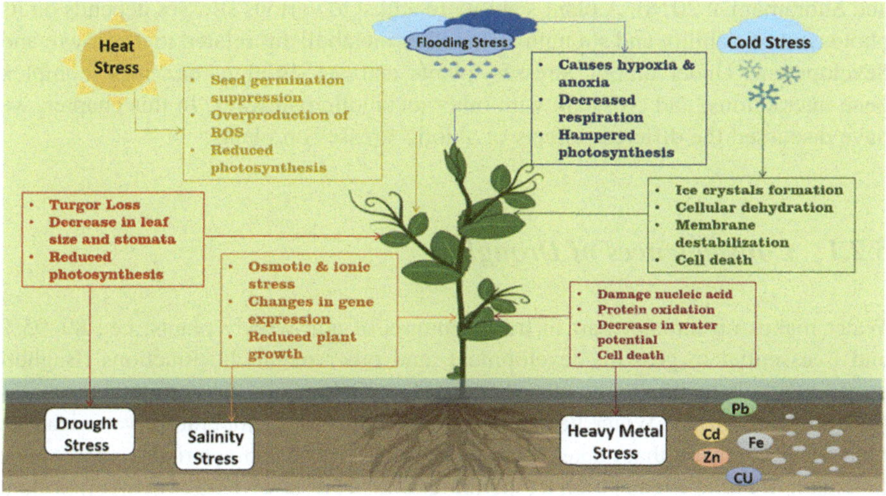

Fig. 5.1 Diverse abiotic stress factors affecting plants

deterioration, and salt have damaged 3.6 billion hectares of dryland agriculture on the planet's 5.2 billion hectares. Half of the world's irrigated land is expected to be damaged by salt-affected soils, resulting in a US$12 billion economic loss (Ansari 2018). Similarly, the global annual cost of land degradation by salt in irrigated countries might be US$ 27.3 billion owing to crop loss. Salinity has been shown to have a negative impact on plant growth. The number of hectares of irrigated land that is becoming increasingly salinized has nearly surpassed 34 million. Although it is impossible to make an exact assessment of agricultural loss (decrease in crop yield and soil health) as a result of abiotic stresses, it is clear that such pressures affect large areas of land and have a major impact on crop quality and quantity.

The soil biodiversity is the multiplex of biological communities on the Earth, with even more richness than tropical forests. Plant-associated microbes are critical to the health of the soil ecosystem and fertility. Soil is a significant organic carbon sink. In the soil, organic leftovers from diverse sources are converted by microbes into humus, a dark-colored, complex organic substance that is degraded by numerous types of bacteria, releasing CO_2 and water (Pandey et al. 2018). Also, it contains nutrients that are also supplied to plants by mineralization (Fig. 5.1).

5.2 Abiotic Stresses and Plants

A variety of abiotic stresses are experienced by plants such as drought flooding, extreme temperatures, salinity, nutrition deficiency, heavy metals that affect the productivity of plants. Because of the sessile nature of plants, these stresses impose major restrictions such as limiting crop production and food security worldwide (Jha and Subramanian 2018). A plant's ability to adjust to various stresses depends on its photosynthetic ability and strength about other metabolisms related to its growth and development. Under abiotic stresses, plants activate different enzymes, complex gene interactions, and crosstalk with other metabolic pathways. In this chapter, we have discussed the different effects of abiotic stresses on plants.

5.2.1 Consequences of Drought

Water makes up a major part of fresh biomass of non-woody plants, i.e., 80–95% and is essential in growth, development, and other metabolic functions (Bagheri et al. 2019). It maintains plant cell turgor thus helping in the respiration process (Khalid et al. 2019). Water has characteristics like high-temperature vaporization, high surface tension that makes it a good solvent. These characteristics allow water to remain in liquid even at high temperatures and act as a solvent for many important molecules, minerals, and ions. Scarcity of water is one of the major issues in plant production. Drought causes almost 50% losses in crop yield. There are many causes of water shortage such as low rainfall, high temperatures, high intensity of light, and salinity. Rainfall is the principal source of water because arid and semi-arid land covers roughly 35% of the Earth's surface. Even places with adequate rainfall face this problem because to the erratic distribution of rainfall throughout the year. As a result of changing global climatic conditions, temperature rises, increasing water evaporation from soil and plants (Salehi-Lisar and Bakhshayeshan-Agdam 2016).

Plant productivity is affected in a variety of ways by drought. Cell division, elongation, and differentiation are all necessary for an organism's growth and development (Salehi-Lisar and Bakhshayeshan-Agdam 2016). Drought has a direct impact on all of these stages since it causes turgor loss, enzyme disruption, and a reduced photosynthesis rate due to a lack of energy. Reduced leaf size and number of stomata, cutinization of leaf surface, increase in number of large vessels, cell wall thickening, premature leaf senescence, and submersion of stomata are some of the morphological and anatomical characteristics influenced by water stress in succulent and xerophytes plants. The fundamental variables in optimal photosynthesis in plants are appropriate leaf area development and stomatal openings. As a result of the lower surface area, increased resistance to gas exchange due to fewer stomatal openings, and increased leaf senescence in water stressed plants, the net photosynthetic rate is reduced. Drought stress primarily affects plant morphology by reducing plant size and biomass production (Khan et al. 2021a). To enhance water uptake under water stress, plants stretch their roots and form ramified root systems. During dehydration, the root to shoot ratio increases, while the total biomass of the plants decreases dramatically.

5.2.2 Consequences of Flooding

Flooding, also known as damped, flooded, drenched, or ponded soil, is a severe stress that harms plant growth and metabolism. One of the key drivers of global warming is changes in environmental circumstances induced by the industrial revolution's release of greenhouse gases such as carbon dioxide and methane (Tewari and Mishra 2018). As a result, changing climatic circumstances have influenced a number of abiotic stressors as well as biotic factors. Scientists have several definitions of flooding, which is regarded as one of the multiple stress factors that causes numerous alterations in plant tissues. Climate change is one of the main causes of erratic rainfall, which causes floods in some parts of the world. Flooding has a detrimental impact on about 13% of the total land area and 10% of agricultural land in the United States (Tewari and Mishra 2018). Around 15–80% of the total yield of crops is lost due to waterlogging condition. Despite the fact that a number of reviews had been published on different stress factors, but very less data is available on flooding stress. Plants can be deprived of CO_2, O_2, and sunlight if excess of water is present. Flooding stress disrupts plant processes. Photosynthesis and aerobic respiration are hampered by a lack of CO_2, O_2, and sunshine. Reduced photosynthesis and aerobic respiration result in a shortage of carbohydrates, restricting plant growth and development. This increased water level induces hypoxia (a drop in oxygen levels) in a small duration, causing anoxia to roots. The redox potential of wet soil and plants rises as a result of hypoxia, leading in the production of reactive oxygen species (ROS). Although ROS is essential for intracellular and intercellular signaling, it also oxidizes membrane lipids, pigments, proteins, and nucleic acids, causing normal plant metabolism to be disrupted. Additionally, increased CO_2 levels in plant roots create phytotoxin buildup in decreased soil, limiting respiration (Blom and Voesenek 1996; Fukao and Bailey-Serres 2008).

5.2.3 Consequences of Salinity

The increase in world population demands for an increase in food sources: that is why more arable land is needed to grow more food crops to feed people (Amirjani 2010). This can be accomplished by increasing the amount of farmed land and crop productivity which can be made possible by bringing agriculture to marginal lands. But the agricultural productivity is limited by salt stress. Soil salinity has deleterious effects on crop production all over the world. According to an estimation, salt damage affects 20% of all irrigated farmland and reduce crop yield (Kumar et al. 2020). The total area of salt-affected soil on the planet is estimated to be 833 million hectares (Negrão et al. 2017). Every year vast swaths of agricultural lands are lost and become unfit for agriculture and that is why removed due to increasing soil salinity. The main factors resulting for soil salinity are the irrigation through saline irrigation water and the application of fertilizers.

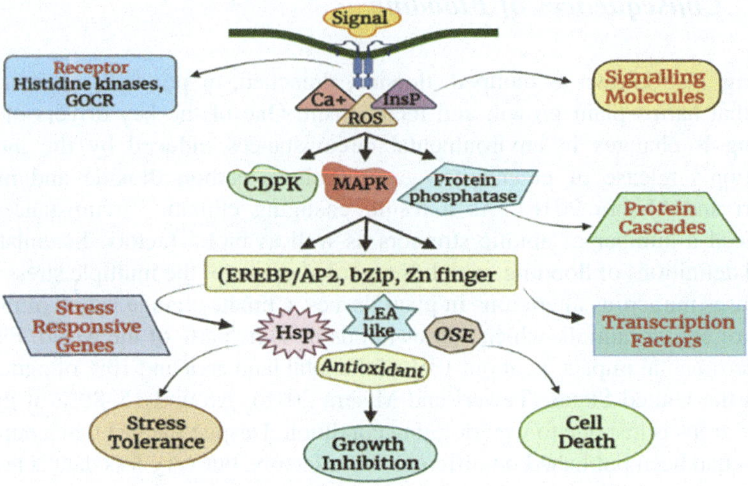

Fig. 5.2 Mechanism induced by salt stress in plants

An excessive amount of salts in soil mainly NaCl has negative results on plant productivity. In the beginning, plant's response to salinity can be divided into two phases depending on the Na^+ sensing; shoot ion-dependent response and shoot ion-independent response (Negrão et al. 2017). The first phase occurs within minutes to days and is followed by stomatal closure and inhibition of leaf expansion. The second phase occurs over a longer period and causes the accumulation of Na^+ in shoot up to toxic concentrations, particularly in old leaves causing premature leaf senses and ultimately reducing plant yield and even plant death.

Salinity alters the morphological, physiological, and biochemical responses of plants, and plants respond to them by various mechanisms (Fig. 5.2). It results in osmotic and ionic stress. On molecular level changes in gene expression can be identified as a result of this effect. Suppressive growth can be seen in plants facing salt stress, but all the plants have different tolerance levels to high salt concentrations. Plants adapt different adaptive strategies to fight different stresses such as production and accumulation of Free Amino Acids (FAA) such as proline by plant tissues (Amirjani 2010). Proline adjusts the osmotic potential in the cytoplasm by acting as a compatible solute. Because there is a major imbalance between reactive oxygen species (ROS) and antioxidant defense synthesis, oxidative stress is a critical signal in the case of biotic and abiotic stressors. Because they cause significant damage to lipids, nucleic acids, and proteins, ROS have been labeled as harmful chemicals (Gratão et al. 2005). They cause variety of injuries to plant metabolism and damage photosynthetic pigments, inactivate proteins, enzymes, and permeabilize membranes by causing lipid peroxidation.

5.2.4 Consequences of Extreme Temperature

The temperature around the plant determines the rate of (Hatfield 2015) growth and development, and each species has a temperature range with a minimum, maximum, and optimum temperature. Extreme events that occur throughout the summer would have the greatest impact on plant productivity. Given the predicted global warming of 1.5–2.0 °C by 2050, heat stress is now regarded as one of the most essential elements affecting plant growth and development. Many elements of vegetative processes, such as growth, yield, and generative development, are influenced by heat exposure. Extreme temperatures cause several physiological and metabolic changes. If we specifically talk about photosynthesis many plant species are reported for disruptive photosynthetic rate due to extreme temperatures. Some of them are reported in Table 5.1. Lower growth, reduced photosynthesis, seed germination suppression, incorrect development, changed secondary metabolism, and overproduction of reactive oxygen species (ROS) leading in oxidative stress have all been linked to higher temperatures.

In cell cultures of *Perilla frutescens* and strawberry, the ideal temperature (25 °C) enhances anthocyanin output. Lowering the temperature encourages the accumulation of anthocyanins while restricting cell development. Strawberry cell culture yielded the highest anthocyanin concentration at 15 °C, which was around 13-fold higher than that at 35 °C. At the relatively high temperature of 28 °C, anthocyanin synthesis was dramatically reduced in *Perilla frutescens* suspension cultures, whereas 25 °C was ideal for pigment productivity. In *Daucus carota* cell suspension cultures, similar findings on optimal anthocyanin productivity have been found (Akula and Ravishankar 2011).

5.2.5 Consequences of Heavy Metals

Higher plants need trace levels of heavy metals like zinc (Zn) and copper (Cu) to live, but others like cadmium (Cd) aren't required. All heavy metals, however, are dangerous at large doses (Riesen and Feller 2005). Because metals cannot be decomposed, their remediation necessitates both immobilization and toxicity reduction or elimination. Scientists have been evaluating solutions including the use of plants to clean contaminated areas in recent years. Phytoextraction (uptake) is the process of using living green plants to extract inorganic toxins from polluted soils and concentrate them in roots and easily harvestable shoots (Suman et al. 2018).

Metal ions (lanthanum, europium, silver, and cadmium) and oxalate have a role in secondary metabolite synthesis. Because it is a major component of the urease enzyme, the trace metal nickel (Ni) is necessary for plant development. Plant development is stifled by Ni concentrations above a particular threshold. In response to cadmium and lead stress, plant growth, pigment content, metabolic parameters, and heavy metal uptake in *Brassica juncea* L. were investigated. Cd and Pb

Table 5.1 Temperature effects on photosynthesis in several plant species

Plant species	Temperature stress	Effect on photosynthetic systems	References
L. esculentum	Cold temperature	Chlorophyll content, leaf area, Pn, Gs, Et, Ci, and Fv/Fm all decreased A decrease in the expression of photosynthesis-related proteins	Ahanger et al. (2020)
Saccharum officinarum	Cold temperature	Chloroplast structural disruption, chloroplast microtubule and grana lamella reduction, and overall photosynthetic efficiency decline	Xalxo et al. (2020)
T. aestivum	Cold temperature	Photosynthesis declines due to thylakoid membrane damage	Djanaguiraman et al. (2020)
	High-temperature environment	Reduced net photosynthetic rate and thylakoid membrane degradation	Djanaguiraman et al. (2020)
	High temperature	Decline in CO_2 assimilation rate and stomatal conductance	Djanaguiraman et al. (2020)
Camellia sinensis	Low temperature	Reduction in Pn, Gs, Et, Ci, and maximal photochemical efficiency of PS II	Zhang et al. (2020)
Populus simonii	High temperature	Reduced electron transport, photosystem disruptions, glycolate pathway activation caused H_2O_2 production, damaging complete photosynthetic apparatus	Sharma et al. (2019)
L. esculentum	Low temperature	Damage to the thylakoid membrane, resulting in a decrease in light energy distribution and a decrease in electron transport rate	Yang et al. (2018)
T. aestivum	High temperature	Reduction in chlorophyll content and photochemical activity of PS II	Liu et al. (2019)
Cucumis sativus	High temperature	Decreased net photosynthetic rate, actual photochemical efficiency, photochemical quenching coefficient, and starch content	He et al. (2018)
Pisum sativum	High temperature	Decreased CO_2 assimilation, stomatal conductance, and water use efficiency	Parvin et al. (2019)
Glycine max	High temperature	Photosynthesis rates, stomatal and mesophyll conductance, photosystem II quantum yield, carboxylation rate, and electron transfer across CO_2 levels have all increased	Xu et al. (2016)
	Cold temperature	Due to a photo-biochemical process, photosynthetic restriction occurs, followed by stomatal and mesophyll limitations. Photosynthesis is inhibited by the build-up of carbohydrates and organic acids	Gago et al. (2020)
Citrullus lanatus	Cold environment	Reduction chlorophyll content, Pn, Gs, Et, and Ci	Wang et al. (2018)
Prosopis chilensis &	Hot environment	Photochemical efficiency and photosynthetic rate are both decreasing	Wang et al. (2018)

(continued)

Table 5.1 (continued)

Plant species	Temperature stress	Effect on photosynthetic systems	References
Prosopis tamarugo			
Hibiscus rosa-sinensis	Cold environment	PS II has a lower photochemistry efficiency, electron transport, and quantum yield	Banerjee and Roychoudhury (2019)
Zea mays	Hot environment	Decreased activities of RUBISCO activase	Kim et al. (2021)
Triticum aestivum	Hot environment	Enzymatic activities are destroyed because of reduced chlorophyll production	Zhou et al. (2021)
Gossypium barbadense	Hot environment	Overall photosynthesis is diminishing due to limited activity of ribulose-1, 5 bisphosphates and photosynthetic electron transport	Osei-Bonsu (2020)

treatment reduced plant growth, chlorophyll content, and carotenoids, while Cd was found to be more detrimental than Pb treatment. This is exemplified by *B. juncea*. During the flowering stage, Cd (900 M) reduced protein level by 94%, whereas Pb (1500 M) reduced protein content by 44%. Proline levels increased as Cd and Pb levels decreased but declined at higher levels. *B. juncea* roots accumulated more Cd and Pb than the shoots. Cd accumulates at a faster rate than Pb, while higher Pb concentrations prevent Cd absorption.

5.2.6 Consequences of Nutrition Deficiency

When plants are stressed, secondary metabolite production may rise since growth is often inhibited more than photosynthesis, and secondary metabolites are the main source of carbon fixation. The relationship between soil pH, air and soil temperature, available moisture, nutrient excesses, soil organic matter, and soil mineral content is intricate. However, in some cases, visual signs may be used to indicate potential difficulties. Deficiencies should be assessed through soil testing and plant tissue analysis. Soil testing is the most exact technique of determining whether primary nutrient levels are optimal for plant survival and growth. Nitrogen is normally a unique test, but it is routinely applied at uniform rates to most crops (Hatfield 2015).

Crop health and productivity are harmed by nutrient shortages and toxicities, which may manifest as unusual visual symptoms. Knowing the function and mobility of each essential nutrient in the plant will aid in determining which one is causing a deficiency or toxicity symptom. Stunted development, chlorosis, interveinal chlorosis, purple or red staining, and necrosis are all signs of general insufficiency. Mobile nutrient shortages arise initially in older, lower leaves, while immobile nutrient deficiencies appear first in younger, upper leaves. Overapplication of nutrients is the most common cause of nutrient poisoning, which causes abnormal growth (excessive or stunted), chlorosis, leaf discoloration, and necrotic patches. Many nutrients may impede the absorption of other nutrients when ingested in excess, thus leading to deficiency symptoms.

5.3 Microbiome

Microbes are the collective communities of microorganisms in a particular environment. They are fundamental for the maintenance of life on Earth, as they play a critical role in combating different diseases. We don't know much about their distribution, but we do know that they can be found in a variety of places, including soils, oceans, the atmosphere, and even within our bodies (Berg et al. 2014). When it comes to plant-associated microorganisms, it has a long history that dates back to Lorenz Hiltner's concept of the rhizosphere in 1904 (Berg et al. 2014). They colonize a wide range of plant parts, from the surface to the roots and deep tissues, and play a critical role in plant health, productivity, community composition, and ecosystem functioning (Singh et al. 2019). This microbial diversity can resist a variety of abiotic stressors (Table 5.2). Resistance and resiliency are two words that come to mind when describing this. These microbial communities can assist plants in coping with a variety of stressors.

5.3.1 Role of Microbiome in Relieving Drought

As the world population tends to reach nine billion by 2025, worldwide food security is threatened (Hanjra and Qureshi 2010). As a result, 70% more food must be produced to meet the growing population's demand. Crop productivity is, however, severely limited by difficult environmental circumstances. Drought stress is an environmental limitation that poses a severe danger to agricultural growth and productivity in most parts of the world. To cope up with this issue scientists are focusing on improving drought resistance traits in crops with the help of molecular breeding and transgenic approaches (Hussain et al. 2018). But these techniques have certain limitations that make it difficult to develop drought resistance crops. Therefore, there is need to explore and utilize the naturally occurring root microflora that confer resistance in plants against drought and many other environmental stresses.

Because of its environmentally favorable nature and low-cost input, the use of naturally existing microflora to combat drought stress is gaining traction around the world. Microbes can be found in practically every component of the plant (phyllosphere, endosphere, and rhizosphere), and the rhizosheath, which refers to the "root-adhering soil fraction" and is also considered part of the plant microbiome, has received less attention. Rhizosheaths are a mechanism that improves drought tolerance in plants. They are made up of mucilage that is released from plants around

Table 5.2 Abiotic stress tolerance in common crop plants is mediated by Rhizobacteria species

Crop	Stress	Rhizobacteria	References
Solanum tuberosum	Salt/ drought/ HMs	*Bacillus pumilus* DH 11, *Bacillus firmus* 40	Khan et al. (2021b)
Helianthus annuus	Drought	*Achromobacter xylosoxidans* (SF2) *Bacillus pumilus* (SF3 and SF4)	Raghuwanshi and Prasad (2018)
Lycopersicon esculentum	Heat	*Bacillus cereus*	Patel et al. (2017)
Oryza sativa	Drought	*Azospirillum brasilense* Az.39	Tisarum et al. (2019)
Triticum aestivum	Heat	*Bacillus velezensis* 5113	Abd El-Daim et al. (2019)
Triticum aestivum	Heat	*Pseudomonas brassicacearum, Bacillus thuringiensis, Bacillus subtilis*	Khan et al. (2021b)
Cucurbita pepo	Drought	*Bacillus circulans* ML2, *Bacillus megaterium* ML3	Khan et al. (2021b)
Lycopersicon esculentum	Heavy metal	*Pseudomonas aeruginosa, Burkholderia gladioli*	Khanna et al. (2019)
Zea mays	Drought	*Klebsiella variicola* F2, *Pseudomonas fluorescens* YX2, *Raoultella planticola* YL2	Razad et al. (2021)
Abelmoschus esculentus	Salinity	*Enterobacter* spp.	Sagar et al. (2020)
Stevia rebaudiana	Salinity	*Streptomyces* spp.	Khan et al. (2021b)
Phaseolus vulgaris	Salinity	*Aneurinibacillus aneurinilyticus, Paenibacillus* spp.	Gupta and Pandey (2020)
Arachis hypogea	Salinity	*Bacillus licheniformis* K11	Subiramani et al. (2020)

soil particles and forms a cylinder around the roots of drought-stressed plants. They are known to host bacterial and fungal populations, as well as microbes that promote plant growth (PGP) (Etesami 2021). Microbes that live in areas that are frequently subjected to drought stress are more adaptable and resistant than those that do not. *Brassica rapa* plants were exposed to recurrent drought stress in one study, and they were better able to regulate bacterial abundance than control plants (Naylor and Coleman-Derr 2018).

Plant responses to microbe-induced drought stress are poorly known. Rolli et al. investigated grapevine and *Arabidopsis* growth in drought-stricken environments. They put eight isolates to the test in vivo for 510 different strains, and the results demonstrated that plant growth-promoting action is stress-dependent, not strain-specific. However, it was observed that several of the microbes studied play a role in plant adaptation to drought stress (Hussain et al. 2018). Plants' mechanisms for resisting tolerance are still unknown. (1) Production of phytohormones like gibberellic acid, abscisic acid (ABA), and indole-3-acetic acid (IAA) to lower ethylene levels in the roots, which is possible when plant ACC is broken down by bacterial ACC-deaminase; (2) production of 1-aminocyclopro pane-1-carboxylate (ACC) deaminase to lower ethylene levels in the roots, which is possible during the breakdown of plant.

Cell division, stem and root elongation, lateral and adventitious root initiation, and root and shoot orientation in response to light and gravity are all influenced by IAA. Some bacteria, such as *Azospirillum* spp. and *Bacillus thuringiensis*, have been discovered to produce IAA, which aids plant survival by increasing root and root hair formation. Similarly, *Azospirillum lipoferum* strains can affect stomatal closure in their host plant maize by generating ABA and gibberellins, which can influence water loss regulation (Vurukonda et al. 2016). Some PGPRs that can produce ACC-deaminase can also help in drought tolerance. For example, treatment of tomato and pepper seedlings with *Achromobacter piechaudii ARV8* resulted in reduced ET concentration as these bacteria are capable of producing ACC-deaminase which is an enzyme capable of cleaving ACC into amine and alpha-ketobutyrate instead of ET (Mayak et al. 2004). Epoxy polysaccharides having architectural characteristics impart functionality to biofilms. These biofilms can improve plants performance in drought stress. *Pseudomonas* spp. and *Acinetobacter* spp. are EPS producing species which promoted drought tolerance in pepper plant by imparting protection against desiccation by the formation of hydrophilic biofilm around the roots (Kaushal et al. 2016).

All these mechanisms induced by PGPR and plant itself are reported in the literature, but ample research is available to advocate the fact that soil microorganisms can improve drought tolerance in plants.

5.3.2 Role of the Microbiome in Relieving Flooding

Flood frequency and severity are predicted to grow as a result of global climate change, which has had a variety of effects on climatic conditions around the world. Plants undergo numerous physiological and biological changes as a result of prolonged submersion under soggy circumstances. Under soggy conditions, the soil in which plants grow undergoes numerous physiological changes. The pores in the soil allow gaseous exchange between the atmosphere, the soil, and the soil microorganisms. This increased the synthesis of the stress hormone ethylene. By manufacturing the enzyme 1-aminocyclopropane-1-carboxylate (ACC) deaminase, many bacterial populations can control ethylene levels. ACC is a plant-produced ethylene precursor that is cleaved by ACC-deaminase, reducing ethylene synthesis. Basil plants infected with microorganisms that produce ACC-deaminase grew faster than control plants. *Pseudomonas putida* UW4, an ACC-deaminase-producing bacterium was used for proteomic profiling of the cucumber plant, and this plant was grown in hypoxic conditions, which showed a shift in the protein profile of the plant toward defense stress. Although this data showed the ability of microbiome in flood stress tolerance, only very few studies are available. Continued study is needed to unravel the mechanisms and plant microbiome interaction.

5.3.3 Role of the Microbiome in Relieving Salinity

Fulfilling the world's food demand is the major challenge as the world population is rapidly growing. So, for this purpose soil salinity is the major focus as it limits crop production and growth in many areas of the world. Several approaches including soil reclamation and management practices can be used to address this issue, but these techniques are expensive and not so practical. Another environment-friendly approach that uses plant microbiome as inoculum is gaining momentum (Hussain et al. 2018). Microorganisms that are exposed to high levels of stress are thought to develop characteristics that allow them to withstand the stress. In the case of salinity stress, the same strategy is employed. Microorganisms that are salinity tolerant can withstand a variety of salt concentrations and overcome them via various salinity tolerant processes. Microorganisms that are saline tolerant can improve the health of salinity-affected soils, sustain ecological functions, and encourage plant growth.

The accumulation of different solutes for osmoregulation, production of extracellular proteases, and production of Na+/K+ antiporters, which is an integral membrane protein and a cotransporter involved in the active transport of two or more molecules, are all reported mechanisms by which salt tolerant bacteria overcome the effect of salt (Mayak et al. 2004). Such salt tolerant bacteria can be used as an inoculum in salt effected soil that will help crops in overcoming the stress. In one of the study 130 rhizobacterial strains were isolated from wheat plants that were sown in saline conditions. Out of these 130 isolates, 24 isolates tolerated the high stress level of NaCl. The authors added that this stress tolerance is due to different genes, hormones, and proteins (Mayak et al. 2004). According to some reported data, arbuscular mycorrhizal fungi (AMF) can help plants develop tolerance to salt stress. Experiments showed that AMF plants inoculated with *Glomus* sp. have increased growth in soil as compared to un-inoculated plants which might be possibly due to increased K^+ and decreased Na^+ concentration in shoots (Giri and Mukerji 2004). This treatment has shown improved salt tolerance in maize, mung bean, cucumber, tomato, and clover plants. Hence the application of a salt-tolerant microbiome can be beneficial for plants to grow in salt-affected soil. However, there is a need to explore that which hidden mechanisms of microbiome are involved in salt tolerance.

5.3.4 Role of the Microbiome in Relieving Extreme Temperatures

Bacteria play an important role in the survival of most types of life on the planet. It's crucial to understand how people and other living things on Earth can adapt to climate change by combining knowledge of many microbiomes. We need to understand not just how microbiomes drive temperature change (via GHG production and consumption), but also how climate change and other anthropogenic activities will affect them. The impact of climate change is predicted to be heavily influenced by microbiome reactions, which are critical for achieving a green future. Despite the fact that microbiomes are critical in the regulation of climate change, they are rarely the focus of climate change research (Cavicchioli et al. 2019; Bakken and Frostegård 2017; Itakura et al. 2013; Ritchie et al. 2018).

Finding the genes that control agriculturally beneficial nitrogen, potassium, and sulfur metabolism, and then improving the bacteria that carry these genes is one strategy (Bakken and Frostegård 2017; Itakura et al. 2013). N_2O emissions from soybean have decreased as a result of the use of microbial strains with increased N_2O reductase activity, and both genetically modified and wild strains with increased N_2O reductase activity give prospects for reducing N_2O emissions (Itakura et al. 2013). The programs that control the stomach microbiota and breeding of cattle, aim for host genetic elements that change microbiome's responses are options for lessening methane emission from livestock.

Microbiomes have the greatest impact on carbon sequestration. As a result, ecological changes that affect aquatic photosynthesis of microbiomes and, as a result, carbon storage—fixed in waters—are a focus for the carbon cycle. Microbiomes contribute significantly to GHG emissions through the biochemical cycles of nitrogen, carbon dioxide, and methane. The local environment, the biome, the food chain, and, most importantly, climate change all have an impact on the balance of microbial GHG uptake versus release. GHG emissions, pollution, agriculture, and finally population increase all have an impact on microbiomes, and this has a direct impact on climate change, hazardous waste, agricultural practices, and disease spread (Venkatramanan et al. 2020a, b). Positive feedbacks will occur as a result of anthropogenic activity that alters the carbon uptake-to-release ratio, hastening climate change. Microbiomes, on the other hand, have great promise for easing anthropogenic concerns through enhanced agricultural results, biofuel production, and pollution reduction. Targeted laboratory study using model bacteria will be used to address specific microbiome challenges (Cavicchioli et al. 2019).

5.3.5 Role of the Microbiome in Relieving Heavy Metals

Heavy metal pollution of soil poses a threat to the natural environment and food safety. Environmental pollution and social disclosure associated with heavy metals are attributed to a variety of human-caused actions such as trade production and mining. Around 50 million locations of soil pollution by heavy metals or metalloids exist worldwide, with existing concentrations exceeding legal thresholds. Heavy metal contamination of soil poses several risks to the environment and individuals, as well as affecting food chain safety and the ability to use the land for agricultural purposes, all of which have a significant impact on food security (Wuana and Okieimen 2011).

Heavy metal pollution of soil poses a threat to the natural environment and food safety. Environmental pollution and social disclosure associated with heavy metals are attributed to a variety of human-caused actions such as trade production and mining. Around 50 million locations of soil pollution by heavy metals or metalloids exist worldwide, with existing concentrations exceeding legal thresholds. Heavy metal contamination of soil poses several risks to the environment and individuals, as well as affecting food chain safety and the ability to use land for agricultural purposes, all of which have a significant impact on food security. The natural environment and food safety are both threatened by heavy metal poisoning of soil. Heavy metal contamination and social disclosure are linked to several human-caused activities, including trade production and mining. Around 50 million sites worldwide have been contaminated by heavy metals or metalloids, with concentrations exceeding legal limits. Heavy metal pollution of soil provides a multitude of threats to the environment and persons, as well as influencing food chain safety and land use for agricultural reasons, all of which have a substantial impact on food security. Bacterial processes can be divided into two categories: molecular and biochemical. The first group consists of heavy metal resistance genes found on chromosomes or plasmids. Microbes interact with extracellular soluble metals and have a role in microbial tolerance through diverse chemical reactions, according to the latter.

The remediation of metals in soil is important to safeguard the natural environment and shield living organisms (Prasad 2021). Conventional remediation techniques include chemical and physical methods. Correspondingly, biological approaches are measured as an active technique for metal remediation, which comprises: bioremediation, and phytoremediation (Suman et al. 2018). Bioremediation by microorganisms is an effective, efficient, and eco-friendly procedure that lowers the cost of the cleaning process associated with the contamination of heavy metals (Mishra 2017; Prasad et al. 2021). Contrary, phytoremediation, depends on the conditions of water, soil, and climate, demonstrating to be a very sluggish and seasonally effective method (Prasad 2022). In addition, interactions of plant–microorganism also perform a vital part in the adaptation to environments polluted with heavy metals and therefore can be examined in depth to enhance microbe-assisted methods of phytoremediation (Ianeva 2009; Spain and Alm 2003).

5.3.6 Role of the Microbiome in Relieving Nutrient Deficiency

Plants are part of a varied ecological unit in their natural habitat, which contains a large number of microorganisms. However, research into a broad range of plant-associated bacteria and their potential to replace synthetic agricultural inputs is still in its early stages. The composition and dynamics of rhizosphere microbiomes have lately made significant advances. Plants impact microbiome architecture, probably through root exudates, and bacteria have evolved to thrive in the rhizosphere niche in a variety of ways, according to evidence.

Plants contribute a considerable amount of carbon to the rhizosphere, which helps soil microbes survive. So, what do the plants get in exchange? Although the N_2 molecule contains a substantial amount of nitrogen in the atmosphere, the majority of nitrogen, potassium, and sulfur atoms are biologically bonded in soils. These nutrient sources are only slightly accessible to plants due to differences in metabolic capacity between plants and bacteria, but they can be digested by a variety of soil microbes. This suggests that in natural ecosystems, microorganisms' nitrogen fixation and nutrient mineralization activities are critical for plant nutrition, as these reactions break down refractory forms of nitrogen, phosphorous, and sulfur to release these elements for plant nutrition (Rovira 1965; van der Heijden et al. 2008). This long-held belief has recently been thrown into question, as many studies have showed that plants may directly absorb various types of organic nitrogen. Despite this, it is often considered that bacteria are the top competitors for organic nitrogen molecules in soil due to their poor diffusivity. Furthermore, isotope labeling tests often confirm the notion that most organic nitrogen is used first by microbial species and then by plants (Kuzyakov and Xu 2013).

For years, researchers have established a list of microbial metabolic pathways relevant to plant nutrition in terms of nitrogen, potassium, and sulfur in the literature on soil microorganisms. The symbiotic connection between bacteria and legumes is often used economically when inoculating field crops with nitrogen-fixing rhizobia strains. But how can these phenomena be strengthened and improved for long-term use in agricultural systems for nitrogen fixation in legumes and nutrition in non-legume crops in terms of nitrogen, potassium, and sulfur? Assume that several bacterial strains have varied metabolic capacities (Timm et al. 2015), along with the vast amount of soil microbial genome sequencing data generated recently (Müller et al. 2016). Finding the genes that control agriculturally beneficial nitrogen, potassium, and sulfur metabolism, and then improving the bacteria that carry these genes is one strategy.

5.4 Current Insights and Future Prospectives of Research

Plant-associated microbiomes are expanding in relevance as a major field of research that enhances plant health, growth, and production. Plant-associated microbiomes are intimately implicated in mechanisms that enable plants survive stressful circumstances, as explained in this chapter. Because of the heavy reliance on it, scientists are now viewing the microbial genome as a plant genome. However, these conclusions are preliminary, and additional research is needed to properly comprehend the highly complex phenomenon of microbial communities. Stress management in plants would be greatly aided by plant-associated microbiomes. One method could be to create transgenic crops with microbe-derived beneficial genes. However, molecular procedures are time-consuming, and there are a slew of other difficulties that make it less accessible. The other strategy that could be more cost-effective and environment-friendly is using microbial inoculants. Several studies have shown beneficial and positive results in field conditions, while others find unsatisfied and negative results (Nadeem et al. 2014). Another strategy could be the development of a microbial consortium for ready-to-use formulations against specific abiotic stresses. Therefore, the mechanisms by which microbes confer stress resistance to plants need more research and understanding. If we develop our knowledge and if we can decode and mimic the biological procedures of microorganisms and be able to regulate them, then we can control these abiotic stresses. Recent trend provides data on understanding the microbe–microbe and plant–microbe interactions that will help us to use PGPM as a sustainable tool against these limiting factors.

5.5 Conclusion

Abiotic elements such as heavy metal, flooding, temperature, light, and drought all have an impact on agricultural output around the world. Abiotic stresses cause plants to respond in a variety of ways, but photosynthetic activity is one of the most significant regulators of overall plant growth and productivity. Reactive oxygen species (ROS) are formed in plants as a result of all of these abiotic stressors, producing oxidative damage, and membrane instability. When RUBISCO activity is decreased, plant photosynthetic efficiency is also diminished. Heavy metals have a significant negative impact on pigment production and accumulation due to enzymatic degradation. Meanwhile, the temperature is progressively rising as a result of several environmental circumstances, such as global warming. Temperatures beyond the permissible threshold stress the plant, causing it to lose its cellular equilibrium and limit its growth, development, and metabolism. As a result, plants' and microorganisms' ability to cope with these stressors differs substantially between species. Agricultural yields still have a lot of room for improvement, despite substantial advances in genetic technologies such as QTL mapping and transgenic approaches. For example, genetic and environmental relationships are thought to be poorly understood. Likewise, finding QTLs for one type of stress does not work for other types of stress. As a result, problems with transgenic plants designed to combat stress persist.

Acknowledgments The authors are thankful to Khola Tazeen for her help in the literature collection and proofreading of this manuscript.

References

Abd El-Daim IA, Bejai S, Meijer J (2019) Bacillus velezensis 5113 induced metabolic and molecular reprogramming during abiotic stress tolerance in wheat. Sci Rep 9(1):1–18

Ahanger MA, Aziz U, Alsahli AA, Alyemeni MN, Ahmad P (2020) Influence of exogenous salicylic acid and nitric oxide on growth, photosynthesis, and ascorbate-glutathione cycle in salt stressed *Vigna angularis*. Biomol Ther 10(1):42

Akula R, Ravishankar GA (2011) Influence of abiotic stress signals on secondary metabolites in plants. Plant Signal Behav 6(11):1720–1731. https://doi.org/10.4161/psb.6.11.17613

Amirjani M (2010) Effect of salinity stress on growth, mineral composition, proline content, antioxidant enzymes of soybean. Am J Plant Physiol 5(6):350–360

Ansari MI (2018) Plant microbiome and its functional mechanism in response to environmental stress. Int J Green Pharm 12:81–92

Bagheri N, Alizadeh O, Zadeh SS, Aref F, Ordookhani K (2019) Biochemical investigation of potential of plant growth-promoting bacteria and auxin treatments to alleviate drought stress in wheat (Triticum aestivum L.). Eurasia J Biosci 13(2):1225–1230

Bakken LR, Frostegård Å (2017) Sources and sinks for N2O, can microbiologist help to mitigate N2O emissions? In: Microbiology E (ed) vol 19. Wiley Online Library, pp 4801–4805

Banerjee A, Roychoudhury A (2019) Cold stress and photosynthesis. In: Photosynthesis, productivity and environmental stress. Wiley, Chichester, pp 27–37

Barea JM (2015) Future challenges and perspectives for applying microbial biotechnology in sustainable agriculture based on a better understanding of plant-microbiome interactions. J Soil Sci Plant Nutr 15(2):261–282

Berg G, Grube M, Schloter M, Smalla K (2014) The plant microbiome and its importance for plant and human health. Front Microbiol 5:1

Blom C, Voesenek L (1996) Flooding: the survival strategies of plants. Trends Ecol Evol 11(7): 290–295

Cavicchioli R, Ripple WJ, Timmis KN, Azam F, Bakken LR, Baylis M et al (2019) Scientists' warning to humanity: microorganisms and climate change. Nat Rev Microbiol 17(9):569–586

Djanaguiraman M, Narayanan S, Erdayani E, Prasad PV (2020) Effects of high temperature stress during anthesis and grain filling periods on photosynthesis, lipids and grain yield in wheat. BMC Plant Biol 20(1):1–12

Etesami H (2021) Potential advantage of rhizosheath microbiome, in contrast to rhizosphere microbiome, to improve drought tolerance in crops. Rhizosphere 20:100439

Fukao T, Bailey-Serres J (2008) Submergence tolerance conferred by Sub1A is mediated by SLR1 and SLRL1 restriction of gibberellin responses in rice. Proc Natl Acad Sci USA 105(43): 16814–16819

Gago J, Daloso DM, Carriquí M, Nadal M, Morales M, Araújo WL et al (2020) The photosynthesis game is in the "inter-play": mechanisms underlying CO2 diffusion in leaves. Environ Exp Bot 178:104174. https://doi.org/10.1016/j.envexpbot.2020.104174

Giri B, Mukerji KG (2004) Mycorrhizal inoculant alleviates salt stress in *Sesbania aegyptiaca* and Sesbania grandiflora under field conditions: evidence for reduced sodium and improved magnesium uptake. Mycorrhiza 14(5):307–312

Gratão PL, Polle A, Lea PJ, Azevedo RA (2005) Making the life of heavy metal-stressed plants a little easier. Funct Plant Biol 32(6):481–494

Gupta S, Pandey S (2020) Enhanced salinity tolerance in the common bean (Phaseolus vulgaris) plants using twin ACC deaminase producing rhizobacterial inoculation. Rhizosphere 16:100241

Hanjra MA, Qureshi ME (2010) Global water crisis and future food security in an era of climate change. Food Policy 35(5):365–377

Hatfield J (2015) Temperature extremes: effect on plant growth and development. Weather Climate Extremes 10:4–10. https://doi.org/10.1016/j.wace.2015.08.001

He L, Yu L, Li B, Du N, Guo S (2018) The effect of exogenous calcium on cucumber fruit quality, photosynthesis, chlorophyll fluorescence, and fast chlorophyll fluorescence during the fruiting period under hypoxic stress. BMC Plant Biol 18(1):1–10

Hussain SS, Mehnaz S, Siddique KH (2018) Harnessing the plant microbiome for improved abiotic stress tolerance. In: Plant microbiome: stress response. Springer, New York, pp 21–43

Ianeva O (2009) Mechanisms of bacteria resistance to heavy metals. Mikrobiolohichnyi Zhurnal 71(6):54–65

Itakura M, Uchida Y, Akiyama H, Hoshino YT, Shimomura Y, Morimoto S et al (2013) Mitigation of nitrous oxide emissions from soils by *Bradyrhizobium japonicum* inoculation. Nat Clim Chang 3(3):208–212

Jha Y, Subramanian R (2018) From interaction to gene induction: an eco-friendly mechanism of PGPR-mediated stress management in the plant. In: Plant microbiome: stress response. Springer, New York, pp 217–232

Kaushal M, Wani S, Ecosystems P, Environment (2016) Rhizobacterial-plant interactions: strategies ensuring plant growth promotion under drought and salinity stress. Agric Ecosyst Environ 231:68–78

Khalid MF, Hussain S, Ahmad S, Ejaz S, Zakir I, Ali MA et al (2019) Impacts of abiotic stresses on growth and development of plants. In: Plant tolerance to environmental stress. CRC Press, West Palm Beach, FL, pp 1–8

Khan N, Ali S, Shahid M, Mustafa A, Sayyed R, Curá J (2021a) Insights into the interactions among roots, rhizosphere and rhizobacteria for improving plant growth and tolerance to abiotic stresses: a review. Cells 10:1551. s note: MDPI stays neu-tral with regard to jurisdictional claims in ...

Khan N, Ali S, Shahid MA, Mustafa A, Sayyed R, Curá JA (2021b) Insights into the interactions among roots, rhizosphere, and rhizobacteria for improving plant growth and tolerance to abiotic stresses: a review. Cell 10(6):1551

Khanna K, Jamwal VL, Gandhi SG, Ohri P, Bhardwaj R (2019) Metal resistant PGPR lowered Cd uptake and expression of metal transporter genes with improved growth and photosynthetic pigments in Lycopersicon esculentum under metal toxicity. Sci Rep 9(1):1–14

Kim SY, Slattery RA, Ort DR (2021) A role for differential Rubisco activase isoform expression in C4 bioenergy grasses at high temperature. GCB Bioenergy 13(1):211–223. https://doi.org/10.1111/gcbb.12768

Kumar A, Singh S, Gaurav AK, Srivastava S, Verma JP (2020) Plant growth-promoting bacteria: biological tools for the mitigation of salinity stress in plants. Front Microbiol 11:1216

Kuzyakov Y, Xu X (2013) Competition between roots and microorganisms for nitrogen: mechanisms and ecological relevance. New Phytol 198(3):656–669. https://doi.org/10.1111/nph.12235

Liu H, Brettell LE, Qiu Z, Singh BK (2020) Microbiome-mediated stress resistance in plants. Trends Plant Sci 25(8):733–743

Liu Y, Xu Q, Li W, Yang X, Zheng Q, Li B et al (2019) Long-term high light stress induces leaf senescence in wheat (Triticum aestivum L.). Photosynthetica 57(3):830–840

Mayak S, Tirosh T, Glick BR (2004) Plant growth-promoting bacteria that confer resistance to water stress in tomatoes and peppers. Plant Sci 166(2):525–530

Mishra GK (2017) Microbes in heavy metal remediation: a review on current trends and patents. Recent Pat Biotechnol 11:88–196

Müller DB, Vogel C, Bai Y, Vorholt JA (2016) The plant microbiota: systems-level insights and perspectives. Annu Rev Genet 50:211–234. https://doi.org/10.1146/annurev-genet-120215-034952

Nadeem SM, Ahmad M, Zahir ZA, Javaid A, Ashraf M (2014) The role of mycorrhizae and plant growth promoting rhizobacteria (PGPR) in improving crop productivity under stressful environments. Biotechnol Adv 32(2):429–448

Naylor D, Coleman-Derr D (2018) Drought stress and root-associated bacterial communities. Front Plant Sci 8:2223

Negrão S, Schmöckel S, Tester M (2017) Evaluating physiological responses of plants to salinity stress. Ann Bot 119(1):1–11

Osei-Bonsu I (2020) Thermotolerance of photosynthesis in cowpea (*Vigna unguiculata*). (Ph.D.), Michigan State University, Ann Arbor. https://www.proquest.com/dissertations-theses/thermotolerance-photosynthesis-cowpea-em-vigna/docview/2394838223/se-2?accountid=135034. ProQuest dissertations & theses global database. 27833893.

Pandey SN, Abid M, Khan M (2018) Diversity, functions, and stress responses of soil microorganisms. In: Egamberdieva D, Ahmad P (eds) Plant microbiome: stress response. Microorganisms for sustainability, vol 5. Springer, Singapore, pp 1–19

Parvin S, Uddin S, Fitzgerald GJ, Tausz-Posch S, Armstrong R, Tausz M (2019) Free air CO2 enrichment (FACE) improves water use efficiency and moderates drought effect on N2 fixation of Pisum sativum L. Plant Soil 436(1):587–606

Patel KS, Naik JH, Chaudhari S, Amaresan N (2017) Characterization of culturable bacteria isolated from hot springs for plant growth promoting traits and effect on tomato (Lycopersicon esculentum) seedling. C R Biol 340(4):244–249

Prasad R (2021) Environmental pollution and remediation. Springer. ISBN: 978-981-15-5499-5. https://www.springer.com/gp/book/9789811554988

Prasad R (2022) Phytoremediation for environmental sustainability. Springer, Singapore. ISBN 978-9811656200. https://doi.org/10.1007/978-981-16-5621-7

Prasad R, Nayak SC, Kharwar RN, Dubey NK (2021) Mycoremediation and environmental sustainability, vol 3. Springer International Publishing. 978-3-030-54421-8. https://www.springer.com/gp/book/9783030544218

Raghuwanshi R, Prasad JK (2018) Perspectives of rhizobacteria with ACC deaminase activity in plant growth under abiotic stress. Root Biol:303–321

Razad A, Sutariati GAK, Rakian TC, Mamangkey J, Silalahi M (2021) Bio-priming using indigenous Pseudomonas fluorescens to improve local upland rice (Padi Gogo) production of Paebiu Kolopua and Pae Wuna cultivars. Int J Sci Technol Manag 2(5):1885–1899

Riesen O, Feller U (2005) Redistribution of nickel cobalt manganese zinc and cadmium via the phloem in young and maturing wheat. J Plant Nutr 28(3):421–430

Ritchie H, Reay DS, Higgins P (2018) Potential of meat substitutes for climate change mitigation and improved human health in high-income markets. Front Sustain Food Syst 2:16

Rovira AD (1965) Interactions between plant roots and soil microorganisms. Annu Rev Microbiol 19:241–266. https://doi.org/10.1146/annurev.mi.19.100165.001325

Sagar A, Sayyed R, Ramteke P, Sharma S, Marraiki N, Elgorban AM, Syed A (2020) ACC deaminase and antioxidant enzymes producing halophilic Enterobacter sp. PR14 promotes the growth of rice and millets under salinity stress. Physiol Mol Biol Plants 26(9):1847–1854

Salehi-Lisar SY, Bakhshayeshan-Agdam H (2016) Drought stress in plants: causes, consequences, and tolerance. In: Drought stress tolerance in plants, vol 1. Springer, New York, pp 1–16

Sharma A, Kumar V, Shahzad B, Ramakrishnan M, Sidhu GPS, Bali AS et al (2019) Photosynthetic response of plants under different abiotic stresses: a review. J Plant Growth Regul:1–23

Singh D, Raina TK, Kumar A, Singh J, Prasad R (2019) Plant microbiome: a reservoir of novel genes and metabolites. Plant Gene. https://doi.org/10.1016/j.plgene.2019.100177

Spaepen S, Vanderleyden J, Okon Y (2009) Plant growth-promoting actions of rhizobacteria. World J Microbiol Biotechnol 51:283–320

Spain A, Alm E (2003) Implications of microbial heavy metal tolerance in the environment. Rev Undergraduate Res 2:1–6

Subiramani S, Ramalingam S, Muthu T, Nile SH, Venkidasamy B (2020) Development of abiotic stress tolerance in crops by plant growth-promoting. Phyto-Microb Stress Regul 125

Suman J, Uhlik O, Viktorova J, Macek T (2018) Phytoextraction of heavy metals: a promising tool for clean-up of polluted environment? Front Plant Sci 9:1476

Tewari S, Mishra A (2018) Flooding stress in plants and approaches to overcome. In: Plant metabolites and regulation under environmental stress. Elsevier, Amsterdam, pp 355–366

Timm CM, Campbell AG, Utturkar SM, Jun SR, Parales RE, Tan WA, Pelletier DA (2015) Metabolic functions of Pseudomonas fluorescens strains from Populus deltoides depend on rhizosphere or endosphere isolation compartment. Front Microbiol 6:1118. https://doi.org/10.3389/fmicb.2015.01118

Tisarum R, Theerawitaya C, Samphumphuang T, Phisalaphong M, Singh HP, Cha-Um S (2019) Promoting water deficit tolerance and anthocyanin fortification in pigmented rice cultivar (Oryza sativa L. subsp. indica) using arbuscular mycorrhizal fungi inoculation. Physiol Mol Biol Plants 25(4):821–835

Tosi M, Mitter EK, Gaiero J, Dunfield K (2020) It takes three to tango: the importance of microbes, host plant, and soil management to elucidate manipulation strategies for the plant microbiome. Can J Microbiol 66(7):413–433

van der Heijden MG, Bardgett RD, van Straalen NM (2008) The unseen majority: soil microbes as drivers of plant diversity and productivity in terrestrial ecosystems. Ecol Lett 11(3):296–310. https://doi.org/10.1111/j.1461-0248.2007.01139.x

van der Heijden MG, Hartmann M (2016) Networking in the plant microbiome. PLoS Biol 14(2): e1002378

Venkatramanan V, Shah S, Prasad R (2020a) Global climate change and environmental policy: resilient and smart agriculture. Springer, Singapore. ISBN: 978-981-329-855-2. https://www.springer.com/gp/book/9789813298552

Venkatramanan V, Shah S, Prasad R (2020b) Global climate change and environmental policy: agriculture perspectives. Springer, Singapore. ISBN: 978-981-13-9569-7. https://doi.org/10.1007/978-981-13-9570-3

Vurukonda SSKP, Vardharajula S, Shrivastava M, SkZ A (2016) Enhancement of drought stress tolerance in crops by plant growth promoting rhizobacteria. Microbiol Res 184:13–24

Wang Y, Li J, Gu W, Zhang Q, Tian L, Guo S, Wei S (2018) Exogenous application of 5-aminolevulinic acid improves low-temperature stress tolerance of maize seedlings. Crop Pasture Sci 69(6):587–593. https://doi.org/10.1071/CP17401

Wuana RA, Okieimen FE (2011) Heavy metals in contaminated soils: a review of sources, chemistry, risks and best available strategies for remediation. Int Schol Res Notices 2011:1–20

Xalxo R, Yadu B, Chandra J, Chandrakar V, Keshavkant S (2020) Alteration in carbohydrate metabolism modulates thermotolerance of plant under heat stress. In: Wani SH, Kumar V (eds) Heat stress tolerance in plants: physiological, molecular and genetic perspectives. Wiley, pp 77–115

Xu G, Singh SK, Reddy VR, Barnaby JY, Sicher RC, Li T (2016) Soybean grown under elevated CO2 benefits more under low temperature than high temperature stress: varying response of photosynthetic limitations, leaf metabolites, growth, and seed yield. J Plant Physiol 205:20–32. https://doi.org/10.1016/j.jplph.2016.08.003

Yang X, Xu H, Li D, Gao X, Li T, Wang R (2018) Effect of melatonin priming on photosynthetic capacity of tomato leaves under low-temperature stress. Photosynthetica 56(3):884–892

Zhang F, Lu K, Gu Y, Zhang L, Li W, Li Z (2020) Effects of low-temperature stress and brassinolide application on the photosynthesis and leaf structure of tung tree seedlings. Front Plant Sci 10:1767

Zhou Z, Wei C, Liu H, Jiao Q, Li G, Zhang J et al (2021) Exogenous ascorbic acid application alleviates cadmium toxicity in seedlings of two wheat (Triticum aestivum L.) varieties by reducing cadmium uptake and enhancing antioxidative capacity. Environ Sci Pollut Res Int 29:21739. https://doi.org/10.1007/s11356-021-17371-z

Chapter 6
Functional Potential of Plant Microbiome for Sustainable Agriculture in Conditions of Abiotic Stresses

Sudhakar Srivastava and Divya Singh

Abstract The growth and yield of crop plants are threatened by a variety of factors like soil and irrigation water quality, climate change, and abiotic and biotic stresses. Among these, abiotic stresses like drought, salinity and metal(loid)s, such as arsenic (As) and cadmium (Cd) constitute three major constraints inducing huge yield losses and quality deterioration of crop produce throughout the world. To enable plants to combat these stresses and to obtain sustainable increases in crop yields, the help is required from its associated microorganisms. Microbiome constitutes all microorganisms intricately linked to plants in above and below-ground part of plants. Microbiomes affect plant growth, development, and yield both in normal and under-stressed conditions via several mechanisms. These effects include changes in bioavailability of essential and toxic metal(loid)s, nitrogen fixation, phosphate solubilization, phytohormone synthesis, and water uptake. The microbiome functions like an extended genome of plants and these microbial associations change dynamically in presence of stress and also influence the effect of stresses on plants in reverse. The research on this dynamic microbe–plant interaction has yielded valuable mechanistic information and identified potential microbes that can be used to achieve sustainable crop production in coming years. The present chapter discusses microbiome and its role in abiotic stresses.

Keywords Climate change · Drought · Salinity · Heat · Metal(loid)s · Microbes

6.1 Introduction

The plant and microbiome are intricately connected to each other and plant growth and development are influenced strongly by the microbiome. The microbiome of plant can be obtained from root surface, rhizosphere, of above ground parts like

S. Srivastava (✉) · D. Singh
Plant Stress Biology Laboratory, Institute of Environment and Sustainable Development, Banaras Hindu University, Varanasi, UP, India
e-mail: sudhakar.iesd@bhu.ac.in

S. Chhabra et al. (eds.), *Plant Microbiome for Plant Productivity and Sustainable Agriculture*, Microorganisms for Sustainability 37,
https://doi.org/10.1007/978-981-19-5029-2_6

leaves and also from inside the plant tissues. These microbiomes can be termed as rhizosphere, phyllosphere, and endosphere (Berg et al. 2014; Bulgarelli et al. 2012; Lundberg et al. 2012). Rhizosphere is studied with respect to microbiomes, plant–soil, and plant–organisms interactions, which include archaea, bacteria, and fungi and also protozoans, nematodes, and viruses; all playing important roles toward plant growth and development.

The microbiome–plant association extends potential pool of proteins and metabolites and helps both mutualistic group of organisms in growth, development, and stress tolerance. Microorganisms help in nutrient bioavailability, uptake, and assimilation in plants, in providing stress-fighting metabolites, hormones, and proteins to combat abiotic and biotic stresses and diseases (Berg et al. 2014). The plant microbiome also assists in the remediation of contaminant present and can help in phytoremediation efforts by enhancing its effectiveness and rate (Frossard et al. 2018). Nonetheless, abiotic and biotic stresses also influence the microbiome composition and functions and in turn influence their own effects on plants. Further effects on microbiome are from agricultural use of chemicals that influence nutrient composition of soil and thus the microbiome. Crop rotation, irrigation pattern, weather changes, etc. also influence plant microbiome. Thus, plant–microbiome interactions are highly dynamic in nature and therefore the research on this aspect has increased in recent past. This is because it is now well understood that amidst rising abiotic stresses, decreasing crop productivity and need of higher crop production, microbiome interventions can be fruitful to sustain crop yields (Ambrosini et al. 2016; Frossard et al. 2018; Lareen et al. 2016; Naylor et al. 2017; Thiem et al. 2018; Young et al. 2018; Prasad et al. 2018).

6.2 Role of Plant Microbiome in Metal(loid) Stress Tolerance

Metals are present ubiquitously and life of all living beings is dependent on the optimum presence and intake of some essential or beneficial ones. However, some metals like cadmium (Cd), arsenic (As), and chromium (Cr) are toxic even in minute quantities. The contamination of toxic metals influences plant microbiome in terms of microbial diversity and operational taxonomic unit (OTU) counts (Gołębiewski et al. 2014; Hur et al. 2011; Jiao et al. 2019; Sheik et al. 2012) or in terms of abundance of specific bacterial species and genera (Berg et al. 2012). The effects of beneficial, essential, and toxic metals on microbiome vary and there is variation in effects due to metal species and crop plant (Gołębiewski et al. 2014; Hur et al. 2011). In a study on Cd, it was found that microbial diversity decreased initially but increased later on demonstrating that microbiome can adjust, tolerate, and revert to normal or stress–tolerance roles within a crop cycle (Jiao et al. 2019). Microbiome can consist of microbes that not only provide the stress tolerance but also comprise of plant growth promoting microbes (PGPMs) and the appropriate application of

PGPMs (Table 6.1) and stress–tolerant microbes can allow to attain sustainable and quality crop yields even in stressed conditions (Chen et al. 2016; Gil-Martínez et al. 2018; Mnasri et al. 2017; Wang et al. 2016). Several researchers have identified PGPMs with high tolerance to metal(loid)s like strains of *Pantoea agglomerans, Alcaligenes faecalis*, and *Bacillus cereus* for cadmium (Cd), arsenate [As(V)], silver (Ag), copper (Cu), lead (Pb), and zinc (Zn) (Luziatelli et al. 2020; El-Meihy et al. 2019; Abo-Amer et al. 2015).

6.3 Role of Plant Microbiome in Drought Stress Tolerance

Drought is a major environmental stressor in agriculture, which results in huge losses in crop yields every year (Lesk et al. 2016). The research on drought has been going on for several decades and a lot of information about its effects on plants, mode of action, and tolerance mechanisms has been revealed. However, the drought-induced changes in plant microbiome are yet to be understood (Bastida et al. 2017; Naylor and Coleman-Derr 2018). As in case of other stresses, the interactions between drought and microbiome are mutual, both affecting each other (Naylor et al. 2017) (Table 6.2). Nonetheless, some microbes possess tolerance to drought and therefore, their abundance is increased in drought conditions (Naylor et al. 2017). The members of Proteobacteria, Bacteroidetes, and Firmicutes have been found to be normally drought tolerant (Soussi et al. 2016). Drought tolerant bacteria have been identified from several drought affected regions (Armada et al. 2018) and focus has been on to find bacteria that are able to not only grow themselves but also protect the crop plant from drought effects. In a study, *Acinetobacter* and *Pseudomonas* bacteria reduced growth inhibition of grapevines in drought conditions by minimizing effects on photosynthesis through the synthesis of 1-aminocyclopropane-1-carboxylate (ACC) deaminase (Rolli et al. 2015). One of the major influences of drought tolerant bacteria on plants for reducing their proneness to drought is augmentation of plant root growth and root water uptake (Armada et al. 2018; Gagne-Bourque et al. 2016; Marasco et al. 2013; Rolli et al. 2015) that can be achieved through the synthesis of various hormones.

6.4 Role of Plant Microbiome in Salinity Stress Tolerance

Salinity is another very important stress factor that causes major agricultural crop losses. The presence of high salinity in agricultural field leads to high osmotic values inhibiting water and nutrient uptake by the plants and thus decreasing plant photosynthesis and biomass (Yaish and Kumar 2015). The microbiome associated to plants, especially roots, is also affected due to less water availability in prolonged stress conditions. However, there are several environments, which are highly saline like sea water and coastal areas and these are home to a number of microbes. Thus,

Table 6.1 Studies demonstrating the role of microorganisms in mitigation of metal(loid) toxicity in plants

S no.	Name of the microorganisms	Plant	Effect	References
1.	*Achromobacter xylosoxidans*	*Brassica juncea*	Microbe inhibited the damaging effect of Cu toxicity by increasing length and dry weight of shoot and root	Ma et al. (2009)
2.	*Burkholderia* sp.	*Lycopersicon esculentum*	This microbe protected the host plant against Cd stress and promoted relative plant growth by reducing the Cd absorption	Dourado et al. (2013)
3.	*Psychrobacter* sp.	*Ricinus communis* and *Helianthus annuus*	This microbe helped the energy crops against Ni toxicity, enhanced fresh and dry weight of plants; it improved the availability of phosphorus and promoted the iron content in both plants	Ma et al. (2010)
4.	*Klebsiella pneumoniae*	*Oryza sativa*	The microbial strain helped in improving the overall growth of the plant by reducing the accumulation of Cd in seedling	Pramanik et al. (2017)
5.	*Bacillus* spp.	*Cicer arietinum*	*Bacillus* helped host plants to fight against Cr toxicity, promoted the growth of all parts of the plant including nodules, maximum increment in chlorophyll and leghemoglobin at 136 mg Cr/kg soil was 23% and 143%, respectively	Wani and Khan (2010)
6.	*Pseudomonas aeruginosa* and *Burkholderia gladioli*	*Lycopersicom esculentum*	These microbes fight against Cd toxicity by enhancing the osmolytes and phenolic compounds in the host plant. The activities of antioxidative enzymes were enhanced in presence of microbes	Khanna et al. (2019)
7.	*Penicillium janthinellum*	*Solanum lycopersicum*	This endophytic fungus significantly improved the growth by improving the shoot length, chlorophyll content, shoot dry weight, and stomatal conductance in plant subjected to Cd stress	Khan et al. (2014)

(continued)

Table 6.1 (continued)

S no.	Name of the microorganisms	Plant	Effect	References
8.	*Achromobacter xylosoxidans*	*Oryza sativa*	*A. xylosoxidans* excreted indole-3-acetic acid, which promoted overall growth of the rice and reduced the effects of As toxicity; it reduced the concentration of As in stem, leaves, bran and grain of rice	Wang et al. (2020)
9.	*Phyllobacterium myrsinacearum*	*Sedum plumbizincicola*	This microbe helped in mobilization of high concentration of Cd, Zn, and Pd in soil and thus helped in preventing the adverse effects of their toxicity	Ma et al. (2013)
10.	*Enterobacter* sp. (in combination with zeolite)	*Brassica napus*	This microorganism helped in alleviating the Cd toxicity in combination with zeolite. The combination improved the physiological attributes such as chlorophyll content, transpiration rate, or stomata conductance of *B. napus*	Saeed et al. (2019)
11.	*Bacillus* sp. and *Stenotrophomoas* sp.	*Raphanus sativus*	These microbes reduced the negative effects of Ni-toxicity by improving plant growth parameters, total chlorophyll, and shoot nitrogen content	Akhtar et al. (2018)
12.	Arbuscular mycorrhizal fungus (with compost)	*Medicago sativa*	This mixture helped in mitigating the stress caused by Zn and Cd and reduced oxidative stress, leading to better plant development; the application of mixture improved the stomatal conductance and sugar content in the host and decreased metal accumulation	Raklami et al. (2020)
13.	*Enterobacter aerogenes*	*Oryza sativa*	This microbe possessed high degree of resistance toward Cd, Pb, and As. This microbe also exhibited some important traits of PGPM under high Cd stress. It reduced the oxidative stress through antioxidants and decreased Cd uptake	Pramanik et al. (2018)

(continued)

Table 6.1 (continued)

S no.	Name of the microorganisms	Plant	Effect	References
14.	*Proteobacter, Alphaproteobacter, Gammaproteobacter, Rhizobiales, Rhizobium,* and Acidobacteria, etc.	*Miscanthus sinensis*	These rhizospheric and endophytic microbes helped in growth of host plants under metal stress. *M. sinensis* showed resistance toward As and Sb in presence of these microbes due to improved carbohydrate metabolism and biosynthesis of secondary metabolites	Sun et al. (2021)
15.	*Bacillus pumilus* and *Pseudomonas* sp. (in combination with leonardite)	*Oryza sativa*	These microbes with leonardite decreased the accumulation of As in the host plant. These microbes also reduced oxidative stress	Dolphen and Thiravetyan (2019)

salt-tolerant and halophytic microbes are known and have been found in highly saline soils (Yan et al. 2015). The saline conditions have also been found to negatively impact plant-mycorrhizal association; mycorrhizae are fungal associates helping plants gain nutrients from soil (Thiem et al. 2018). Bacterial abundance and diversity were also found to reduce in saline soils (Thiem et al. 2018; Yaish et al. 2016). The research over the years has screened, identified, and characterized potential salinity tolerant PGPMs that have been applied either singly or in form of different consortia (Table 6.3). Salt tolerant PGPMs have been found to possess potential for production of exopolysaccharides, hormones (gibberellic acid, indole-3-acetic acid), and siderophores (Fatima et al. 2020). A salt tolerant bacterium *Cellulomonas pakistanensis* sp. nov. strain NCCP-11T was isolated from paddy rice that was found to enhance nutrient availability to plants by the degradation of organic matter via the production of cellulases and hemicellulases (Ahmed et al. 2014; Duy et al. 2016).

6.5 Sustainable Agriculture in the Future Scenarios

The global population has been continuously increasing over past several decades and trend is proposed to be the same in near future. The same comment holds true for climatic changes that have been worsening and projected increase in temperature and disturbance in rainfall patterns are threatening. The need of sustainable crop plant production becomes challenging in the face of these major pressures; population growth and extreme climatic conditions (Lesk et al. 2016). The population growth has also reduced the available lands and has increased the rate of depletion of natural

Table 6.2 Studies demonstrating the role of microorganisms in mitigation of drought and water stress in plants

S. no	Name of microorganism	Name of plant	Effect	References
1.	*Achromobacter xylosoxidans* and *Bacillus pumilus*	*Helianthus annuus*	This is an endophytic bacterium which helped in promoting growth of sunflower seedling under the influence of water stress; it also produced silicic acid and inhibited the growth of pathogenic fungi	Forchetti et al. (2010)
2.	*Klebsiella variicola*, *Raoultella planticola*, and *Pseudomonas fluorescens*	*Zea mays*	These microbes induced a positive effect on dry matter weight of plants under drought stress and increased relative water content of leaves	Gou et al. (2015)
3.	*Micrococcus luteus*	*Zea mays*	*M. luteus* was able to produce exopolysaccharides which helped in surviving under water stress; this microbe showed positive enhancement for root, shoot, and number of leaves of maize plant	Raza and Faisal (2013)
4.	*Bacillus subtilis*	*Arabidopsis thaliana* and *Brassica campestris*	This microbe helped in tolerating water stress by enhancing soil nutrients, nitrogen fixation, and key elements such as potassium, phosphorus, and iron	Woo et al. (2020)
5.	*Azospirillum lipoferum*	*Triticum aestivum*	The microbe helped in alleviating the drought stress, improved growth, relative water content, leaf photosynthetic pigment, antioxidant enzymes, and grain yield	Agami et al. (2017)
6.	*Enterobacter* sp. and *Bacillus* sp.	*Triticum aestivum*	These microbes mitigated the drought stress by making the root system more branched and increased total root length, surface area, and number of root tips	Jochum et al. (2019)
7.	*Euphorbia trignisa*, *Bacillus* sp., *Lysinibacillus*, *Enterobacter*, *Stenotrophomonas*, *Lelliottia*, and *Pseudomonas*	*Solanum lycopersicom*	Bacterization of the host plant by these microbes led to overall biomass increment and greater water retention in foliage	Eke et al. (2019)

(continued)

Table 6.2 (continued)

S. no	Name of microorganism	Name of plant	Effect	References
8.	*Streptomyces laurentii* and *Penicillium* sp.	*Sorghum bicolour*	Both of these are P—Solubilizing bacteria which increased growth in all parameters and also increased chlorophylls, proline, and glycine betaine	Kour et al. (2020)
9.	*Pseudomonas flourescense* and *Bacillus amyloliquefaciens*	*Mentha piperita*	Plants treated with these microbes had higher content of phenol, osmolytes, and antioxidants and showed change in the root system	Chiappero et al. (2019)
10.	*Pseudomonas* sp., *Stenotrophomonas* sp., *Undibacterium* and *Providencia* sp.	*Alhagi sparsifolia*	All these microbes helped in improving the growth capacity of inoculated plants under drought stress measured in terms of length and weight of shoot and root	Zhang et al. (2020)

resources like water, nutrients, and clean air. All such factors reduce the crop growth and productivity. Further stress on the agriculture comes from the xenobiotics, metal (loid) contamination, and other abiotic stresses. The situation overall is quite gloomy and sustainable crop production needs intensive research and inputs from all sectors in near future. In this scenario, the role of microbiome has been considered to be vital. Microbes are known to exist in extremes of environment and thrive. Even in agricultural fields subjected to abiotic stress conditions, microorganisms thrive in abundance and such microorganisms also possess key properties for augmenting the growth of plants. Therefore, the use of such microbes, microbial consortia and microbiome interactions, holds the key to the sustainable agriculture in future.

The increasing temperature is the most obvious effect of changing climate and high temperature stress is known to restrict the growth and yield of plants owing to multifaceted effects on water availability, rate of transpiration, stomatal movement, photosynthetic efficiency, and optimal enzyme activities (Asseng et al. 2013; Wahid and Close 2007; Ashraf and Hafeez 2004). The high temperature has been found to decrease biomass of plants, spike and floret numbers, and seed setting, induce scorching of leaves and twigs and early senescence of leaves (Ebrahim et al. 1998; Fahad et al. 2016; Mitra et al. 2021). At $\geq$30 °C, the floret sterility has been found to increase owing to poor germination and shedding of pollens (Fahad et al. 2015, 2016). At cellular level as well, high temperature can cause protein breakdown (Xie et al. 2003), and impair cell membrane structure and properties (Beney and Gervais 2001). In agricultural scenario, the microorganisms are supposed to be prone to high temperatures, which can change the community and functions of microbes and reduce PGPMs in the soil (Singh et al. 2019). The high temperature of the soil and air disturbs evapotranspiration and water relations of soil and plants as well as

Table 6.3 Studies demonstrating the role of microorganisms in mitigation of salinity stress in plants

S. no.	Name of microorganism	Name of plant	Effect	References
1.	*Pseudomonas fluorescens* *Pseudomonas putida* *Bacillus subtilis*	*Glycine max*	These microbes helped in alleviating salt stress by reducing germination time and enhancing the length and fresh weight of the upper plant body. They promoted antioxidant enzymes activity	Abulfaraj and Jalal (2021)
2.	*Bacillus amyloliquefaciens*	*Mentha piperita*	This microbe helped plant in harsh saline conditions by enhancing total chlorophyll content. It released acetoin as one of the main components which helped in increment in length and dry weight of root	Cappellari and Banchio (2020)
3.	Arbuscular mycorrhizal fungi (AMF) and *Pseudomonas* sp. (coinoculation)	*Sulla coronaria*	They helped in the mitigation of salt stress via various mechanisms like reducing shoot Na accumulation, electrolyte leakage, and by lowering level of malondialdehyde in plants	Hidri et al. (2019)
4.	*Stenotrophomonas maltophilia*	*Arachis hypogaea*	It helped in escaping the salt stress by protecting photosynthetic pigments in peanut plants. It also increased auxin like growth enhancing hormones	Alexander et al. (2020)
5.	*Bacillus subtilis*	*Cicer arietinum*	It helped in alleviating the salt stress by reducing accumulation of sodium and promoting the accumulation of potassium, calcium, and magnesium in plant	Abd-Allah et al. (2018)
6.	*Bacillus licheniformis*	*Triticum aestivum*	This microorganism directly helped in plant growth by supporting increment in length of root, shoot, fresh weight, and dry weight and by reducing sodium ion content	Singh and Jha (2016)
7.	*B. subtilis* and Arbuscular mycorrhizal fungi (AMF)	*Acacia benth*	The microbiome helped the plant to survive in saline stress by increasing the accumulation of important elements like nitrogen, phosphorus, potassium, magnesium, and calcium	Hashem et al. (2016)

(continued)

Table 6.3 (continued)

S. no.	Name of microorganism	Name of plant	Effect	References
8.	*Piriformospora indica*	*Hordeum vulgare*	This mutualistic fungus helped in alleviating salt stress by modulating ion accumulation in colonized plants by increasing the foliar potassium (K^+)/sodium (Na^+) ratio. It promoted Ca^+ accumulation	Alikhani et al. (2013)
9.	*Trichoderma harzianum*	*Cucumis sativus*	This microbe helped in mitigating the salt stress by affecting antioxidant enzymes including peroxidase, polyphenol oxidase phenylalanine ammonia-lyase, catalase, superoxide dismutase, ascorbate peroxidase and glutathione reductase, proline, and ascorbic acid. This microbe increased the concentration of K(+) ion and decreased that of Na(+) ion and ethylene	Zhang et al. (2019)
10.	*Pseudomonas putida*	*Arabidopsis thaliana*	It significantly improved the germination rate of host plants under salt stress. It also upregulated the stress tolerance gene like *LOX2* (jasmonic acid synthesis)	Chu et al. (2019)
11.	*Bacillus licheniformis* and *Pseudomonas plecoglossicida*	*Helianthus annuus*	These PGPMs helped in reducing oxidative stress and improving height and root length of plants. They also upregulated the physiological and biochemical functions at high salinity levels	Yasmeen et al. (2020)
12.	*Azospirillum lipoferum*	*Cicer arietinum*	It alleviated inhibitory impacts of salinity on chickpea growth via modulating osmolytes, antioxidants machinery, and stress related gene expression	El-Esawi et al. (2019)
13.	*Azospirillum brasilense*	*Trifolium repens*	This microbe helped in improvement of overall growth of plant, leaf area and chlorophyll content. It increased the ratio of K^+/Na^+ by decreasing the level of Na^+	Khalid et al. (2017)

(continued)

Table 6.3 (continued)

S. no.	Name of microorganism	Name of plant	Effect	References
14.	*Bacillus licheniformis* and *Enterobacter asburiae*	*Chenopodium quinoa*	These microbes helped in mitigating the salt stress by producing IAA, siderophores, hydrogen cyanide, ammonia, and many extracellular enzymes. They decreased the concentration of Na ion under saline stress	Mahdi et al. (2020)

microbes and affects overall crop growth and yield (Lamaoui et al. 2018). Microbiome functions carry potential to deal with heat stress and other climatic changes to sustain high crop yields in near future (Mahmud et al. 2021). Soil microbes can be engineered to address soil fertility and crop production problems, such as by enhancing nutrient availability. This is because microbes can evolve, adapt, and influence to the climatic changes in real time and can also allow host plants also to adapt slowly (Qiao et al. 2017). The future research needs to study microbiome with respect to heat, drought, salinity, and other abiotic stresses and climatic changes in isolation and in conjunction to engineer potential microbial consortia for sustainable crop production.

Acknowledgments S.S. acknowledges funding support from Institute of Eminence, Banaras Hindu University (Scheme no. 6031).

References

Abd-Allah EF, Alqarawi AA, Hashem A, Radhakrishnan R, Al-Huqail AA, Al-Otibi FON et al (2018) Endophytic bacterium *Bacillus subtilis* (BERA 71) improves salt tolerance in chickpea plants by regulating the plant defense mechanisms. J Plant Interact 13:37–44

Abo-Amer AE, El-Shanshoury AER, Alzahrani OM (2015) Isolation and molecular characterization of heavy metal-resistant *Alcaligenes faecalis* from sewage wastewater and synthesis of silver nanoparticles. Geomicrobiol J 32:836–845

Abulfaraj AA, Jalal RS (2021) Use of plant growth-promoting bacteria to enhance salinity stress in soyabean (*Glycine max* L.) plants. Saudi J Biol Sci 28:3823–3834

Agami RA, Ghramh HA, Hasheem M (2017) Seed inoculation with *Azospirillum lipoferum* alleviates the adverse effects of drought stress on wheat plants. J Appl Bot Food Qual 90: 165–173

Ahmed I, Kudo T, Abbas S, Ehsan M, Iino T, Fujiwara T et al (2014) *Cellulomonas pakistanensis* sp. nov., a moderately halotolerant Actinobacteria. Int J Syst Evol Micrbiol 64:2305–2311

Akhtar MJ, Ullah S, Ahmad I, Abdul R, Nadeem SM, Khan MY et al (2018) Nickel phytoextraction through bacterial inoculation in *Raphanus sativus*. Chemosphere 190:234–242

Alexander A, Singh VK, Mishra A (2020) Halotolerant PGPR *Stenotrophomonas maltophilia* BJ01 induces salt tolerance by modulating physiology and biochemical activities of *Arachis hypogaea*. Front Microbiol 11:568289

Alikhani M, Khatabi B, Sepehri M, Nekouei MK, Mardi M, Salekdeh GH (2013) A proteomics approach to study the molecular basis of enhanced salt tolerance in barley (*Hordeum vulgare* L) conferred by root mutualistic fungus *Piriformospora indica*. Mol BioSyst 9:1498–1510

Ambrosini A, de Souza R, Passaglia LMP (2016) Ecological role of bacterial inoculants and their potential impact on soil microbial diversity. Plant Soil 400:193–207

Armada E, Leite MFA, Medina A, Azcon R, Kuramae EE (2018) Native bacteria promote plant growth under drought stress condition without impacting the rhizomicrobiome. FEMS Microbiol Ecol 94:fiy092

Ashraf M, Hafeez M (2004) Thermotolerance of pearl millet and maize at early growth stages: growth and nutrient relations. Biol Plant 48:81–86

Asseng S, Ewert F, Rosenzweig C, Jones JW, Hatfield JL, Ruane AC et al (2013) Uncertainty in simulating wheat yields under climate change. Nat Clim Chang 3:827–832

Bastida F, Torres IF, Andres-Abellan M, Baldrian P, Lopez-Mondejar R, Vetrovský T, Richnow HH et al (2017) Differential sensitivity of total and active soil microbial communities to drought and forest management. Glob Chang Biol 23:4185–4203

Beney L, Gervais P (2001) Influence of the fluidity of the membrane on the response of microorganisms to environmental stresses. Appl Microbiol Biotechnol 57:34–42

Berg J, Brandt KK, Al-Soud WA, Holm PE, Hansen LH, Sørensen SJ et al (2012) Selection for Cu-tolerant bacterial communities with altered composition, but unaltered richness, via long-term Cu exposure. Appl Environ Microbiol 78:7438–7446

Berg G, Grube M, Schloter M, Smalla K (2014) Unraveling the plant microbiome: looking back and future perspectives. Front Microbiol 5:1–7

Bulgarelli D, Rott M, Schlaeppi K, van Themaat EVL, Ahmadinejad N, Assenza F et al (2012) Revealing structure and assembly cues for Arabidopsis root-inhabiting bacterial microbiota. Nature 488:91–95

Cappellari LR, Banchio E (2020) Microbial volatile organic compounds produced by *Bacillus amyloliquefaciens* GB03 ameliorate the effects of salt stress in *Mentha piperita* principally through acetoin emission. J Plant Growth Regul 39:764–775

Chen L, He L, Wang Q, Sheng X (2016) Synergistic effects of plant growth-promoting Neorhizobium huautlense T1-17 and immobilizers on the growth and heavy metal accumulation of edible tissues of hot pepper. J Hazard Mater 312:123–131

Chiappero J, Cappellari LR, Alderete LGS, Palermo TB, Banchio E (2019) Plant growth promoting rhizobacteria improves the antioxidant status in *Mentha piperita* grown under drought stress leading to enhancement of plant growth and phenolic content. Ind Crop Prod 139:111553

Chu TN, Tran BTH, Bui LV, Hoang MTT (2019) Plant growth promoting rhizobacterium *Pseudomonas* PS01 induces salt tolerance in *Arabidopsis thaliana*. BMC Res Notes 12:11

Dolphen R, Thiravetyan P (2019) Reducing arsenic in rice grains by leonardite and arsenic-resistant endophytic bacteria. Chemosphere 223:448–454

Dourado MN, Martins PF, Quecine MC, Piotto FA, Souza LA, Franco MR et al (2013) *Burkholderia* sp. SCMS54 reduces cadmium toxicity and promotes growth in tomato. Ann Appl Biol 163:494–507

Duy MV, Hoi NT, Ve NB, Thuc LV, Trang NQ (2016) Influence of *Cellulomonas flavigena*, *Azospirillum* sp. and *Pseudomonas* sp. on rice growth and yield grown in submerged soil amended with rice straw. In: Sayyed RZ, Reddy MS, Al-Turki AI (eds) Recent trends in PGPR research for sustainable crop productivity. Scientific Publishers, Delhi, pp 238–242

Ebrahim MK, Zingsheim O, El-Shourbagy MN, Moore PH, Komor E (1998) Growth and sugar storage in sugarcane grown at temperatures below and above optimum. J Plant Physiol 153: 593–602

Eke P, Kumar A, Sahu KP, Wakam LN, Sheoran N, Ashajyothi M, Patel A et al (2019) Endophytic bacteria of desert cactus (*Euphorbia trigonas* Mill) confer drought tolerance and induce growth promotion in tomato (*Solanum lycopersicum* L.). Microbiol Res 228:126302

El-Esawi MA, Al-Ghamdi AA, Ali HM, Alayafi AA (2019) *Azospirillum lipoferum* FK1 confers improved salt tolerance in chickpea (*Cicer arietinum* L.) by modulating osmolytes, antioxidant machinery and stress-related gene expression. Environ Exp Bot 159:55–65

El-Meihy RM, Abou-Aly HE, Youssef AM, Tewfike TA, El-Alkshar EA (2019) Efficiency of heavy metals-tolerant plant growth promoting bacteria for alleviating heavy metals toxicity on sorghum. Environ Exp Bot 162:295–301

Fahad S, Hussain S, Bano A, Saud S, Hassan S, Shan D et al (2015) Potential role of phytohormones and plant growth-promoting rhizobacteria in abiotic stresses: consequences for changing environment. Environ Sci Pollut Res 22:4907–4921

Fahad S, Hussain S, Saud S, Khan F, Hassan S, Nasim W et al (2016) Exogenously applied plant growth regulators affect heat-stressed rice pollens. J Agron Crop Sci 202:139–150

Fatima T, Mishra I, Verma R, Arora NK (2020) Mechanisms of halotolerant plant growth promoting Alcaligenes sp. Involved in salt tolerance and enhancement of the growth of rice under salinity stress. 3 Biotech 10:361

Forchetti G, Masciarelli O, Izaguirre MJ, Alemano S, Alvarez D, Abdala G (2010) Endophytic bacteria improve seedling growth of sunflower under water stress, produce salicylic acid, and inhibit growth of pathogenic fungi. Curr Microbiol 61:485–493

Frossard A, Donhauser J, Mestrot A, Gygax S, Baath E, Frey B (2018) Long- and short-term effects of mercury pollution on the soil microbiome. Soil Biol Biochem 120:191–199

Gagne-Bourque F, Bertrand A, Claessens A, Aliferis KA, Jabaji S (2016) Alleviation of drought stress and metabolic changes in timothy (*Phleum pratense* L.) colonized with *Bacillus subtilis* B26. Front Plant Sci 7:584

Gil-Martínez M, Lopez-García A, Domínguez MT, Navarro-Fernandez CM, Kjøller R, Tibbett M et al (2018) Ectomycorrhizal fungal communities and their functional traits mediate plant-soil interactions in trace element contaminated soils. Front Plant Sci 9:1682

Gołębiewski M, Deja-Sikora E, Cichosz M, Tretyn A, Wrobel B (2014) 16S rDNA pyrosequencing analysis of bacterial community in heavy metals polluted soils. Microbial Ecol 67:635–647

Gou W, Zheng P, Chen F, Zhang L, Cui Z, Cao M et al (2015) Accumulation of choline and gylcinebetaine and drought stress tolerance induced in maize (*Zea mays*) by three plant growth promoting rhizobacteria (PGPR) strains. Pak J Bot 47:581–586

Hashem A, Abd-Allah EF, Alqarawi AA, Shah MA (2016) Induction of osmoregulation and modulation of salt stress in *Acacia gerrardii* Benth. by arbuscular mycorrhizal fungi and *Bacillus subtilis* (BERA 71). BioMed Res Int 2016:6294098

Hidri R, Mahmoud OMB, Farhat N, Cordero I, Pueyo JJ, Debez A et al (2019) Arbuscular mycorrhizal fungus and rhizobacteria affect the physiology and performance of *Sulla coronaria* plants subjected to salt stress by mitigation of ionic imbalance. J Plant Nutr Soil Sci 182:451–462

Hur M, Kim Y, Song H-R, Kim JM, Choi YI, Yi H (2011) Effect of genetically modified poplars on soil microbial communities during the phytoremediation of waste mine tailings. Appl Environ Microbiol 77:7611–7619

Jiao S, Chen W, Wei G (2019) Resilience and assemblage of soil microbiome in response to chemical contamination combined with plant growth. Appl Environ Microbiol 85:e02523–e02518

Jochum MD, McWilliams KL, Borrego EJ, Kolomiets MV, Niu G, Pierson A et al (2019) Bioprospecting plant growth-promoting rhizobacteria that mitigate drought stress in grasses. Front Microbiol 10:2106

Khalid M, Bilal M, Hassani D, Iqbal HMN, Wang H, Huang D (2017) Mitigation of salt stress in white clove (*Trifolium repens*) by *Azospirillum brasilense* and its inoculation effect. Bot Stud 58:5

Khan AL, Waqas M, Hussain J, Al-Harrasi A, Lee IJ (2014) Fungal endophyte *Penicillium janthinellum* LK5 can reduce cadmium toxicity in *Solanum lycopersicum* (Sitiens and Rhe). Biol Fertil Soil 50:75–85

Khanna K, Jamwal VL, Sharma A, Gandhi SG, Ohari P, Bhardwaj R et al (2019) Supplementing with plant growth promoting rhizobacteria (PGPR) alleviates cadmium toxicity in *Solanum lycopersicum* by modulating the expression of secondary metabolites. Chemosphere 217:463–474

Kour D, Rana KL, Kaur T, Sheikh I, Yadav AN, Kumar V, Dhaliwal HS, Saxena AK (2020) Microbe-mediated alleviation of drought stress and acquisition of phosphorus in great millet (*Sorghum bicolour* L.) by drought-adaptive and phosphorus-solubilizing microbes. Biocatal Agricult Biotechnol 23:101501

Lamaoui M, Jemo M, Datla R, Bekkaoui F (2018) Heat and drought stresses in crops and approaches for their mitigation. Front Chem 6:26

Lareen A, Burton F, Schafer P (2016) Plant root-microbe communication in shaping root microbiomes. Plant Mol Biol 90:575–587

Lesk C, Rowhani P, Ramankutty N (2016) Influence of extreme weather disasters on global crop production. Nature 529:84–87

Lundberg DS, Lebeis SL, Paredes SH, Yourstone S, Gehring J, Malfatti S et al (2012) Defining the core Arabidopsis thaliana root microbiome. Nature 488:86–90

Luziatelli F, Ficca AG, Cardarelli M, Melini F, Cavalieri A, Ruzzi M (2020) Genome sequencing of Pantoea agglomerans C1 provides insights into molecular and genetic mechanisms of plant growth-promotion and tolerance to heavy metals. Microorganisms 8:153

Ma Y, Rajkumar M, Freitas H (2009) Inoculation of plant growth promoting bacterium *Achromobacter xylosoxidans* strain Ax10 for the improvement of copper phytoextraction by *Brassica juncea*. J Environ Manag 90:831–837

Ma Y, Rajkumar M, Vicente JAF, Freitas H (2010) Inoculation of Ni-resistance plant growth promoting bacterium *Psychrobacter* sp. strain SRS8 for the improvement of phytoextraction by energy crops. Int J Phytoremediation 13:126–139

Ma Y, Rajkumar M, Luo Y, Freitas H (2013) Phytoextraction of heavy metals polluted soil using *Sedum plumbizincicola* inoculated with metal mobilization *Phyllobacterium myrsinacearum* RC6B. Chemosphere 93:1386–1392

Mahdi I, Fahsi N, Hafidi M, Allaoui A, Biskri L (2020) Plant growth enhancement using rhizospheric halotolerant phosphate solubilizing bacterium *Bacillus licheniformis* QA1 and *Enterobacter asburiae* QF11 isolated from *Chenopodium quinoa* Wild. Microorganisms 8:948

Mahmud K, Missaoui A, Lee K, Ghimire B, Presley HW, Makaju S (2021) Rhizosphere microbiome manipulation for sustainable crop production. Curr Plant Biol 27:100210

Marasco R, Rolli E, Vigani G, Borin S, Sorlini C, Ouzari H et al (2013) Are drought-resistance promoting bacteria cross-compatible with different plant models? Plant Signal Behav 8:e26741

Mitra D, Rodriguez AMD, Cota FIP, Khoshru B, Panneerselvam P, Moradi S et al (2021) Amelioration of thermal stress in crops by plant growth-promoting rhizobacteria. Physiol Mol Plant Pathol 115:101679

Mnasri M, Janouskova M, Rydlova J, Abdelly C, Ghnaya T (2017) Comparison of arbuscular mycorrhizal fungal effects on the heavy metal uptake of a host and a non-host plant species in contact with extraradical mycelial network. Chemosphere 171:476–484

Naylor D, Coleman-Derr D (2018) Drought stress and root-associated bacterial communities. Front Plant Sci 8:2223

Naylor D, DeGraaf S, Purdom E, Coleman-Derr D (2017) Drought and host selection influence bacterial community dynamics in the grass root microbiome. ISME J 11:2691–2704

Pramanik K, Mitra S, Sarkar A, Maiti TK (2018) Alleviation of phytotoxic effects of cadmium on rice seedlings by cadmium resistant PGPR strain *Enterobacter aerogenes* MCC 3092. J Hazard Mater 351:317–329

Pramanik K, Mitra S, Sarkar A, Soren T, Maiti TK (2017) Characterization of cadmium resistance *Klebsiella pneumoniae* MCC 3091 promoted rice seedling growth by alleviating phytotoxicity of cadmium. Environ Sci Pollut Res 24:24419–24437

Prasad R, Gill SS, Tuteja N (2018) Crop improvement through microbial biotechnology. Elsevier. ISBN: 9780444639882

Qiao Q, Wang F, Zhang J, Chen Y, Zhang C, Liu G et al (2017) The variation in the rhizosphere microbiome of cotton with soil type, genotype and developmental stage. Sci Rep 7:3940

Raklami A, El Gharmali A, Rahou Y, Oufdou K, Meddich A (2020) Compost and mycorrhizae application as a technique to alleviate Cd and Zn stress in *Medicago sativa*. Int J Phytoremediation 23:190–201

Raza FA, Faisal M (2013) Growth promotion of maize by desiccation tolerant *Micrococcus luteus*-chp37 isolated from Cholistan desert, Pakistan. Aust J Crop Sci 7:1693–1698

Rolli E, Marasco R, Vigani G, Ettoumi B, Mapelli F, Deangelis ML et al (2015) Improved plant resistance to drought is promoted by the root associated microbiome as a water stress-dependent trait. Environ Microbiol 17:316–331

Saeed Z, Naveed M, Imran M, Bashir MA, Sattar A, Mustafa A et al (2019) Combined use of *Enterobacter* sp. MN17 and zeolite reverts the adverse effects of cadmium on growth, physiology and antioxidant activity of *Brassica napus*. PLoS One 14:e0213016

Sheik CS, Mitchell TW, Rizvi FZ, Rehman Y, Faisal M, Hasnain S et al (2012) Exposure of soil microbial communities to chromium and arsenic alters their diversity and structure. PLoS One 7: e40059

Singh RP, Jha PN (2016) A halotolerant bacterium *Bacillus licheniformis* HSW-16 augments induced systemic tolerance to salt stress in wheat plant (*Triticum aestivum*). Front Plant Sci 7: 1890

Singh RK, Masurkar P, Pandey SK, Kumar S (2019) Rhizobacteria–plant interaction, alleviation of abiotic stresses. In: Sayyed R, Arora N, Reddy M (eds) Plant growth promoting rhizobacteria for sustainable stress management. Springer, Singapore, pp 345–353

Soussi A, Ferjani R, Marasco R, Guesmi A, Cherif H, Rolli E et al (2016) Plant-associated microbiomes in arid lands: diversity, ecology and biotechnological potential. Plant Soil 405: 357–370

Sun X, Song B, Xu R, Zhang M, Gao P, Lin H, Sun W (2021) Root-associated (rhizosphere and endosphere) microbiomes of the *Miscanthus sinensis* and their response to the heavy metal contamination. J Environ Sci 104:387–398

Thiem D, Gołębiewski M, Hulisz P, Piernik A, Hrynkiewicz K (2018) How does salinity shape bacterial and fungal microbiomes of *Alnus glutinosa* roots? Front Microbiol 9:651

Wahid A, Close TJ (2007) Expression of dehydrins under heat stress and their relationship with water relations of sugarcane leaves. Biol Plant 51:104–109

Wang Q, Chen L, He L-Y, Sheng X-F (2016) Increased biomass and reduced heavy metal accumulation of edible tissues of vegetable crops in the presence of plant growth-promoting *Neorhizobium huautlense* T1-17 and biochar. Agric Ecosyst Environ 228:9–18

Wang KT, Li YP, Wu YC, Qui ZQ, Ding Z, Wang X et al (2020) Improved grain yield and lowered arsenic accumulation in rice plants by inoculating with arsenite-oxidizing *Achromobacter xylosoxidans* GD03. Ecotoxicol Environ Saf 206:111229

Wani PA, Khan MS (2010) *Bacillus* species enhance growth parameters of chickpea (*Cicer arietinum* L.) in chromium stressed soils. Food Chem Toxicol 48:3262–3267

Woo OG, Kim H, Kim J-S, Keum H-L, Lee KC, Sul WJ, Lee JH (2020) *Bacillus subtilis* strain GOT9 confers enhanced tolerance to drought and salt stresses in *Arabidopsis thaliana* and *Brassica campestris*. Plant Physiol Biochem 148:359–367

Xie C, Li YQ, Tang W, Newton RJ (2003) Study of dynamical process of heat denaturation in optically trapped single microorganisms by near-infrared Raman spectroscopy. J Appl Phys 94: 6138–6142

Yaish MW, Kumar PP (2015) Salt tolerance research in date palm tree (*Phoenix dactylifera* L.), past, present, and future perspectives. Front Plant Sci 6:348

Yaish MW, Al-Lawati A, Jana GA, Vishwas Patankar H, Glick BR (2016) Impact of soil salinity on the structure of the bacterial endophytic community identified from the roots of caliph medic (*Medicago truncatula*). PLoS One 11:e0159007

Yan N, Marschner P, Cao W, Zuo C, Qin W (2015) Influence of salinity and water content on soil microorganisms. Int Soil Water Cons Res 3:316–323

Yasmeen T, Ahmad A, Arif MS, Mubin M, Rehman K, Shahzad SM et al (2020) Biofilm forming rhizobacteria enhance growth and salt tolerance in sunflower plants by stimulating antioxidant enzymes activity. Plant Physiol Biochem 156:242–256

Young E, Carey M, Meharg AA, Meharg C (2018) Microbiome and ecotypic adaption of *Holcus lanatus* (L.) to extremes of its soil pH range, investigated through transcriptome sequencing. Microbiome 6:48

Zhang F, Wang Y, Liu C, Chen F, Ge H, Tian F et al (2019) *Trichoderma harzianum* mitigates salt stress in cucumber via multiple responses. Ecotoxicol Environ Saf 170:436–445

Zhang L, Zhang W, Li Q, Cui R, Wang Z, Wang Y et al (2020) Deciphering the root endosphere microbiome of the desert plant *Alhagi sparsifolia* for desert resistance-promoting bacteria. Appl Environ Microbiol 86:e02863-19

Chapter 7
The Beneficial Plant Microbial Association for Sustainable Agriculture

Sivakumar Natesan, Shyamkumar Rajaram, Devaprakash Manoharan, and Thirumalaivasan Ramachandran

Abstract Microbes are ubiquitous and can associate to colonize plants and exhibits different modes of interactions. Plant beneficial microbes could colonize both the phyllosphere and rhizosphere to promote the various aspect of plant growth and other various compartments in plants. These beneficial microbes are generally called plant growth-promoting microbes (PGPMs), they can become an excellent alternative to remove or reduce the use of various toxic agrochemicals including synthetic chemical fertilizers and biocides. The association of PGPMs provides nutrients, protection against pathogens as well as various environmental stress responses either direct or indirect mechanisms. The soil and rhizosphere microbes beneficially associate either the root surface or phyllosphere region of the plant and influence the growth and health fitness of crops. Some microbes directly interact with the plant to develop a symbiotic relationship (e.g., *Rhizobium*, mycorrhizal fungi), and few can interact at the surface of the root with either associative symbiosis (*Azospirillum*) or nonsymbiotic beneficial interactions such as nutrient acquisition, solubilization, and translocation of minerals and water, vitamin and growth hormone synthesis, mineralization of soil organic residues, inhibits harmful pathogens and nematodes, production of iron siderophores to chelate ions, and provide induced resistance against various biotic and abiotic stresses. This chapter describes the basic beneficial microbial interactions on the rhizosphere, phyllosphere, and their beneficial effects on the host for sustainable agriculture, specifically, bacterial nodulation, mycorrhizal infection, microbial endophytes, development of bioinoculums, and their benefits to the plant. Further, the functions of beneficial microbes to the plants and the soil have been discussed. Besides, rhizosphere microbiome engineering and its role in sustainable agriculture have been also discussed.

S. Natesan (✉) · D. Manoharan · T. Ramachandran
Department of Molecular Microbiology, School of Biotechnology, Madurai Kamaraj University, Madurai, Tamil Nadu, India

S. Rajaram
Department of Biotechnology, Kamaraj College of Engineering and Technology, Virudhunagar, Tamil Nadu, India

S. Chhabra et al. (eds.), *Plant Microbiome for Plant Productivity and Sustainable Agriculture*, Microorganisms for Sustainability 37,
https://doi.org/10.1007/978-981-19-5029-2_7

Keywords Beneficial microbes · Microbiome engineering · Phyllosphere microbial association · Rhizosphere microbes · Sustainable agriculture

7.1 Introduction

Increasing the human population often drives-up demand for food, especially since almost one in seven people around the world is continually hungry, lacking enough food to be healthy. The Food and Agriculture Organization (FAO) projects that by 2050, the world population will reach 9.2 billion, hence it is an urgent need to increase the agricultural productivity up to 60–70% from the current levels (Pawlak and Kołodziejczak 2020). In order to supply the growing demand for food, certain factors are necessary to reach this goal including a suitable environment with the climatic condition, good adequate fertile soil, and also suitable cultivation practices. From the mid twentieth century to date, the majority of agricultural farmers use chemical fertilizers and other agrochemicals to increase crop productivity. The chemical fertilizers as well as the synthetic agrochemicals give sudden effects on the yield; however, they continuously eradicate the soil beneficial microbes as well as the soil health (Meena et al. 2020). In addition, extensive and improper use of such chemicals causes environmental issues as well as health issues for humans and animals.

The beneficial role of soil-borne microbes especially rhizosphere microbes has been emphasized in many cases. They can play a vital role in earth biogeochemical cycling that can maintain soil fertility and soil health (Basu et al. 2021). Hence, last few decades scientists give more attention to screen, select, and identify the most beneficial microbes from the environment of the various plant species, and studied the real benefits toward the plant and our environment (Pawlak and Kołodziejczak 2020). There are various microorganisms including bacteria, actinobacteria, fungi, and cyanobacteria have beneficially interacted with a wide range of plants either symbiotically or nonsymbiotically. The major benefits of such microbial interactions are usually either direct or indirect benefits such as mineral supply phosphorus (P), nitrogen (N), potassium (K), and sulfur (S). Mineralization of organic waste and inorganic substances provides enhanced nutrient uptake, synthesis of plant growth hormone-like substances, quench pathogenic microbes including soil nematodes and other insect vectors (biocontrol), decreases either biotic or abiotic stresses, detoxification of chemicals, reduction of heavy metal toxicity, and also reduces the impact in global climatic change. Hence, scientists have begun to direct their interests toward ensuring soil microorganisms and their sustainability in agricultural use, as bio-inoculants as a suitable alternative for chemical fertilizers and biocides (Abdul Rahman et al. 2021).

Among the various environment, rhizosphere soil and the phyllosphere have more beneficially interacted with a large number of diverse microorganisms (Prasad and Zhang 2022). Symbiotic, associative symbiotic, and nonsymbiotic modes microbes are largely exploited the rhizosphere and phyllosphere region for their surveillance (Sivakumar et al. 2020). For example, the most prominent and well-known symbiotic microorganism that fixes atmospheric N_2 in the form of ammonia is Rhizobium in legume nodules; *Azotobacter*, *Azospirillum*, and Phosphobacter are other well-known beneficial microbes being considered as biofertilizers for various crops. In addition to the mycorrhizal fungi, Frankia can also mobilize various minerals to the host plant and give protection against environmental stresses (Oleńska et al. 2020).

Rhizosphere health is important for plant growth and development; it depends on the diversity and abundance of beneficial microbes. For the sustainable agriculture, it is necessary to maintain the rhizosphere microbiome through effective soil management, which can define as the process of improving the nutrient efficiency in the soil to enhance the nutrient availability to improve beneficial microbes for the growth and yield of crops (Tahat et al. 2020). Generally, beneficial soil microbes enhance the rhizosphere health through the multidimensional mechanism, which includes fixing the N_2, hydrolyzing complex organic waste residues to simple molecules, production of siderophores, hydrogen cyanide, and organic acids, secreting enzymes to solubilize phosphate and potassium, and production of vitamins and growth hormones, ultimately it improves soil fertility. There are several research works undertaken to isolate many of such beneficial microorganisms from the soil of both agriculture and forest that can exhibit plant growth-promoting activities called, plant growth-promoting rhizobacteria (PGPR) (Prasad et al. 2015). PGPR and mycorrhiza fungi are currently used as a tool for food security and sustainable agriculture (Diagne et al. 2020).

Recent scientific discoveries gaining more and more information related to the soil and rhizosphere microbiome and its complex architecture are used to design a new and healthier environment for a specific type of crop in a particular soil is possible. Through such microbial manipulation, the rhizosphere gets more benefits which overcome the drawbacks of the single type of microbial supplement. This technology is now called microbiome engineering or designing, specifically rhizosphere microbiome engineering (Albright et al. 2021). This chapter describes the important beneficial microbial associations in both the rhizosphere and phyllosphere region of plants and their beneficial role mainly PGPR functions. In addition, the role of beneficial microbes for the use of biofertilizers and biocontrol agents in sustainable agriculture is analyzed. Further, microbiome engineering with beneficial microbes to manipulate the rhizosphere microbiome for agricultural sustainability is also discussed.

7.2 Beneficial Microbial Interactions in Plants

Microbes are ubiquitous, they can survive and develop adaptation to any environment and establish their colonies in different niches. Plants have different microhabitats that hold several complex microbial communities such as bacteria, archaea, and fungi, named microbiome (Bang et al. 2018). This microbiome of microbial consortia develops a unique community structure in the various microhabitats of the plants, generally by either epiphytic or endophytic association. Various studies revealed this complex microbial colonization on different structural components of

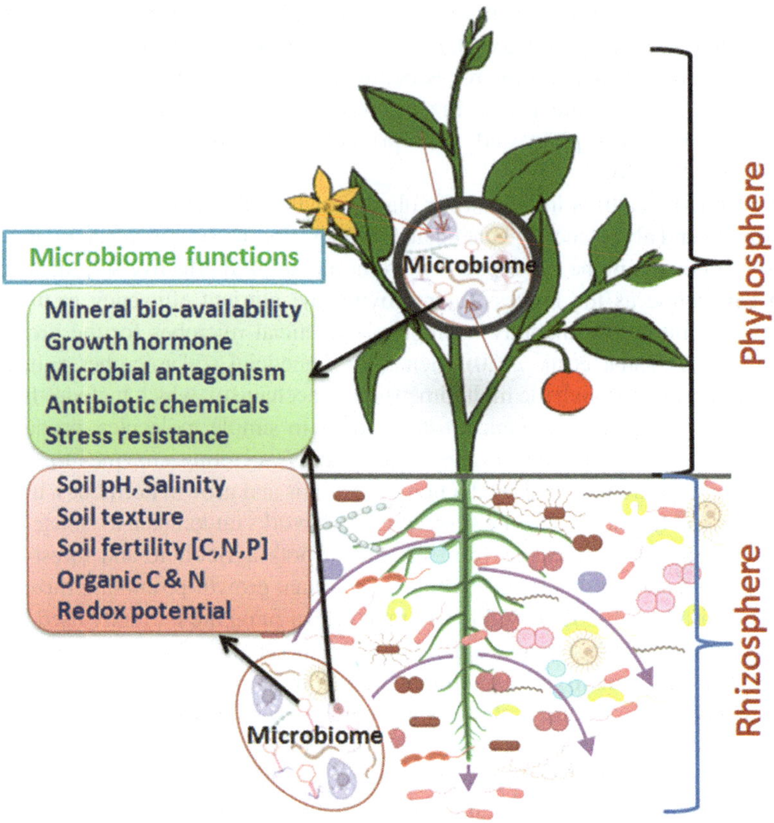

Fig. 7.1 Microbial associations on plant

plants such as phyllosphere (caulosphere (stems), phylloplane (leaves), anthosphere (flowers), and carposphere (fruits)), and rhizosphere regions (Fig. 7.1), besides, the spermosphere, a small area which surrounds a germination seed in the soil which has a unique microenvironment for microbial diversity (van der Heijden et al. 2007; Hodge and Fitter 2010).

7.3 Rhizosphere Microbiome Interaction

The soil is the natural and largest microbial reservoir for plants, contains 10^6 to 10^{11} microbial cells per gram and consists of 30,000 and above prokaryotic diversity (Berendsen et al. 2012). The rhizosphere is the region where the plant roots are firmly attached with soil particles for their stability, nutrients, and water. This region is densely populated with both prokaryotic and eukaryotic microbes and nematodes. In this region, plants secrete root exudates called rhizodeposits, consists of sugars,

Fig. 7.2 Factors determination of microbial interactions on plant

amino acids, organic and inorganic acids, siderophores, vitamins, nucleosides, and polysaccharide (Cooper 2007; Odelade and Babalola 2019; Khan et al. 2021), which chemotactically attract the microorganism to establish their associative/symbiotic colonies on the surface (epiphytes) and inner tissues (endophytes) of root (Canarini et al. 2019). The rhizosphere microbes heavily influence the soil nutrients recycling (decomposition, N_2 fixation, and solubilization of rock minerals) and make them available for plants. Hence, the abundance and diversity of microbes in the rhizosphere region are comparatively ~100 times greater than the nearby other soil habitat (Backer et al. 2018). Several biotic and abiotic factors have determined the microbial abundance and interaction with plant roots. Factors which include native microbes, temperature, wind flow, light intensity, oxygen availability, soil fertility, heavy metals, and agrochemicals (Fig. 7.2) have direct influences on beneficial microbial interaction (Dastogeer et al. 2020). Further, the composition of rhizodeposits varied in different plant species, which determines the microbial composition in the rhizosphere soil (McNear 2013; Mendes et al. 2013; Paterson et al. 2007). For example, legume plants' rhizosphere has a significant number of microbial diversity than cereal crops due to its specialized chemoattractants (flavonoids and organic acids), which interact with the symbiotic association of N_2 fixing *Rhizobium* sp. that makes nodules on its root (Cooper 2007). The exudates of roots are important to mobilize and modulate the community of the microbiome in the rhizosphere, along with other biotic and abiotic factors (Mendes et al. 2013; Korenblum et al. 2020). However, some selected microbial taxa are common in specific plant species, which are always detected in the similar plant species in the different environment called "core microbes" (Jones et al. 2019). The core microbe of a specific plant species is usually not disturbed by various agricultural practices including the long-term tillage and crop rotations (Bziuk et al. 2021). Besides, the seed contains beneficial microbes, after germination, it develops an endophytic association with plants (Shao et al. 2021) and also increases the sustainability of the crop (Zhang et al. 2019). In root, *Rhizobium*, mycorrhizal fungi, and *Frankia* are symbiotically associated with a wide range of plants and promote growth and yield. Further, *Azotobacter*, *Azospirillum*, *Phosphobacter,* and other rhizosphere microbes beneficially assemble in the rhizosphere microbiome and provide various health benefits to the host plant.

7.3.1 *Rhizobium Nodulation: A Beneficial Microbe–Plant Interaction*

Nodulation is an important event for the biological N_2 fixation in legumes. *Rhizobium* is a genus of Gram-negative soil bacteria, the well-known nitrogen-fixing root nodule bacterial symbiont of legumes and other flowering plants. The legume-rhizobia symbioses mainly occur in three Leguminosae subfamilies, the Caesalpinioideae, Mimosoideae, and Papilionoideae. According to the species-specific interaction of Rhizobia in legumes further named *Bradyrhizobium*, *Ochrobactrum*, *Rhizobium*, *Ensifer*, *Mesorhizobium*, *Burkholderia*, *Neorhizobium*, *Allorhizobium*, *Devosia*, *Phyllobacterium*, *Microvirga*, *Ochrobactrum*, and *Pararhizobium* (Andrews and Andrews 2017). The bacteria colonize plant cells within root nodules, where they convert atmospheric nitrogen into ammonia by a nitrogenase enzyme and then provide organic nitrogenous compounds such as glutamine or ureides to the plant. Plant provides photosynthetically fixed organic carbon for *Rhizobium* growth. During this association, both partners have mutually benefited (Varma et al. 2020).

The root-nodulation process starts with a signal transduction process between the host plant and its microsymbionts (Oldroyd 2013; Wang et al. 2018a). Under nitrogen starvation, legume roots secrete a complex mixture of exudates (infection flavonoids) into the rhizosphere, which stimulates nodulation (*nod*) genes to synthesize nodulation factors, a lipochitooligosaccharidic signal that is essential for initiating symbiotic development in most legumes (Oldroyd et al. 2011). Further, rhizobia polysaccharides such as exo- (EPS), capsular- (KPS), and lipo-

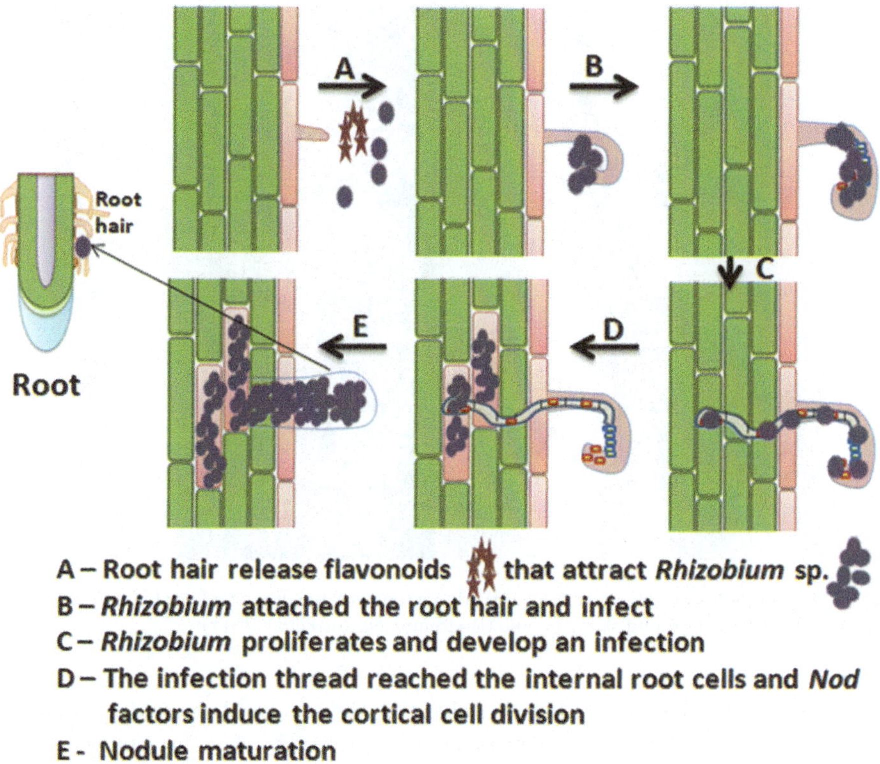

Fig. 7.3 Stages of Rhizobium root nodule formation. (Adapted from Farssi et al. 2018)

polysaccharides (LPS) also have significant role in establishing symbiotic relationship with legumes (Gibson et al. 2008). Exopolysaccharides have been shown to be required for a rhizobial infection that can suppress plant defense and promote nodulation. This mechanism was reported in different nodulating rhizobium species such as *Rhizobium leguminosarum*, *Sinorhizobium meliloti* (Peck et al. 2006), *Bradyrhizobium japonicum*, *R. alamii*, *R. lentis*, *R. japonicum*, *R. metallidurans*, *R. smilacinae*, *R. phaseoli*, and *R. trifolii* (Breakspear et al. 2014).

Exopolysaccharides have been shown to be required for rhizobial infection that can suppress plant defense and nodulation. Figure 7.3 described the mechanism of root nodulation. Infection starts with recognition of the *Nod* signals in the root hair of the specified host and induces the modification on the root hair growth, which looks like curling at the tip of root hair results in a cork-screw structure. Rhizobia colonize at the curled surface of root hair and infect, then enter into the root hair by endocytosis. The invaded rhizobium is confined within a specialized infection tube called infection thread which further progresses intracellularly to the cortex, where it develops into a primordium, a newly forming nodule. Bacteria multiply within a membrane-bound compartment named symbiosome, a unit of biological N_2 fixation where bacteria make symbiotic interaction with host cell for nutrients and ammonia

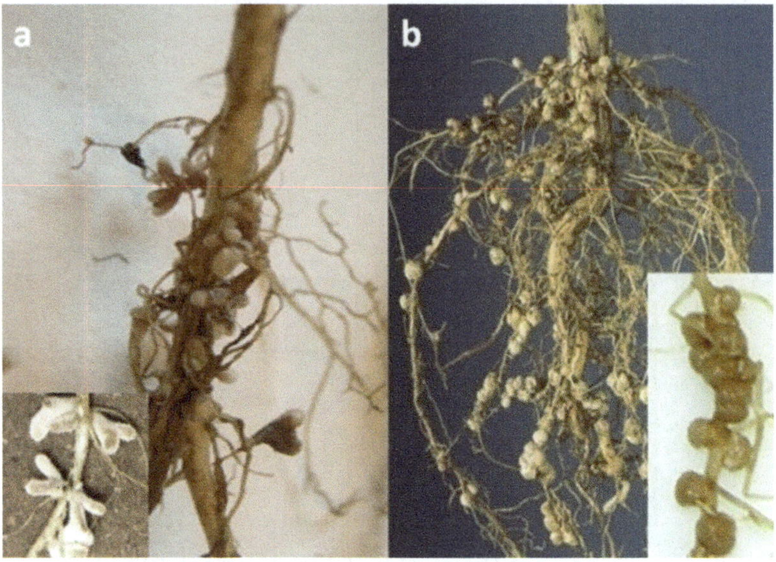

Fig. 7.4 Morphology of legume nodule. (**a**) Alfalfa root with fingerlike nodule (Jensen et al. 2012); (**b**) soybean root with round shape nodule (Rodríguez-Navarro et al. 2011)

exchange, within the nodule bacteria changing their shape (bacteroids). There are two types of nodules identified in plants (Fig. 7.4). The first one appears as the modified lateral root with a terminal apical meristem along with lateral vascular cells called intermediate nodule (e.g., alfalfa, fingerlike in shape); the other one is the determinate nodule of soybean (round shape), where the cortical cells divide and expand to develop globular nodule structure. The development of this nodule is in a radial pattern (Sulieman and Tran 2013).

The symbiotic nitrogen-fixation by the rhizobia-legumes comprises 14 genera and more than 98 species (Berrada and Fikri-Benbrahim 2014). The N_2 fixing ability of the rhizobia in the nodule should vary with the associated species, nodule shape, growth pattern, and also acidic or alkaline N compounds (Sprent 2009). There are a number of regulatory processes involved in the N_2 fixation in root nodules. the following are some important components to regulate the N_2 fixation; (1) enzymatic regulation of C- and N-metabolism, (2) O_2 availability regulated by a physical barrier and leghemoglobin (Lb), (3) molecular control of nodule number, (4) production of reactive oxygen species and reactive nitrogen species (RNS), and (5) signal mediated regulation of nitrogen-fixing (*nif*) gene expression (Sulieman and Tran 2013). Within the nodules, N_2 is reduced to ammonia that is used to incorporate into the plant component.

The quantity of N_2 fixed by *Rhizobium* varies approximately 50–100 kg N/hectare with the species of legumes under different environmental conditions. Legumes bioinoculated with *Rhizobium* increases the yield up to 35% (Amat et al. 2020). Generally, for the formulation of commercial biofertilizer, Rhizobium is mixed with suitable carriers such as lignite and charcoal peat, or in a liquid-based formulation used with methylcellulose, polyethylene glycol (0.5%), gum arabic, and polyvinyl pyrrolidone (0.5%) (Tittabutr et al. 2007) consist of 10^{10} to 10^{12} cfu/g. It is generally used for economically important legume plants, and the dosage depends on the type of crop, soil type, and environmental conditions (Barman et al. 2017). Table 7.1 represents the importance of Rhizobium inoculums for sustainable agriculture.

7.3.2 *Azotobacter*

Azotobacter is a free-living, nonsymbiotic N_2-fixing bacterium abundant in soil and plays an important role in the nitrogen (N_2) and phosphorous (P) cycle (Velmourougane et al. 2019; Rana et al. 2020). Several studies revealed that the rates of N_2-fixation in soil range from 15 kg/ha/year to 60 kg/ha/year (Romero-Perdomo et al. 2017; Saha et al. 2017). In addition, *Azotobacter* accelerates the mineralization of soil organic residues and also reduces heavy metals (Kizilkaya 2009; Din et al. 2019) and has the ability to produce vitamins. *Azotobacter* contributes ~0.06% to the rhizosphere microbiome (Hassen et al. 2020). Species of *Azotobacter* such as *A. chroococcum* and *A. vinelandii* are most abundant in tropical soil, *A. insignis*, *A. beijerinckii*, and *A. macrocytogenes* are present in a wide range of soil pH, both acidic and alkaline soil (Kennedy et al. 2015). Azotobacter species is used as a suitable biofertilizer for non-leguminous crops such as cotton, rice, vegetables, sorghum, sugarcane, and potato (Ritika and Utpal 2014). Several reports suggested that *Azotobacter* inoculum increases the yield of various crops (Wani and Ali 2013), rapeseed and mustard (Dutta and Singh 2002), sugarcane (Kizilkaya 2009), etc. Moreover, *Azotobacter* species are able to synthesize plant growth hormones such as indole acetic acid (IAA), gibberellins, and cytokinins. Besides, *Azotobacter* strains can also protect host plants indirectly from phytopathogens, prevent seedling mortality, and regulate other beneficial rhizosphere microorganisms (Arora et al. 2018; Ansari and Mahmood 2019a, b). Through the available literature, *Azotobacter* influences overall plant growth, but the exact mechanism is not yet fully understood (Sumbul et al. 2020). In addition, to fix nitrogen, they can solubilize phosphates and also mineralize organic sulfur (El-Badry et al. 2016). The exopolysaccharide of *Azotobacter* sp. solubilizes tricalcium phosphate (Yi et al. 2008).

However, soil organic matter acts as limiting components for *Azotobacter* surveillance, poor growth of Azotobacter was reported in a reduced amount of soil organic matter with other factors such as temperature, moisture content, salinity, and soil pH (Andjelković et al. 2018). *Azotobacter* grows optimally at a pH of 7–7.5, however, *A. chroococcum* survived at pH 9 (Andjelković et al. 2018). Further, the plant growth-promoting activity of *Azotobacter* is mainly affected by soil salinity; but *Azotobacter salinestris* is known to tolerate 8% of NaCl. *Azotobacter* is a mesophilic bacterium that can grow at 25–30 °C, but it can develop cyst under unfavorable temperature (45 °C), that germinate under favorable conditions (Saribay 2003).

Table 7.1 Applications of Rhizobium for sustainable agriculture

Strain	Host plant	Function	References
Rhizobium leguminosarum bv. viceae	*Pisum sativum* L.	Soil inoculum increased pea nodulation under short season conditions	Begum et al. (2001a)
Rhizobium leguminosarum	*Pisum sativum* L. and *Lens culinaris*	Flavonoids induced nod gene expression and preactivated *nod* genes	Begum et al. (2001b)
Rhizobium spp. PchDMS and Pch43	Chickpea	Increased yield and antioxidant activities and induced resistance against *Fusarium oxysporum* f. sp. *ciceris* (Foc)	Arfaoui et al. (2005)
Rhizobium strains BARI RPs-2001 and BARI RPs-2002	*Pisum sativum*	Soil inoculation increases pod weight and length	Ahmed et al. (2007)
Rhizobium sp.	*Lens culinaris* Medik	Increased yield and seed quality	Al-Karaki (2008)
Sinorhizobium meliloti	*Medicago sativa* L.	The tripartite symbiosis of Rhizobium, AM, and Lucerne can improve the performance of Lucerne in organic farming and under dry conditions	Ardakani et al. (2009)
Rhizobium sp.	*Pisum sativum*	Indigenous rhizobia increase the seed quality	Agarwal and Choure (2011)
Rhizobium japonicum	Soybean	Biocontrol agent against soil-borne pathogens *Fusarium solani* and *Macrophomina phaseolina*	Al-Ani et al. (2012)
Mesorhizobium spp.	*Cicer arietinum* L	Increased nutrient uptake and yield of chickpea	Verma et al. (2013)
Rhizobium sp.	*Glycine max* (L.) Merril	Reproductive growth stages on shoot biomass and yield of soybean	Lamptey et al. (2014)
Rhizobium trifolii, *R. phaseoli*, *R. leguminosarum* and *Bradyrhizobium japonicum*	Legume plant	Inoculation of solid and liquid biofertilizers resulted into enhanced plant growth by providing balanced nutrient supply	Datta et al. (2015)
Rhizobium spp.	*Pisum sativum* L.	*Rhizobium* coupled with soil application of organic manures could increase yield in pea	Olle et al. (2016)
Rhizobium sp.	*Lens culinaris* Medik.	Symbiotic efficiency of native and exotic Rhizobium strains nodulating lentil	Tena et al. (2016)
Rhizobium sp.	*Phaseolus vulgaris* L.	Higher plant-growth-promoting effects	Assefa et al. (2017)
Rhizobium leguminosarum bv. *viceae*	*Pisum sativum*	Disease resistance against *Didymella pinodes*	Ranjbar Sistani et al. (2017)

(continued)

Table 7.1 (continued)

Strain	Host plant	Function	References
Rhizobium sp.	*Glycine max* (L) *Merril*	Increased growth and yield	Herliana et al. (2019)
Rhizobium sp. and *Bradyrhizobium* spp.	*Phaseolus vulgaris* L. and *Glycine max* L.	Rhizobial inoculation improved drought tolerance, biomass and grain yields	Aserse et al. (2020)
Rhizobium spp.	*Vicia faba* L.	Combined use of inoculant and chemical fertilizer increased growth and yield	Genetu et al. (2021)

Azotobacter produces siderophores (Ansari et al. 2017), an iron (Fe) chelating molecule, that can use the utilization of extracellular "Fe" molecules, and they act as an important iron resource in the environment (Wichard et al. 2009). For example, *Azotobacter* expresses iron-rich nitrogenases (Baars et al. 2016), through which they can reduce nitrogen (N_2 fixation). The Fe-siderophore complex of *Azotobacter* shows antiphytopathogenic activities and can directly improve plant growth by protecting plants from the pathogens attack (Hayat et al. 2010). In addition, it can use molybdenum (Mo) or vanadium (V) for its nitrogenase activity and absorb heavy metals like W and Zn (Kraepiel et al. 2009; Sumbul et al. 2020). Abo-Amer et al. (2014) demonstrated heavy metals (Co^{2+}, Ni^{2+}, Zn^{2+}, and Cu^{2+}) absorbing efficiency of *Azotobacter* using heavy metal contaminated soil. In addition, *Azotobacter* provided good protection to the plants against pathogenic infections caused by *Rhizoctonia solani* and *Xanthomonas campestris* and also inhibits root-knot nematode *Meloidogyne incognita* (Akram et al. 2016; Sumbul et al. 2020). Besides, *Azotobacter* can produce antibiotic-like substance anisomycin, a fungicidal compound that inhibits the growth of *Alternaria*, *Aspergillus*, *Curvularia*, *Fusarium*, *Helminthosporium*, *Macrophomina*, and *Rhizoctonia* (Jnawali et al. 2015). Further, *Azotobacter* sp. has also been reported to degrade toxic pesticide compounds such as 2-chlorophenol, 4-chlorophenol, 2,6-dichlorophenol, 2,4-6-trichlorophenol, and 2,4-dichlorophenoxyacetic acid (2,4-D) (Gaofeng et al. 2004; Kumar et al. 2016). All such properties of *Azotobacter* sp. could be a promising character for considering a potent biofertilizer for the benefits of crops, a few are highlighted in Table 7.2.

7.3.3 *Azospirillum*

A genus *Azospirillum* is a group of gram-negative, flagellated, slightly curved, microaerophilic (or) facultative anaerobic diazotrophs, nonfermentative, and nitrogen-fixing bacteria from the family of Rhodospirillaceae, the best-characterized PGPR, able to colonize the roots of several plant species. Twenty-one species of *Azospirillum* have been isolated from the rhizosphere of many types of grass and cereals in tropical as well as in temperate climates, namely *Azospirillum brasilense*,

Table 7.2 Effect of *Azotobacter*-based biofertilizers on yields of different crops

Strain	Crop	Effect on yield	References
Azotobacter sp.	Cabbage	Yield increase 13.9%	Sarkar et al. (2010)
Azotobacter sp.	*Capsicum annum* L. and *Pisum sativum* L.	Yield increase 11.4; decrease Fusarium root rot infection	Jaipaul et al. (2011)
Azotobacter	Wheat	Yield increase 14.3%	Milošević et al. (2012)
Azotobacter sp.	Cluster Bean	Yield increases up to 16.6%	Deshmukh et al. (2014)
Azotobacter chroococcum	Potato	Yield increases 62.3%	El-Sayed (2014)
Azotobacter sp.	Broccoli	Yield increase 17.3%	Singh et al. (2014)
Azotobacter sp.	Rice	Yield increase up to 26.9%	Zayadan et al. (2014)
Azotobacter sp.	Cucumber	Yield increase up to 21%	Saeed et al. (2015)
Azotobacter sp.	Chickpea	Yield increase 35.5%	Ansari et al. (2015)
Azotobacter sp.	Tomato	Yield increase 23.8%	Singh et al. (2015a)
Azotobacter sp.	Carrot	Yield increase 34.2%	Sarma et al. (2015)
Azotobacter sp.	Cotton	Yield increase 13%	Romero-Perdomo et al. (2017)

A. amazonense, *A. irakense*, *A. lipoferum*, *A. largimobile*, *A. halopraeferens*, *A. oryzae*, *A. canadensis*, *A. doeberinerens*, *A. melinis* (Peng et al. 2006; Mehnaz et al. 2007; Saharan and Nehra 2011), *A. formosense* (Lin et al. 2012), *A. fermentarium* (Lin et al. 2013), *A. humicireducens* (Zhou et al. 2013), *A. soli* (Lin et al. 2015), *A. agricola* (Lin et al. 2016), *A. griseum* (Yang et al. 2019), *A. palustre* (Tikhonova et al. 2019), *A. ramasamyi* (Anandham et al. 2019), *A. thermophilum* (Zhao et al. 2020), *A. oleiclasticum* (Wu et al. 2021), and *A. baldaniorum* (dos Santos Ferreira et al. 2020). Among them, *A. brasilense* and *A. lipoferum* are well-known species for their physiological and genetic characteristics (Fibach-Paldi et al. 2012). The bacteria have a wide range of growth temperature (5–42 °C) and pH (5.0–9.0) (Steenhoudt and Vanderleyden 2000). The *Azospirillum*–plant root interactions are explained by four different aspects such as habitat, plant root association, N_2 fixation, and biosynthesis of phytohormones.

The beneficial growth-promoting role of *Azospirillum* ensures the biofertilizer applications in various commercial crops. Other than the N_2 fixation, *Azospirillum* relies on the synthesis of phytohormones such as auxins (Spaepen and Vanderleyden 2015), abscisic acid (Cohen et al. 2009), cytokinins (Tien et al. 1979), ethylene (Strzelczyk et al. 1994), gibberellins (Bottini et al. 1989), and salicylic acid (Sahoo et al. 2014). Like *Azotobacter* sp., *Azospirillum* sp. can solubilize inorganic phosphates (Turan et al. 2012) and alleviate several abiotic stresses, such as drought and salinity (Rodríguez-Salazar et al. 2009; García et al. (2017)), and also the heavy metals (Bashan and de-Bashan 2010). The mechanisms involved in this process are quite complex, where *Azospirillum* inoculum under high NaCl enhances K, Ca uptake, and also increase chlorophyll *a* and *b* biosynthesis, and photoprotective pigments such as carotenoids, violaxanthin, zeaxanthin, antheraxanthin, lutein, and neoxanthin resulting in the production of soluble saccharides, proteins and other osmolytes like proline and betaine, which improve salt stress resistance (Mehnaz 2015; Singh et al. 2015b). *Azospirillum* inhibits Na uptake, enhances the uptake of Ca and K, and triggers nitrate reductase and nitrogenase activity in shoot and roots (Hamdia et al. 2004). *Azospirillum* improves drought resistance in plants, through the induction of jasmonic acid (JA) and salicylic acid (SA), which improves the tolerance to drought and salinity stress (Kaushal and Wani 2016; Fukami et al. 2018). Further, *Azospirillum* lowers the synthesis of ethylene, which is related to senescence processes induced by stress conditions (Kumari et al. 2018), and adjustment of water loss by stomatal closure control (Kaushal and Wani 2016). *Azospirillum* develops biofilm on the root surface (Molina-Favero et al. 2008), which is another possible way to resist stress responses. In addition, *Azospirillum* can reduce the biotic stress against the plant pathogens through a siderophore-induced systemic resistance mechanism (Tortora et al. 2011) or cause alteration of host physiology to induce several secondary metabolite synthesis that mediated resistance against pathogens (Bashan and De-Bashan 2010). Some important biofertilizer applications of *Azospirillum* sp. are listed in Table 7.3.

7.3.4 Actinorhizal (Frankia–Plants) Interaction

Actinorhizae are a symbiotic association between the actinomycete genus *Frankia* and various dicotyledonous plant families including Betulaceae, *Casuarinaceae*, *Coriariaceae*, *Datiscaceae*, Elaeagnaceae, Myricaceae, Rhamnaceae, and Rosaceae (Diagne et al. 2013). In this association, Frankia develops nodules on actinorhizal plants (morphologically distinct from legume-rhizobia nodules), able to fix nitrogen up to 300 N_2 kg/hectare/year (Shantharam and Mattoo 1997). Frankia is a gram-positive, filamentous free-living, and nitrogen-fixing actinobacterium that usually forms a symbiotic association with actinorhizal plants or in soil (Fig. 7.3a). *Frankia* forms hyphae and multilocular sporangia which are located on hyphae either terminal or intercalary (Obertello et al. 2003). The morphology of the *Frankia* in nodule varies according to the host plant, and it forms the nodule in a perennial root, wherein bacteria are hosted and nitrogen is fixed (Santi et al. 2013). In the genus *Frankia*, 11 species such as *Frankia alni*, *F. asymbiotica*, *F. canadensis*, *F. casuarinae*, *F. coriariae*, *F. discariae*, *F. elaeagni*, *F. inefficax*, *F. irregularis*, *F. saprophytica*, and *F. Torreyi* (Nouioui et al. 2016, 2019) are able to make two types of an actinorhizal nodule; they are (1) intracellular and (2) extracellular

Table 7.3 Use of *Azospirillum* sp. for sustainable agriculture

Strain	Host plant	Function	References
Azospirillum brasilense ssp. 248	Triticum aestivum	Improved nitrogen content and counteract the effects of salinity	Alamri and Mostafa (2009)
Azospirillum brasilense	*Triticum aestivum*	Increasing grain yield	Ardakani and Mafakheri (2011)
Azospirillum brasilense	*Capsicum annuum*	Ameliorate the deleterious effect of NaCl in a sweet pepper	Amor and Cuadra-Crespo (2012)
Azospirillum brasilense sp. 245	*Lactuca sativa* L. cv. Crimor INTA	Overcome the negative effects of aging on lettuce seeds	Carrozzi et al. (2012)
Azospirillum lipoferum	*Zea mays* L.	Increased shoot and root fresh and dry weight; shoot and root length	Bano et al. (2013)
Azospirillum brasilense sp. 245	*Lactuca sativa* L.	Seed inoculation improved freshly product quality but also extend storage life in lettuce grown under salt stress	Fasciglione et al. (2015)
Azospirillum brasilense sp. 245	*Arabidopsis thaliana*	Improved plants seed yield, plants survival, proline levels, and relative leaf water content under drought	Cohen et al. (2015)
Azospirillum brasilense	Wheat	Increased antioxidant activity	Méndez-Gómez et al. (2015)
Azospirillum brasilense	*Urochloa ruziziensis*	Makes the plant more efficient at removing reactive oxygen species and protecting chlorophyll *a*	Bulegon et al. (2016)
Azospirillum brasilense	*Zea mays* and *Triticum aestivum*	Seed inoculation improves plant growth and yield	Fukami et al. (2016)
Azospirillum brasilense + *Bradyrhizobium*	*Glycine max* (L.)	Coinoculation stimulates an early nodulation and maintenance of nodule under drought; increased the grain yield	Cerezini et al. (2016)
A. brasilense Az39	*Glycine max*	Microbial IAA increases plant growth	Puente et al. (2017)
Azospirillum brasilense	*Trifolium repens*	Salt stress resistance	Khalid et al. (2017)
Azospirillum sp.	*Zea mays*	Increased tolerance to osmotic and salt stress	García et al. (2017)
Azospirillum brasilense	*Zea mays*	Seed or leaf spray inoculum increases plant growth through phytohormones and eliciting genes related to plant-stress	Fukami et al. (2017)

(continued)

Table 7.3 (continued)

Strain	Host plant	Function	References
		tolerance and defense against pathogens	
Azospirillum brasilense	*Zea mays*	Coculture-effective strategy to mitigate salinity stress	Fukami et al. (2018)
Azospirillum brasilense	*Fragaria ananassa*	Activation of systemic acquired resistance against pathogens	Elías et al. (2018)

infection (Pawlowski and Demchenko 2012). In this actinorhizal endosymbiotic nodule, nitrogen is fixed and can be transported to the host, and reduced carbon from the plant can be transferred to the Frankial partner (Franche and Bogusz 2012; Gonzalez et al. 2015). However, little is known about the plant growth-promoting (PGP) properties of *Frankia*, some of these strains have been found to solubilize inorganic phosphate, and synthesize plant hormones, and siderophores (Tisa et al. 2016).

Morphology of actinorhizal nodule is varied in different host species, septate, pear-shaped without stalk, or septate, elongated, or club-shaped (Berg 1994). The symbiotic interaction starts with Frankial infection at the surface of root hair. The intracellular infection starts with the deformation of root hairs with a signaling molecule produced by Frankia. With Rhizobium infection, actinorhizal hyphae are tangled by curled root hairs and few can enter into the infection site of folding, then it invaded to develop infection thread formation. Within the structure, Frankial filaments are encapsulated in plant cell-wall-based components. Frankial invasion triggers cell divisions in the root cortex near to the infected root hair, forming a mitotically active zone called the pre-nodule, penetrating the thin wall of the recently expanded cortical cells, and enlarging wherein N_2 can fix. During the infection process, when the infected cell matures, the tips of the hyphae in due course differentiate vesicles that will fix nitrogen (Newcomb and Wood 1987). Nitrogenase, which is an oxygen-labile enzyme, is exclusively expressed in these vesicles, and fixes the N_2. Moreover, the periderm of the nodule acts as an oxygen barrier, wherein the bacterial hemoglobin could also support shuttle oxygen to the sites of respiration (Pawlowski et al. 2007). Figure 7.5 illustrates the structure and physiology of the *Frankia* nodule.

Actinorhizal symbiosis generally gives resistance to plants against abiotic stresses. Several reports suggested the impact of Actinorhizal nodule on dicotyledons. Frankia is highly tolerant to salinity, hence it can be used for rehabilitation of soil affected by salt, heavy metal, oil, etc. (Diagne et al. 2013). For example, the symbiotic association between *C. equisetifolia* and *Frankia* can be widely used for the recovery of saline soil (Ngom et al. 2016). The effect of bio-inoculation of isolated *Frankia* sp. strains has been demonstrated to improve drought stress tolerance in tomato (*Solanum lycopersicum* L. cv. Pusa ruby) (Mohan Gupta et al. 2020), which give an idea to use *Frankia* for sustainable agriculture.

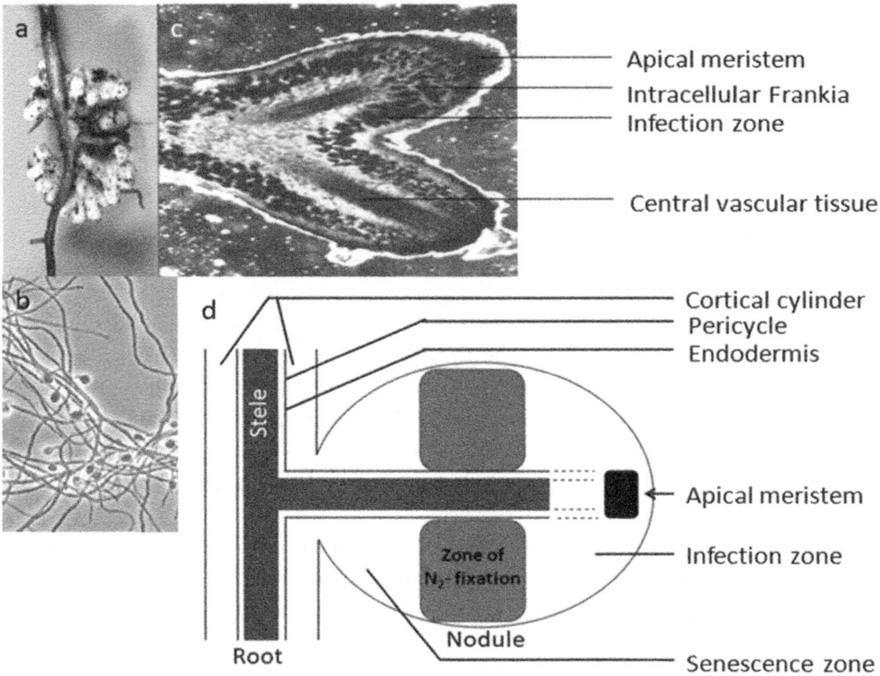

Fig. 7.5 Actinorhizal nodule. (**a**) Frankia nodule; (**b**) microscopic image of Frankial hyphae and spore; (**c**) cross-section of young nodule; (**d**) schematic drawing of a mature actinorhizal nodule lobe. (Adapted from Rascio and La Rocca 2008. Biological Nitrogen Fixation. Encyclopedia of Ecology)

7.3.5 Mycorrhizal Interaction

Fungi are the second most diverse and abundant microbe in soil, which could exploit very diverse substrates for their nutritional purpose, especially they do saprophytic mode of life. A few fungal groups establish the mutualistic symbiont with its associated host. Mycorrhizal fungi are a heterogeneous group of fungi that can carry out their life cycle as free-living and also associate with the roots of higher plants (over 90% of plant species). In mycorrhizal association both partners are benefited, fungal partner improves the nutrient status of the host plant especially for the mineral nutrients, water absorption, disease resistance, and growth, whereas the host plant provides space and nutrients for the fungal growth and propagation (Martin et al. 2008; Varma et al. 2017a, b, c). The fungal partners do an essential role in mineral recycling through their mycelium, absorb soil nutrients, and feed them to the plant. Moreover, fungal partners develop an extensive hyphal network in the soil named web which can effectively connect the host plant and supply the nutrients efficiently. Species of three major groups of fungi such as Ascomycota,

Basidiomycota (septate), and Glomeromycota (aseptate) develop either ecto- or endo–mycorrhizas (Javot et al. 2007a, b).

7.3.5.1 Ectomycorrhizae

Ectomycorrhizas occur mainly in certain woody plants of gymnosperms and angiosperms, and it contributes 30% of the total microbial biomass in forest soil (Johnson and Gehring 2007). More than 6000 species of ascomycetes (*Cenococcum geophilum*, *Tuber borchii*, *Scleroderma hypogaeum*), basidiomycetes (*Amanita muscaria*, *Hebeloma cylindrosporum*, *Laccaria bicolor*, *Paxillus involutus*, *Pisolithus tinctorius*, *Suillus bovinus*, *Xerocomus badius*), and zygomycetes take part in this ectomycorrhizal (EM) assemblage, where they colonize roots of trees, such as pine (Pinaceae) and beech (Fagaceae) of the forest ecosystem. These fungi form symbioses with lateral roots of plants and create a specialized characteristic structure called "Hartig net" (labyrinthine structure), a network of inward-growing hyphae that extends into the root penetrating between the epidermis and cortex cells of its host plants. Due to this association, fungi can induce altered root architecture including root hair modification or inhibition (Fig. 7.6a). The EM fungal hyphae occasionally join together into macroscopic structures called sporocarps, similar to the xylem, which serve in water uptake (Smith and Read 1997; Agerer 2001). EM fungi are generally carried out saprotrophic and symbiotic modes of life. During saprophytic life, EM fungi release many extracellular hydrolytic enzymes that can use organic nutrient sources. EM fungi have an instance loss of plant cell wall degrading enzymes, hence it restricts penetrating the intracellular space of plant root (Martin et al. 2008).

7.3.5.2 Endomycorrhizae

Endomycorrhizas are further divided into orchid, ericoid, and arbuscular mycorrhizas (AMs). The arbuscular mycorrhizal fungi are the most abundant of the endomycorrhiza (Smith and Read 2008), they are obligate biotrophic symbionts the hyphae penetrate the cells of the inner root cortex to form specialized branched structures called "arbuscules" (Javot et al. 2007a, b), a specific nutrient transfer interface of the symbiosis (Genre et al. 2005, 2008). However, arbuscules do not penetrate the protoplast of plant cells, which can develop a new and distinct periarbuscular membrane (PAM) continuous with the plasma membrane of the cortical cell (Genre et al. 2005) (Fig. 7.6b). Several AM fungi form vesicles as lipid-rich storage organelle, called vesicular-arbuscular mycorrhiza (VAM). AM fungi are classified under the separate phylum *Glomeromycota* consisting of four genera, *Acaulospora*, *Gigaspora*, *Glomus*, and *Sclerocystis*. In the case of Orchid and Ericoid mycorrhiza, they can specifically make symbiotic colonization in Ericales and Orchidaceae respectively. In orchids, the flow of carbon is from the

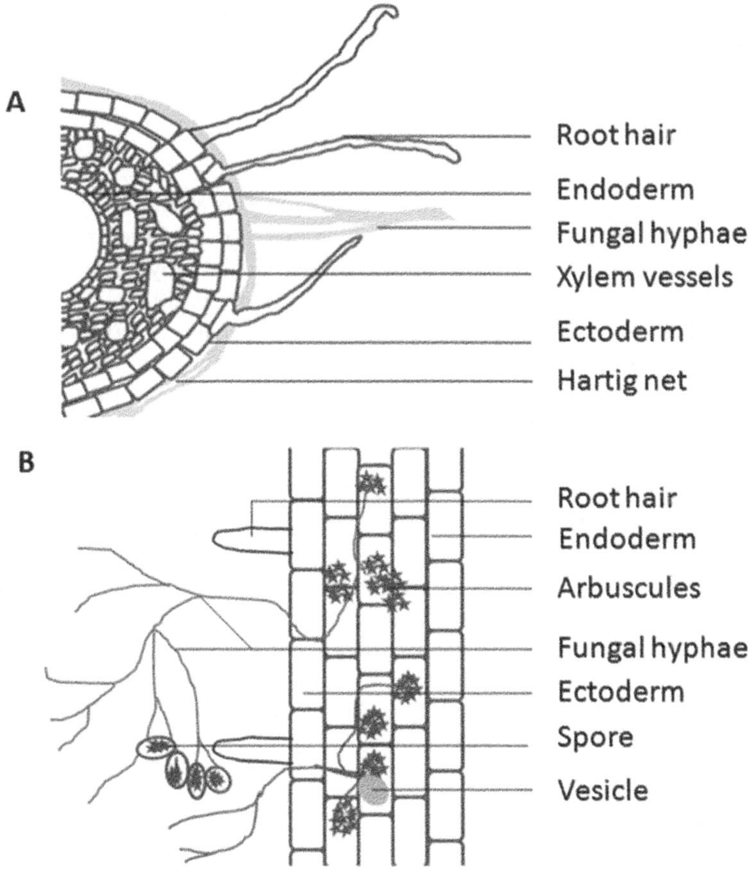

Fig. 7.6 Schematic diagram of the main structural features of ectomycorrhiza (**a**) and AM mycorrhiza (**b**). (Adapted from, Marschner 2012, Rhizosphere Biology, In: Marschner's Mineral Nutrition of Higher Plants)

fungus to the host which is entirely different from the other mycorrhiza, this is essential for orchid seed germination (Smith and Read 2008).

7.3.5.3 Fungal Endophytes

Endophytic fungi are a group of fungi that live in almost all plant species to maintain symbiotic life without causing symptoms of disease (Gazis and Chaverri 2010; Porras-Alfaro and Bayman 2011). Most endosymbiotic fungi live at least part of their life cycle away from the plant (Rodriguez et al. 2009). The members of *Ascomycota*, *Basidiomycota*, and *Mucoromycota* are the major group of endosymbionts of plants, including *Moeniiplasma glomeromycotorum*, *Rhizopus microspores*, *Mortierella elongate*, *Mycoavidus cysteinexigens*, *Piriformospora indica*,

Paracoinothyrium variabile (Bärenstrauch et al. 2020). However, fungi such as *Cladosporium, Curvularia, Colletotrichum, Cordana, Deightoniella, Fusarium, Guignardia, Nigrospora, Periconiella, Phoma*, and *Verticillium* are identified as pathogenic endophytes (Photita et al. 2005; Cui et al. 2021). The endosymbiotic colonization as well as the nutrient transfer mechanisms of such beneficial fungi is not clearly understood when compared with mycorrhizal fungi (Saikkonen 2007; Behie et al. 2012; Behie and Bidochka 2014).

In mutualistic symbiosis, both partners of endosymbionts benefit from this association (Jia et al. 2016), where host plants provide shelters and other proliferation facilities to fungi, and fungi can alter the metabolic activities of host, such as enhance growth, promote nutrient acquisition, and improve metal and drought tolerance (Mejia et al. 2014; Poveda et al. 2021). In addition, they also enhance the defense efficiency of host plants against pests as well as pathogenic microorganisms (Cui et al. 2021).

7.3.6 Benefits of Fungal–Plant Interactions

Beneficial plant-associated fungi are broadly distributed and lead to benefits for the plant nitrogen transfer, phosphate mobilization, drought resistance, salt stress, and also the alleviator of biotic stresses. The major benefit of mycorrhizal association to the plant is listed in Fig. 7.7.

7.3.6.1 Soil Health

Healthy soil is more productive; in sustainable agriculture management of soil is an important event that increases the crop productivity. Healthy soil is nutrient rich and abundant in microbial diversity with good texture and aeration. Soil is rich in fungal diversity, especially mycorrhizal fungi form a complex networking between plant and soil, which provide nutrient transport, decomposition of organic residues in soil, regulation of microbial community, increase water holding ability, and aeration of soil (Fig. 7.5). Arbuscular mycorrhizal (AM) fungi are the important microorganism in agricultural soil, and they can significantly increase crop productivity (Bagyaraj and Ashwin 2017; Frąc et al. 2018). The positive effect of fungi in the soil is to maintain soil quality and plant health along with other PGPRs from soil microbiome. However, a human activity heavily affects soil microbiome and affects the soil texture, aeration, and water-holding capability. In grassland soil, basidiomycetes dominate and actively involve in the decomposition and mineral recycling process (Cassman et al. 2016; Yang et al. 2017). In a forest, ectomycorrhizal associations strongly regulate soil texture and health. Moreover, soil organic matter could regulate the diversity and abundance of fungi and other microorganisms. They are the major energy source of soil microorganisms and maintain the nutritional quality of soil (Jiménez-Morillo et al. 2016).

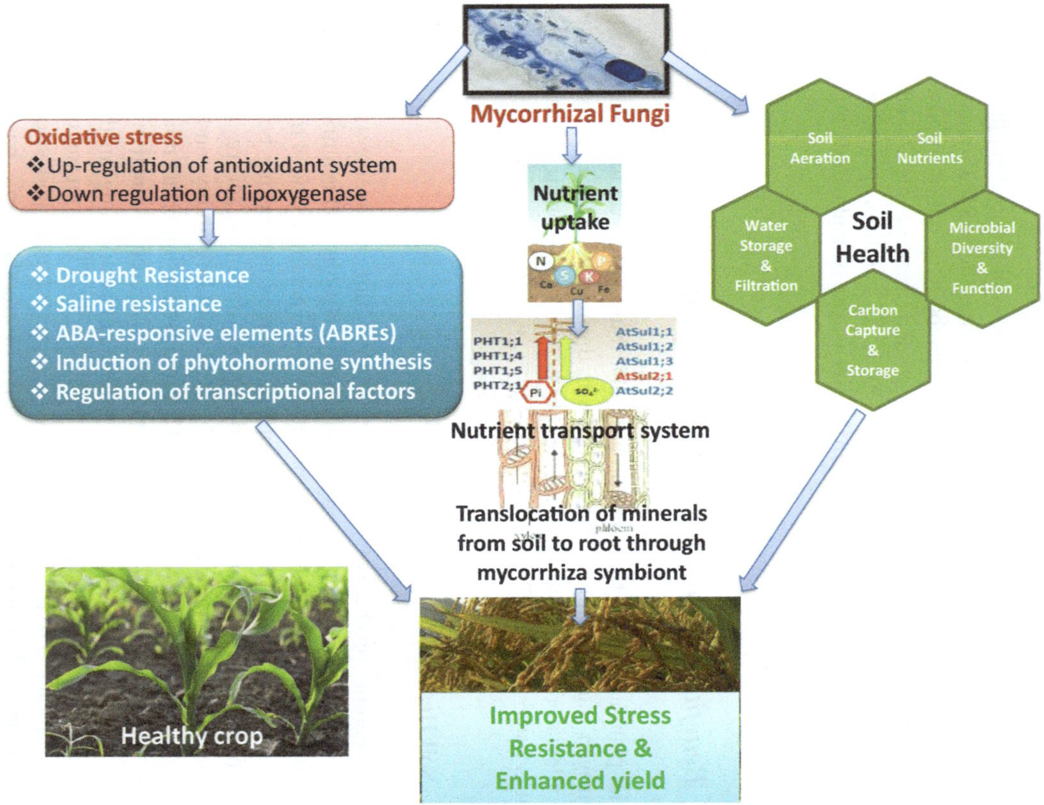

Fig. 7.7 A diagrammatic representation of mycorrhizal functions to regulate various processes in the ecosystem and plant growth promotion under abiotic stress conditions

7.3.6.2 Nitrogen Uptake

Nitrogen is the most important nutrient needed for the primary productivity of the plant, which is essential for plant growth, health, and yield. Generally, roots absorb nitrogen in their inorganic forms such as nitrate (NO_3^-) and ammonium (NH_4^+) ions. In the atmosphere, a large amount of nitrogen exist, some can be readily incorporated by soil containing nitrogen-fixing organisms, and some quantity may be immobilized in organic forms like peptides, proteins, and amino acids. Through the fungal decomposition of organic substances, nitrogen can be mineralized and available for plant nutrition. Several soil fungi such as mycorrhizal fungi break down the organic nitrogen by their extracellular proteases for their energy requirements and provide nitrogen reserve to their host plant (Sinsabaugh 2010). Earlier studies have reported that approximately 20–75% of the total N uptake of AM plants can be transferred through the AM fungi to their hosts (Ahanger et al. 2014; Hashem et al. 2018a). For example, AM fungi and ectomycorrhizal fungi release proteases and peptidases to hydrolyze polymers to simple monomers like nitrates (NO_3) or ammonium (NH_4^+), then absorbed by the extraradical mycelia and further synthesis of amino acids (arginine) into intracellular mycelia. After its migration, the amino acid arginine is broken down into ammonia via the urea cycle and transported to the plant root (Ngwene et al. 2013). This mechanism helps to absorb a large amount of NH_4^+ and NO_3 via the fungal hyphae (Mian et al. 2009). This kind of nutrient transport is mediated with a specialized cell structure developed by arbuscular mycorrhizae-root in the symbiosis called PAM, which contains AMT family of ammonium transporters (Kobae et al. 2010; López-Pedrosa et al. 2006) and high-affinity nitrate transporters (NRT) (Kemppainen and Pardo 2013). Several studies highlighted the impact of AM fungi on the ability to absorb and transfer N to the nearby plants or host plants (Zhu et al. 2016; Battini et al. 2017; Turrini et al. 2018; Wang et al. 2018a, b; Liu et al. 2018a, b). Few examples are listed in Table 7.4.

7.3.6.3 Phosphate Transfer

Phosphorous is one of the growth-limiting nutrients for plants. Plants are unable to use insoluble rock or organic phosphates from the soil (Ruttenberg 2003). In nature, weathering process releases phosphate ions, orthophosphates, $H_2PO_4^-$, and HPO_4^{2-} are the primary forms of phosphorus taken up by plants and or root colonized mycorrhiza (Yang et al. 2012). Fungal mycelium directly absorbs phosphates by a translocation process and is transported into the cortical cells of the root in the form of inorganic phosphates (Smith and Smith 2011). In addition, mycorrhiza secretes enzymes such as phosphomonoesterases and phosphatases, they hydrolyze organic P compounds in soil, and through this mechanism, phosphate can be a cycle in plant nutrition. Unlike nitrogen transport, the absorption and transport of phosphate are also mediated with a number of phosphate transporters in both mycorrhizal fungi and plants, which are located in PAM of arbuscular mycorrhizal symbiosis (Benedetto

Table 7.4 Mycorrhizal transporters in nutrient transport to plant

Nutrients	Transporter	Mycorrhizal species	References
Nitrogen	GintAMT1, GintAMT2, GintAMT3	*Glomus intraradices*	López-Pedrosa et al. (2006), Pérez-Tienda et al. (2011), Calabrese et al. (2016)
	RiPTR2	*Rhizophagus irregularis*	Belmondo et al. (2014)
Phosphorus	GiPT	*Glomus intraradices*	Maldonado-Mendoza et al. (2001)
	GmosPT	*Glomus mosseae*	Balestrini et al. (2007)
	GigmPT	*Gigaspora margarita*	Xie et al. (2016)
	GiPT	*Rhizophagus irregularis*	Yang et al. (2012)
Zinc	GintZnT1	*Glomus intraradices*	González-Guerrero et al. (2005)
	GintZnT1	*Rhizophagus irregularis*	
Sugar	GpMST1	Glomeromycota	Schüßler et al. (2007)
	RiMST2	*Glomus* sp.	Helber et al. (2011)
	RiMST2	*Rhizophagus irregularis*	Helber et al. (2011)
	RiMST5		Ait Lahmidi et al. (2016)
Arsenic	RiArsAB	*Rhizophagus irregularis*	Maldonado-Mendoza and Harrison (2018)
	RiMT-11		González-Chávez et al. (2014)
Iron	*RiFRE1* *RiFTR1-2*	*Rhizophagus irregularis*	Tamayo et al. (2014)

et al. 2005). Several studies reported that the expression of transporters in mycorrhizal fungi is high at symbiotic association and also under low phosphorous conditions (Krajinski et al. 2014; Wang et al. 2014a, b). AM fungi inoculation improved P contents in plant tissues (Jixiang et al. 2017; Garcés-Ruiz et al. 2017; Wang et al. 2018a, b; Liu et al. 2018a, b; Chhabra and Dowling 2017; Chhabra 2019) by the expression of phosphate transporters during the mycorrhizal association in plants, for example, *Piriformospora indica* and *H. cylindrosporum* association in plant root under low soil phosphate conditions (Tatry et al. 2009; Yadav et al. 2010; Gill et al. 2016).

7.3.6.4 Other Soil Nutrients Transport

Along with nitrogen and phosphate transfer, AM and ectomycorrhizal fungi are able to transfer potassium, sulfur, zinc, copper, and cesium (Cartmill et al. 2008; Nord and Lynch 2009). The above nutrients are generally rich in most soil; however, the plant cannot easily access these elementary nutrients in some soil. Like phosphate mineralization, AM fungi and other mycorrhizal fungi solubilize and transfer such soil nutrients to their host plant. Several gene expression studies disclosed the efficiency of mineral uptake by the AM fungi, for example, the upregulation of sulfate transporters in *M. truncatula* is due to fungal colonization (Casieri et al. 2012). Sulfur uptake in plant roots is directly connected with the phosphate content of the soil, whereas low soil phosphate with fungal colonization increases both phosphate and sulfate uptake (Garcia et al. 2014). Moreover, various reports suggested that mycorrhizae have the ability to transfer both macro and micronutrients such as N, P, K, Ca, Mg, Fe, Cu, and Zn from soil (Asrar et al. 2012; Berruti et al. 2016; Bati et al. 2015; Zaefarian et al. 2013; Ali et al. 2015). Mycorrhizae can mobilize such minerals at a high level under stress conditions (Asrar et al. 2012), for example, under salinity stress increases P, Ca, and K transport in *Euonymus japonica* (Gómez-Bellot et al. 2015), and under drought stress P and Fe transport in *Pelargonium graveolens* L. (Amiri et al. 2017). Under salt stress conditions AM fungi-treated *Cucumis sativus* plants have absorbed a high amount of P, Ca^{2+}, N, Mg^{2+}, and K^{+} (Hashem et al. 2018b). Few examples are listed in Table 7.4. However, little information is available on the ability of endophytes to transfer soil nutrients (Usuki and Narisawa 2007). Usually, fungal endophytes make asymptomatic interactions with the plant without any beneficial nutrient transport (Behie and Bidochka 2014).

7.3.6.5 Mutual Exchange of Minerals

AM fungi develop symbiosis with roots to obtain essential nutrients from the host plant and reciprocally provide mineral nutrients, for example, N, P, K, Ca, Zn, and S. Normally, readily available free carbon in the soil ecosystem is a limited one. Primary carbon metabolites of plants fixed through photosynthesis are often simply metabolized by fungi, which may be the reason fungi are able to colonize on the root to reciprocally exchange the minerals (Smith et al. 2009). For example, the mycorrhizal fungi *Glomus intraradices* transfers phosphorus and nitrogen to the roots of wild carrot (*Daucus carota*) in a reciprocal exchange of carbon (Liao et al. 2013; Fang and St Leger 2010). In contrast, the decrease in available carbon delays the overall rate of phosphate transfer, which indicates the need for the mechanism of fungal interaction with plant roots. Moreover, different abiotic conditions can also influence the nutrient exchange in between mycorrhizal symbiosis (Hoysted et al. 2021), where the atmospheric (CO_2) influences the carbon-nutrient exchanges between Mucoromycotina “fine root endophyte” (MFRE) fungi and vascular plant. As a result, reciprocal nutrient exchange stabilizes the cooperative relationship between the symbiotic partners because of the nutritionally mutualistic relation (Field et al. 2015; Orchard et al. 2017). Several fungal associations on plant roots essentially transfer nonlimiting nutrients from the soil through the reciprocal exchange of ions and nutrients. In AM fungi, *Glomus* spp., the amount of carbon received from the symbiotic association is directly related to the phosphate transfer

(Bücking and Shachar-Hill 2005). However, endophytes take the photosynthetic carbon from the plant without reciprocal exchange of nutrients. Very limited information is available in the nonmycorrhizal fungal symbionts nutrient exchange process (Behie and Bidochka 2014). These findings offer new insights into the regulatory mechanism of mineral nutrient uptake by host plants from AM fungi.

7.3.6.6 Drought Resistance

Drought stress is multidimensional stress that can affect plant life and causes changes in the morphological, physiological, biochemical, and molecular properties of plants. Shortage of water induces oxidative stress (Impa et al. 2012; Hasanuzzaman et al. 2013), which leads to giving deleterious effects on plant growth by affecting enzyme activity, nutrient uptake, and assimilation (Ahanger and Agarwal 2016; Ahanger et al. 2017a, b). Numerous studies reported that mycorrhizal association alleviates drought stress in several crop plants such as barley, maize, onion, soybean, strawberry, and wheat (Yooyongwech et al. 2016; Moradtalab et al. 2019). One such mechanism of drought resistance is enlarging the root surface by the association of mycorrhizal fungi. Such a kind of symbiotic association regulates a variety of physiological processes including osmotic regulation, increase proline and glutathione levels, and stomatal conductance, and decreases transpiration rate which plants get resistant to drought (Zhang et al. 2016; Rani 2016; Li et al. 2019). Table 7.5 contains some representative evidence for fungal-induced drought resistance in plants.

7.3.6.7 Salinity Stress Tolerance

Salts occur naturally within soils and water, salinization is the process of increasing the salt content in the soil, which is an increasing environmental problem that creates a cruel risk to global food safety. Salinity induces the generation of high-level expression of reactive oxygen species that affects plant growth and productivity (Ahanger et al. 2017a, b, 2018). However, mycorrhizal association mitigates salinity-associated adverse effects on plants (Santander et al. 2019) and increases crop productivity (Elhindi et al. 2017; Ait-El-Mokhtar et al. 2019) through the modification of plant physiology and biochemical activities such as photosynthetic rate, stomatal conductance, leaf water relations, and water use efficiency (El-Nashar 2017; Wang et al. 2018b) in *Ocimum basilicum* L. (Elhindi et al. 2017) and *Allium sativum* (Borde et al. 2010).

In addition, support with mineral nutrient transport, mycorrhizal symbionts enhance the synthesis of some plant metabolites, for example, jasmonic acid and salicylic acid. Under salt stress, AM fungi inoculated plants showed higher resistance to saline stress and improved nutrient transport, which may be due to enhanced mineral absorption by the mycorrhizal symbiont (Hashem et al. 2018b), for example, AM inoculation increases chlorophyll contents and regulate Na^+ transport with

Table 7.5 Mycorrhizal association on plant drought resistance

Fungus species	Plant	Responses	References
Glomus deserticola	*Antirrhinum majus* L.	Increased shoot and root length, leaf number, leaf area, water content, Chl content, and proline	Asrar et al. (2012)
Funneliformis mosseae, *F. geosporus*	*Fragaria ananassa*	Increased shoot and root biomass, water use efficiency (WUE), and plant survival	Boyer et al. (2015)
F. mosseae and *Rhizophagus intraradices*	*Robinia pseudoacacia* L.	Increased dry biomass, WUE, and photosynthesis	Yang et al. (2014)
R. irregularis, *G. intraradices*	*Lettuce and tomato*	Increased biomass, strigolactone production, and ABA accumulation	Ruiz-Lozano et al. (2015)
R. intraradices, *F. mosseae*	*Pelargonium graveolens*	Increase plant biomass, and essential oil content	Amiri et al. (2015)
Septoglomusconstrictum, *Glomus* spp. *G. aggregatum*	*Glycine max*	Improved mineral (N &P) uptake and water content	Grümberg et al. (2014)
G. intraradices, *Gigaspora gregaria*, *Scutellospora gregaria*	*Vigna subterranea*	Increased mineral absorption, soluble sugars, and acid but reduced proline content	Tsoata et al. (2015)
G. intraradices	*Hordeum vulgare*	Increased root volume and P content	Bayani et al. (2015)
R. intraradices	*Zea mays*	Increased uptake of P, N, K, and Mg, plant dry weight and WUE	Zhao et al. (2015)
G. mosseae, *G. fasciculatum*, *Gigaspora decipiens*	*Triticum aestivum* L.	Increased plant growth and total chlorophyll content	Pal and Pandey (2016)
R. intraradices	*Triticum durum*	Increased biomass, mineral nutrition (Cu, Fe, Mn, & Zn), and gliadins in grains	Goicoechea et al. (2016)
Glomus spp.	*Ipomoea batatas*	Proline and soluble sugars adjust osmotic potential	Yooyongwech et al. (2017)
F. mosseae, *Paraglomus occultum*	*Poncirus trifoliata*	Increased mineral absorption, and leaf water potential	Zhang et al. (2018a)
R. intraradices, *Claroideoglomusclaroideum*, *F. mosseae*	*Hordeum vulgare*, *Hordeum vulgare* ssp. *spontaneum*	Increase root biomass alleviation of drought stress	Sendek et al. (2019)

(continued)

Table 7.5 (continued)

Fungus species	Plant	Responses	References
AMF	*Triticum aestivum* L.	Increased phosphorus uptake, pigment biosynthesis, and accumulation of plant metabolites	Metwally et al. (2019)
R. irregularis	*Malus hupehensis*	Enhanced drought by regulating genes in the MAPK pathway	Huang et al. (2020)
R. irregularis	*Zea mays*	Improved photosystem II activities, biomass, and drought tolerance in seedlings	Hu et al. (2020)
Glomus spp.	*Saccharum arundinaceum* Retz.	Increased water uptake, phenolics, ascorbic acid, glutathione, antioxidant enzymes, and plant biomass	Mirshad and Puthur (2017)

support of Mg^{2+} and N nutrients in *Capsicum annuum* (Çekiç et al. 2012), and higher biomass production, increased synthesis of proline, increased N uptake in lettuce (Santander et al. 2019). In addition, the mycorrhizal fungi can improve some organic acid synthesis in plants under saline stress, which increases betaine production and substantiates osmoregulation (Begum et al. 2019) in maize plants (Sheng et al. 2011). Table 7.6 presents few examples of mycorrhizal association on the adaptation of plant under saline stress.

7.3.6.8 Heavy Metal(s) Tolerance

Heavy metals are elements that generally produce common toxic effects on plants, such as low biomass productivity, inhibition of growth and photosynthesis, chlorosis, altered water balance and nutrient assimilation, and senescence, which ultimately cause plant death (Singh et al. 2016; Moghadam et al. 2016). Some heavy metals (Co, Cu, Fe, Mn, Mo, Ni, V, and Zn) may require trace amounts for the normal functioning of cells, when their concentration is excess it can become harmful. Moreover, excess of such heavy metals affects soil properties, in particular, soil pH, water-holding ability, mineral properties, soil aeration, and also microbial diversity. When plants grow in such soil, growth and yield may be severally affected, and accumulate toxic heavy materials in its root, leaf, fruits, and seeds (Yousaf et al. 2016). However, mycorrhizal association specifically, AM fungi reduces the toxic effects of such heavy metals and inhibits their transport to the plant. Heavy metal toxicity is typically reduced by chelation through immobilizing the metals in their hyphae or hyphal mantle and storing them in vacuoles. Hence, the concentration of

Table 7.6 Effect of mycorrhizal association on the adaptation of plant under saline and heavy metal stresses

Fungus species	Plant	Responses	References
Glomus fasciculate	*Acacia nilotica*	Improved root and shoot biomass as well as P, Zn, and Cu contents	Giri et al. (2007)
Glomus intraradices	*Solanum lycopersicum* L.	Improved dry matter, ion uptake, growth parameters, and chlorophyll content	Hajiboland et al. (2010)
Glomus mosseae and *Paraglomus occultum*	*Citrus tangerine*	Higher antioxidant defense systems response to salinity	Wu et al. (2010)
Glomus intraradices	*Trigonella foenumgraecum*	Better plant growth, lower leaf senescence, and decreased lipid peroxidation, higher osmolyte (glycine betaine, sugars), and polyamines concentration, and more and bigger plastoglobules formation	Evelin et al. (2012, 2013)
Glomus viscosum	*Medicago sativa*	Reduce salt toxicity, maintain the yield attributes, and regulate anthocyanins enrichment in the pericarp of grains	Campanelli et al. (2013)
G. mosseae	*Kosteletzkya virginica*	Increased mineral nutrition and antioxidant enzymes	Zhang et al. (2014)
Claroideoglomus etunicatum	*Oryza sativa* (L)	Improved quantum yield of PSII photochemistry, net photosynthetic rate, stomatal conductance	Porcel et al. (2015)
Claroideoglomus etunicatum	*Aeluropus littoralis*	Increased shoot and root dry mass, stomatal conductance, soluble sugars, free α-amino acids, and Na+ and K^+uptake	Hajiboland et al. (2015)
Claroideoglomus etunicatum	*Oryza sativa*	Root and shoot P, Na, and K content and expression of genes with a role in the uptake, transport, or compartmentation of Na/K	Porcel et al. (2016)
Funneliformis mosseae	*Cicer arietinum*	Increased antioxidant enzymes and modulated ascorbate-glutathione (ASA-GSH) cycle	Garg and Bhandari (2016)
Rhizophagus irregularis	*Solanum lycopersicum* L.	Enhanced shoot fresh weight (FW), leaf area, leaf number, root FW, and levels of growth hormones	Khalloufi et al. (2017)
G. mosseae	*Leymus chinensis*	Increased colonization rate, seedling weight, water contents, and both P and N	Jixiang et al. (2017)
Glomus etunicatum, *Glomus intraradices*, *Glomus mosseae*	*Cucumis sativus* L.	Increased biomass, photosynthetic pigment synthesis, and enhanced antioxidant enzymes	Hashem et al. (2018b)
R. irregularis	*Elaeagnus angustifolia*	Higher activities of superoxide dismutase (SOD), catalase (CAT), and ascorbate peroxidase (APX) in the leaves of the mycorrhizal seedlings in response to salinity	Chang et al. (2018)

(continued)

Table 7.6 (continued)

Fungus species	Plant	Responses	References
R. intraradices	*Cicer arietinum* L.	Improving root biomass, root to shoot ratio, and nutrient acquisition, maintaining ion equilibrium, and modulating carbohydrate metabolism and reproductive yield	Garg and Bharti (2018)

heavy metals is diluted to the point of being inaccessible to the plants (Punamiya et al. 2010; Garg and Chandel 2011; Kapoor et al. 2013; Audet 2014). In addition, at the high concentration of heavy metals in soil, mycorrhizae increase their morphology and biomass as well as uptake of maximum quantity of heavy metals and fix their cells, hence it reduces their toxicity to plants (Miransari 2017). Several researchers found the impact of AM fungal symbionts in various plant growth under heavy metal stress. Generally, AM fungal mycelia show a high cation-exchanging process through which it uptakes a large quantity of heavy metals especially in the extra-radical mycelium and also detoxifies the heavy metals (Kamal et al. 2010). For example, restoration of Cd in the extra-radical mycelium of AM fungi binds to glomalin glycoprotein (Janoušková and Pavlíková 2010), further, lowering the Cd level in both the vacuoles and cell wall, which brought about Cd detoxification, observed in rice (Li et al. 2016) and give tolerance in alfalfa (Wang et al. 2012). A few examples of heavy metal adaptation by mycorrhizal fungi are listed in Table 7.7.

7.3.6.9 Adaptation Under High and Low Temperature

Soil temperature promotes either heat stress or cold stress to plants; it certainly affects the growth and yield of crops through several physiology receptive stresses including oxidative stress, abscission, and senescence, wilting and burning of leaves, and inhibition of seed germination and growth (Bunn et al. 2009; Chen et al. 2013; Hasanuzzaman et al. 2013). However, mycorrhizal association alleviates the stressful effect of heat or cold through improving water–plant relationships strengthening plant immunity, and also promoting sufficient protein expression (Zhu et al. 2010; Begum et al. 2019). For example, AM fungi such as *Rhizophagus intraradices*, *Funneliformis mosseae*, and *F. geosporum* increase plant growth by regulation of stomatal conductance, and transpiration rate under high temperature in *Zea mays* (Mathur et al. 2016), similarly, *R. irregularis* in *Solanum lycopersicum* (Calvo-Polanco et al. 2016), *Rhizophagus irregularis*, *Funneliformis mosseae*, *F. geosporum*, and *Claroideoglo musclaroideumi* on *Triticum aestivum* L. (Cabral et al. 2016).

Table 7.7 Heavy metal adsorption by AM fungi

Metal	Plant	Fungi	Activity	References
Heavy metal	*Sesbania rostrata*	*Glomus mosseae*	Stimulated formation of root nodules, and increased N and P contents	Lin et al. (2007)
Cd and Zn	*Cajanus cajan* L.	*Rhizophagus irregularis*	Improved root biomass, nutrient status (P, N, Mg, Fe), and proline biosynthesis	Garg and Singh (2018)
Zn	*Medicago truncatula*	*Rhizophagus irregularis*	Increased expression of transporters and decreased absorption of Zn	Nguyen et al. (2019)
Cd	*Cajanus cajan* L.	*Glomus mosseae*	Increase in biothiols (NPSH) and glutathione (GSH), with a larger pool of NP-SH which strongly induced accumulation of phytochelatins	Garg and Aggarwal (2012)
	Lonicera japonica	*Glomus versiforme (Gv) and Rhizophagus intraradices (Ri)*	Increased several antioxidative enzymes and decreased of shoot Cd concentrations	Jiang et al. (2016)
	Trigonella foenum-graecum L.	*Glomus monosporum, G. clarum, Gigaspora nigra*, and *Acaulospora laevis*	Increased antioxidant enzyme activities and malondialdehyde content.	Abdelhameed and Metwally (2019)

7.4 Beneficial Microbial Association on Phyllosphere

The phyllosphere is defined as "the aerial part of the plant or the part of a plant above the ground (leaves, stems, buds, flowers, and fruits), generally, the surface of leaves, thought to provide a home for microorganisms," a habitat for a variety of microorganisms including bacteria, yeasts, and fungus (Yadav and Yadav 2019). Bacteria are the most common microbial occupants exceeding 10^8 cells/cm^2 of the phyllosphere followed by archaea, filamentous fungus, and yeasts play a role in phyllosphere microbial community formation, either epiphytic or endophytic (Leveau 2015; Whipps et al. 2008). The epiphytic phyllosphere microbial communities are extremely complex, containing both culturable and unculturable organisms (Müller and Ruppel 2014). Microbes in the phyllosphere zone have evolved to insensible environmental conditions with epiphytic microbes, particularly vulnerable to temperature, light, UV radiation, and a lack of water or nutrition availability, which influences the richness and composition of the phyllosphere microbial community (Vorholt 2012). The interaction between plants and invading microbes are also influencing the development and maintenance of plant–microbe or microbe–microbe relationships such as commensalism, mutualism, and or antagonism (Rastogi et al. 2013; Barrera et al. 2019). Depending on the association microbes may have either beneficial or harmful effects on the host (Colston and Jackson 2016).

Recent years, extensive research work has analyzed the health status of animals based on individual microbiomes, emphasizing their need for microbiome analysis in various ecosystems, such as soils, water, and plants (Alivisatos et al. 2015; Dubilier et al. 2015; Thaiss et al. 2016). Figure 7.8 represents the various limiting factors regulating the formation of the phyllosphere microbial community. In the phyllosphere, microbes from air and soil, factors such as environmental conditions, rainfall, agricultural practice, human intervention, seasonal variations, and plant species are regulating microbial growth on the leaf surface (Beattie 2002). Moreover, the cuticle layer of the leaf keeps moisture leaving from the interior of the leaf tissues and regulates the amount of water that remains on the leaf surface (Neinhuis and Barthlott 1997). Though, bacteria overcome the water demand by building aggregates to develop biofilms and secrete extracellular polymeric substances (EPS) that can withstand desiccation (Morris and Kinkel 2002; Afzal et al. 2019).

7.4.1 Phyllosphere Microbiome

The phyllosphere's microbial communities are varied from one plant to another, composed with different kinds of bacteria, filamentous fungus, yeasts, and algae, in addition to cyanobacteria, protozoans, and nematodes (Morris and Kinkel 2002; Lindow and Brandl 2003). Several recent cultivation-independent investigations have provided new information about the phyllosphere microbial communities. It is clear that these communities do not reflect random assembly of microorganisms, but are rather subjected to selection, resulting in predictable microbial communities with few dominating phyla and other subgroups (Vorholt 2012). Bacteria are the dominant epiphytes (e.g., methylobacterium, methylophilus, methylibium, and hyphomicrobium) followed by yeast and certain filamentous fungi, mostly present as latent spores rather than active mycelia range between 10^2 and 10^8 CFU/g leaf (Delmotte et al. 2009).

7.4.1.1 Phyllosphere Bacteria

The phyllosphere represents the largest bacterial habitat on Earth and maintained either more prevalent epiphytic or endophytic life. Compare with the rhizosphere, the phyllosphere is an extremely flexible habitat, where epiphytes are frequently exposed to various growth-limiting environmental factors and antimicrobial compounds (Laforest-Lapointe et al. 2016). However, endophytes complete their life within the host plant with limited multiplication due to host defense (Singh and Dubey 2018). The core phyllosphere bacterial biome comprised Proteobacteria, Firmicutes, Actinobacteria, and Bacteroidetes (Vorholt 2012; Laforest-Lapointe et al. 2016; Müller et al. 2016a). The microbiome assemblage on the phyllosphere of distinct plant species could depend on the interaction of several environmental factors (Andreote et al. 2014), and also the host (Müller et al. 2016b). Generally, they

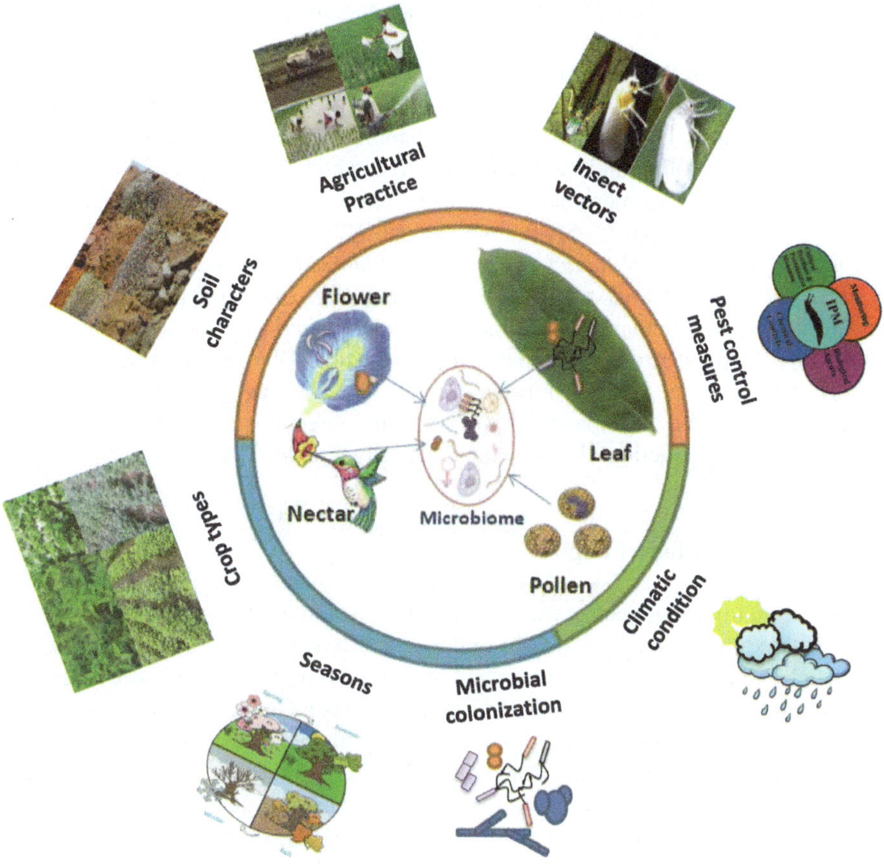

Fig. 7.8 Phyllosphere microbiome. Factors controlling the phyllosphere microbiome formation

are derived from the air as well as from the seeds, soil, and water (Lemanceau et al. 2017). Several works highlighted that Alphaproteobacteria (*Sphingomonas*, *Methylobacterium*), and Gammaproteobacteria (*Pseudomonas*) are the most abundant phyla, followed by Firmicutes, and Actinobacteria (Müller et al. 2016b; Durand et al. 2018; Carvalho et al. 2017). For example, *Bacillus* (Venkatachalam et al. 2016; Thapa and Prasanna 2018), *Citrobacter* (Kecskeméti et al. 2016), *Curtobacterium* (Steven et al. 2018), *Enterobacteria* (Thapa et al. 2017; Dong et al. 2019), *Erwinia* (Thapa and Prasanna 2018), Exiguobacterium (Venkatachalam et al. 2016), *Pseudomonas* (Cid et al. 2016), *Microbacterium* (Toju et al. 2019; Knief et al. 2012), *Methylobacterium* (Janakiev et al. 2020), *Flavobacterium* (Bodenhausen et al. 2013), *Firmicutes* (Mina et al. 2020), *Rhizobium*, *Xanthomonas* (Ottesen et al. 2013), *Weissella* sp. (Dong et al. 2019), and Streptomyces (Thapa and Prasanna 2018) are identified from the phyllosphere of crop plants. In addition, diazotrophic bacteria such as *Azotobacter*, *Beijerinckia*, and *Klebsiella* and Cyanobacteria such as

Nostoc, *Scytonema*, and *Stigonema* are also the inhabitant of the phyllosphere (Bashir et al. 2022).

Compared with the rhizosphere the phyllosphere is a nutrient-poor environment with a more dynamic structure. On the phyllosphere, plants release a very less quantity of sugars generally fructose, methanol, sucrose, and also trace levels of volatile organic compounds and amino acids, which can support the growth of bacteria and develop microbial assemblage on a leaf (Tecon and Leveau 2016; Quan and Liang 2017; Sivakumar et al. 2020). Methylotrophs such as *Hyphomicrobium*, *Methylobacterium*, *Methylibium*, *Methylophilus*, *Methylocapsa*, *Methylocella*, and *Methylocysts* are the dominant flora of the phyllosphere environment that can use methanol or methane as a sole source of carbon and energy and also helps to plant growth (Abanda-Nkpwatt et al. 2006; Krishnamoorthy et al. 2018). The presence of such nutrients on the phyllosphere surface can facilitate diverse bacteria; they are spatially distributed and interact with one another through direct physical interaction (Tecon et al. 2018; Sivakumar et al. 2020).

7.4.1.2 Phyllosphere Fungi

Fungi are an important part of the leaf microbiome, having a significant role on their host, and exhibit both epiphytic and endophytic life with a wide range of metabolic activities including mineral nutrition and recycling, biotic and abiotic stress-resistant and adaptation (Guerreiro et al. 2018; Yao et al. 2019). On the phyllosphere, yeast is the most common epiphytic fungus followed by filamentous fungi. *Cladosporium*, *Colletotrichum*, *Chaetomium*, *Alternaria*, *Penicillium*, *Acremonium*, *Mucor*, and *Aspergillus* are the dominant group of filamentous fungi that make epiphytic and also an endophytic association with phyllosphere (Inácio et al. 2002; Ripa et al. 2019; Dhayanithy et al. 2019). Sreekanth et al. (2017) isolated *Alternaria alternata*, *Setosphaeria* sp., *Cochliobolus* sp., *Alternaria* sp. *Phomaherbarum*, *Davidiellatassiana*, *Botryosphaeria dothidea*, *Ulocladium alternariae*, *Phoma macrostoma* var. incolorata, *Phoma exigua* var. exigua, *Cladosporium cladosporioides*, *Botryospaeria* sp., *Guignardia mangiferae*, *Pyrenophora tritici-repentis*, *Guignardia alliacea*, and *Rhizopus oryzae* from *Catharanthus roseus*. This multifungal group of such associations plays an essential role in the operation of ecosystem function and maintaining complex food webs in the phyllosphere. Fungal endophytes synthesize several metabolites including volatile organic compounds, and secondary metabolites, which directly provide antipathogenic effects to the host as well as induce plant immune functions.

7.4.1.3 Phyllosphere Actinomycetes

Along with bacteria, actinobacteria contribute considerable attention to the phyllosphere. Some actinobacteria interact with the host plant to maintain symbiotic life residing in plant tissues giving huge significance to the host through their secondary metabolites (Singh and Dubey 2018). Various studies revealed that plants at arid, semiarid, and mangrove contains a vast diversity of actinomycetes such as *Actinoplanes missouriensis*, *Actinoallomurus acaciae*, *A. coprocola*, *A. oryzae*, *Actinomaduraglauciflava*, *Amycolatopsistolypomycina*, *Jishengella endophytica*, *Kribbella* sp., *Microbisporamesophila*, *Microbispora* sp., *Micromonospora* sp., *Nocardioides* sp., *Nocardia alba*, *Nonomura earubra*, *Nonomuraea* sp., *Pseudonocardia* sp., *planotetraspora* sp., *Pseudonocardia endophytica*, *P. halophobica*, *Streptomyces* sp., and *Streptomyces javensis* (Chen et al. 2011; Xie et al. 2011; Yadav 2017; Yadav and Yadav 2018; Yadav and Yadav 2019; Sivakumar et al. 2020). The diversity and distribution of endophytic actinomycetes in the phyllosphere are determined by the host plant physiology and their environment (Du et al. 2013; Dinesh et al. 2017; Nalini and Prakash 2017). Endophytic actinomycetes have wide functional properties in their associated plants, including protection and mineral recycling.

7.4.2 *Functions of Phyllosphere Microorganism for Sustainable Agriculture*

7.4.2.1 Plant Nutrition Acquisition and Growth

The phyllosphere microbiome is being considered a vital component in plant nutrient acquisition and has little impact on the host plant growth (Stone et al. 2018). Conversely, nitrogen-fixing bacteria present in the phyllosphere facilitate microbial diversity formulation and improve the health status of the plant. Mechanistic investigations revealed the essentials of plant phyllosphere nitrogen dynamics in stress resistance and also the growth (Delmotte et al. 2009). Plant cells can absorb the end products of nitrogen fixation (ammonia, nitrate, and nitrogen dioxide) more easily in warm conditions because nitrogenous end products readily diffuse into the leaf's inner tissue. In the phyllosphere, numerous plant species colonized by chemoautotrophic nitrogen-fixing bacteria have been found, particularly archaea (Abadi et al. 2021; Qin et al. 2019). In humid tropical conditions, epiphytic microbes can fix nitrogen, generally, tropical epiphytes grow in the plant phyllosphere to fix nitrogen and are able to meet a significant portion of their nitrogen needs (Hietz et al. 2002; Abril et al. 2005; Stanton et al. 2019). Several studies reported that α-proteobacteria, β-proteobacteria, and γ-proteobacteria along with cyanobacteria (*Oscillatoria Microcoleus*, *Scytonema*, and *Stigonema*) are significant nitrogen fixers of phyllosphere (Freiberg 1998; Fürnkranz et al. 2008; Bao et al. 2020). In addition, the presence of nitrifying bacteria in the phyllosphere of tropical forest plants (Whipps et al. 2008; Maignien et al. 2014; Carrell and Frank 2014; Guerrieri et al. 2020) and crop plants has a significant impact on plant growth (Miyamoto et al. 2004). But the actual mechanism of nitrogen fixation on the phyllosphere is not yet clear.

Other than the nitrogen acquisition, phyllosphere microbes can solubilize phosphates and potassium. Various phyllosphere inhabitants such as *Acetobacter* sp. *Arthrobacter humicola*, *Bacillus alcalophilus*, *B. aryabhattai*, *B. thuringiensis*, *Chryseobacterium* sp., *Cellulosimicrobium*, *Enterobacter* sp., *Klebsiella terrigena*, *Methylobacterium extorquens*, *M. mesophilicum*, *M. radiotolerans*, *Paenibacillus amylolyticus*, *Phyllobacterium*, *Pseudomonas fuscovaginae*, *P. fluorescens CHA0*, *Psychrobacterfozii*, *Rhodococcus*, and *Serratia* have been documented as phosphate solubilizers used to promote plant growth and development (Kumar et al. 2013; Verma et al. 2014; de Souza et al. 2013; Thapa et al. 2017). In order to solubilize phosphates, some phyllosphere bacteria such as *Bacillus mucilagenosus*, *P. amylolyticus*, and *Psychrobacter* sp. solubilize potassium (K) (Verma et al. 2014). In the same way, some phyllosphere microbiota takes part in the uptake of micronutrients like zinc (Esitken et al. 2010), sulfur (Yu et al. 2013), and copper (Esitken et al. 2010). Moreover, the contribution of the phyllosphere microbiome for the growth of plant is comparably less than that of the microbiome of the rhizosphere. Benefits of phyllosphere microbes to the growth of plant begin germination of seeds. Many of the phyllosphere bacteria produce plant growth regulators including cytokines (CKs), auxins. Microbial production of CKs on leaf surface facilitates the transport of nutrients from the surface to inner tissues especially nitrogen transport (Lemanceau et al. 2017). In these instances, the phyllosphere microbiome may influence the total plant's capacity to get nutrients, rather than just the foliage (Bram et al. 2018).

7.4.2.2 Biological Control

Microbial association on phyllosphere protects their residing host through various mechanisms. One such mechanism is changing nutrient distribution on leaf surfaces can manipulate microbial communities against pathogens (Williams et al. 2013; Manching et al. 2014). In some cases, phyllosphere microbes produce higher biomass through which competitive mechanism can quench the pathogenic growth (Pusey and Wend 2012), which was observed in rice (Ren et al. 2016) and lettuce (Williams et al. 2013), and also fix nitrogen (Feng et al. 2003), this could be important for agricultural sustainability (Gupta et al. 2018). *Pseudomonas fluorescens* A506 crumb leaf blight disease-causing *Erwinia amylovora* through competition for nutrients and space (Wilson and Lindow 1993). However, in general, phyllosphere microbes produce antagonistic substances (antibiotic), which can reduce both bacterial and fungal pathogenesis. Bacterial flora such as *Pseudomonas* spp. and *Bacillus* spp. residing in the phyllosphere could produce antibiotic-like peptides, which induce systemic resistance of plants against various fungal pathogens (Vorholt 2012; Ceballos et al. 2012; Griffin and Carson 2015). The use of *Bacillus* spp. isolated from phyllosphere and soil has been largely exploited as a biocontrol agent in agriculture (Fernando et al. 2007), due to its broad-spectrum antagonistic activity it has been used as a biocontrol for sugar beet (Collins and Jacobsen 2003), citrus (Huang et al. 2012), cotton, rice and amaranth leaves (Wang et al. 2014a, b), strawberry (Wei et al. 2016), and cacao (Villamil Carvajal et al. 2015). The ability to produce endospores, secondary metabolites (Mendes et al. 2013), and proteins (Zhang et al. 2018b) by *Bacillus* sp. and the capability to induce systemic resistance in a host are considered to be useful biocontrol agents for phyllosphere pathogens (Lahlali and Peng 2014; Wei et al. 2016).

In addition, the interaction of microbes on the phyllosphere stimulates gene expression of its residing host by signaling process, through this mechanism plants can recognize (molecular recognition pattern) and distinguish the presence of pathogens (Hunter et al. 2010) and other beneficial microbes (Rastogi et al. 2012). Similarly, the expression of pathogenesis-related proteins and antimicrobial proteins by *Sphingomonas melonis* on the phyllosphere controls the potent plant pathogen *Pseudomonas syringae* (Innerebner et al. 2011). Moreover, some phyllosphere microbes can produce siderophores, by which epiphytic bacteria inhibit pathogens, and reduce oxidative stresses (Fu et al. 2016; Thapa et al. 2017; Gupta et al. 2018; Santos Kron et al. 2020). For example, *Pseudomonas syringae* pv. syringae has pathogen quenching activity through siderophore production (Wensing et al. 2010). *Pseudomonas protegens* CS1 produces enantio-pyochelin siderophore has the antagonistic activity against the plant pathogen *Xanthomonas citri* (Michavila et al. 2017). In addition, some microorganisms could induce the systemic resistance of plants by changing upregulation of various antioxidant enzymes, such as ascorbate peroxidase (APX), catalase (CAT), peroxidase (POD), polyphenol oxidase (PPO), and superoxide dismutase (SOD) (Abd El-Gawad et al. 2015). Meanwhile, researchers could suggest the use of native microorganisms as biocontrol agents due to their host-specific association, colonization, and adaptation to local environmental conditions (Kumar and Gopal 2015; Cruz-Martín et al. 2017), and shows multiple modes of activity against pathogenic microbes. For example, representatives of the Bacillaceae family isolated from *Mussa* spp. phyllosphere showed antifungal activity against black Sigatoka disease (*Mycosphaerella fijiensis*) (Poveda et al. 2010; Cruz-Martín et al. 2017), also in *Mussa* spp., *Bacillus pumilus* showed chitinolytic and glucanolytic activity against *M. fijiensis* (Cruz-Martín et al. 2018).

7.4.2.3 Anti-insect Activity

Endophytic microbes produce various secondary metabolites (alkaloids), within the plant tissues, which can reduce or kill the plant-feeding insects and pest. The alkaloids exhibit strong insecticidal activity mainly to inhibit the developmental stages of insect larvae, affect feeding activities, reduce survival, and also the egg-laying ability (Akutse et al. 2014; Martinuz et al. 2012; Resquín-Romero et al. 2016; Sánchez-Rodríguez et al. 2018). Endophytic alkaloids are toxic to plant-feeding insects, for example, endophytic fungi such as *Acremonium coenophialum* and *Epichloe typhina* interact with *Lolium* and *Festuca* grasses to produce multiple loline, peramine, ergovaline, which have toxic effect to aphids (Siegel et al. 1990); *Tripterygium wilfordii* produce cytotoxic alkaloid, cytochalasin (Wagenaar et al. 2000); *Phomopsis* sp. synthesize Phomapsichalasin (antimicrobial agent) (Horn et al. 1995); *Streptomyces* sp. produces 1-N-methyl albonoursin (antibiotic) in perennial ryegrass (Gureny and Mantle 1993) are other few examples. In addition, *Beauveria bassiana* reduces the damage caused by a poppy stem gall wasp (*Iraella luteipes*) (Quesada-Moraga et al. 2009), *Helicoverpa zea* (Powell et al. 2009), and *Tuta absoluta* (Klieber and Reineke 2016; Qayyum et al. 2015). Endophytic fungal entomopathogens, *B. bassiana* and *Metarhizium brunneum* kill aphid such as *Aphidius colemani* and *Myzus persicae*, hence it could be used as a biocontrol agent to reduce the herbivorous damage on several crops such as tomato, potato, sugarcane, and pea (De Sassi et al. 2006; Akutse et al. 2014; Qayyum et al. 2015; Jaber and Araj 2018). Moreover, *B. bassiana* offers protection against plant viral pathogens, Zucchini yellow mosaic virus (ZYMV) transmitted by aphid vector (Jaber and Salem 2014). Hence, in agriculture, the pest management practices the beneficial endophytic bacteria and fungi have been used to control disease transmission and insect pest-mediated crop damage.

7.4.2.4 Host Stress Tolerance

Microbial association in phyllosphere provides stress tolerance against various environmental stresses such as extreme drought, UV rays, and temperature (Sun et al. 2010; Meena et al. 2015). UV radiation causes damage to the genetic material, which alters the physiology and growth of plants (Yurimoto et al. 2021). Some of the plant's defenses may be provided by bacterial EPS, which may protect it from desiccation and UV radiation. It has been shown that the gel-like EPS matrix of aquatic *Pseudomonas aeruginosa* biofilm helps to transport water and nutrients to the plant. Biofilms presence in the phyllosphere may provide able to minimize the loss of water to get more resistant to desiccation than biofilms in water (Chaturvedi et al. 2016). *Piriformospora indica*, endophytic fungi alleviate drought stress-induced oxidative stress in many plants (Waller et al. 2005; Sun et al. 2010). *Pseudomonas* sp. is often a major element of the phyllosphere, indicating that naturally existing biofilms give protection against desiccation and UV light. Pigmented bacteria provide increased UV resistance than other bacteria. It is plausible that phyllosphere microorganisms may give some UV protection to the plant host via the creation of EPS and the pigmentation by UV-absorbing pigment (Jacobs and Sundin 2001; Compant et al. 2019). In arid conditions, often result in an increasing C:N ratios in leaves suggesting that one of the secondary impacts of drought is nitrogen restriction, implying a possible function for the phyllosphere microbiome alleviating drought-induced nitrogen limitation (Rico et al. 2014). Other environmental stresses may also be tolerated by plants because of microbially acquired tolerance (Vacher et al. 2016).

Ecological stress stimulates the plant to produce various reactive oxygen species (ROS), hydrogen peroxide, hydroxyl radicals, and superoxide radicals, causing oxidative damage to various cellular components such as proteins, nucleic acids, and membranes (Das and Roychoudhury 2014; Lata et al. 2018; Huang et al. 2019). Beneficial plant endophytes alleviate oxidative damages induced by both biotic and abiotic stresses; they can stimulate the plant defense mechanism to protect themselves from such defects. There are several research outcomes highlighting the endophytic or endosymbiotic association of both bacteria and fungi in phyllosphere and rhizosphere induces plant defense responses through the expression of peroxidase enzymes like superoxide dismutase and glutathione reductase (Egamberdieva et al. 2017; Lata et al. 2018; Oukala et al. 2021). Endophytes could also produce stress-responsive enzymes, which can also protect the plant cell from oxidative damages, for example, the endophytic *Enterobactes* sp. contain several superoxide genes, peroxidase, catalase, and hyperperoxide reductases (Taghavi et al. 2005). In some cases, it can produce various antioxidant metabolites and accumulates in plant tissues, for example, the presence of the endophytic fungus *Epichloë coenophiala* in tall fescue grasses accumulates mannitol an antioxidative fungal metabolite, which provides oxidative stress-related protection in plant (Lata et al. 2018); *Pestalotiopsis microspora* produces pestacin and isopestacin in endosymbiotic interaction with plant (Strobel and Daisy 2003).

Above all significant merits provide the use of beneficial phyllosphere microbiome interaction on plant growth and its productivity guide us to use those microbes for sustainable agriculture.

7.5 PGBs Bioinoculant Formulation for Sustainable Agriculture

Bioinoculants are ecofriendly microbes having the ability to promote plant growth and yield in a variety of mechanisms (Prasad et al. 2019). It is an excellent alternative of chemical fertilizers, improves the soil nutrients, antagonist against phytopathogens and soil nematodes, facilitator, stress relieving agent producer, and detoxifier, as the whole, it enhances the crop production more economically and ecofriendly. Several researches revealed the exploitation of PGPR as bioinoculant imparts success in sustainable agriculture through the low-cost process. There are different bioinoculant formulation are in agricultural practice which contains PGPBs such as solid (e.g., powder) or liquid formulations, metabolite formulations, solid carrier-based formulations, and synthetic polymer-based formulations. Efficiency of the bioinoculant is based on the type of inoculums, their appropriate formulation, production, and storage condition of microbial cells, where in some cases the application is more efficient and it always maintains rich colony-forming units (CFUs), either in the soil or in the phyllosphere (Orozco-Mosqueda et al. 2021), and it shows some detectable antipathogenic activity against soil-borne pathogens. On foliar application of such bioinoculants may interfere with the growth of pathogens causing infections in the aerial part of plant (Owen et al. 2015). Commonly, the bioinoculant formulation process depends on the type of applications in the larger field or in the small greenhouse conditions. Moreover, the time of inoculums given, quantity, and the type of organism used along with environmental factors are influencing the efficiency of inoculums, which could kill or inhibits the propagation of pathogens. Mostly they are applied to the field as prophylaxis, before observing disease symptoms in crops to prevent the pathogenic attack and managing to protect even the products during post-harvest (Sridhar et al. 2014).

Commercially available bioinoculants contain the beneficial microorganisms that have plant growth-promoting activities and biocontrol along with fertilizer effects to the plants. Some bioinoculants can be used to kill several larvae of harmful insects and pest. For example, *Bacillus thuringiensis* control the larvae of several moths and butterflies, *Bacillus popilliae* kill Japanese beetles, *Nosema locustae* control grasshoppers, and entomopathogenic fungi such as *Beauveria bassiana*, *Metharhizum anisopliae,* and *Isaria fumosorosea* kill various insects (Maina et al. 2018). *Trichoderma* sp. controls the growth of various plant fungal pathogens such as *Rhizoctonia*, *Fusarium*, *Verticillium*, *Pythium*, and *Phytophthora.* The best choice of bioinoculum application is on seed coat (seedlings), hence the beneficial microbes start its interaction at the early stage of plant growth. For example, Bacillus and Pseudomonas show its direct effect on plant growth and also eliminate pathogens from its rhizosphere region (Hernández-León et al. 2015).

However, in the use of various beneficial microbes as bioinoculants, several barriers could affect the effect and application including loss of biological activity, physiological and agronomical factors affecting the survival of inoculants, lifetime effect, etc. For example, when the spore form of *Bacillus* sp. is used for soil application, which prolongs its shelf life (Santoyo et al. 2012). Some non-sporulating traits required more specific and special formulations. Once it has been inoculated, the bioinoculant maintains its persistence in the soil and makes them colonize in the rhizosphere. Hence, it is our need to select a more appropriate bioinoculum which can compete rhizosphere or phyllosphere microflora and also able to colonize and perform beneficial activities in the microenvironment (Hartmann et al. 2008; Berendsen et al. 2012; Liu et al. 2020).

For this reason, more research is required to evaluate abiotic factors on the beneficial properties of each bioinoculant. However, prior to the development of commercial bioinoculants, the following properties should be properly evaluated which include (1) pathogenecity, (2) specificity, (3) interactions with other soil microbes, (4) shelf life of the preparation, (5) longevity time in soil/crop, (6) performance, (7) dosage level, and (8) cost effect (Arora et al. 2010). Some important biocontrol and biofertilizer effects of beneficial microbes are listed in Table 7.8.

7.5.1 Microbial Consortium as Bioinoculum

In recent years, crop quality with higher nutrition and minerals is of increasing concern with the prolonged demands from consumers (Timmusk et al. 2017). A soil-native microbe significantly contributes to the health and development of plants against nutritional deficiency and abiotic/biotic stress factors (Bakker et al. 2018; Oyserman et al. 2018). Commercially available bioinoculants have limited success

Table 7.8 Microbial consortia and its beneficial role

Microbial consortia	Beneficial effect	References
Bacillus subtilis, *R. tropici*	Reduced disease severity caused by *Fusarium solani*, increase the yield of bean	De Jensen et al. (2002)
Bradyrhizobium, *G. mosseae*	Inhibits *Cylindrocladium parasiticum*, increased production of soybean	Yao et al. (2002)
AMF, *Rhizobium*, *Azospirillum*, *Bacillus*, *Pseudomonas*	Increased nodulation, higher plant growth, increased yield, increases systemic resistance against stress responses, increased nutrient uptake	Singh et al. (2010)
Pseudomonas, *Azotobacter*, *Azospirillum*	High nutrient intake, shoot, and root wet and dry biomass, enhanced lycopene production, increased antioxidants in tomato	Ordookhani et al. (2010)
Bacillus cereus, *B. subtilis*, *Serratia*	Enhanced photosynthetic efficiency, reduce wilt symptoms, reduce mono-dehydroascorbate, increased coat proline content, increase systemic resistance to plant	Wang (2009)
Pseudomonas aeruginosa, *Trichoderma harzianum*, *B. subtilis*	Increased defense against sclerotinia rot, induced systemic resistance	Jain et al. (2012)
P. aeruginosa, *Trichoderma* sp., *Rhizobium* sp.	Activated physiological defense response against color rot pathogen in chickpea	Singh et al. (2013)
Pseudomonas fluorescens, *Trichoderma*, *Rhizobium*	Increased antioxidant enzyme activity, induced resistance of plant against biotic stress	Singh et al. (2013)
Streptomyces spp. *Pseudomonas* sp., *P. flurescens*, *G. intraradices*	Increased plant growth, yield, and increased drought resistance in Finger millet	Khadka et al. (2016)

due to their formulation of either one or few microbial taxa. It is unlikely that it never consolidates our entire basic requirement for commercial crop production (Hart et al. 2018). Hence, using the synthetic method to formulate microbial consortia (MC) that could potentially replace and/or redesign the structure and function of plant microbiome is a remedial measure to solve the disadvantages of conventional biofertilizers (Qin et al. 2016; Hart et al. 2018; Kong et al. 2018). Several reports highlighted the merits of such consortia of bioinoculants, coinoculation of rhizobia with other PGPR enhances nodulation ability, nitrogen fixation, phytohormone production, and growth of legumes (Bansal and Srivastava 2012; Gupta et al. 2015). Similarly, diazotrophic such as *Azospirillum*, *Azotobacter*, *Bacillus*, *Pseudomonas*, *Enterobacter*, and *Serratia* are a few PGPR that have been successfully coinoculated with rhizobium (Tajini et al. 2012; Sekar et al. 2016). The bioinoculums of microbial consortia often target to enhance the metabolite content (sugars, ascorbic acid, folic acid, vitamin, and anthocyanin) and nutrients (N, Ca, P, Mg, K, Na, Fe, Mn, Cu, Zn, and B) in crops. Hence, the inoculums of consortia are formulated with PGPRs and AM fungi, which promotes the highest nutraceuticals in the crops (Battini et al. 2016; Torres et al. 2016; Avio et al. 2017; Bona et al. 2015).

The efficiency of the beneficial microbes and their consortia on growth promotion and productivity in the field condition in various crops have been accounted for in the formulation studies (Table 7.8). The following are the major components that need to be evaluated for the consortia design and development: (a) compatible strains with a different pattern of plant colonization, (b) compatible strains combined with a broad spectrum of activity against pathogens, (c) compatible strains combined with different mode of action, and (d) compatible strains combined with a genetically diverse group to adapt in a wide range of pH, temperature, moisture, and relative humidity (Sekar et al. 2016).

7.6 Engineering Host Microbiome for Sustainable Agriculture

The microbiome is defined as the collective genomes of the microbes such as bacteria, fungi, protozoa, and viruses that live in a specified environment. Diverse microbial community in the microbiome gives either beneficial or harmful effect on its host and environment. Harmful microbial domination in a microbiome causes diseases and disorders. Microbiome engineering is a kind of process to manipulate the microbial abundance with the inoculation of one or more new beneficial microbes in an adequate amount (Albright et al. 2021). However, the engineering of microbiome can be used to modify microbial diversity and community structure and could restore ecological balance, and has been employed for improving human health and agricultural productivity. Researchers understand the relationships between microbe–microbe and microbes–host, which creates novel insights into microbiome engineering, where the new microbial community with a small number of species can create a stable ecosystem. Plant microbiome is represented as the microbiome in root and root-associated area. In agriculture, rhizosphere microbiome engineering will impose green revolution to develop high yield in crops, which greatly improved food security worldwide (Bano et al. 2021).

Microbiome engineering is an attempt to find and improve the function of an ecosystem by manipulating the required composition of microbes (Albright et al. 2021). For sustainable agriculture, currently, selective beneficial microbes with desirable properties could blend and formulate synthetically to establish complex beneficial microbiome at the rhizosphere and phyllosphere to promote the growth and fitness of the plant under various ecoclimatic conditions. Like animal probiotic supplements, soil has been enriched with plant beneficial microbes that can colonize and modulate the rhizosphere community through various interactions and signaling processes. In the following section, the agricultural importance of rhizosphere microbiome engineering has been discussed.

7.6.1 Rhizosphere Microbiome Engineering

The root-associated microbiome consists of different compositions of the microbial communities represent in plant root and the rhizosphere they could share the nutrients and space together with plants. In rhizospheres, several thousand bacterial and fungal taxa including beneficial and harmful species are maintaining a complex food web to develop a complex microbial community structure. Functionally, the root-associated microbiome influences plant physiology and plays anti-pathogen roles (Agler et al. 2016; Foo et al. 2017). However, the manipulation of root-associated microbial composition and networking for highly efficient beneficial microbes of PGPRs (nutrient mobilizers, antibiotics, siderophore, and plant growth hormone producers) that promote growth, provide fitness and health of the plant as well as enhance the yield.

In the microbiome engineering process, beneficial microbes are isolated from healthy crops and analyzed for their microbial composition, nutritional and physiological characteristics, role in the rhizosphere community, and environmental importance. This information has accounted for the effective formulation of rhizosphere microbiome engineering. Rhizosphere microbiome engineering is a hot topic in agriculture studies because of its unpredictable benefits to the agricultural crops (Kumar and Dubey 2020). In natural environment, rhizosphere microbiome functions are continuously changing due to unstable environmental conditions. In this status, microbial communities both adapt and survive or failed to survive, that is the reason behind the limited level of success of bioinoculants (biofertilizers and biocontrol agents) in various environmental conditions. Thus, there is a need to formulate and engineer the microbiome with highly adaptable traits with optimum PGPR functions and maintain a healthy microbiome in various soil and environmental conditions. This system of agriculture is called "microbiome-mediated smart agriculture system (MiMSAS)," where the engineered microbiome has the ability to withstand the various environmental stresses (Bano et al. 2021; Kaul et al. 2021).

Using the applications of new emerging techniques in the synthetic biotechnology, microbes can be constructed based on the required characteristics capable to adapt and effectively perform plant growth (Peng et al. 2016; Dubey et al. 2019). Genetic manipulation of the desirable traits could be based on the conventional features of plant-microbe intercellular communications, or intracellular metabolic pathways, and also the composition of microbial communities that enhances plant performance (Dubey et al. 2019). Figure 7.9 illustrated the microbiome engineering for sustainable agriculture. The potential property used to engineer the microorganism is the desirable mutualistic character of the PGPR microbes. In addition, engineered microbes make positive beneficial communication with unmodified microbes there in the natural environment, which can facilitate syntrophic nutrition within the microbiome makes enhanced plant growth (Ren et al. 2016; Muleta 2017; Charubin and Papoutsakis 2019). After strengthening the characters through genetic modification, large-scale field experiments can be performed with control experiments in various soil and environmental stress responses such as drought, salinity,

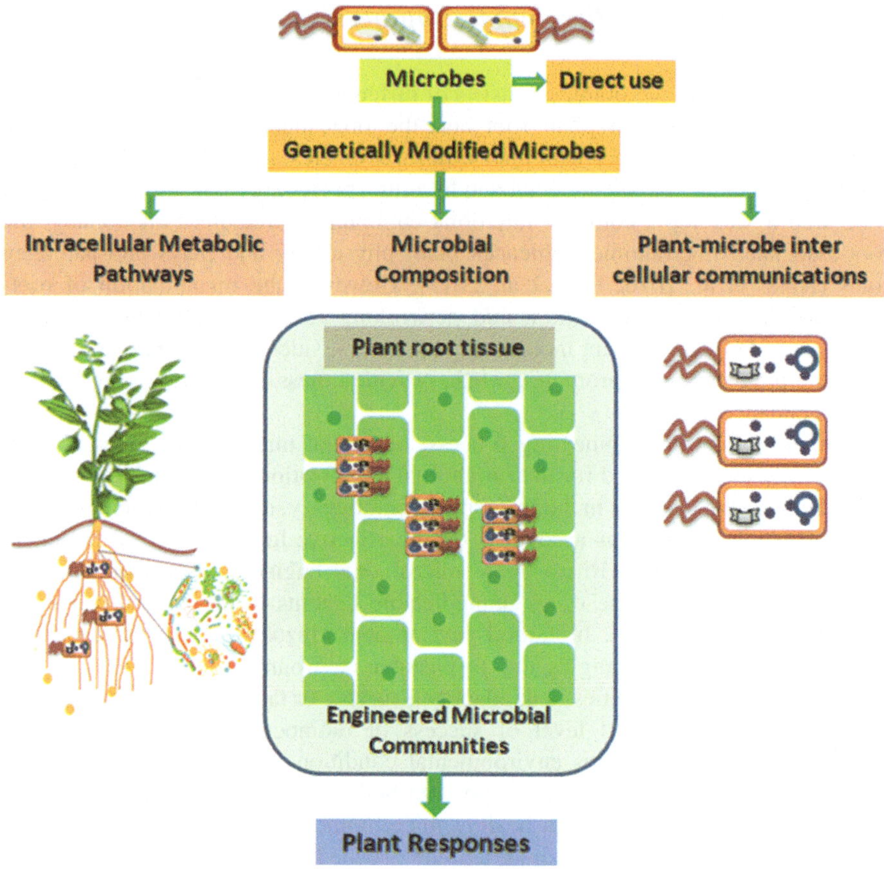

Fig. 7.9 A diagrammatic representation of microbiome engineering. The microbes are engineered genetically in order to modify the characters of the plant–microbe intercellular interactions, or microbial intracellular metabolic pathways, and to change the microbial composition. (Adapted from Dubey et al. 2019, *Biodivers Conserv.* 28, 2405–2429)

pH, elevated level of carbon dioxide, and light on microbial community composition and mutualistic benefits of inoculations (Dubey et al. 2019).

Rhizosphere microbiome engineering can alleviate the drought stress in plants with the association of rhizospheric, endophytic, and symbiotic microbes, and increases soil fertility which can reduce the use of commercial chemical fertilizers for crop yield (Zolla et al. 2013). Reduced use of such chemicals and fertilizers could retain the biodiversity in the soil or the engineered microbes synergistically interact with other soil microbes to re-establish the healthy rhizosphere microbiome. The main advantage of microbiome engineering is that their microbial composition is controllable and easily detectable. In addition, engineered microbes can compete the pathogen by multiple mechanisms and protect the plant against the infections. For example, for the control of potato common scab and tobacco black root rot (Gopal et al. 2013), five root-associated bacteria have been shown to protect *Nicotiana attenuata* from a sudden-wilt disease (Santhanam et al. 2015). However, the development of engineered microbiomes limits our understanding of the core microbiome and only culturable microbes can be included in microbiome engineering (Table 7.9).

7.7 Conclusion

Beneficial microbes that are associated with the plants are having plant growth-promoting activities (PGPMs), as growth promoters, biofertilizers, or biocontrol agents for sustainable agricultural practices. Microorganisms sustain a specific microenvironment and provide various benefits to the soil, plant, and environment. In the rhizosphere environment, many bacterial beneficial interactions have been established by bacteria, fungi, actinomycetes, cyanobacteria, etc. Most significantly, the symbiotic association between Rhizobium and mycorrhizal fungi provides various nutrients through nitrogen fixation, transportation, solubilization of phosphates, and other insoluble minerals. Saprophytic microbes could decompose organic waste material to enrich the soil nutrient and recycling process. The beneficial microbes that are associated with the plant as endophytic and epiphytes of the phyllosphere and rhizosphere have various biological functions to control pathogenic infection and alleviate different environmental stresses (drought, salinity, heavy metal, oxidative stresses) and control pest and nematode infections. Studies of the microbiome in the phyllosphere and rhizosphere instigate the production of bioinoculants. To overcome the setbacks of single conventional bioinoculants combined microbes with better attributes could be used for species-specific microbiome engineering mainly in rhizosphere region. Moreover, the knowledge generated through modern technologies used to understand the functional role of individual microbes in a microbiome and its impact on soil and plant health guide us to redesign beneficial rhizosphere microbiomes that will improve soil health, and crop productivity through sustainable agriculture.

Acknowledgments The authors would like to thank the DST-PURSE facility, Madurai Kamaraj University for providing internet facility and RUSA (MHRD) programs.

Table 7.9 Functions of microbiome transfer for sustainable agriculture

Microbiome method	Functions and limitations	References
Native root-associated microbiota transplant	Inhibit plant diseases, resist environmental stresses, and promote growth; Easy to manipulate; Limited by the availability of functional native microbiome	Gopal et al. (2013), Zolla et al. (2013)
Synthetic microbiome: root-associated microbiota transplant	Inhibit plant diseases, resist environmental stresses, promote growth; Controllable customization of microbial composition for beneficial effects; Limited by the understanding of the core microbiome; Only applicable to culturable microbes	Santhanam et al. (2015), Glick (2012)
Signaling molecules: Administration of root extrudates, e.g., salicylic acid	Resist environmental stresses, promote growth; Promote assembly of balanced microbiome; Limited by the availability of signaling molecules	Lebeis et al. (2015)
Agricultural management: Organic farming	Improve soil fertility	Navarrete et al. (2015)
Plant-based methods: Plant breeding and cultivar selection	Enhanced production of exudates; Influences the microbial diversity by enhancing the growth of some selected microbes present in the rhizosphere	Dubey et al. (2019)
	Alteration of plant resistance to disease and environmental factors; Improved tolerance toward to resist adverse environmental conditions (edaphic, biological, and climatic)	Ryan et al. (2009), Lynch and Neufeld (2015)
Genetic modification: Change in the amount of signaling molecules, and organic exudates in soil	Plants are engineered to secrete exudates that direct specific microbial diversity for providing; Plant induces microbiome for beneficial functional traits like production of siderophore, antifungal, antimicrobial, antibiotics acts as a biocontrol agent. Improving resistance toward adverse environment conditions. Use in bioremediation of contaminants	Bakker et al. (2012), Imam et al. (2016), Sharma et al. (2013)

(continued)

Table 7.9 (continued)

Microbiome method	Functions and limitations	References
Genetic modification: Plants are engineered for producing exudates which modify properties	Plant growth is enhanced at acidic or low pH, resistance salinity, alkalinity and water stress. Enhanced resistance towards Al^{3+}. Enhanced phosphate solublization. Increase in shoot biomass, longer and larger root hairs	Ryan et al. (2009), Yang et al. (2007)
Meta-organism based management and selection of complementary microbiomes and plants	Crop rotation; Managing soil diversity by induction of suppressive soils; Improving physico-chemical characteristics of the soil; Elevation in organic carbon content and higher level of nutrients cycling	Kumar et al. (2015)

Author Contributions All authors listed have made a considerable intellectual contribution to this chapter and approved it for publication.

References

Abadi VAJM, Sepehri M, Rahmani HA, Dolatabad HK, Shamshiripour M, Khatabi B (2021) Diversity and abundance of culturable nitrogen fixing bacteria in the phyllosphere of maize. J Appl Microbiol 131(2):898–912. https://doi.org/10.1111/jam.14975

Abanda-Nkpwatt D, Musch M, Tschiersch J, Boettner M, Schwab W (2006) Molecular interaction between *Methylobacterium extorquens* and seedlings: growth promotion, methanol consumption, and localization of the methanol emission site. J Exper Botany 57(15):4025–4032. https://doi.org/10.1093/jxb/erl173

Abd El-Gawad HG, Ibrahim MFM, Abd El-Hafez AA, Abou El-Yazied A (2015) Contribution of pink pigmented facultative Methylotrophic bacteria in promoting antioxidant enzymes, growth and yield of Snap Bean. Synthesis 2:4. https://doi.org/10.5829/idosi.aejaes.2015.15.7.12709

Abdelhameed R, Metwally R (2019) Alleviation of cadmium stress by arbuscular mycorrhizal symbiosis. Int J Phytoremed 21(7):663–671. https://doi.org/10.1080/15226514.2018.1556584

Abdul Rahman N, Abdul Hamid N, Nadarajah K (2021) Effects of abiotic stress on soil microbiome. Int J Mol Sci 22(16):9036. https://doi.org/10.3390/ijms22169036

Abo-Amer AE, Abu-Gharbia MA, Soltan ESM, Abd El-Raheem WM (2014) Isolation and molecular characterization of heavy metal-resistant Azotobacter chroococcum from agricultural soil and their potential application in bioremediation. Geomicrobiol J 31(7):551–561. https://doi.org/10.1080/01490451.2013.850561

Abril G, Guérin F, Richard S, Delmas R, Galy-Lacaux C, Gosse P, Tremblay A, Varfalvy L, Dos Santos M, Matvienko B (2005) Carbon dioxide and methane emissions and the carbon budget of a 10-year old tropical reservoir (Petit Saut, French Guiana). Wiley, London. https://doi.org/10.1029/2005GB002457

Afzal I, Shinwari Z, Sikandar S, Shahzad S (2019) Plant beneficial endophytic bacteria: mechanisms, diversity, host range and genetic determinants. Microbiol Res 221:36–49. https://doi.org/10.1016/j.micres.2019.02.001

Agarwal S, Choure RG (2011) Use of indigenous Rhizobia as effective bioinoculants for *Pisum sativum*. Int J Biotech Biosci 1(1):89–96. https://doi.org/10.3389/fpls.2017.01961

Agerer R (2001) Exploration types of ectomycorrhizae. Mycorrhiza 11(2):107–114. https://doi.org/10.1007/s005720100108

Agler M, Ruhe J, Kroll S, Morhenn C, Kim S, Weigel D, Kemen E (2016) Microbial hub taxa link host and abiotic factors to plant microbiome variation. PLoS Biol 14(1):1002352. https://doi.org/10.1371/journal.pbio.1002352

Ahanger M, Agarwal R (2016) Potassium up-regulates antioxidant metabolism and alleviates growth inhibition under water and osmotic stress in wheat (*Triticum aestivum* L). Protoplasma 254(4):1471–1486. https://doi.org/10.1007/s00709-016-1037-0

Ahanger M, Hashem A, Abd-Allah E, Ahmad P (2014) Arbuscular mycorrhiza in crop improvement under environmental stress. Emer Technol Manag Crop Stress Tol 2:69–95. https://doi.org/10.1016/B978-0-12-800875-1.00003-X

Ahanger M, Tittal M, Mir R, Agarwal R (2017a) Alleviation of water and osmotic stress-induced changes in nitrogen metabolizing enzymes in Triticum aestivum L. cultivars by potassium. Protoplasma 254(5):1953–1963. https://doi.org/10.3389/fpls.2019.01068

Ahanger M, Tomar N, Argal S, Agarwal R (2017b) Plant growth under water/salt stress: ROS production; antioxidants and significance of added potassium under such conditions. Physiol Mol Biol Plants 23(4):731–744. https://doi.org/10.1007/s12298-017-0462-7

Ahanger M, Alyemeni M, Wijaya L, Alamri S, Alam P, Ahmad AM (2018) Potential of exogenously sourced kinetin in protecting *Solanum lycopersicum* from NaCl-induced oxidative stress through up-regulation of the antioxidant system, ascorbate-glutathione cycle and glyoxalase system. PLoS One 13(9):0202175. https://doi.org/10.1371/journal.pone.0202175

Ahmed R, Solaiman ARM, Halder NK, Siddiky MA, Islam MS (2007) Effect of inoculation methods of *Rhizobium* on yield attributes, yield and protein content in seed of pea. J Soil Nature 1(3):30–35

Ait Lahmidi N, Courty P, Brulé D, Chatagnier O, Arnould C, Doidy J, Berta G, Lingua G, Wipf D, Bonneau L (2016) Sugar exchanges in arbuscular mycorrhiza: RiMST5 and RiMST6, two novel *Rhizopha gusirregularis* monosaccharide transporters, are involved in both sugar uptake from the soil and from the plant partner. Plant Physiol Biochem 107:354–363. https://doi.org/10.1016/j.plaphy.2016.06.023

Ait-El-Mokhtar M, Laouane R, Anli M, Boutasknit A, Wahbi S, Meddich A (2019) Use of mycorrhizal fungi in improving tolerance of the date palm (*Phoenix dactylifera* L.) seedlings to salt stress. Scia Hortic 253:429–438. https://doi.org/10.1016/j.scienta.2019.04.066

Akram M, Rizvi R, Sumbul A, Ansari RA, Mahmood I (2016) Potential role of bio-inoculants and organic matter for the management of root-knot nematode infesting chickpea. Cogent Food Agricul 2(1):1183457. https://doi.org/10.1080/23311932.2016.1183457

Akutse KS, Fiaboe KK, Van den Berg J, Ekesi S, Maniania NK (2014) Effects of endophyte colonization of Vicia faba (Fabaceae) plants on the life—history of leaf minerparasitoids *Phaedroto mascabriventris* (Hymenoptera: Braconidae) and Diglyphusisaea (Hymenoptera: Eulophidae). PLoS One 9(10):109965

Alamri SA, Mostafa YS (2009) Effect of nitrogen supply and Azospirillumbrasilense Sp-248 on the response of wheat to seawater irrigation. Saudi J Biol Sci 16(2):101–107. https://doi.org/10.1016/j.sjbs.2009.10.009

Al-Ani RA, Adhab MA, Mahdi MH, Abood HM (2012) *Rhizobium japonicum* as a biocontrol agent of soybean root rot disease caused by *Fusarium solani* and *Macrophomina phaseolina*. Plant Prot Sci 48(4):149–155. https://doi.org/10.17221/16/2012-PPS

Albright MB, Louca S, Winkler DE, Feeser KL, Haig SJ, Whiteson KL, Emerson JB, Dunbar J (2021) Solutions in microbiome engineering: prioritizing barriers to organism establishment. ISME J 16:1–8

Ali N, Masood S, Mukhtar T, Kamran MA, Rafique M, Munis MFH, Chaudhary HJ (2015) Differential effects of cadmium and chromium on growth, photosynthetic activity, and metal uptake of *Linumus itatissimum* in association with Glomus intraradices. Environ Monit Asses 187(6):1–11. https://doi.org/10.1007/s10661-015-4557-8

Alivisatos AP, Blaser MJ, Brodie EL, Chun M, Dangl JL, Donohue TJ, Dorrestein PC, Gilbert JA, Green JL, Jansson JK, Knight R (2015) A unified initiative to harness Earth's microbiomes. Science 350(6260):507–508. https://doi.org/10.1126/science.aac8480

Al-Karaki GN (2008) *Rhizobium* and phosphorus influence on lentil seed protein and lipid. J Plant Nutr 22:351–358. https://doi.org/10.1080/01904169909365632

Amat D, Thakur JK, Mandal A, Sahu A, Kiran Kumar Reddy K (2020) Production and utilization of legume inoculants (*Rhizobium*) in India. Harit Dhara 3(2):28–31

Amiri R, Nikbakht A, Etemadi N (2015) Alleviation of drought stress on rose geranium [*Pelargonium graveolens* (L.) Herit.] in terms of antioxidant activity and secondary metabolites by mycorrhizal inoculation. Scia Hortic 197:373–380. https://doi.org/10.1016/j.scienta.2015.09.062

Amiri R, Nikbakht A, Etemadi N, Sabzalian MR (2017) Nutritional status, essential oil changes and water-use efficiency of rose geranium in response to arbuscular mycorrhizal fungi and water deficiency stress. Symbiosis 73(1):15–25. https://doi.org/10.1007/s13199-016-0466-z

Amor FMD, Cuadra-Crespo P (2012) Plant growth-promoting bacteria as a tool to improve salinity tolerance in sweet pepper. Funct Plant Biol 39(1):82–90. https://doi.org/10.1071/FP11173

Anandham R, Heo J, Krishnamoorthy R, Senthil Kumar M, Gopal NO, Kim SJ, Kwon SW (2019) *Azospirillum ramasamyi* sp. nov., a novel diazotrophic bacterium isolated from fermented bovine products. Int J Syst Evol Microbiol 69(5):1369–1375. https://doi.org/10.1099/ijsem.0.003320

Andjelković S, Vasića T, Radovića J, Babića S, Markovića J, Zornića V (2018) Abundance of azotobacter in the soil of natural and artificial grasslands. Solut Proj Sustain Soil Manag 172. https://orcid.org/0000-0003-2312-0251

Andreote F, Gumiere T, Durrer A (2014) Exploring interactions of plant microbiomes. Scientia Agricola 71(6):528–539. https://doi.org/10.1590/0103-9016-2014-0195

Andrews M, Andrews M (2017) Specificity in legume-rhizobia symbioses. Int J Mol Sci 18(4):705. https://doi.org/10.3390/ijms18040705

Ansari RA, Mahmood I (eds) (2019a) Plant health under biotic stress: volume 1: organicstrategies. Springer, Singapore. https://doi.org/10.1007/978-981-13-6043-5_8

Ansari RA, Mahmood I (eds) (2019b) Plant health under biotic stress: volume 2: microbial interactions. Springer, Singapore

Ansari MF, Tipre DR, Dave SR (2015) Efficiency evaluation of commercial liquid biofertilizers for growth of Cicer aeritinum (chickpea) in pot and field study. Biocat Agricult Biotechnol 4(1): 17–24. https://doi.org/10.1016/j.bcab.2014.09.010

Ansari RA, Rizvi R, Sumbul A, Mahmood I (2017) PGPR: current vogue in sustainable crop production. In: Probiotics and plant health. Springer, Singapore, pp 455–472. https://doi.org/10.1016/j.sjbs.2020.08.004

Ardakani MR, Mafakheri S (2011) Designing a sustainable agroecosystem for wheat (Triticum *aestivum* L.) production. J Appl Environ Biol Sci 1(10):401–413

Ardakani MR, Pietsch G, Moghaddam A, Raza A, Friedel JK (2009) Response of root properties to tripartite symbiosis between lucerne (Medicago sativa L.), rhizobia and mycorrhiza under dry organic farming conditions. Am J Agricul Biol Sci 4(4):266–277. https://doi.org/10.3844/ajabssp.2009.266.277

Arfaoui A, Sifi B, El Hassni M, El Hadrami I, Boudabbous A, Chérif M (2005) Biochemical analysis of chickpea protection against Fusarium wilt afforded by two Rhizobium isolates. Plant Pathol J 4(1):35–42. https://doi.org/10.3923/ppj.2005.35.42

Arora NK, Khare E, Maheshwari DK (2010) Plant growth promoting rhizobacteria: constraints in bioformulation, commercialization, and future strategies. In: Plant growth and health promoting bacteria. Springer, Berlin, pp 97–116. https://doi.org/10.1007/978-3-642-13612-2_5

Arora M, Saxena P, Abdin MZ, Varma A (2018) Interaction between Piriformospora indica and Azotobacter chroococcum governs better plant physiological and biochemical parameters in

Artemisia annua L. plants grown under in vitro conditions. Symbiosis 75(2):103–112. https://doi.org/10.1007/s13199-017-0519-y

Aserse AA, Markos D, Getachew G, Yli-Halla M, Lindström K (2020) Rhizobial inoculation improves drought tolerance, biomass and grain yields of common bean (*Phaseolus vulgaris* L.) and soybean (*Glycine max* L.) at Halaba and Boricha in Southern Ethiopia. Arch Agron Soil Sci 66(4):488–501. https://doi.org/10.1080/03650340.2019.1624724

Asrar AA, Abdel-Fattah GM, Elhindi KM (2012) Improving growth, flower yield, and water relations of snapdragon (Antirhinum majus L.) plants grown under well-watered and water-stress conditions using arbuscular mycorrhizal fungi. Photosynthetica 50(2):305–316. https://doi.org/10.1007/s11099-012-0024-8

Assefa AH, Amsalu B, Tana T (2017) Response of common Bean (*Pharsalus vulgaris* L.) cultivars to combined application of rhizobium and NP fertilizer at Melkassa, central Ethiopia. Int J Plant Soil Sci 14(1):1–10. https://doi.org/10.9734/IJPSS/2017/30864

Audet P (2014) Arbuscular mycorrhizal fungi and metal phytoremediation: ecophysiological complementarity in relation to environmental stress. In: Emerging technologies and management of crop stress tolerance. Academic, Amsterdam, pp 133–160

Avio L, Sbrana C, Giovannetti M, Frassinetti S (2017) Arbuscular mycorrhizal fungi affect total phenolics content and antioxidant activity in leaves of oak leaf lettuce varieties. Scia Hortic 224: 265–271. https://doi.org/10.1016/j.scienta.2017.06.022

Baars O, Zhang X, Morel FM, Seyedsayamdost MR (2016) The siderophore metabolome of Azotobacter vinelandii. App Env Microbiol 82(1):27–39. https://doi.org/10.1128/aem.03160-15

Backer R, Rokem JS, Ilangumaran G, Lamont J, Praslickova D, Ricci E, Subramanian S, Smith DL (2018) Plant growth-promoting rhizobacteria: context, mechanisms of action, and roadmap to commercialization of biostimulants for sustainable agriculture. Front Plant Sci 9:1473. https://doi.org/10.3389/fpls.2018.01473

Bagyaraj DJ, Ashwin R (2017) Soil biodiversity: role in sustainable horticulture. Biodivers Hortic Crops 5:1–18

Bakker MG, Manter DK, Sheflin AM, Weir TL, Vivanco JM (2012) Harnessing the rhizosphere microbiome through plant breeding and agricultural management. Plant Soil 360:1–13. https://doi.org/10.1146/annurev.arplant.57.032905.105159

Bakker PAHM, Pieterse CMJ, de Jonge R, Berendsen RL (2018) The soil-borne legacy. Cell 172: 1178–1180. https://doi.org/10.1016/j.cell.2018.02.024

Balestrini R, Gómez-Ariza J, Lanfranco L, Bonfante P (2007) Laser microdissection reveals that transcripts for five plant and one fungal phosphate transporter genes are contemporaneously present in arbusculated cells. Mol Plant Microbe Interact 20(9):1055–1062. https://doi.org/10.1094/mpmi-20-9-1055

Bang C, Dagan T, Deines P, Dubilier N, Duschl WJ, Fraune S, Hentschel U, Hirt H, Hülter N, Lachnit T, Picazo D (2018) Metaorganisms in extreme environments: do microbes play a role in organismal adaptation? Zoology 127:1–19. https://doi.org/10.1016/j.zool.2018.02.004

Bano Q, Ilyas N, Bano A, Zafar N, Akram A, Hassan F (2013) Effect of Azospirillum inoculation on maize (Zea mays L.) under drought stress. Pak J Bot 45(S1):13–20

Bano S, Wu X, Zhang X (2021) Towards sustainable agriculture: rhizosphere microbiome engineering. Appl Microbiol Biotechnol 105(19):7141–7160. https://doi.org/10.1016/j.jare.2020.04.014

Bansal R, Srivastava JP (2012) Antioxidative defense system in pigeonpea roots under waterlogging stress. Acta Physiol Plant 34(2):515–522. https://doi.org/10.1093/femsec/fiaa017

Bao L, Gu L, Sun B, Cai W, Zhang S, Zhuang G, Bai Z, Zhuang X (2020) Seasonal variation of epiphytic bacteria in the phyllosphere of Gingko biloba, Pinus bungeana and Sabina chinensis. FEMS Microbiol Ecol 96(3):17. https://doi.org/10.1093/femsec/fiaa017

Bärenstrauch M, Mann S, Jacquemin C, Bibi S, Sylla OK, Baudouin E, Buisson D, Prado S, Kunz C (2020) Molecular crosstalk between the endophyte Paraconiothyrium variabile and the phytopathogen Fusarium oxysporum—modulation of lipoxygenase activity and beauvericin production during the interaction. Fungal Genet Biol 139:103383. https://doi.org/10.1016/j.fgb.2020.103383

Barman M, Paul S, Choudhury AG, Roy P, Sen J (2017) Biofertilizer as prospective input for sustainable agriculture in India. Int J Curr Microbiol Appl Sci 6(11):1–177. https://doi.org/10.20546/ijcmas.2017.611.141

Barrera SE, Sarango-Flóres SW, Montenegro-Gómez SP (2019) The phyllosphere microbiome and its potential application in horticultural crops. A review. Rev Colomb Cienc Hortic 13(3): 384–396. https://doi.org/10.17584/rcch.2019vl3i3.8405

Bashan Y, De-Bashan LE (2010) How the plant growth-promoting bacterium Azospirillum promotes plant growth—a critical assessment. Adv Agron 108:77–136. https://doi.org/10.1016/S0065-2113(10)08002-8

Bashir I, War AF, Rafiq I, Reshi ZA, Rashid I, Shouche YS (2022) Phyllosphere microbiome: diversity and functions. Microbioll Res 254:126888. https://doi.org/10.1016/j.micres.2021.126888

Basu S, Kumar G, Chhabra S, Prasad R (2021) Role of soil microbes in biogeochemical cycle for enhancing soil fertility. In: Verma JP, Macdonald C, Gupta VK, Podile AR (eds) New and future developments in microbial biotechnology and bioengineering: phytomicrobiome for sustainable agriculture. Elsevier, pp 149–157

Bati CB, Santilli E, Lombardo L (2015) Effect of arbuscular mycorrhizal fungi on growth and on micronutrient and macronutrient uptake and allocation in olive plantlets growing under high total Mn levels. Mycorrhiza 25:97–108. https://doi.org/10.1007/s00572-014-0589-0

Battini F, Bernardi R, Turrini A, Agnolucci M, Giovannetti M (2016) Rhizophagus intraradices or its associated bacteria affect gene expression of key enzymes involved in the rosmarinic acid biosynthetic pathway of basil. Mycorrhiza 26(7):699–707. https://doi.org/10.1007/s00572-016-0707-2

Battini F, Grønlund M, Agnolucci M, Giovannetti M, Jakobsen I (2017) Facilitation of phosphorus uptake in maize plants by mycorrhizosphere bacteria. Sci Rep 7(1):1–11. https://doi.org/10.1038/s41598-017-04959-0

Bayani R, Saateyi A, Faghani E (2015) Influence of arbuscular mycorrhiza in phosphorus acquisition efficiency and drought-tolerance mechanisms in barley. Int J Biosci 7(1):86–94. https://doi.org/10.12692/ijb/7.1.86-94

Beattie GA (2002) Leaf surface waxes and the process of leaf colonization by microorganisms. Phyllosphere Microbiol:3–26. https://doi.org/10.1094/MPMI.2002.15.12.1236

Begum AA, Leibovitch S, Migner P, Zhang F (2001a) Inoculation of pea (*Pisum sativum* L.) by *Rhizobium leguminosarum bv.* viceae preincubated with naringenin and hesperetin or application of naringenin and hesperetin directly into soil increased pea nodulation under short season conditions. Plant Soil 237(1):71–80

Begum AA, Leibovitch S, Migner P, Zhang F (2001b) Specific flavonoids induced nod gene expression and pre-activated nod genes of Rhizobium leguminosarum increased pea (*Pisums ativum* L.) and lentil (*Lens culinaris* L.) nodulation in controlled growth chamber environments. J Expl Bot 52(360):1537–1543. https://doi.org/10.1093/jexbot/52.360.1537

Begum N, Qin C, Ahanger MA, Raza S, Khan MI, Ashraf M, Ahmed N, Zhang L (2019) Role of arbuscular mycorrhizal fungi in plant growth regulation: implications in abiotic stress tolerance. Front Plant Sci 10:1068. https://doi.org/10.3389/fpls.2019.01068

Behie SW, Bidochka MJ (2014) Nutrient transfer in plant–fungal symbioses. Trends Plant Sci 19(11):734–740

Behie SW, Zelisko PM, Bidochka MJ (2012) Endophytic insect-parasitic fungi translocate nitrogen directly from insects to plants. Science 336(6088):1576–1577. https://doi.org/10.1126/science.1222289

Belmondo S, Fiorilli V, Pérez-Tienda J, Ferrol N, Marmeisse R, Lanfranco L (2014) A dipeptide transporter from the arbuscular mycorrhizal fungus Rhizophagus irregularis is upregulated in the intraradical phase. Front Plant Sci 5:436. https://doi.org/10.3389/fpls.2014.00436

Benedetto A, Magurno F, Bonfante P, Lanfranco L (2005) Expression profiles of a phosphate transporter gene (GmosPT) from the endomycorrhizal fungus Glomus mosseae. Mycorrhiza 15(8):620–627

Berendsen R, Pieterse C, Bakker P (2012) The rhizosphere microbiome and plant health. Trends Plant Sci 17(8):478–486

Berg RH (1994) Symbiotic vesicle ultrastructure in high pressure-frozen, freeze-substituted actinorhizae. Protoplasma 183(1-4):37–48

Berrada H, Fikri-Benbrahim K (2014) Taxonomy of the rhizobia: current perspectives. Microbiol Res J Int 4(6):616–639

Berruti A, Lumini E, Balestrini R, Bianciotto V (2016) Arbuscular mycorrhizal fungi as natural biofertilizers: let's benefit from past successes. Front Environ Sci 6:1559. https://doi.org/10.3389/fmicb.2015.01559

Bodenhausen N, Horton M, Bergelson J (2013) Bacterial communities associated with the leaves and the roots of *Arabidopsis thaliana*. PLoS One 8(2):e56329

Bona E, Lingua G, Manassero P, Cantamessa S, Marsano F, Todeschini V, Copetta A, D'Agostino G, Massa N, Avidano L, Gamalero E (2015) AM fungi and PGP pseudomonads increase flowering, fruit production, and vitamin content in strawberry grown at low nitrogen and phosphorus levels. Mycorrhiza 25(3):181–193

Borde M, Dudhane M, Jite PK (2010) AM fungi influences the photosynthetic activity, growth and antioxidant enzymes in Allium sativum L. under salinity condition. Notulae Sci Biol 2(4):64–71

Bottini R, Fulchieri M, Pearce D, Pharis RP (1989) Identification of gibberellins A1, A3, and iso-A3 in cultures of *Azospirillum lipoferum*. Plant Physiol 90(1):45–47

Boyer LR, Brain P, Xu XM, Jeffries P (2015) Inoculation of drought-stressed strawberry with a mixed inoculum of two arbuscular mycorrhizal fungi: effects on population dynamics of fungal species in roots and consequential plant tolerance to water deficiency. Mycorrhiza 25(3): 215–227

Bram WGS, Eric AW, Colin RJ (2018) The role of the phyllosphere microbiome in plant health and function. Annual Plant Rev 1:1–24. https://doi.org/10.1002/9781119312994.apr0614

Breakspear A, Liu C, Roy S, Stacey N, Rogers C, Trick M, Morieri G, Mysore KS, Wen J, Oldroyd GE, Downie JA (2014) The root hair "infectome" of Medicago truncatula uncovers changes in cell cycle genes and reveals a requirement for auxin signaling in rhizobial infection. Plant Cell 26(12):4680–4701. https://doi.org/10.1105/tpc.114.133496

Bücking H, Shachar-Hill Y (2005) Phosphate uptake, transport and transfer by the arbuscular mycorrhizal fungus Glomus intraradices is stimulated by increased carbohydrate availability. New Phytol 165(3):899–912. https://doi.org/10.1111/j.1469-8137.2004.01274.x

Bulegon LG, Guimarães VF, Laureth JCU (2016) *Azospirillum brasilense* affects the antioxidant activity and leaf pigment content of Urochloaruziziensis under water stress. Pesqui Agro Pecu Bras 46:343–349. https://doi.org/10.1590/1983-40632016v4641489

Bunn R, Lekberg Y, Zabinski C (2009) Arbuscular mycorrhizal fungi ameliorate temperature stress in thermophilic plants. Ecology 90(5):1378–1388. https://doi.org/10.1890/07-2080.1

Bziuk N, Maccario L, Straube B, Wehner G, Sørensen SJ, Schikora A, Smalla K (2021) The treasure inside barley seeds: microbial diversity and plant beneficial bacteria. Environ Microbiome 16:20. https://doi.org/10.1186/s40793-021-00389-8

Cabral C, Ravnskov S, Tringovska I, Wollenweber B (2016) Arbuscular mycorrhizal fungi modify nutrient allocation and composition in wheat (*Triticum aestivum* L.) subjected to heat-stress. Plant Soil 408(1-2):385–399. https://doi.org/10.1007/S11104-016-2942-X

Calabrese S, Pérez-Tienda J, Ellerbeck M, Arnould C, Chatagnier O, Boller T, Schüßler A, Brachmann A, Wipf D, Ferrol N, Courty PE (2016) GintAMT3—a low-affinity ammonium transporter of the arbuscular mycorrhizal *Rhizopha gusirregularis*. Front Plant Sci 7:679. https://doi.org/10.3389/fpls.2016.00679

Calvo-Polanco M, Sánchez-Romera B, Aroca R, Asins M, Declerck S, Dodd I, Martínez-Andújar C, Albacete A, Ruiz-Lozano J (2016) Exploring the use of recombinant inbred lines in combination with beneficial microbial inoculants (AM fungus and PGPR) to improve drought stress tolerance in tomato. Environ Exp Bot 131:47–57. https://doi.org/10.1016/j.envexpbot.2016.06.015

Campanelli A, Ruta C, De Mastro G, Morone-Fortunato I (2013) The role of arbuscular mycorrhizal fungi in alleviating salt stress in Medicago sativa L. var. icon. Symbiosis 59(2):65–76. https://doi.org/10.1007/s13199-012-0191-1

Canarini A, Kaiser C, Merchant A, Richter A, Wanek W (2019) Root exudation of primary metabolites: mechanisms and their roles in plant responses to environmental stimuli. Front Plant Sci 10:157. https://doi.org/10.3389/fpls.2019.00157

Carrell AA, Frank AC (2014) Pinus flexilis and Picea engelmannii share a simple and consistent needle endophyte microbiota with a potential role in nitrogen fixation. Front Microbiol 5:333. https://doi.org/10.3389/fmicb.2014.00333

Carrozzi LE, Creus CM, Barassi CA, Monterubbianesi G, Di Benedetto A (2012) Reparation of aged lettuce (*Lactuca sativa*) seeds by osmotic priming and *Azospirillum brasilense* inoculation. Botany 90(12):1093–1102. https://doi.org/10.1139/b2012-087

Cartmill AD, Valdez-Aguilar LA, Bryan DL, Alarcón A (2008) Arbuscular mycorrhizal fungi enhance tolerance of vinca to high alkalinity in irrigation water. Scia Hortic 115(3):275–284. https://doi.org/10.1016/j.scienta.2007.08.019

Carvalho M, Ma Y, Oliveira R, Freitas H (2017) Endophytic actinobacteria for sustainable agricultural applications. In: Endophytes: crop productivity and protection. Springer, Cham, pp 163–189. https://doi.org/10.1007/978-3-319-66541-2_8

Casieri L, Gallardo K, Wipf D (2012) Transcriptional response of Medicago truncatula sulphate transporters to arbuscular mycorrhizal symbiosis with and without sulphur stress. Planta 235(6): 1431–1447. https://doi.org/10.1007/s00425-012-1645-7

Cassman NA, Leite MF, Pan Y, de Hollander M, van Veen JA, Kuramae EE (2016) Plant and soil fungal but not soil bacterial communities are linked in long-term fertilized grassland. Sci Rep 6(1):1–11. https://doi.org/10.1038/srep23680

Ceballos I, Mosquera S, Angulo M, Mira JJ, Argel LE, Uribe-Velez D, Romero-Tabarez M, Orduz-Peralta S, Villegas V (2012) Cultivable bacteria populations associated with leaves of banana and plantain plants and their antagonistic activity against Mycosphaerella fijiensis. Microbial Ecol 64(3):641–653. https://doi.org/10.1007/s00248-012-0052-8

Çekiç FÖ, Ünyayar S, Ortaş İ (2012) Effects of arbuscular mycorrhizal inoculation on biochemical parameters in *Capsicum annuum* grown under long term salt stress. Turkish J Bot 36(1):63–72. https://doi.org/10.3906/Bot-1008-32

Cerezini P, Kuwano BH, dos Santos MB, Terassi F, Hungria M, Nogueira MA (2016) Strategies to promote early nodulation in soybean under drought. Field Crops Res 196:160–167. https://doi.org/10.1016/j.fcr.2016.06.017

Chang W, Sui X, Fan XX, Jia TT, Song FQ (2018) Arbuscular mycorrhizal symbiosis modulates antioxidant response and ion distribution in salt-stressed Elaeagnus angustifolia seedlings. Front Microbiol 9:652. https://doi.org/10.3389/fmicb.2018.00652

Charubin K, Papoutsakis ET (2019) Direct cell-to-cell exchange of matter in a synthetic Clostridium syntrophy enables CO2 fixation, superior metabolite yields, and an expanded metabolic space. Metab Eng 52:9–19. https://doi.org/10.1016/j.ymben.2018.10.006

Chaturvedi H, Singh V, Gupta G (2016) Potential of bacterial endophytes as plant growth promoting factors. J Plant Pathol Microbiol 7(9):1–6. https://doi.org/10.4172/2157-7471.1000376

Chen M, Zhang L, Zhang X (2011) Isolation and inoculation of endophytic actinomycetes in root nodules of Elaeagnus angustifolia. Mod Appl Sci 5(2):264. https://doi.org/10.5539/mas.v5n2p264

Chen S, Jin W, Liu A, Zhang S, Liu D, Wang F, Lin X, He C (2013) Arbuscular mycorrhizal fungi (AMF) increase growth and secondary metabolism in cucumber subjected to low temperature stress. Scia Hortic 160:222–229. https://doi.org/10.1016/j.scienta.2013.05.039

Chhabra S (2019) Phosphorus management in agroecosytems and role and relevance of microbes in environmental sustainability. Sustain Green Technol Environ Manag:53–66. https://doi.org/10.1007/978-981-13-2772-8_3

Chhabra S, Dowling DN (2017) Endophyte-promoted nutrient acquisition: phosphorus and iron. In: Functional importance of the plant microbiome: implications for agriculture, forestry and bioenergy, vol 2017. Springer, Cham, pp 21–42. https://doi.org/10.1007/978-3-319-65897-1_3

Cid F, Inostroza N, Graether S, Bravo L, Jorquera M (2016) Bacterial community structures and ice recrystallization inhibition activity of bacteria isolated from the phyllosphere of the Antarctic vascular plant Deschampsia antarctica. Polar Biol 40(6):1319–1331. https://doi.org/10.1007/s00300-016-2036-5

Cohen AC, Travaglia CN, Bottini R, Piccoli PN (2009) Participation of abscisic acid and gibberellins produced by endophytic Azospirillum in the alleviation of drought effects in maize. Botany 87(5):455–462. https://doi.org/10.1139/B09-023

Cohen AC, Bottini R, Pontin M, Berli FJ, Moreno D, Boccanlandro H, Travaglia CN, Piccoli PN (2015) Azospirillum brasilense ameliorates the response of Arabidopsis thaliana to drought mainly via enhancement of ABA levels. Physiol Plant 153(1):79–90. https://doi.org/10.1111/ppl.12221

Collins DP, Jacobsen BJ (2003) Optimizing a *Bacillus subtilis* isolate for biological control of sugar beet Cercospora leaf spot. Biol Cont 26(2):153–161. https://doi.org/10.1016/S1049-9644(02)00132-9

Colston TJ, Jackson CR (2016) Microbiome evolution along divergent branches of the vertebrate tree of life: what is known and unknown? Mol Ecol 25(16):3776–3800. https://doi.org/10.1111/mec.13730

Compant S, Samad A, Faist H, Sessitsch A (2019) A review on the plant microbiome: ecology, functions, and emerging trends in microbial application. J Adv Res 19:29–37. https://doi.org/10.1016/j.jare.2019.03.004

Cooper JE (2007) Early interactions between legumes and rhizobia: disclosing complexity in a molecular dialogue. J Appl Microbiol 103(5):1355–1365. https://doi.org/10.1111/j.1365-2672.2007.03366.x

Cruz-Martín M, Acosta-Suárez M, Mena E, Roque B, Pichardo T, Alvarado-Capó Y (2017) Antifungal activity of Musa phyllosphere *Bacillus pumilus* strain against Mycosphaerella fijiensis. Trop Plant Pathol 42(2):121–125. https://doi.org/10.1007/s40858-017-0139-3

Cruz-Martín M, Acosta-Suárez M, Mena E, Roque B, Pichardo T, Alvarado-Capó Y (2018) Effect of *Bacillus pumilus* CCIBP-C5 on *Musa-Pseudocercospora fijiensis* interaction. 3 Biotech 8(2):122. https://doi.org/10.1007/s13205-018-1152-z

Cui R, Lu X, Chen X, Malik WA, Wang D, Wang J, Wang S, Guo L, Chen C, Wang X, Wang X (2021) A novel raffinose biological pathway is observed by symbionts of cotton≡Verticillium dahliae to improve salt tolerance genetically on cotton. J Agron Crop Sci 207(6):956–969. https://doi.org/10.1111/jac.12556

Das K, Roychoudhury A (2014) Reactive oxygen species (ROS) and response of antioxidants as ROS-scavengers during environmental stress in plants. Front Env Sci 2:53. https://doi.org/10.3389/fenvs.2014.00053

Dastogeer KMG, Tumpa FH, Sultana A, Akter MA, Chakraborty A (2020) Plant microbiome—an account of the factors that shape community composition and diversity. Current Plant Biol 23:100161. https://doi.org/10.1016/j.cpb.2020.100161

Datta A, Singh RK, Tabassum S (2015) Isolation, characterization and growth of Rhizobium strains under optimum conditions for effective biofertilizer production. Int J Pharm Sci Rev Res 32(1):199–208

De Jensen CE, Percich JA, Graham PH (2002) Integrated management strategies of bean root rot with Bacillus subtilis and Rhizobium in Minnesota. Field Crops Res 74(2-3):107–115. https://doi.org/10.1016/S0378-4290(01)00200-3

De Sassi C, Müller CB, Krauss J (2006) Fungal plant endosymbionts alter life history and reproductive success of aphid predators. Proc R Soc B Biol Sci 273(1591):301–1306. https://doi.org/10.1098/rspb.2005.3442

de Souza R, Beneduzi A, Ambrosini A, Da Costa PB, Meyer J, Vargas LK, Schoenfeld R, Passaglia LM (2013) The effect of plant growth-promoting rhizobacteria on the growth of rice (Oryza sativa L.) cropped in southern Brazilian fields. Plant Soil 366:585–603. https://doi.org/10.1007/s11104-012-1430-1

Delmotte N, Knief C, Chaffron S, Innerebner G, Roschitzki B, Schlapbach R, von Mering C, Vorholt JA (2009) Community proteogenomics reveals insights into the physiology of phyllosphere bacteria. Proc Natl Acad Sci 106(38):16428–16433. https://doi.org/10.1073/pnas.0905240106

Deshmukh R, Sonah H, Patil G, Chen W, Prince S, Mutava R, Vuong T, Valliyodan B, Nguyen HT (2014) Integrating omic approaches for abiotic stress tolerance in soybean. Front Plant Sci 5:244. https://doi.org/10.3390/d12100370

Dhayanithy G, Subban K, Chelliah J (2019) Diversity and biological activities of endophytic fungi associated with Catharanthus roseus. BMC Microbiol 19:22. https://doi.org/10.1186/s12866-019-1386-x

Diagne N, Arumugam K, Ngom M, Nambiar-Veetil M, Franche C, Narayanan KK, Laplaze L (2013) Use of Frankia and actinorhizal plants for degraded lands reclamation. Bio Med Res Int 2013:1–9. https://doi.org/10.1155/2013/948258

Diagne N, Ngom M, Djighaly PI, Fall D, Hocher V, Svistoonoff S (2020) Roles of arbuscular mycorrhizal fungi on plant growth and performance: importance in biotic and abiotic stressed regulation. Diversity 12(10):370. https://doi.org/10.3390/d12100370

Din M, Nelofer R, Salman M, Khan FH, Khan A, Ahmad M, Jalil F, Din JU, Khan M (2019) Production of nitrogen fixing *Azotobacter* (SR-4) and phosphorus solubilizing *Aspergillus* niger and their evaluation on Lagenaria siceraria and *Abelmoschus esculentus*. Biotechnol Rep 22:e00323. https://doi.org/10.1016/j.btre.2019.e00323

Dinesh R, Srinivasan V, TE S, Anandaraj M, Srambikka H (2017) Endophytic actinobacteria: diversity, secondary metabolism and mechanisms to unsilence biosynthetic gene clusters. Crit Rev Microbiol 43(5):546–566. https://doi.org/10.1080/1040841x.2016.1270895

Dong CJ, Wang LL, Li Q, Shang QM (2019) Bacterial communities in the rhizosphere, phyllosphere and endosphere of tomato plants. PLoS One 14(11):e0223847. https://doi.org/10.1371/journal.pone.0223847

dos Santos Ferreira N, Hayashi Sant' Anna F, Massena Reis V et al (2020) Genome-based reclassification of *Azospirillum brasilense* Sp245 as the type strain of *Azospirillum baldaniorum* sp. nov. Int J Syst Evol Microbiol 70(12):6203–6212. https://doi.org/10.1099/ijsem.0.004517

Du HJ, Zhang YQ, Liu HY, Su J, Wei YZ, Ma BP, Guo BL, Yu LY (2013) Allonocardiopsis opalescens gen. nov., sp. nov., a new member of the suborder Streptosporangineae, from the surface-sterilized fruit of a medicinal plant. Int J Syst Evol Microbiol 63:900–904. https://doi.org/10.1099/ijs.0.041491-0

Dubey A, Mall MA, Khan F, Chowdhary K, Yadav S, Kumar A, Sharma S, Khare PK, Khan ML (2019) Soil microbiome: a key player for conservation of soil health under changing climate. Biodivers Conserv 28(8):2405–2429. https://doi.org/10.1007/s10531-019-01760-5

Dubilier N, McFall-Ngai M, Zhao L (2015) Microbiology: create a global microbiome effort. Nat News 526(7575):631

Durand A, Maillard F, Alvarez-Lopez V, Guinchard S, Bertheau C, Valot B, Blaudez D, Chalot M (2018) Bacterial diversity associated with poplar trees grown on a Hg-contaminated site: community characterization and isolation of Hg-resistant plant growth-promoting bacteria. Sci Total Environ 622-623:1165–1177. https://doi.org/10.1016/j.scitotenv.2017.12.069

Dutta S, Singh MS (2002) Mustard and rapeseed response to azotobacter. Indian J Hill Farming 15:44–46

Egamberdieva D, Wirth SJ, Alqarawi AA, Abd_Allah EF, Hashem A (2017) Phytohormones and beneficial microbes: essential components for plants to balance stress and fitness. Front Microbiol 8:2104. https://doi.org/10.3389/fmicb.2017.02104

El-Badry MA, Elbarbary TA, Ibrahim IA, Abdel-Fatah YM (2016) *Azotobacter vinelandii* evaluation and optimization of Abu Tartur Egyptian phosphate ore dissolution. Saudi J Pathol Microbiol 1(3):80–93. https://doi.org/10.21276/sjpm.2016.1.3.2

Elhindi KM, El-Din AS, Elgorban AM (2017) The impact of arbuscular mycorrhizal fungi in mitigating salt-induced adverse effects in sweet basil (*Ocimum basilicum* L.). Saudi J Biol Sci 24(1):170–179. https://doi.org/10.1016/j.sjbs.2016.02.010

Elías JM, Guerrero-Molina MF, Martínez-Zamora MG, Díaz-Ricci JC, Pedraza RO (2018) Role of ethylene and related gene expression in the interaction between strawberry plants and the plant growth-promoting bacterium Azospirillum brasilense. Plant Biol 20(3):490–496. https://doi.org/10.1111/plb.12697

El-Nashar YI (2017) Response of snapdragon (Antirrhinum majus L.) to blended water irrigation and arbuscular mycorrhizal fungi inoculation: uptake of minerals and leaf water relations. Photosynthetica 55(2):201–209. https://doi.org/10.1007/s11099-016-0650-7

El-Sayed MT (2014) The response of Fusarium solani to Cd (II) and Cu (II) in pure culture. Egypt J Microbiol 5:99–117. https://doi.org/10.21608/ejm.2014.244

Esitken A, Yildiz HE, Ercisli S, Donmez MF, Turan M, Gunes A (2010) Effects of plant growth promoting bacteria (PGPB) on yield, growth and nutrient contents of organically grown strawberry. Scia Hortic 124(1):62–66. https://doi.org/10.1007/s00572-011-0392-0

Evelin H, Giri B, Kapoor R (2012) Contribution of Glomus intraradices inoculation to nutrient acquisition and mitigation of ionic imbalance in NaCl-stressed Trigonella foenum-graecum. Mycorrhiza 22(3):203–217. https://doi.org/10.1007/s00572-011-0392-0

Evelin H, Giri B, Kapoor R (2013) Ultrastructural evidence for AMF mediated salt stress mitigation in Trigonella foenum-graecum. Mycorrhiza 23(1):71–86. https://doi.org/10.1007/s00572-012-0449-8

Fang W, St Leger RJ (2010) Mrt, a gene unique to fungi, encodes an oligosaccharide transporter and facilitates rhizosphere competency in Metarhizium robertsii. Plant Physiol 154(3):1549–1557. https://doi.org/10.1104/pp.110.163014

Farssi O, Mouradi M, Ghoulam C, Bouizgaren A, Berrougui H, Farissi M (2018) Genes and molecular signals involved in legumes-rhizobia symbiosis. Ann West Univ Timiş Ser Biol 21(2):193–206

Fasciglione G, Casanovas EM, Quillehauquy V, Yommi AK, Goni MG, Roura SI, Barassi CA (2015) Azospirillum inoculation effects on growth, product quality and storage life of lettuce plants grown under salt stress. Scia Hortic 195:154–162. https://doi.org/10.1016/j.scienta.2015.09.015

Feng Y, Shen D, Dong X, Song W (2003) In vitro symplasmata formation in the rice diazotrophic endophyte *Pantoea agglomerans* YS19. Plant Soil 255(2):435–444

Fernando WGD, Nakkeeran S, Zhang Y, Savchuk S (2007) Biological control of *Sclerotiniasclerotiorum* (Lib.) de Bary by *Pseudomonas* and *Bacillus* species on canola petals. Crop Protect 26(2):100–107. https://doi.org/10.1016/j.cropro.2006.04.007

Fibach-Paldi S, Burdman S, Okon Y (2012) Key physiological properties contributing to rhizosphere adaptation and plant growth promotion abilities of Azospirillumbrasilense. FEMS Microbiol Lett 326(2):99–108. https://doi.org/10.1111/j.1574-6968.2011.02407.x

Field KJ, Rimington WR, Bidartondo MI, Allinson KE, Beerling DJ, Cameron DD, Duckett JG, Leake JR, Pressel S (2015) First evidence of mutualism between ancient plant lineages (H. aplomitriopsida liverworts) and M. ucoromycotina fungi and its response to simulated P. alaeozoic changes in atmospheric CO2. New Phytol 205(2):743–756. https://doi.org/10.1111/nph.13024

Foo JL, Ling H, Lee YS, Chang MW (2017) Microbiome engineering: current applications and its future. Biotechnol J 12(3):1600099. https://doi.org/10.1002/biot.201600099

Frąc M, Hannula SE, Bełka M, Jędryczka M (2018) Fungal biodiversity and their role in soil health. Front Microbiol 9:707. https://doi.org/10.3389/fmicb.2018.00707

Franche C, Bogusz D (2012) Signalling and communication in the actinorhizal symbiosis. In: Signaling and communication in plant symbiosis. Springer, Berlin, pp 73–92. https://doi.org/10.1007/978-3-642-20966-6_4

Freiberg E (1998) Microclimatic parameters influencing nitrogen fixation in the phyllosphere in a Costa Rican premontane rain forest. Oecologia 117(1-2):9–18. https://doi.org/10.1007/s004420050625

Fu SF, Sun PF, Lu HY, Wei JY, Xiao HS, Fang WT, Cheng BY, Chou JY (2016) Plant growth-promoting traits of yeasts isolated from the phyllosphere and rhizosphere of Droseraspatulata Lab. Fungal Biol 120(3):433–448. https://doi.org/10.1016/j.funbio.2015.12.006

Fukami J, Nogueira MA, Araujo RS, Hungria M (2016) Accessing inoculation methods of maize and wheat with *Azospirillum brasilense*. Amb Express 6(1):1–13. https://doi.org/10.1186/s13568-015-0171-y

Fukami J, de la Osa C, Ollero FJ, Megías M, Hungria M (2017) Co-inoculation of maize with Azospirillum brasilense and Rhizobium tropici as a strategy to mitigate salinity stress. Funct Plant Biol 45(3):328–339. https://doi.org/10.1071/fp17167

Fukami J, Cerezini P, Hungria M (2018) *Azospirillum*: benefits that go far beyond biological nitrogen fixation. AMB Expr 8:73. https://doi.org/10.1186/s13568-018-0608-1

Fürnkranz M, Wanek W, Richter A, Abell G, Rasche F, Sessitsch A (2008) Nitrogen fixation by phyllosphere bacteria associated with higher plants and their colonizing epiphytes of a tropical lowland rainforest of Costa Rica. ISME J 2(5):561–570. https://doi.org/10.1038/ismej.2008.14

Gaofeng W, Hong X, Mei J (2004) Biodegradation of chlorophenols, a review. Chem J Internet 10: 67. www.chemistrymag.on/cji/2004/

Garcés-Ruiz M, Calonne-Salmon M, Plouznikoff K, Misson C, Navarrete-Mier M, Cranenbrouck S, Declerck S (2017) Dynamics of short-term phosphorus uptake by intact mycorrhizal and non-mycorrhizal maize plants grown in a circulatory semi-hydroponic cultivation system. Front Plant Sci 8:1471. https://doi.org/10.3389/fpls.2017.01471

Garcia K, Delteil A, Conéjéro G, Becquer A, Plassard C, Sentenac H, Zimmermann S (2014) Potassium nutrition of ectomycorrhizal P inus pinaster: overexpression of the Hebeloma cylindrosporum H c T rk1 transporter affects the translocation of both K+ and phosphorus in the host plant. New Phytol 201(3):951–960. https://doi.org/10.1111/nph.12603

García JE, Maroniche G, Creus C, Suárez-Rodríguez R, Ramirez-Trujillo JA, Groppa MD (2017) In vitro PGPR properties and osmotic tolerance of different Azospirillum native strains and their effects on growth of maize under drought stress. Microbiol Res 202:21–29. https://doi.org/10.1016/j.micres.2017.04.007

Garg N, Aggarwal N (2012) Effect of mycorrhizal inoculations on heavy metal uptake and stress alleviation of *Cajanus cajan* (L.) Millsp. genotypes grown in cadmium and lead contaminated soils. Plant Growth Regul 66:9–26. https://doi.org/10.1007/s10725-011-9624-8

Garg N, Bhandari P (2016) Silicon nutrition and mycorrhizal inoculations improve growth, nutrient status, K+/Na+ ratio and yield of Cicer arietinum L. genotypes under salinity stress. Plant Growth Regul 78(3):371–387. https://doi.org/10.1007/s10725-015-0099-x

Garg N, Bharti A (2018) Salicylic acid improves arbuscular mycorrhizal symbiosis, and chickpea growth and yield by modulating carbohydrate metabolism under salt stress. Mycorrhiza 28(8): 727–746. https://doi.org/10.1007/s00572-018-0856-6

Garg N, Chandel S (2011) Role of Arbuscular Mycorrhizal (AM) fungi on growth, cadmium uptake, osmolyte, and phytochelatin synthesis in Cajanus cajan (L.) Millsp. under NaCl and Cd stresses. J Plant Growth Regul 31(3):292–308. https://doi.org/10.3389/fpls.2019.01068

Garg N, Singh S (2018) Arbuscular mycorrhiza *Rhizophagus irregularis* and silicon modulate growth, proline biosynthesis and yield in Cajanus cajan L. Millsp. (pigeon pea) genotypes under cadmium and zinc stress. J Plant Growth Regul 37(1):46–63. https://doi.org/10.1007/s00344-017-9708-4

Gazis R, Chaverri P (2010) Diversity of fungal endophytes in leaves and stems of wild rubber trees (Hevea brasiliensis) in Peru. Fungal Ecol 3(3):240–254. https://doi.org/10.1016/j.funeco.2009.12.001

Genetu G, Yli-Halla M, Asrat M, Alemayehu M (2021) Rhizobium inoculation and chemical fertilisation improve faba bean yield and yield components in Northwestern Ethiopia. Agriculture 11(7):678. https://doi.org/10.3390/agriculture11070678

Genre A, Chabaud M, Timmers T, Bonfante P, Barker DG (2005) Arbuscular mycorrhizal fungi elicit a novel intracellular apparatus in Medicago truncatula root epidermal cells before infection. Plant Cell 17(12):3489–3499. https://doi.org/10.1105/tpc.105.035410

Genre A, Chabaud M, Faccio A, Barker DG, Bonfante P (2008) Prepenetration apparatus assembly precedes and predicts the colonization patterns of arbuscular mycorrhizal fungi within the root cortex of both Medicago truncatula and Daucus carota. Plant Cell 20(5):1407–1420. https://doi.org/10.1105/tpc.108.059014

Gibson K, Kobayashi H, Walker G (2008) Molecular determinants of a symbiotic chronic infection. Annu Rev Genet 42(1):413–441. https://doi.org/10.1146/annurev.genet.42.110807.091427

Gill SS, Gill R, Trivedi DK, Anjum NA, Sharma KK, Ansari MW, Ansari AA, Johri AK, Prasad R, Pereira E, Varma A, Tuteja N (2016) *Piriformospora indica*: potential and significance in plant stress tolerance. Front Microbiol 7:332. https://doi.org/10.3389/fmicb.2016.00332

Giri B, Kapoor R, Mukerji KG (2007) Improved tolerance of Acacia nilotica to salt stress by arbuscular mycorrhiza, Glomus fasciculatum may be partly related to elevated K/Na ratios in root and shoot tissues. Microbial Ecol 54(4):753–760. https://doi.org/10.1007/s00248-007-9239-9

Glick BR (2012) Plant growth-promoting bacteria: mechanisms and applications. Scientifica 2012: 963401. https://doi.org/10.6064/2012/963401

Goicoechea N, Bettoni MM, Fuertes-Mendizabal T, González-Murua C, Aranjuelo I (2016) Durum wheat quality traits affected by mycorrhizal inoculation, water availability and atmospheric CO2 concentration. Crop Pasture Sci 67(2):147–155. https://doi.org/10.1071/CP15212

Gómez-Bellot M, Ortuño M, Nortes P, Vicente-Sánchez J, Bañón S, Sánchez-Blanco M (2015) Mycorrhizal euonymus plants and reclaimed water: biomass, water status and nutritional responses. Scia Hortic 186:61–69. https://doi.org/10.1016/j.scienta.2015.02.022

Gonzalez AJ, Larraburu EE, Llorente BE (2015) *Azospirillum brasilense* increased salt tolerance of jojoba during in vitro rooting. Indusl Crops Prod 76:41–48. https://doi.org/10.1016/j.indcrop.2015.06.017

González-Chávez MDCA, Miller B, Maldonado-Mendoza IE, Scheckel K, Carrillo-González R (2014) Localization and speciation of arsenic in Glomus intraradices by synchrotron radiation spectroscopic analysis. Fungal Biol 118(5-6):444–452. https://doi.org/10.1016/j.funbio.2014.03.002

González-Guerrero M, Azcon-Aguilar C, Mooney M, Valderas A, MacDiarmid CW, Eide DJ, Ferrol N (2005) Characterization of a *Glomus intraradices* gene encoding a putative Zn transporter of the cation diffusion facilitator family. Fungal Gen Biol 42(2):130–140. https://doi.org/10.1016/j.fgb.2004.10.007

Gopal M, Gupta A, Thomas GV (2013) Bespoke microbiome therapy to manage plant diseases. Front Microbiol 4:355. https://doi.org/10.3389/fmicb.2013.00355

Griffin EA, Carson WP (2015) The ecology and natural history of foliar bacteria with a focus on tropical forests and agroeco systems. Bot Rev 81(2):105–149. https://doi.org/10.1007/s12229-015-9151-9

Grümberg B, Urcelay C, Shroeder M, Vargas-Gil S, Luna C (2014) The role of inoculum identity in drought stress mitigation by arbuscular mycorrhizal fungi in soybean. Biol Ferti Soils 51(1): 1–10. https://doi.org/10.1007/s00374-014-0942-7

Guerreiro S, Dawson R, Kilsby C, Lewis E, Ford A (2018) Future heat-waves, droughts and floods in 571 European cities. Environ Res Let 13(3):034009

Guerrieri R, Lecha L, Mattana S, Cáliz J, Casamayor EO, Barceló A et al (2020) Partitioning between atmospheric deposition and canopy microbial nitrification into throughfall nitrate fluxes in a Mediterranean forest. J Ecol 108(2):626–640

Gupta G, Parihar SS, Ahirwar NK, Snehi SK, Singh V (2015) Plant growth promoting rhizobacteria (PGPR): current and future prospects for development of sustainable agriculture. J Microb Biochem Technol 7(2):96–102. https://doi.org/10.4172/1948-5948.1000188

Gupta N, Vats S, Bhargava P (2018) Sustainable agriculture: role of metagenomics and metabolomics in exploring the soil microbiota. In: Choudhary D, Kumar M, Prasad R, Kumar V (eds) Silico approach for sustainable agriculture. Springer, Singapore. https://doi.org/10.1007/978-981-13-0347-0_11

Gureny KA, Mantle PG (1993) Biosynthesis of 1-N-methylalbonoursin by an endophytic Streptomyces sp. isolated from perennial ryegrass. J Nat Prod 56(7):1194–1198

Hajiboland R, Aliasgharzadeh N, Laiegh SF, Poschenrieder C (2010) Colonization with arbuscular mycorrhizal fungi improves salinity tolerance of tomato (*Solanum lycopersicum* L.) plants. Plant Soil 331(1):313–327. https://doi.org/10.1007/s11104-009-0255-z

Hajiboland R, Dashtebani F, Aliasgharzad N (2015) Physiological responses of halophytic C4 grass Aeluropus littoralis to salinity and arbuscular mycorrhizal fungi colonization. Photosynthetica 53(4):572–584. https://doi.org/10.1007/s11099-015-0131-4

Hamdia MAES, Shaddad MAK, Doaa MM (2004) Mechanisms of salt tolerance and interactive effects of Azospirillum brasilense inoculation on maize cultivars grown under salt stress conditions. Plant Growth Regul 44(2):165–174. https://doi.org/10.1023/B:GROW.0000049414.03099.9b

Hart M, Antunes P, Chaudhary VB, Abbott L (2018) Fungal inoculants in the field: is the reward greater than the risk? Funct Ecol 32:126–135. https://doi.org/10.1111/1365-2435.12976

Hartmann A, Rothballer M, Schmid M (2008) Lorenz Hiltner, a pioneer in rhizosphere microbial ecology and soil bacteriology research. Plant Soil 312(1):7–14. https://doi.org/10.1007/s11104-007-9514-z

Hasanuzzaman M, Gill SS, Fujita M (2013) Physiological role of nitric oxide in plants grown under adverse environmental conditions. In: Plant acclimation to environmental stress. Springer, New York, pp 269–322. https://doi.org/10.1007/978-1-4614-5001-6_11

Hashem A, Abd-Allah EF, Alqarawi AA, Egamberdieva D (2018a) Arbuscular mycorrhizal fungi and plant stress tolerance. In: Plant microbiome: stress response. Springer, Singapore, pp 81–103. https://doi.org/10.1007/978-981-10-5514-0_4

Hashem A, Alqarawi AA, Radhakrishnan R, Al-Arjani, ABF, Aldehaish HA, Egamberdieva D, Abd_Allah EF (2018b) Arbuscular mycorrhizal fungi regulate the oxidative system, hormones and ionic equilibrium to trigger salt stress tolerance in Cucumis sativus L. Saudi J Biol Sci 25(6): 1102–1114. https://doi.org/10.1016/j.sjbs.2018.03.009

Hassen AI, Pierneef R, Swanevelder ZH, Bopape FL (2020) Microbial and functional diversity of Cyclopia intermedia rhizosphere microbiome revealed by analysis of shotgun metagenomics sequence data. Data Brief 32:106288. https://doi.org/10.1016/j.dib.2020.106288

Hayat R, Ali S, Amara U, Khalid R, Ahmed I (2010) Soil beneficial bacteria and their role in plant growth promotion: a review. Ann Microbiol 60(4):579–598. https://doi.org/10.1007/s13213-010-0117-1

Helber N, Wippel K, Sauer N, Schaarschmidt S, Hause B, Requena N (2011) A versatile monosaccharide transporter that operates in the arbuscular mycorrhizal fungus Glomus sp is crucial for the symbiotic relationship with plants. Plant Cell 23(10):3812–3823. https://doi.org/10.1105/tpc.111.089813

Herliana O, Harjoso T, Anwar AHS, Fauzi A (2019) The effect of rhizobium and N fertilizer on growth and yield of black soybean (*Glycine max* (L) Merril). IOP Conf Ser Earth Environ Sci 255:012015

Hernández-León R, Rojas-Solís D, Contreras-Pérez M, del Carmen Orozco-Mosqueda M, Macías-Rodríguez LI, Reyes-de la Cruz H, Valencia-Cantero E, Santoyo G (2015) Characterization of the antifungal and plant growth-promoting effects of diffusible and volatile organic compounds produced by Pseudomonas fluorescens strains. Biol Cont 81:83–92. https://doi.org/10.1016/j.biocontrol.2014.11.011

Hietz P, Wanek W, Wania R, Nadkarni NM (2002) Nitrogen-15 natural abundance in a montane cloud forest canopy as an indicator of nitrogen cycling and epiphyte nutrition. Oecologia 131(3):350–355. https://doi.org/10.1007/s00442-002-0896-6

Hodge A, Fitter AH (2010) Substantial nitrogen acquisition by arbuscular mycorrhizal fungi from organic material has implications for N cycling. Proc Nat Acad Sci 107(31):13754–13759. https://doi.org/10.1073/pnas.1005874107

Horn WS, Simmonds MSJ, Schwartz RE, Blaney WM (1995) Phomopsichalasin, a novel antimicrobial agent from an endophytic *Phomopsis* sp. Tetrahedron 51(14):3969–3978. https://doi.org/10.1016/0040-4020(95)00139-Y

Hoysted GA, Kowal J, Pressel S, Duckett JG, Bidartondo MI, Field KJ (2021) Carbon for nutrient exchange between Lycopodiellainundata and Mucoromycotina fine root endophytes is unresponsive to high atmospheric CO_2. Mycorrhiza 31:1–10. https://doi.org/10.1007/s00572-021-01033-6

Hu Y, Xie W, Cen B (2020) Arbuscular mycorrhiza improved drought tolerance of maize seedlings by altering photosystem II efficiency and the levels of key metabolites. Chem Biol Technol Agricul 7(1):1–14. https://doi.org/10.1007/s00572-021-01033-6

Huang TP, Tzeng DDS, Wong AC, Chen CH, Lu KM, Lee YH et al (2012) DNA polymorphisms and biocontrol of Bacillus antagonistic to citrus bacterial canker with indication of the interference of phyllosphere biofilms. PLoS One 7(7):e42124. https://doi.org/10.1371/journal.pone.0042124

Huang L, Yu LJ, Zhang X, Fan B, Wang FZ, Dai YS, Qi H, Zhou Y, Xie LJ, Xiao S (2019) Autophagy regulates glucose-mediated root meristem activity by modulating ROS production in Arabidopsis. Autophagy 15(3):407–422. https://doi.org/10.1080/15548627.2018.1520547

Huang D, Ma M, Wang Q, Zhang M, Jing G, Li C, Ma F (2020) Arbuscular mycorrhizal fungi enhanced drought resistance in apple by regulating genes in the MAPK pathway. Plant Physiol Biochem 149:245–255. https://doi.org/10.1016/j.plaphy.2020.02.020

Hunter PJ, Hand P, Pink D, Whipps JM, Bending GD (2010) Both leaf properties and microbe-microbe interactions influence within-species variation in bacterial population diversity and structure in the lettuce (Lactuca species) phyllosphere. Appl Environ Microbiol 76(24):8117–8125. https://doi.org/10.1128/AEM.01321-10

Imam J, Singh PK, Shukla P (2016) Plant microbe interactions in post genomic era: perspectives and applications. Front Microbiol 7:1488. https://doi.org/10.3389/fmicb.2016.01488

Impa SM, Nadaradjan S, Jagadish SVK (2012) Drought stress induced reactive oxygen species and anti-oxidants in plants. In: Abiotic stress responses in plants. Springer, New York, pp 131–147. https://doi.org/10.1007/978-1-4614-0634-1_7

Inácio J, Pereira P, Carvalho DM, Fonseca A, Amaral-Collaco MT, Spencer-Martins I (2002) Estimation and diversity of phylloplane mycobiota on selected plants in a Mediterranean-type ecosystem in Portugal. Microbial Ecol 44(4):344–353. https://doi.org/10.1007/s00248-002-2022-z

Innerebner G, Knief C, Vorholt JA (2011) Protection of Arabidopsis thaliana against leaf-pathogenic Pseudomonas syringae by Sphingomonas strains in a controlled model system. Appl Environ Microbiol 77(10):3202–3210

Jaber LR, Araj S-E (2018) Interactions among endophytic fungal entomopathogens (Ascomycota: Hypocreales), the green peach aphid Myzus persicae Sulzer (Homoptera: Aphididae), and the aphid endoparasitoid Aphidius colemani Viereck (Hymenoptera: Braconidae). Biol Control 116:53–61. https://doi.org/10.1016/j.biocontrol.2017.04.005

Jaber LR, Salem NM (2014) Endophytic colonisation of squash by the fungal entomopathogen Beauveria bassiana (Ascomycota: Hypocreales) for managing Zucchini yellow mosaic virus in cucurbits. Biocont Sci Technol 24(10):1096–1109. https://doi.org/10.1080/09583157.2014.923379

Jacobs JL, Sundin GW (2001) Effect of solar UV-B radiation on a phyllosphere bacterial community. Appl Environ Microbiol 67(12):5488–5496. https://doi.org/10.1128/AEM.67.12.5488-5496.2001

Jain A, Singh S, Kumar Sarma B, Bahadur Singh H (2012) Microbial consortium-mediated reprogramming of defence network in pea to enhance tolerance against Sclerotinia sclerotiorum. J Appl Microbiol 112(3):537–550. https://doi.org/10.1111/j.1365-2672.2011.05220.x

Jaipaul SS, Dixit AK, Sharma AK (2011) Growth and yield of capsicum (Capsicum *annum*) and garden pea (Pisum sativum) as influenced by organic manures and biofertilizers. Indian J Agri Sci 81(7):637–642

Janakiev T, Dimkić I, Bojić S, Fira D, Stanković S, Berić T (2020) Bacterial communities of plum phyllosphere and characterization of indigenous antagonistic Bacillus thuringiensis R3/3 isolate. J Appl Microbiol 128(2):528–543

Janoušková M, Pavlíková D (2010) Cadmium immobilization in the rhizosphere of arbuscular mycorrhizal plants by the fungal extraradical mycelium. Plant Soil 332(1):511–520

Javot H, Penmetsa RV, Terzaghi N, Cook DR, Harrison MJ (2007a) A Medicago truncatula phosphate transporter indispensable for the arbuscular mycorrhizal symbiosis. Proc Nat Acad Sci 104(5):1720–1725

Javot H, Pumplin N, Harrison MJ (2007b) Phosphate in the arbuscular mycorrhizal symbiosis: transport properties and regulatory roles. Plant Cell Environ 30(3):310–322. https://doi.org/10.1111/j.1365-3040.2006.01617.x

Jensen ES, Peoples MB, Boddey RM, Gresshoff PM, Hauggaard-Nielsen H, Alves BJR, Morrison MJ (2012) Legumes for mitigation of climate change and the provision of eedstock for biofuels and biorefineries. A review. Agron Sustain Dev 32(2):329–364

Jia M, Chen L, Xin HL, Zheng CJ, Rahman K, Han T, Qin LP (2016) A friendly relationship between endophytic fungi and medicinal plants: a systematic review. Front Microbiol 7:906

Jiang Q, Zhuo F, Long S, Zhao H, Yang D, Ye Z, Li S, Jing Y (2016) Can arbuscular mycorrhizal fungi reduce Cd uptake and alleviate Cd toxicity of *Lonicera japonica* grown in Cd-added soils? Scie Rep 6:21805. https://doi.org/10.1038/srep21805

Jiménez-Morillo NT, González-Pérez JA, Jordán A, Zavala LM, de la Rosa JM, Jiménez-González MA, González-Vila FJ (2016) Organic matter fractions controlling soil water repellency in sandy soils from the Doñana National Park (Southwestern Spain). Land Degrad Dev 27(5): 1413–1423

Jixiang L, Yingnan W, Shengnan S, Chunsheng M, Xiufeng Y (2017) Effects of arbuscular mycorrhizal fungi on the growth, photosynthesis and photosynthetic pigments of *Leymus chinensis* seedlings under salt-alkali stress and nitrogen deposition. Sci Total Environ 576: 234–241. https://doi.org/10.1016/j.scitotenv.2016.10.091

Jnawali AD, Ojha RB, Marahatta S (2015) Role of *Azotobacter* in soil fertility and sustainability—a review. Adv Plants Agric Res 2(6):1–5

Johnson NC, Gehring CA (2007) Mycorrhizas: Symbiotic mediators of Rhizosphere and ecosystem processes. Rhizosphere:73–100. https://doi.org/10.1016/b978-012088775-0/50006-9

Jones P, Garcia BJ, Furches A, Tuskan GA, Jacobson D (2019) Plant host-associated mechanisms for microbial selection. Front Plant Sci 10:862. https://doi.org/10.3389/fpls.2019.00862

Kamal S, Prasad R, Varma A (2010) Soil microbial diversity in relation to heavy metals. In: Sherameti I, Varma A (eds) Soil heavy metals, vol 19. Springer-Verlag, Berlin, pp 31–64

Kapoor R, Evelin H, Mathur P, Giri B (2013) Arbuscular mycorrhiza: approaches for abiotic stress tolerance in crop plants for sustainable agriculture. In: Tuteja N, Gill SS (eds) Plant acclimation to environmental stress. Springer Science+Business Media, Berlin, pp 359–401. https://doi.org/10.1007/978-1-4614-5001-6_14

Kaul S, Choudhary M, Gupta S, Dhar MK (2021) Engineering host microbiome for crop improvement and sustainable agriculture. Front Microbiol 12:1125

Kaushal M, Wani SP (2016) Plant-growth-promoting rhizobacteria: drought stress alleviators to ameliorate crop production in drylands. Ann Microbiol 66(1):35–42

Kecskeméti E, Berkelmann-Löhnertz B, Reineke A (2016) Are epiphytic microbial communities in the carposphere of ripening grape clusters (Vitis vinifera L.) different between conventional, organic, and biodynamic grapes? PLoS One 11(8):e0160852. https://doi.org/10.1371/journal.pone.0160852

Kemppainen MJ, Pardo AG (2013) LbNrt RNA silencing in the mycorrhizal symbiont Laccariabicolor reveals a nitrate-independent regulatory role for a eukaryotic NRT2-type nitrate transporter. Environ Microbiol Rep 5:353–366

Kennedy C, Rudnick P, MacDonald ML, Melton T (2015) Azotobacter. In: Trujillo ME, Dedysh S, DeVos P, Hedlund B, Kämpfer P, Rainey FA, Whitman WB (eds) Bergey's manual of systematics of archaea and bacteria. Atlanta, GA, American Cancer Society, pp 1–33. https://doi.org/10.1002/9781118960608.gbm01207

Khadka K, Rachana D, Asis S, Laxmi R, Khem C, Shreeram S, Shrimala Y, Deepak U, Swikar K, Ram R, Pashupati C (2016) Constraints and opportunities for promotion of finger millet in Nepal. https://doi.org/10.13140/RG.2.2.13997.69606

Khalid M, Bilal M, Hassani D, Iqbal N, Wang H, Huang D (2017) Mitigation of salt stress in white clover (*Trifolium repens*) by *Azospirillum brasilense* and its inoculation effect. Bot Stud 58:5

Khalloufi M, Martínez-Andújar C, Lachaâl M, Karray-Bouraoui N, Pérez-Alfocea F, Albacete A (2017) The interaction between foliar GA3 application and arbuscular mycorrhizal fungi inoculation improves growth in salinized tomato (*Solanum lycopersicum* L.) plants by modifying the hormonal balance. J Plant Physiol 214:134–144. https://doi.org/10.1016/j.jplph.2017.04.012

Khan N, Ali S, Shahid MA, Mustafa A, Sayyed RZ, Curá JA (2021) Insights into the interactions among roots, rhizosphere, and rhizobacteria for improving plant growth and tolerance to abiotic stresses: a review. Cells 10(6):1551. https://doi.org/10.3390/cells10061551

Kizilkaya R (2009) Nitrogen fixation capacity of *Azotobacter* spp. strains isolated from soils in different ecosystems and relationship between them and the microbiological properties of soils. J Environ Biol 30:73–82

Klieber J, Reineke A (2016) The entomopathogen *Beauveria bassiana* has epiphytic and endophytic activity against the tomato leaf miner *Tutaabsoluta*. J Appl Entomol 140:580–589. https://doi.org/10.1111/jen.12287

Knief C, Delmotte N, Chaffron S, Stark M, Innerebner G, Wassmann R, von Mering C, Vorholt JA (2012) Metaproteogenomic analysis of microbial communities in the phyllosphere and rhizosphere of rice. ISME J 6(7):1378–1390. https://doi.org/10.1038/ismej.2011.192

Kobae Y, Tamura Y, Takai S, Banba M, Hata S (2010) Localized expression of arbuscular mycorrhiza-inducible ammonium transporters in soybean. Plant Cell Physiol 51:411–1425

Kong Z, Hart M, Liu H (2018) Paving the way from the lab to the field: using synthetic microbial consortia to produce high-quality crops. Front Plant Sci 9:1467

Korenblum E, Dong Y, Szymanski J, Panda S, Jozwiak A, Massalha H, Meir S, Rogachev I, Aharoni A (2020) Rhizosphere microbiome mediates systemic root metabolite exudation by root-to-root signaling. Proc Natl Acad Sci 117(7):3874–3883

Kraepiel AML, Bellenger JP, Wichard T, Morel FM (2009) Multiple roles of siderophores in free-living nitrogen-fixing bacteria. Biometals 22(4):573

Krajinski F, Courty P-E, Sieh D, Franken P, Zhang H, Bucher M, Gerlach N, Kryvoruchko I, Zoeller D, Udvardi M, Hause B (2014) The H+-ATPase HA1 of Medicago truncatula is essential for phosphate transport and plant growth during arbuscular mycorrhizal symbiosis. Plant Cell 26(4):1808–1817. https://doi.org/10.1105/tpc.113.120436

Krishnamoorthy R, Kwon S, Kumutha K, Senthilkumar M, Ahmed S, Sa T, Anandham R (2018) Diversity of culturable methylotrophic bacteria in different genotypes of groundnut and their potential for plant growth promotion. 3 Biotech 8(6):275. https://doi.org/10.1007/s13205-018-1291-2

Kumar A, Dubey A (2020) Rhizosphere microbiome: engineering bacterial competitiveness for enhancing crop production. J Adv Res 24:337–352. https://doi.org/10.1016/j.jare.2020.04.014

Kumar B, Gopal D (2015) Effective role of indigenous microorganisms for sustainable environment. 3 Biotech 5(6):867–876. https://doi.org/10.1007/s13205-015-0293-6

Kumar V, Singh P, Jorquera M, Sangwan P, Kumar P, Verma A, Agrawal S (2013) Isolation of phytase-producing bacteria from Himalayan soils and their effect on growth and phosphorus uptake of Indian mustard (*Brassica juncea*). World J Microbiol Biotechnol 29(8):1361–1369

Kumar A, Dames JF, Gupta A, Sharma S, Gilbert JA, Ahmad P (2015) Current developments in arbuscular mycorrhizal fungi research and its role in salinity stress alleviation: a biotechnological perspective. Crit Rev Biotechnol 35:461–474

Kumar A, Trefault N, Olaniran AO (2016) Microbial degradation of 2, 4-dichlorophenoxyacetic-acid: insight into the enzymes and catabolic genes involved, their regulation and biotechnological implications. Crit Rev Microbiol 42(2):194–208

Kumari MER, Gopal AV, Lakshmipathy R (2018) Effect of stress tolerant plant growth promoting rhizobacteria on growth of blackgram under stress condition. Int J Curr Microbiol App Sci 7(1): 1479–1487

Laforest-Lapointe I, Messier C, Kembel S (2016) Host species identity, site and time drive temperate tree phyllosphere bacterial community structure. Microbiome 4(1):27. https://doi.org/10.1186/s40168-016-0174-1

Lahlali R, Peng G (2014) Suppression of clubroot by C lonostachys rosea via antibiosis and induced host resistance. Plant Pathol 63(2):447–455

Lamptey S, Ahiabor BDK, Yeboah S, Osei D (2014) Effect of Rhizobium inoculants and reproductive growth stages on shoot biomass and yield of soybean (*Glycine max* (L.) Merril). J Agricul Sci 6(5):44–54. https://doi.org/10.5539/jas.v6n5p44

Lata R, Chowdhury S, Gond SK, White JF (2018) Induction of abiotic stress tolerance in plants by endophytic microbes. Let Appl Microbiol 66:268–276

Lebeis SL, Paredes SH, Lundberg DS, Breakfield N et al (2015) Plant microbiome. Salicylic acid modulates colonization of the root microbiome by specific bacterial taxa. Science 349:860–864

Lemanceau P, Barret M, Mazurier S, Mondy S, Pivato B, Fort T, Vacher C (2017) Plant communication with associated microbiota in the spermosphere, rhizosphere and phyllosphere. Adv Bot Res 82:101–133. https://doi.org/10.1016/bs.abr.2016.10.007. How plants communicate with their biotic environment

Leveau JH (2015) Life of microbes on aerial plant parts. In: Principles of plant-microbe interactions. Springer, Cham, pp 17–24

Li H, Luo N, Zhang LJ, Zhao HM, Li YW, Cai QY et al (2016) Do arbuscular mycorrhizal fungi affect cadmium uptake kinetics, subcellular distribution and chemical forms in rice? Sci Total Environ 571:1183–1190. https://doi.org/10.1016/j.scitotenv.2016.07.124

Li C, Zhou J, Wang X, Liao H (2019) A purple acid phosphatase, GmPAP33, participates in arbuscule degeneration during arbuscular mycorrhizal symbiosis in soybean. Plant Cell Environ 42(6):2015–2027

Liao X, Fang W, Lin L, Lu H-L, St. Leger RJ (2013) Metarhizium robertsii produces and extracellular invertase (MrINV) that plays a pivotal role in rhizospheric interactions and root colonization. PLoS One 8:1–11. https://doi.org/10.1371/journal.pone.0078118

Lin A-J, Zhang X-H, Wong M-H, Ye Z-H, Lou L-Q, Wang Y-S, Zhu Y-G (2007) Increase of multimetal tolerance of three leguminous plants by arbuscular mycorrhizal fungi colonization. Environ Geochem Health 29:473–481

Lin SY, Shen FT, Young LS, Zhu ZL, Chen WM, Young CC (2012) *Azospirillum formosense* sp. nov., a diazotroph from agricultural soil. Int J syst Evol Microbiol 62(5):1185–1190

Lin SY, Liu YC, Hameed A, Hsu YH, Lai WA, Shen FT, Young CC (2013) *Azospirillum fermentarium* sp. nov., a nitrogen-fixing species isolated from a fermenter. Int J syst Evol Microbiol 63(10):3762–3768

Lin SY, Hameed A, Liu YC, Hsu YH, Lai WA, Shen FT, Young CC (2015) *Azospirillum soli* sp. nov., a nitrogen-fixing species isolated from agricultural soil. Int J Syst Evol Microbiol 65(12):4601–4607

Lin SY, Liu YC, Hameed A, Hsu YH, Huang HI, Lai WA, Young CC (2016) *Azospirillum agricola* sp. nov., a nitrogen-fixing species isolated from cultivated soil. Int J Syst Evol Microbiol 66(3): 1453–1458

Lindow SE, Brandl MT (2003) Microbiology of the phyllosphere. Appl Environ Microbiol 69: 1875–1883. https://doi.org/10.1128/AEM.69.4.1875-1883

Liu C, Ravnskov S, Liu F, Rubæk GH, Andersen MN (2018a) Arbuscular mycorrhizal fungi alleviate abiotic stresses in potato plants caused by low phosphorus and deficit irrigation/partial root-zone drying. J Agric Sci 156:46–58. https://doi.org/10.1017/S0021859618000023

Liu Z, Beskrovnaya P, Melnyk RA, Hossain SS, Khorasani S, O'Sullivan LR, Haney CH (2018b) A genome-wide screen identifies genes in rhizosphere-associated *Pseudomonas* required to evade plant defenses. mBio 9(6):e00433-18. https://doi.org/10.1128/mBio.00433-18

Liu H, Brettell LE, Singh B (2020) Linking the phyllosphere microbiome to plant health. Trends Plant Sci 25:41–844

López-Pedrosa A, Gonzalez-Guerrero M, Valderas A, Azcon-Aguilar C, Ferrol N (2006) GintAMT1 encodes a functional high-affinity ammonium transporter that is expressed in the extraradical mycelium of Glomus intraradices. Fungal Gene Biol 43:102–110. https://doi.org/10.1016/j.fgb.2005.10.005

Lynch MDJ, Neufeld JD (2015) Ecology and exploration of the rare biosphere. Nat Rev Microbiol 13:217–229

Maignien L, DeForce EA, Chafee ME, Eren AM, Simmons SL (2014) Ecological succession and stochastic variation in the assembly of Arabidopsis thaliana phyllosphere communities. mBio 5 (1):e00682-13. https://doi.org/10.1128/mBio.00682-13

Maina UM, Galadima IB, Gambo FM, Zakaria D (2018) A review on the use of entomopathogenic fungi in the management of insect pests of field crops. J Entomol Zool Stud 6(1):27–32

Maldonado-Mendoza IE, Harrison MJ (2018) RiArsB and RiMT-11: two novel genes induced by arsenate in arbuscular mycorrhiza. Fungal Biol 122(2-3):121–130

Maldonado-Mendoza I, Dewbre G, Harrison M (2001) A phosphate transporter gene from the extra-radical mycelium of an arbuscular mycorrhizal fungus. Mol Plant Microbe Interact 14(10): 1140–1148. https://doi.org/10.1094/MPMI.2001.14.10.1140

Manching HC, Balint-Kurti PJ, Stapleton AE (2014) Southern leaf blight disease severity is correlated with decreased maize leaf epiphytic bacterial species richness and the phyllosphere bacterial diversity decline is enhanced by nitrogen fertilization. Front Plant Sci 5:403. https://doi.org/10.3389/fpls.2014.00403

Marschner P (2012) Chapter 15—rhizosphere biology. In: Marschner's mineral nutrition of higher plants, 3rd edn. Academic, San Diego, pp 369–388

Martin F, Aerts A, Ahren D, Brun A, Danchin EGJ, Duchaussoy F, Gibon J, Kohler A, Lindquist E, Pereda V et al (2008) The genome of Laccariabicolor provides insights into mycorrhizal symbiosis. Nature 452:88–92

Martinuz A, Schouten A, Sikora R (2012) Systemically induced resistance and microbial competitive exclusion: implications on biological control. Phytopathology 102:260–266. https://doi.org/10.1094/PHYTO-04-11-0120

Mathur S, Sharma MP, Jajoo A (2016) Improved photosynthetic efficacy of maize *Zea mays* plants with arbuscular mycorrhizal fungi (AMF) under high temperature stress. J Photochem Photobiol B 180:149–154. https://doi.org/10.1016/j.jphotobiol.2018.02.002

McNear DH Jr (2013) The rhizosphere—roots, soil and everything in between. Nat Educ Knowl 4(3):1

Meena RK, Singh RK, Singh NP et al (2015) Isolation of low temperature surviving plant growth–promoting rhizobacteria (PGPR) from pea (Pisum sativum L.) and documentation of their plant growth promoting traits. Biocat Agric Biotechnol 4:806–811

Meena RS, Kumar S, Datta R, Lal R, Vijayakumar V, Brtnicky M, Sharma MP, Yadav GS, Jhariya MK, Jangir CK, Pathan SI, Dokulilova T, Pecina V, Marfo TD (2020) Impact of agrochemicals on soil microbiota and management: a review. Land 9:34. https://doi.org/10.3390/land9020034

Mehnaz S (2015) Azospirillum: a biofertilizer for every crop. In: Plant microbes symbiosis: applied facets. Springer, New Delhi, pp 297–314

Mehnaz S, Weselowski B, Lazarovits G (2007) *Azospirillum zeae* sp. nov., a diazotrophic bacterium isolated from rhizosphere soil of *Zea mays*. Int J Sys Evol Microbiol 57(12):2805–2809

Mejia LC, Herre EA, Sparks JP, Winter K, Garcia MN, Van Bael SA et al (2014) Pervasive effects of a dominant foliar endophytic fungus on host genetic and phenotypic expression in a tropical tree. Front Microbiol 5:479. https://doi.org/10.3389/fmicb.2014.00479

Mendes R, Garbeva P, Raaijmakers JM (2013) The rhizosphere microbiome: significance of plant beneficial, plant pathogenic, and human pathogenic microorganisms. FEMS Microbiol Rev 37(5):634–663. https://doi.org/10.1111/1574-6976.12028

Méndez-Gómez M, Castro-Mercado E, Alexandre G, García-Pineda E (2015) Oxidative and antioxidative responses in the wheat-*Azospirillum brasilense* interaction. Protoplasma 253: 477–486

Metwally A, Azooz M, Nafady N, El-Enany A (2019) Arbuscular mycorrhizal symbiosis alleviates drought stress imposed on wheat plants (*Triticum aestivum* L.). Appl Ecol Environ Res 17 (6):13713–13727

Mian IA, Riaz M, Cresser MS (2009) The importance of ammonium mobility in nitrogen-impacted unfertilized grasslands: a critical reassessment. Environ Pollut 157(4):1287–1293. https://doi.org/10.1016/j.envpol.2008.11.041

Michavila G, Adler C, De Gregorio PR et al (2017) *Pseudomonas protegens* CS1 from the lemon phyllosphere as a candidate for citrus canker biocontrol agent. Plant Biol 19:608–617

Milošević N, Tintor B, Protić R, Cvijanovic G, Dimitrijević T (2012) Effect of inoculation with Azotobacter chroococcum on wheat yield and seed quality. Romanian Biotechnol Lett 17(3): 7352–7357

Mina D, Pereira J, Lino-Neto T, Baptista P (2020) Epiphytic and endophytic bacteria on olive tree phyllosphere: exploring tissue and cultivar effect. Microbial Ecol 80(1):145–157

Miransari M (2017) Arbuscular mycorrhizal fungi and heavy metal tolerance in plants. In: Wu QS (ed) Arbuscular mycorrhizas and stress tolerance of plants. Springer Nature, Singapore, pp 161–174. https://doi.org/10.1007/978-3-319-68867-1_4

Mirshad P, Puthur J (2017) Drought tolerance of bioenergy grass *Saccharum spontaneum* L. enhanced by arbuscular mycorrhizae. Rhizosphere 3(1):1–8. https://doi.org/10.1016/j.rhisph.2016.09.004

Miyamoto T, Kawahara M, Minamisawa K (2004) Novel endophytic nitrogen-fixing clostridia from the grass Miscanthus sinensis as revealed by terminal restriction fragment length polymorphism analysis. Appl Environ Microbiol 70(11):6580–6586

Moghadam SK, Bahadoran Z, Mirmiran P, Tohidi M, Azizi F (2016) Association between dietary acid load and insulin resistance: tehran lipid and glucose study. Prev Nutr Food Sci 21(2): 104–109. https://doi.org/10.3746/pnf.2016.21.2.104

Mohan Gupta S, Kumar K, Kumar Joshi R, Gupta S, Bala M (2020) Frankia: a promising N-fixing plant growth promoting rhizobacteria (PGPR) improved drought tolerance in crops at higher altitude. In: Microbiological advancements for higher altitude agro-ecosystems & sustainability. Springer, Singapore, pp 411–431

Molina-Favero C, Creus CM, Simontacchi M, Puntarulo S, Lamattina L (2008) Aerobic nitric oxide production by Azospirillum brasilense Sp245 and its influence on root architecture in tomato. Mol Plant Microbe Int 21(7):1001–1009

Moradtalab N, Roghieh H, Nasser A, Tobias EH, Günter N (2019) Silicon and the association with an arbuscular-mycorrhizal fungus (*Rhizophagus clarus*) mitigate the adverse effects of drought stress on strawberry. Agronomy 9:41. https://doi.org/10.3390/agronomy9010041

Morris CE, Kinkel LL (2002) Fifty years of phyllosphere microbiology: significant contributions to research in related fields. In: Lindow SE, Hecht-Poinar EI, Elliott VJ (eds) Phyllosphere microbiology. APS Press, St Paul, pp 365–375

Muleta D (2017) Legume response to arbuscular mycorrhizal fungi inoculation in sustainable agriculture. In: Zaidi A, Khan M, Musarrat J (eds) Microbes for legume improvement. Springer, Cham

Müller T, Ruppel S (2014) Progress in cultivation-independent phyllosphere microbiology. FEMS Microbiol Ecol 87:2–17. https://doi.org/10.1111/1574-6941.12198

Müller D, Vogel C, Bai Y, Vorholt J (2016a) The plant microbiota: systems-level insights and perspectives. Annual Rev Gene 50(1):211–234

Müller T, Behrendt U, Ruppel S, von der Waydbrink G, Müller ME (2016b) Fluorescent pseudomonads in the phyllosphere of wheat: potential antagonists against fungal phytopathogens. Curr Microbiol 72(4):383–389

Nalini MS, Prakash HS (2017) Diversity and bioprospecting of actinomycete endophytes from the medicinal plants. Lett Appl Microbiol 64(4):261–270

Navarrete AA, Tsai SM, Mendes LW, Faust K et al (2015) Soil microbiome responses to the short-term effects of *Amazonian deforestation*. Mol Ecol 24:2433–2448

Neinhuis C, Barthlott W (1997) Characterization and distribution of water-repellent, self-cleaning plant surfaces. Annals Bot 79:667–677

Newcomb W, Wood SM (1987) Morphogenesis and fine structure of Frankia (Actinomycetales): the microsymbiont of nitrogen-fixing actinorhizal root nodules. Int Rev Cytol 109:1–88. https://doi.org/10.1016/s0074-7696(08)61719-2. PMID: 3323105

Ngom M, Gray K, Diagne N, Oshone R, Fardoux J, Gherbi H, Valérie H, Sergio S, Laurent L, Tisa Louis S, Sy Mame O, Champion A (2016) Symbiotic performance of diverse Frankia strains on salt-stressed Casuarina glauca and Casuarina equisetifolia plants. Front Plant Sci 7:1331

Nguyen TD, Cavagnaro TR, Watts-Williams SJ (2019) The effects of soil phosphorus and zinc availability on plant responses to mycorrhizal fungi: a physiological and molecular assessment. Sci Rep 9:14880. https://doi.org/10.1038/s41598-019-51369-5

Ngwene B, Elke G, Eckhard G (2013) Influence of different mineral nitrogen sources (NO3-N vs. NH4+–N) on arbuscular mycorrhiza development and N transfer in a Glomus intraradices–cowpea symbiosis. Mycorrhiza 23:107–117

Nord EA, Lynch JP (2009) Plant phenology: a critical controller of soil resource acquisition. J Exp Bot 60:1927–1937

Nouioui I, Ghodhbane-Gtari F, Montero-Calasanz MD, Göker M et al (2016) Proposal of a type strain for *Frankia alni* (Woronin 1866) Von Tubeuf 1895, emended description of *Frankia alni*, and recognition of *Frankia casuarinae* sp. nov. and *Frankia elaeagni* sp. nov. Int J Syst Evol Microbiol 66:5201–5210. https://doi.org/10.1099/ijsem.0.001496

Nouioui I, Cortés-albayay C, Carro L, Castro JF, Gtari M, Ghodhbane-Gtari F, Klenk H-P, Tisa LS, Sangal V, Goodfellow M (2019) Genomic insights into plant growth-promoting potentialities of the genus Frankia. Front Microbiol 10:1457. https://doi.org/10.3389/fmicb.2019.01457

Obertello M, Sy MO, Laplaze L et al (2003) Actinorhizal nitrogen fixing nodules: infection process, molecular biology and genomics. African J Biotechnol. 2(12):528–538

Odelade KA, Babalola OO (2019) Bacteria, Fungi and Archaea domains in rhizospheric soil and their effects in enhancing agricultural productivity. Int J Environ Res Public Health 16(20): 3873. https://doi.org/10.3390/ijerph16203873

Oldroyd GE (2013) Speak, friend, and enter: signalling systems that promote beneficial symbiotic associations in plants. Nat Rev Microbiol 11:252–264. https://doi.org/10.1038/nrmicro2990

Oldroyd GE, Murray JD, Poole PS, Downie JA (2011) The rules of engagement in the legume-rhizobial symbiosis. Annu Rev Genet 45:119–144. https://doi.org/10.1146/annurev-genet-110410-132549

Oleńska E, Małek W, Wójcik M, Swiecicka I, Thijs S, Vangronsveld J (2020) Beneficial features of plant growth-promoting rhizobacteria for improving plant growth and health in challenging conditions: a methodical review. Sci Total Environ 743:140682

Olle M, Lepse L, Williams I (2016) Organic farming of pea in the northern hemisphere—a revieiv. Acta Hortic 1123:137–141. https://doi.org/10.17660/ActaHortic.2016.1123.19

Orchard S, Hilton S, Bending GD, Dickie IA, Standish RJ, Gleeson DB, Jeffery RP, Powell JR, Walker C, Bass D, Monk J, Simonin A, Ryan MH (2017) Fine endophytes (Glomus tenue) are related to Mucoromycotina, not Glomeromycota. New Phytol 213:481–486. https://doi.org/10.1111/nph.14268

Ordookhani K, Khavazi K, Moezzi A, Rejali F (2010) Influence of PGPR and AMF on antioxidant activity, lycopene and potassium contents in tomato. African J Agricul Res 5(10):1108–1116

Orozco-Mosqueda MC, Flores A, Rojas-Sánchez B, Urtis-Flores CA, Morales-Cedeño LR, Valencia-Marin MF, Chávez-Avila S, Rojas-Solis D, Santoyo G (2021) Plant growth-promoting bacteria as bioinoculants: attributes and challenges for sustainable crop improvement. Agronomy 11(6):1167. https://doi.org/10.3390/agronomy11061167

Ottesen A, González Peña A, White J, Pettengill J, Li C, Allard S, Rideout S, Allard M, Hill T, Evans P, Strain E, Musser S, Knight R, Brown E (2013) Baseline survey of the anatomical microbial ecology of an important food plant: *Solanum lycopersicum* (tomato). BMC Microbiol 13(1):114

Oukala N, Aissat K, Pastor V (2021) Bacterial endophytes: the hidden actor in plant immune responses against biotic stress. Plants 10(5):1012

Owen D, Williams A, Griffith G, Withers P (2015) Use of commercial bio-inoculants to increase agricultural production through improved phosphrous acquisition. Appl Soil Ecol 86:41–54

Oyserman BO, Medema MH, Raaijmakers JM (2018) Road MAPs to engineer host microbiomes. Curr Opin Microbiol 43:46–54

Pal A, Pandey S (2016) Role of arbuscular mycorrhizal fungi on plant growth and reclamation of barren soil with wheat (*Triticum aestivum* L.) crop. Int. J Soil Sci 12(1):25–31

Paterson E, Gebbing T, Abel C, Sim A, Telfer G (2007) Rhizodeposition shapes rhizosphere microbial community structure in organic soil. New Phytol 173(3):600–610

Pawlak K, Kołodziejczak M (2020) The role of agriculture in ensuring food security in developing countries: considerations in the context of the problem of sustainable food production. Sustainability 12:5488. https://doi.org/10.3390/su12135488

Pawlowski K, Demchenko KN (2012) The diversity of actinorhizal symbiosis. Protoplasma 249(4): 967–979

Pawlowski K, Jacobsen KR, Alloisio N et al (2007) Truncated hemoglobins in actinorhizal nodules of *Datisca glomerata*. Plant Biol 9:776–785. https://doi.org/10.1055/s-2007-965258

Peck MC, Fisher RF, Long SR (2006) Diverse flavonoids stimulate NodD1 binding to *nod* gene promoters in *Sinorhizobium meliloti*. J Bacteriol 188:5417–5427

Peng G, Wang H, Zhang G, Hou W, Liu Y, Wang ET, Tan Z (2006) Azospirillum melinis sp. nov., a group of diazotrophs isolated from tropical molasses grass. Int J Syst Evol Microbiol 56(6): 1263–1271

Peng XN, Gilmore SP, O'Malley MA (2016) Microbial communities for bioprocessing: lessons learned from nature. Curr Opin Chem Eng 14:103–109. https://doi.org/10.1016/j.coche.2016.09.003

Pérez-Tienda J, Testillano PS, Balestrini R, Fiorilli V, Azcón-Aguilar C, Ferrol N (2011) GintAMT2, a new member of the ammonium transporter family in the arbuscular mycorrhizal fungus *Glomus intraradices*. Fungal Genet Biol 48:1044–1055. https://doi.org/10.1016/j.fgb.2011.08.003

Photita W, Lumyong S, Lumyong P, McKenzie EHC, Hyde KD (2005) Are some fungi isolated as endophytes of *Musa acuminata* latent pathogens? Fungal Divers 16:131–140

Porcel R, Redondogómez S, Mateosnaranjo E, Aroca R, Garcia R, Ruizlozano JM (2015) Arbuscular mycorrhizal symbiosis ameliorates the optimum quantum yield of photosystem II and reduces non-photochemical quenching in rice plants subjected to salt stress. J Plant Physiol 185:75–83. https://doi.org/10.1016/j.jplph.2015.07.006

Porcel R, Aroca R, Azcon R, Ruiz-Lozano J (2016) Regulation of cation transporter genes by the arbuscular mycorrhizal symbiosis in rice plants subjected to salinity suggests improved salt tolerance due to reduced Na+ root-to-shoot distribution. Mycorrhiza 26(7):673–684

Porras-Alfaro A, Bayman P (2011) Hidden fungi, emergent properties: endophytes and microbiomes. Annu Rev Phytopathol 49:291–315

Poveda K, Jiménez MIG, Kessler A (2010) The enemy as ally: herbivore-induced increase in crop yield. Ecol Appl 20(7):1787–1793

Poveda J, Eugui D, Abril-Urías P, Velasco P (2021) Endophytic fungi as direct plant growth promoters for sustainable agricultural production. Symbiosis 85(1):1–19

Powell WA, Klingeman WE, Ownley BH, Gwinn KD (2009) Evidence of endophytic *Beauveria bassiana* in seed-treated tomato plants acting as a systemic entomopathogen to larval *Helicover pazea* (Lepidoptera: Noctuidae). J Entomol Sci 44(4):391–396

Prasad R, Zhang S-H (2022) Benefical microorganisms in agriculture. Springer, Singapore. https://link.springer.com/book/9789811907326

Prasad R, Kumar M, Varma A (2015) Role of PGPR in soil fertility and plant health. In: Egamberdieva D, Shrivastava S, Varma A (eds) Plant growth-promoting rhizobacteria (PGPR) and medicinal plants. Springer International Publishing, Cham, pp 247–260

Prasad R, Kumar V, Kumar M, Choudhary D (2019) Nanobiotechnology in bioformulations. Springer International Publishing. ISBN 978-3-030-17061-5. https://www.springer.com/gp/book/9783030170608

Puente ML, Gualpa JL, Lopez GA, Molina RM, Carletti SM, Cassán FD (2017) The benefits of foliar inoculation with *Azospirillum brasilense* in soybean are explained by an auxin signaling model. Symbiosis 76:41–49

Punamiya P, Datta R, Sarkar D, Barber S, Patel M, Das P (2010) Symbiotic role of glomus mosseae in phytoextraction of lead in vetiver grass [*Chrysopogon zizanioides* (L.)]. J Hazard Mater 177(1–3):465–447

Pusey PL, Wend C (2012) Potential of osmoadaptation for improving *Pantoeaag glomerans* E325 as biocontrol agent for fire blight of apple and pear. Biol Control 62(1):29–37

Qayyum MA, Wakil W, Arif MJ, Sahi ST, Dunlap CA (2015) Infection of *Helicoverpa armigera* by endophytic *Beauveria bassiana* colonizing tomato plants. Biol. Control 90:200–207. https://doi.org/10.1016/j.biocontrol.2015.04.005

Qin Y, Druzhinina IS, Pan X, Yuan Z (2016) Microbially mediated plant salt tolerance and microbiome-based solutions for saline agriculture. Biotechnol Adv 34(7):1245–1259

Qin C, Tao J, Liu T et al (2019) Responses of phyllosphere microbiota and plant health to application of two different biocontrol agents. AMB Expr 9:42. https://doi.org/10.1186/s13568-019-0765-x

Quan M, Liang J (2017) The influences of four types of soil on the growth, physiological and biochemical characteristics of *Lycoris aurea* (L' Her.). Herb Scie Rep 7(1):43284. https://doi.org/10.1038/srep43284

Quesada-Moraga E, Munoz-Ledesma FJ, Santiago-Alvarez C (2009) Systemic protection of *Papaver somniferum* L. against Iraellaluteipes (Hymenoptera: Cynipidae) by an endophytic strain of *Beauveria bassiana* (Ascomycota: Hypocreales). Environ Entomol 38(3):723–730

Rana KL, Kour D, Kaur T, Devi R, Yadav AN, Yadav N et al (2020) Endophytic microbes: biodiversity, plant growth-promoting mechanisms and potential applications for agricultural sustainability. Antonie Van Leeuwenhoek 113:1075–1107

Rani B (2016) Effect of arbuscular mycorrhiza fungi on biochemical parameters in wheat *Triticum aestivum* L. under drought conditions. Front Plant Sci 10:499. https://doi.org/10.3389/fpls.2019.00499. Doctoral dissertation, CCSHAU, Hisardrought stress in C_3 (*Leymus chinensis*) and C_4 (*Hemarthria altissima*) grasses via altering antioxidant enzyme activities and photosynthesis.

Ranjbar Sistani N, Kaul H-P, Desalegn G, Wienkoop S (2017) Rhizobium impacts on seed productivity, quality, and protection of pisum sativum upon disease stress caused by didymellapinodes: phenotypic, proteomic, and metabolomic traits. Front Plant Sci 8:1961. https://doi.org/10.3389/fpls.2017.01961

Rascio N, La Rocca N (2008) Biological nitrogen fixation. In: Jørgensen SE, Fath BD (eds) Encyclopedia of ecology. Academic, Oxford, pp 412–419

Rastogi G, Sbodio A, Tech JJ, Suslow TV, Coaker GL, Leveau JH (2012) Leaf microbiota in an agroecosystem: spatiotemporal variation in bacterial community composition on field-grown lettuce. ISME J 6(10):1812–1822

Rastogi G, Coaker GL, Leveau JHJ (2013) New insights into the structure and function of phyllosphere microbiota through high-throughput molecular approaches. FEMS Microbiol Lett 348(1):1–10. https://doi.org/10.1111/1574-6968.12225

Ren CG, Bai YJ, Kong CC et al (2016) Synergistic interactions between salt-tolerant rhizobia and arbuscular mycorrhizal fungi on salinity tolerance of *Sesbania cannabina* plants. J Plant Growth Regul 35:1098–1107. https://doi.org/10.1007/s00344-016-9607-0

Resquín-Romero G, Garrido-Jurado I, Delso C, Ríos-Moreno A, Quesada-Moraga E (2016) Transient endophytic colonizations of plants improves the outcome of foliar applications of mycoinsecticides against chewing insects. J. Invertebr Pathol 136:23–31. https://doi.org/10.1016/j.jip.2016.03.003

Rico L, Ogaya R, Terradas J, Peñuelas J (2014) Community structures of N_2-fixing bacteria associated with the phyllosphere of a Holm oak forest and their response to drought. Plant Biol 16:586–593

Ripa F, Cao W, Tong S, Sun J (2019) Assessment of plant growth promoting and abiotic stress tolerance properties of wheat endophytic fungi. Bio Med Resh Int 2019:6105865. https://doi.org/10.1155/2019/6105865

Ritika B, Utpal D (2014) Biofertilizer, a way towards organic agriculture: a review. African J Microbiol Res 8:2332–2343. https://doi.org/10.5897/AJMR2013.6374

Rodriguez RJ, Woodward C, Kim YO, Redman RS (2009) Habitat-adapted symbiosis as a defense against abiotic and biotic stresses. In: White JF, Torres MS (eds) Defensive mutualism in microbial symbiosis. CRC Press, Boca Raton, FL, pp 335–346

Rodríguez-Navarro D, Oliver M, Albareda Contreras J, Ruiz-Sainz E (2011) Soybean interactions with soil microbes, agronomical and molecular aspects. Agron Sustain Dev 31(1):173–190. https://doi.org/10.1051/agro/2010023.hal-00930474

Rodríguez-Salazar J, Suárez R, Caballero-Mellado J, Iturriaga G (2009) Trehalose accumulation in Azospirillum brasilense improves drought tolerance and biomass in maize plants. FEMS Microbiol Lett 296(1):52–59

Romero-Perdomo F, Abril J, Camelo M, Moreno-Galván A, Pastrana I, Rojas-Tapias D et al (2017) *Azotobacter chroococcum* as a potentially useful bacterial biofertilizer for cotton (*Gossypium hirsutum*): effect in reducing N fertilization. Rev Argent Microbiol 49:377–383. https://doi.org/10.1016/j.ram.2017.04.006

Ruiz-Lozano J, Aroca R, Zamarreño Á, Molina S, Andreo-Jiménez B, Porcel R, García-Mina J, Ruyter-Spira C, López-Ráez J (2015) Arbuscular mycorrhizal symbiosis induces strigolactone biosynthesis under drought and improves drought tolerance in lettuce and tomato. Plant Cell Environ 39(2):441–452

Ruttenberg KC (2003) The global phosphorus cycle. Treatise Geochem 8:585–643. https://doi.org/10.1016/B0-08-043751-6/08153-6

Ryan P, Dessaux Y, Thomashow L, Weller D (2009) Rhizosphere engineering and management for sustainable agriculture. Plant Soil 321(1-2):363–383

Saeed KA, Ahmed SA, Hassan IA, Ahmed PH (2015) Effect of bio-fertilizer and chemical fertilizer on growth and yield in cucumber (*Cucumis sativus* L.) in green house condition. Am Eurasian J Agric Environ Sci 15:353–358. https://doi.org/10.3923/pjbs.2015.129.134

Saha B, Saha S, Das A, Bhattacharyya P, Basak N, Sinha A et al (2017) Biological nitrogen fixation for sustainable agriculture. In: Meena VS (ed) Agriculturally important microbes for sustainable agriculture. Springer, Singapore, pp 81–128. https://doi.org/10.1007/978-981-10-5343-6_4

Saharan BS, Nehra V (2011) Plant growth promoting rhizobacteria: a critical review. Life Sci Med Res 21(1):30

Sahoo RK, Ansari MW, Pradhan M, Dangar TK, Mohanty S, Tuteja N (2014) Phenotypic and molecular characterization of native Azospirillum strains from rice fields to improve crop productivity. Protoplasma 251(4):943–953

Saikkonen K (2007) Forest structure and fungal endophytes. Fungal Biol Rev. 21:67–74

Sánchez-Rodríguez AR, Raya-Díaz S, Zamarreño ÁM, García-Mina JM, Del Campillo MC, Quesada-Moraga E (2018) An endophytic *Beauveria bassiana* strain increases spike production in bread and durum wheat plants and effectively controls cotton leafworm (*Spodoptera littoralis*) larvae. Biol Control 116:90–102. https://doi.org/10.1016/j.biocontrol.2017.01.012

Santander C, Sanhueza M, Olave J, Borie F, Valentine C, Cornejo P (2019) Arbuscular mycorrhizal colonization promotes the tolerance to salt stress in lettuce plants through an efficient modification of ionic balance. J Soil Sci Plant Nutr 19(2):321–331. https://doi.org/10.1007/s42729-019-00032-z

Santhanam R, Luu VT, Weinhold A, Goldberg J, Oh Y, Baldwin IT (2015) Native root-associated bacteria rescue a plant from a sudden-wilt disease that emerged during continuous cropping. Proc Natl Acad Sci USA 112:E5013–E5020

Santi C, Bogusz D, Franche C (2013) Biological nitrogen fixation in non-legume plants. Annals Bot 111(5):743–767. https://doi.org/10.1093/aob/mct048

Santos Kron A, Zengerer V, Bieri M, Dreyfuss V, Sostizzo T, Schmid M, Lutz M, Remus-Emsermann MN, Pelludat C (2020) *Pseudomonas orientalis* F9 pyoverdine, safracin, and phenazine mutants remain effective antagonists against *Erwinia amylovora* in apple flowers. Appl Environ Microbiol 86(8):e02620-19

Santoyo G, Orozco-Mosqueda MDC, Govindappa M (2012) Mechanisms of biocontrol and plant growth-promoting activity in soil bacterial species of *Bacillus* and *Pseudomonas*: a review. Biocontrol Sci Technol 22:855–872

Saribay GF (2003) Growth and nitrogen fixation dynamic of azotobacter chroococcum in nitrogen-free and OMW containing medium [M.S.- Master of Science]. Middle East Technical University.

Sarkar A, Mandal AR, Prasad PH, Maity TK, Chandra B, Viswavidyalaya K et al (2010) Influence of nitrogen and biofertilizer on growth and yield of cabbage. J Crop Weed 6:72–73

Sarma I, Phookan DB, Boruah S (2015) Influence of manures and biofertilizers on carrot (*Daucus carota* L.) cv. early nantes growth, yield and quality. J Ecofriendly Agric 10:25–27

Schüßler A, Martin H, Cohen D, Fitz M, Wipf D (2007) Arbuscular Mycorrhiza-Studies on the geosiphon symbiosis lead to the characterization of the first *Glomeromycotan* sugar transporter. Plant Signal Behavior 2(5):431–434

Sekar J, Raj R, Vaiyapuri PR (2016) Microbial consortia for sustainable agriculture: commercialization and regulatory issues in India. In: Singh HB et al (eds) Agriculturally important microorganisms. Springer Science, Business Media, Singapore, pp 107–131

Sendek A, Karakoç C, Wagg C et al (2019) Drought modulates interactions between arbuscular mycorrhizal fungal diversity and barley genotype diversity. Sci Rep 9:9650. https://doi.org/10.1038/s41598-019-45702-1

Shantharam S, Mattoo AK (1997) Enhancing biological nitrogen fixation: an appraisal of current and alternative technologies for N input into plants. Plant Soil 194:205–216

Shao J, Miao Y, Liu K, Ren Y, Xu Z, Zhang N, Feng H, Shen Q, Zhang R, Xun W (2021) Rhizosphere microbiome assembly involves seed-borne bacteria in compensatory phosphate solubilization. Soil Biol Biochem 159:108273

Sharma S, Shukla KP, Singh V, Singh J, Devi S, Tewari A (2013) Plant-microbe symbiosis: perspectives and applications. In: Arora NK (ed) Plant microbe symbiosis: fundamentals and advances. Springer, New Delhi, pp 119–145

Sheng M, Tang M, Zhang F, Huang Y (2011) Influence of arbuscular mycorrhiza on organic solutes in maize leaves under salt stress. Mycorrhiza 21:423–430. https://doi.org/10.1007/s00572-010-0353-z

Siegel MR, Latch GCM, Bush LP, Fannin FF, Rowan DD, Tapper BA, Johnson MC (1990) Fungal endophyte-infected grasses: alkaloid accumulation and aphid response. J Chemical Ecol 16(12): 3301–3315

Singh R, Dubey AK (2018) Diversity and applications of endophytic actinobacteria of plants in special and other ecological niches. Front Microbiol 9:1767. https://doi.org/10.3389/fmicb.2018.01767

Singh BP, Hatton BJ, Balwant S, Cowie AL, Kathuria A (2010) Influence of biochars on nitrous oxide emission and nitrogen leaching from two contrasting soils. J Environ Qual 39:1224–1235. https://doi.org/10.2134/jeq2009.0138

Singh A, Sarma BK, Upadhyay RS, Singh HB (2013) Compatible rhizosphere microbes mediated alleviation of biotic stress in chickpea through enhanced antioxidant and phenylpropanoid activities. Microbiol Res 168(1):33–40

Singh A, Maji S, Kumar S (2014) Effect of biofertilizers on yield and biomolecules of anti-cancerous vegetable broccoli. Int J Bioresour Stress Manag 5:262. https://doi.org/10.5958/0976-4038.2014.00565.x

Singh R, Jha P, Jha P (2015a) The plant-growth-promoting bacterium *Klebsiella* sp. SBP-8 confers induced systemic tolerance in wheat (*Triticum aestivum*) under salt stress. J Plant Physiol 184: 57–67

Singh SK, Sharma HR, Shukla A, Singh U, Thakur A (2015b) Effect of biofertilizers and mulch on growth, yield and quality of tomato in mid-hills of Himachal Pradesh. Int J Farm Sci 5:98–110

Singh S, Parihar P, Singh R, Singh VP, Prasad SM (2016) Heavy metal tolerance in plants: role of transcriptomics, proteomics, metabolomics, and ionomics. Front Plant Sci 6:1143. https://doi.org/10.3389/fpls.2015.01143

Sinsabaugh RL (2010) Phenol oxidase, peroxidase and organic matter dynamics of soil. Soil Biol Biochem 42:391–404

Sivakumar N, Sathishkumar R, Selvakumar G, Shyamkumar R, Arjunekumar K (2020) Phyllospheric microbiomes: diversity, ecological significance, and biotechnological applications. Plant Microb Sust Agricult. pp 113–172. https://doi.org/10.1007/978-3-030-38453-1_5

Smith SE, Read DJ (eds) (1997) Mycorrhizal symbiosis. Academic, London, p 13

Smith SE, Read DJ (2008) Mycorrhizal symbiosis, 3rd edn. Academic, London

Smith SE, Smith FA (2011) Roles of arbuscular mycorrhizas in plant nutrition and growth: new paradigms from cellular to ecosystem scales. Annu Rev Plant Biol 62:227–250

Smith FA, Grace EJ, Smith SE (2009) More than a carbon economy: nutrient trade and ecological sustainability in facultative arbuscular mycorrhizal symbioses. New Phytol 182:347–358

Spaepen S, Vanderleyden J (2015) Auxin signaling in *Azospirillum brasilense*: a proteome analysis. In: Biological nitrogen fixation. Wiley, New York, pp 937–940. https://doi.org/10.1002/9781119053095.ch91

Sprent JI (2009) Legume nodulation: a global perspective. Wiley, Hoboken, NJ. 1-4443-1639-7.

Sreekanth D, Kristin IM, Brett AN (2017) Endophytic fungi from Cathranthus roseus: a potential resource for the discovery of antimicrobial polyketides. Nat Prod Chem Res 5:256

Sridhar K, Rajesh V, Omprakash S, Prathyusha C, Devi KS (2014) A critical review on organic farming of vegetables. Int J Appl Biol Pharm Technol 5(1):216–221

Stanton DE, Batterman SA, Von Fischer JC, Hedin LO (2019) Rapid nitrogen fixation by canopy microbiome in tropical forest determined by both phosphorus and molybdenum. Ecology 100(9):e02795

Steenhoudt O, Vanderleyden J (2000) Azospirillum, a free-living nitrogen-fixing bacterium closely associated with grasses: genetic, biochemical and ecological aspects. FEMS Microbiol Rev 24(4):487–506

Steven B, Huntley R, Zeng Q (2018) The influence of flower anatomy and apple cultivar on the apple flower phytobiome. Phytobiom J 2(3):171–179

Stone B, Weingarten E, Jackson C (2018) The role of the phyllosphere microbiome in plant health and function. Annual Plant Rev:533–556

Strobel G, Daisy B (2003) Bioprospecting for microbial endophytes and their natural products. Microbiol Mol Biol Rev 67:491–502

Strzelczyk E, Kampert M, Li CY (1994) Cytokinin-like substances and ethylene production by Azospirillum in media with different carbon sources. Microbiol Res 149(1):55–60

Sulieman S, Tran LSP (2013) Asparagine: an amide of particular distinction in the regulation of symbiotic nitrogen fixation of legumes. Crit Rev Biotechnol 33:309–327

Sumbul A, Ansari RA, Rizvi R, Mahmood I (2020) Azotobacter: a potential bio-fertilizer for soil and plant health management. Saudi J Biol Sci 27:3634–3640. https://doi.org/10.1016/j.sjbs.2020.08.004

Sun C, Johnson JM, Cai D, Sherameti I, Oelmüller R, Lou B (2010) Piriformospora indica confers drought tolerance in Chinese cabbage leaves by stimulating antioxidant enzymes, the expression of drought-related genes and the plastid-localized CAS protein. J Plant Physiol 167(12): 1009–1017. https://doi.org/10.1016/j.jplph.2010.02.013

Taghavi S, Barac T, Greenberg B, Borremans B, Vangronsveld J, van Der Lelie D (2005) Horizontal gene transfer to endogenous endophytic bacteria from poplar improved phyto-remediation of toluene. Appl Environ Microbiol 71:8500–8505

Tahat MMM, Alananbeh KA, Othman YI, Leskovar D (2020) Soil health and sustainable agriculture. Sustainability 12(12):4859. https://doi.org/10.3390/su12124859

Tajini F, Trabelsi M, Drevon JJ (2012) Combined inoculation with *Glomus intraradices* and *Rhizobium tropici* CIAT899 increases phosphorus use efficiency for symbiotic nitrogen fixation in common bean (*Phaseolus vulgaris* L.). Saudi J Biol Scie 19(2):157–163

Tamayo E, Gómez-Gallego T, Azcón-Aguilar C, Ferrol N (2014) Genome-wide analysis of copper, iron and zinc transporters in the arbuscular mycorrhizal fungus *Rhizophagus irregularis*. Front Plant Sci 2014(14):55

Tatry M, Kassis EE, Lambilliotte R, Corratgé C, Van Aarle I, Amenc LK, Alary R, Zimmermann S, Sentenac H, Plassard C (2009) Two differentially regulated phosphate transporters from the symbiotic fungus *Hebeloma cylindrosporum* and phosphorus acquisition by ectomycorrhizal *Pinus pinaster*. Plant J 57:1092–1102

Tecon R, Leveau J (2016) Symplasmata are a clonal, conditional, and reversible type of bacterial multicellularity. Scie Rep 6(1):31914

Tecon R, Ebrahimi A, Kleyer H, Erev Levi S, Or D (2018) Cell-to-cell bacterial interactions promoted by drier conditions on soil surfaces. Proc Nat Acad Sci 115(39):9791–9796

Tena W, Wolde-Meskel E, Walley F (2016) Symbiotic efficiency of native and exotic *rhizobium* strains nodulating lentil (*Lens culinaris* Medik.) in soils of Southern Ethiopia. Agronomy 6(1): 11. https://doi.org/10.3390/agronomy6010011

Thaiss CA, Zmora N, Levy M, Elinav E (2016) The microbiome and innate immunity. Nature 535: 65–74

Thapa S, Prasanna R (2018) Prospecting the characteristics and significance of the phyllosphere microbiome. Ann Microbiol 68(5):229–245. https://doi.org/10.1007/s13213-018-1331-5

Thapa S, Prasanna R (2018) Prospecting the characteristics and significance of the phyllosphere microbiome. Ann Microbiol 68(5):229–245

Thapa S, Prasanna R, Ranjan K, Velmourougane K, Ramakrishnan B (2017) Nutrients and host attributes modulate the abundance and functional traits of phyllosphere microbiome in rice. Microbiol Res 204:55–64

Tien TM, Gaskins MH, Hubbell D (1979) Plant growth substances produced by *Azospirillum brasilense* and their effect on the growth of pearl millet (*Pennisetum americanum* L.). Appl Environ Microbiol 37(5):1016–1024

Tikhonova EN, Grouzdev DS, Kravchenko IK (2019) *Azospirillum palustre* sp. nov., a methylotrophic nitrogen-fixing species isolated from raised bog. Int J Syst Evol Microbiol 69(9):2787–2793

Timmusk S, Behers L, Muthoni J, Muraya A, Aronsson A-C (2017) Perspectives and challenges of microbial application for crop improvement. Front Plant Sci 8:49. https://doi.org/10.3389/fpls.2017.00049

Tisa LS, Oshone R, Sarkar I, Ktari A, Sen A, Gtari M (2016) Genomic approaches toward understanding the actinorhizal symbiosis: an update on the status of the *Frankia* genomes. Symbiosis 70:5–16. https://doi.org/10.1007/s13199-016-0390-2

Tittabutr P, Payakapong W, Teaumroong N, Singleton PW, Boonkerd N (2007) Growth, survival andfield performance of Bradyrhizobial liquid inoculant formulations with polymeric additives. Sci Asia 33:69–77. https://doi.org/10.2306/scienceasia1513-1874.2007.33.069

Toju H, Kurokawa H, Kenta T (2019) Factors influencing leaf- and root-associated communities of bacteria and fungi across 33 plant orders in a Grassland. Front Microbiol 10:241. https://doi.org/10.3389/fmicb.2019.00241

Torres N, Goicoechea N, Morales F, Antolín MC (2016) Berry quality and antioxidant properties in *Vitis vinifera* cv. Tempranillo as affected by clonal variability, mycorrhizal inoculation and temperature. Crop Pasture Sci 67(9):961–977. https://doi.org/10.1071/CP16038

Tortora ML, Díaz-Ricci JC, Pedraza RO (2011) *Azospirillum brasilense* siderophores with antifungal activity against *Colletotrichum acutatum*. Arch Microbiol 193(4):275–286. https://doi.org/10.1007/s00203-010-0672-7

Tsoata E, Njock SR, Youmbi E, Nwaga D (2015) Early effects of water stress on some biochemical and mineral parameters of mycorrhizal *Vigna subterranea* (L.) Verdc (Fabaceae) cultivated in Cameroon. Int J Agron Agric Res 7(2):21–35

Turan M, Gulluce M, von Wirén N, Sahin F (2012) Yield promotion and phosphorus solubilization by plant growth–promoting rhizobacteria in extensive wheat production in Turkey. J Plant Nut Soil Sci 175(6):818–826. https://doi.org/10.1002/jpln.201200054

Turrini A, Avio L, Giovannetti M, Agnolucci M (2018) Functional complementarity of arbuscular mycorrhizal fungi and associated microbiota: the challenge of translational research. Front Plant Sci 9:1407. https://doi.org/10.3389/fpls.2018.01407

Usuki F, Narisawa K (2007) A mutualistic symbiosis between a dark septate endophytic fungus, Heteroconium chaetospira, and a non mycorrhizal plant, Chinese cabbage. Mycologia 99:175–184. https://doi.org/10.3852/mycologia.99.2.175

Vacher C, Hampe A, Porté AJ, Sauer U, Compant S, Morris CE (2016) The phyllosphere: microbial jungle at the plant–climate interface. Annu Rev Ecol Evol Syst 47(1):1–24. https://doi.org/10.1146/annurev-ecolsys-121415-032238

Van Der Heijden MGA, Bardgett RD, van Straalen NM (2007) The unseen majority: soil microbes as drivers of plant diversity and productivity in terrestrial ecosystems. Ecol Lett 11:296–310. https://doi.org/10.1111/j.1461-0248.2007.01139.x

Varma A, Prasad R, Tuteja N (2017a) Mycorrhiza: function, diversity and state-of-art. Springer International Publishing. ISBN: 978-3-319-53064-2. http://www.springer.com/us/book/9783319530635

Varma A, Prasad R, Tuteja N (2017b) Mycorrhiza: eco-physiology, secondary metabolites, nanomaterials. Springer International Publishing. ISBN: 978-3-319-57849-1. http://www.springer.com/us/book/9783319578484

Varma A, Prasad R, Tuteja N (2017c) Mycorrhiza: nutrient uptake, biocontrol, ecorestoration. Springer International Publishing. ISBN: 978-3-319-68867-1. http://www.springer.com/us/book/9783319688664

Varma A, Swati T, Prasad R (2020) Plant microbe symbiosis. Springer International Publishing. ISBN: 978-3-030-36247-8. https://www.springer.com/gp/book/9783030362478

Velmourougane K, Prasanna R, Chawla G, Nain L, Kumar A, Saxena AK (2019) *Trichoderma–Azotobacter* biofilm inoculation improves soil nutrient availability and plant growth in wheat and cotton. J Basic Microbiol 59:632–644. https://doi.org/10.1002/jobm.201900009

Venkatachalam S, Ranjan K, Prasanna R, Ramakrishnan B, Thapa S, Kanchan A (2016) Diversity and functional traits of culturable microbiome members, including cyanobacteria in the rice phyllosphere. Plant Biol 18(4):627–637. https://doi.org/10.1111/plb.12441

Verma JP, Yadav J, Tiwari KN, Kumar A (2013) Effect of indigenous *Mesorhizobium* spp. and plant growth promoting rhizobacteria on yields and nutrients uptake of chickpea (*Cicer arietinum* L.) under sustainable agriculture. Ecol Eng 51:282–286

Verma P, Yadav AN, Kazy SK, Saxena AK, Suman A (2014) Evaluating the diversity and phylogeny of plant growth promoting bacteria associated with wheat (*Triticum aestivum*) growing in central zone of India. Int J Curr Microbiol Appl Sci 3(5):432–447

Villamil Carvajal JE, Viteri Rosero SE, Villegas Orozco WL (2015) Application of microbial antagonists for the biological control of *Moniliophthora roreri* Cif & Par in Theobroma cacao

L. under field conditions. Rev Fac Nac Agron Medellin 68(1):7441–7450. https://doi.org/10.15446/rfnam.v68n1.47830

Vorholt JA (2012) Microbial life in the phyllosphere. Nat Rev 10:828–840. https://doi.org/10.1038/nrmicro2910

Wagenaar MM, Corwin J, Strobel G, Clardy J (2000) Three new cytochalasins produced by an endophytic fungus in the genus Rhinocladiella. J Nat Prod 63(12):1692–1695. https://doi.org/10.1021/np0002942

Waller F, Achatz B, Baltruschat H, Fodor J, Becker K, Fisher M et al (2005) The endophytic fungus *Piriformospora indica* reprograms barley to salt-stress tolerance, disease resistance, and higher yield. PNAS 102:13386–13391

Wang S (2009) Molecular mechanism of plant growth promotion and induced systemic resistance to tobacco mosaic virus by *Bacillus* spp. J. Microbiol Biotechnol 19:1250–1258

Wang Y, Jing H, Gao Y (2012) Arbuscular mycorrhizal colonization alters subcellular distribution and chemical forms of cadmium in *Medicago sativa* L. and resists cadmium toxicity. PLoS One 7:3161–3164. https://doi.org/10.1371/journal.pone.0048669

Wang X, Xue Y, Han M, Bu Y, Liu C (2014a) The ecological roles of Bacillus thuringiensis within phyllosphere environments. Chemosphere 108:258–264. https://doi.org/10.1016/j.chemosphere.2014.01.050

Wang S, Zhang S, Sun C, Xu Y, Chen Y, Yu C, Qian Q, Jiang DA, Qi Y (2014b) Auxin response factor (OsARF12), a novel regulator for phosphate homeostasis in rice (*Oryza sativa*). New Phytologist 201:91–103. https://doi.org/10.1111/nph.12499

Wang P, Niu GY, Fang YH, Wu RJ, Yu JJ, Yuan GF, Pozdniakov SP, Scott RL (2018a) Implementing dynamic root optimization in Noah-MP for simulating phreatophytic root water uptake. Water Reso Res 54(3):1560–1575. https://doi.org/10.1002/2017WR021061

Wang Y, Wang M, Li Y, Wu A, Huang J (2018b) Effects of arbuscular mycorrhizal fungi on growth and nitrogen uptake of *Chrysanthemum morifolium* under salt stress. PLoS One 13(4):e0196408. https://doi.org/10.1371/journal.pone.0196408

Wani S, Ali T (2013) Potential use of *Azotobacter chroococcum* in crop production: an overview. Curr Agric Res J 1:35–38. https://doi.org/10.12944/CARJ.1.1.04

Wei F, Hu X, Xu X (2016) Dispersal of *Bacillus subtilis* and its effect on strawberry phyllosphere microbiota under open field and protection conditions. Scie Rep 6:22611. https://doi.org/10.1038/srep22611

Wensing A, Braun SD, Büttner P, Expert D, Völksch B, Ullrich MS, Weingart H (2010) Impact of siderophore production by *Pseudomonas syringae* pv. syringae 22d/93 on epiphytic fitness and biocontrol activity against *Pseudomonas syringae* pv. glycinea 1a/96. Appl Environ Microbiol 76(9):2704–2711. https://doi.org/10.1128/AEM.02979-09

Whipps JM, Hand P, Pink D, Bending GD (2008) Phyllosphere microbiology with special reference to diversity and plant genotype. J Appl Microbiol 105:1744–1755. https://doi.org/10.1111/j.1365-2672.2008.03906.x

Wichard T, Bellenger JP, Morel FM, Kraepiel AM (2009) Role of the siderophore azotobactin in the bacterial acquisition of nitrogenase metal cofactors. Environ Sci Tech 43(19):7218–7224. https://doi.org/10.1021/es8037214

Williams T, Moyne A, Harris L, Marco M (2013) Season, irrigation, leaf age, and escherichia coli inoculation influence the bacterial diversity in the lettuce phyllosphere. PLoS One 8(7):e68642. https://doi.org/10.1371/journal.pone.0068642

Wilson M, Lindow SE (1993) Interactions between the biological control agent Pseudomonas fluorescens A506 and Erwinia amylovora in pear blossoms. Phytopathology 83(1):117–123. https://doi.org/10.1094/Phyto-83-117

Wu QS, Zou YN, Liu W, Ye XF, Zai HF, Zhao LJ (2010) Alleviation of salt stress in citrus seedlings inoculated with mycorrhiza: changes in leaf antioxidant defense systems. Plant Soil Environ 56:470–475. https://doi.org/10.17221/54/2010-PSE

Wu D, Zhang XJ, Liu HC, Zhou YG, Wu XL, Nie Y et al (2021) *Azospirillum oleiclasticum* sp. nov, a nitrogen-fixing and heavy oil degrading bacterium isolated from an oil production mixture of

Yumen Oilfield. Syst Appl Microbiol 44(1):126171. https://doi.org/10.1016/j.syapm.2020.126171

Xie QY, Wang C, Wang R, Qu Z, Lin HP, Goodfellow M, Hong K (2011) *Jishengella endophytica* gen. nov., sp. nov., a new member of the family Micromonosporaceae. Int J Syst Evol Microbiol 61(5):1153–1159. https://doi.org/10.1099/ijs.0.025288-0

Xie X, Lin H, Peng X, Xu C, Sun Z, Jiang K, Huang A, Wu X, Tang N, Salvioli A, Bonfante P, Zhao B (2016) Arbuscular mycorrhizal symbiosis requires a phosphate transceptor in the Gigaspora margarita fungal symbiont. Mol Plant 9(12):1583–1608. https://doi.org/10.1016/j.molp.2016.08.011

Yadav AN (2017) Agriculturally important microbiomes: biodiversity and multifarious PGP attributes for amelioration of diverse abiotic stresses in crops for sustainable agriculture. Biomed J Sci Tech Res 1(4):861–864. https://doi.org/10.26717/BJSTR.2017.01.000321

Yadav AN, Yadav N (2018) Stress-adaptive microbes for plant growth promotion and alleviation of drought stress in plants. Acta Sci Agric 2(6):85–88

Yadav N, Yadav AN (2019) Actinobacteria for sustainable agriculture. J Appl Biotechnol Bioeng 6(1):38–41. https://doi.org/10.15406/jabb.2019.06.00172

Yadav V, Kumar M, Deep DK, Kumar H, Sharma R, Tripathi T, Tuteja N, Saxena AK, Johri AK (2010) A phosphate transporter from the root endophytic fungus Piriformospora indica plays a role in phosphate transport to the host plant. J Biol Chem 285:26532–26544. https://doi.org/10.1074/jbc.M110.111021

Yang H, Knapp J, Koirala P, Rajagopal D, Peer WA, Silbart LK et al (2007) Enhanced phosphorus nutrition in monocots and dicots over-expressing a phosphorus responsive type I H+-pyrophosphatase. Plant Biotechnol J 5:735–745. https://doi.org/10.1111/j.1467-7652.2007.00281.x

Yang S-Y, Grønlund M, Jakobsen I, Grotemeyer MS, Rentsch D, Miyao A et al (2012) Nonredundant regulation of rice arbuscular mycorrhizal symbiosis by two members of the phosphate transporter1 gene family. Plant Cell 24(10):4236–4251. https://doi.org/10.1105/tpc.112.104901

Yang Y, Tang M, Sulpice R, Chen H, Tian S, Ban Y (2014) Arbuscular mycorrhizal fungi alter fractal dimension characteristics of Robinia pseudoacacia L. seedlings through regulating plant growth, leaf water status, photosynthesis, and nutrient concentration under drought stress. J Plant Growth Reg 33:612–625. https://doi.org/10.1007/s00344-013-9410-0

Yang Y, Zhao J, Jiang Y, Hu Y, Zhang M, Zeng Z (2017) Response of bacteria harboring nirS and nirK genes to different N fertilization rates in an alkaline northern Chinese soil. Eur J Soil Biol 82:1–9. https://doi.org/10.1016/j.ejsobi.2017.05.006

Yang Y, Zhang R, Feng J, Wang C, Chen J (2019) *Azospirillum griseum* sp. nov., isolated from lakewater. Int J Syst Evol Microbiol 69(12):3676–3681. https://doi.org/10.1099/ijsem.0.003460

Yao ZY, Kan FL, Wang ET, Wei GH, Chen WX (2002) Characterization of rhizobia that nodulate legume species of the genus *Lespedeza* and description of *Bradyrhizobium yuanmingense* sp. nov. Int J Syst Evol Microbiol 52:2219–2230. https://doi.org/10.1099/ijs.0.01408-0

Yao H, Sun X, He C, Maitra P, Li X, Guo L (2019) Phyllosphere epiphytic and endophytic fungal community and network structures differ in a tropical mangrove ecosystem. Microbiome 7:57. https://doi.org/10.1186/s40168-019-0671-0

Yi Y, Huang W, Ge Y (2008) Exopolysaccharide: a novel important factor in the microbial dissolution of tricalcium phosphate. World J Microbiol Biotechnol 24:1059–1065. https://doi.org/10.1007/s11274-007-9575-4

Yooyongwech S, Samphumphuang T, Tisarum R, Theerawitaya C, Chaum S (2016) Arbuscular mycorrhizal fungi (AMF) improved water deficit tolerance in two different sweet potato genotypes involves osmotic adjustments via soluble sugar and free proline. Sci Hort. 198: 107–117. https://doi.org/10.1016/j.scienta.2015.11.002

Yooyongwech S, Samphumphung T, Tisaram R, Theerawitaya C, Suriyan CU (2017) Physiological, morphological changes and storage root yield of sweetpotato [*Ipomoea batatas* (L.) Lam.]

under peg-induced water stress. Not Bot Horti Agrobo 45(1):164–171. https://doi.org/10.15835/nbha45110651

Yousaf B, Liu G, Wang R, Zia-ur-Rehman M, Rizwan MS, Imtiaz M, Murtaza G, Shakoor A (2016) Investigating the potential influence of biochar and traditional organic amendments on the bioavailability and transfer of CD in the soil–plant system. Environ Earth Sci 75(5):374. https://doi.org/10.1007/s12665-016-5285-2

Yu X, Lund SP, Scott RA, Greenwald JW, Records AH, Nettleton D, Lindow SE, Gross DC, Beattie GA (2013) Transcriptional responses of *Pseudomonas syringae* to growth in epiphytic versus apoplastic leaf sites. Proc Natl Acad Sci U S A 110(5):E425–E434. https://doi.org/10.1073/pnas.1221892110

Yurimoto H, Shiraishi K, Sakai Y (2021) Physiology of methylotrophs living in the phyllosphere. Microorganisms. 9(4):809. https://doi.org/10.3390/microorganisms9040809

Zaefarian F, Rezvani M, Ardakani MR, Rejali F, Miransari M (2013) Impact of mycorrhizae formation on the phosphorus and heavy-metal uptake of Alfalfa. Comm Soil Sci. Plant Anal 44:1340–1352. https://doi.org/10.1080/00103624.2012.756505

Zayadan BK, Matorin D, Baimakhanova GB, Bolathan K, Oraz GD, Sadanov AK (2014) Promising microbial consortia for producing biofertilizers for rice fields. Microbiol 83(4):391–397. https://doi.org/10.1134/S0026261714040171

Zhang HS, Qin FF, Qin P, Pan SM (2014) Evidence that arbuscular mycorrhizal and phosphate-solubilizing fungi alleviate NaCl stress in the halophyte *Kosteletzkya virginica*: nutrient uptake and ion distribution within root tissues. Mycorrhiza 24(5):383–395. https://doi.org/10.1007/s00572-013-0546-3

Zhang H, Liu Z, Chen H, Tang M (2016) Symbiosis of arbuscular mycorrhizal fungi and *Robinia pseudoacacia* L. improves root tensile strength and soil aggregate stability. PLoS One 11(4):e0153378. https://doi.org/10.1371/journal.pone.0153378

Zhang F, He J, Ni Q, Wu Q, Zou Y (2018a) Enhancement of drought tolerance in trifoliate orange by mycorrhiza: changes in root sucrose and proline metabolisms. Not Bot Horti Agrobot Cluj 46(1):270–276. https://doi.org/10.15835/nbha46110983

Zhang X, Han L, Wang Q, Zhang C, Yu Y, Tian J, Kong Z (2018b) The host actin cytoskeleton channels rhizobia release and facilitates symbiosome accommodation during nodulation in Medicago truncatula. New Phytol 221:1049–1059. https://doi.org/10.1111/nph.15423

Zhang J, Liu YX, Zhang N et al (2019) *NRT1.1B* is associated with root microbiota composition and nitrogen use in field-grown rice. Nat Biotechnol 37:676–684. https://doi.org/10.1038/s41587-019-0104-4

Zhao R, Guo W, Bi N, Guo J, Wang L, Zhao J, Zhang J (2015) Arbuscular mycorrhizal fungi affect the growth, nutrient uptake and water status of maize (Zea mays L.) grown in two types of coal mine spoils under drought stress. Appl Soil Ecol 88:41–49. https://doi.org/10.1016/j.apsoil.2014.11.016

Zhao ZL, Ming H, Ding CL, Ji WL, Cheng LJ, Niu MM et al (2020) *Azospirillum thermophilum* sp. nov., isolated from a hot spring. Int J Syst Evol 70(1):550–554. https://doi.org/10.1099/ijsem.0.003788

Zhou S, Han L, Wang Y, Yang G, Zhuang L, Hu P (2013) *Azospirillum humicireducens* sp. nov., a nitrogen-fixing bacterium isolated from a microbial fuel cell. Int J Syst Evol Microbiol 63:2618–2624. https://doi.org/10.1099/ijs.0.046813-0

Zhu XC, Song FB, Xu HW (2010) Arbuscular mycorrhizae improve low temperature stress in maize via alterations in host water status and photosynthesis. Plant Soil 331:129–137. https://doi.org/10.1007/s11104-009-0239-z

Zhu XC, Song FB, Liu SQ, Liu FL (2016) Arbuscular mycorrhiza improves growth, nitrogen uptake, and nitrogen use efficiency in wheat grown under elevated CO_2. Mycorrhiza 26:133–140. https://doi.org/10.1007/s00572-015-0654-3

Zolla G, Badri DV, Bakker MG, Manter DK, Vivanco JM (2013) Soil microbiomes vary in their ability to confer drought tolerance to *Arabidopsis*. Appl Soil Ecol 68:1–9. https://doi.org/10.1016/j.apsoil.2013.03.007

Chapter 8
Microbiome of Plants: The Diversity, Distribution, and Their Potential for Sustainable Agriculture

Ajay Kumar Pandey, Shweta Mehta, Priyanka Bhati, and Sagar Chhabra

Abstract Plants contain associated microbiomes, which facilitate the plant's productivity by supporting their healthy growth in native niche. Globally, sustainable food production is quintessential, wherein the growth-promoting abilities of plant-associated microbes become vital. Recent literatures reveal significant structure and dynamics on plant microbiome, which recognizes the regulation of plant fitness by native microbiome. Henceforth, it is essential to practice modern microbial developments for strengthening crop production. Many of the environmental factors modulate the microbiome organization. This chapter emphasizes the comprehensive characterization of different niche's microbiome and impact of differential influential factors such as genotype, microbial interactions, agricultural habits, bioinoculants, and abiotic factors to microbial diversity and sustainable plant health. Moreover, conventional and advanced approaches (omics/multi-omics) for studying plant microbiome diversity and plant microbiome interactions are also discussed, which would be helpful in developing novel microbial formulations for sustainable agriculture.

Keywords Plant microbiome · Microbiome diversity · Niche · Agriculture · Bioinoculants

---

A. K. Pandey (✉)
Department of Life Sciences and Biotechnology, Chhatrapati Shahu Ji Maharaj University (CSJMU), Kanpur, India, Uttar Pradesh

Division of Life Sciences, School of Basic & Applied Sciences, Galgotias University, Greater Noida, Uttar Pradesh, India
e-mail: ajaykp4@csjmu.ac.in

S. Mehta · P. Bhati · S. Chhabra
Department of Biosciences, School of Basic and Applied Sciences, Galgotias University, Greater Noida, Uttar Pradesh, India

S. Chhabra et al. (eds.), *Plant Microbiome for Plant Productivity and Sustainable Agriculture*, Microorganisms for Sustainability 37,
https://doi.org/10.1007/978-981-19-5029-2_8

8.1 Introduction

Term "plant microbiome" describes "the microorganism's community with the potential of harboring a space in plant body as symbionts, commensal or pathogen." Plants harbor microbial hotspots along with soil as major natural microbial resource. This microbial plethora is vital in promoting plant's growth, health, and productivity. However, food productivity enhancement strategies (use of fertilizers, pesticides, and agrochemicals), industrialization, and anthropogenic interventions led to an enforced soil and microenvironment deterioration and attenuation in agriculture lands (Compant et al. 2019). These factors cumulatively exert pressure on agroecosystems and challenge the high-yielding economical and sustainable agriproducts. In this background, plant microbiomes are potential tool for enhancing agricultural productivity while maintaining sustainable ecosystem. The plant microbiomes not only promote the growth of plants but also enhance the soil fertility for facilitating sustainable agriculture. Moreover, microbiome aids in seed germination, growth, productivity, yield, adaptation, and stress resistance/tolerance of plants, which impacts the fitness of the plant leading to the sustainable agriculture (Aamir et al. 2021).

8.2 Plant Microbiome: Diversity, Composition, and Distribution

Microbiomes possess high degree of diversity and are dispersed on soil, water, and air, which can associate with the plant organs (Gupta et al. 2021). The microbiome distribution on plant is majorly categorized into phyllospheric (present on aerial parts such as stem, leaves, and flowers), endospheric (present on inner plant tissues), and rhizospheric (present on below ground components such as roots and surrounding soil) (Fig. 8.1) (Aamir et al. 2021; Santos and Olivares 2021). In addition, microbes also reside as epiphyte on plant surface (Compant et al. 2019). As endophyte microbes penetrate the epidermis and colonize inter- and intracellularly. In general, ectomycorrhizal fungi penetrate root spaces intracellularly after developing a mantle, which surrounds root tip. However, endomycorrhizal fungi develop arbuscules for intracellular colonization (Santos and Olivares 2021). Naturally, healthy plant harbors multifaceted diversity of microorganisms such as bacteria, and fungi are foremost constituent associated with plants (Trivedi et al. 2020). Some examples of plant-associated bacterial and fungal strains and their functional role in normal/stressed environment are listed in Tables 8.1 and 8.2, respectively.

Generally, microbial dynamics is associated with the types of plant species, its developmental stages, or genetic makeup. And, changes in host plant environment during the different life stages play a crucial role and arbitrate the microbial diversity by innate defense system stimulation of plant (Dastogeer et al. 2020; Gupta et al. 2021). Additionally, abiotic and biotic stresses-driven and -compromised defense

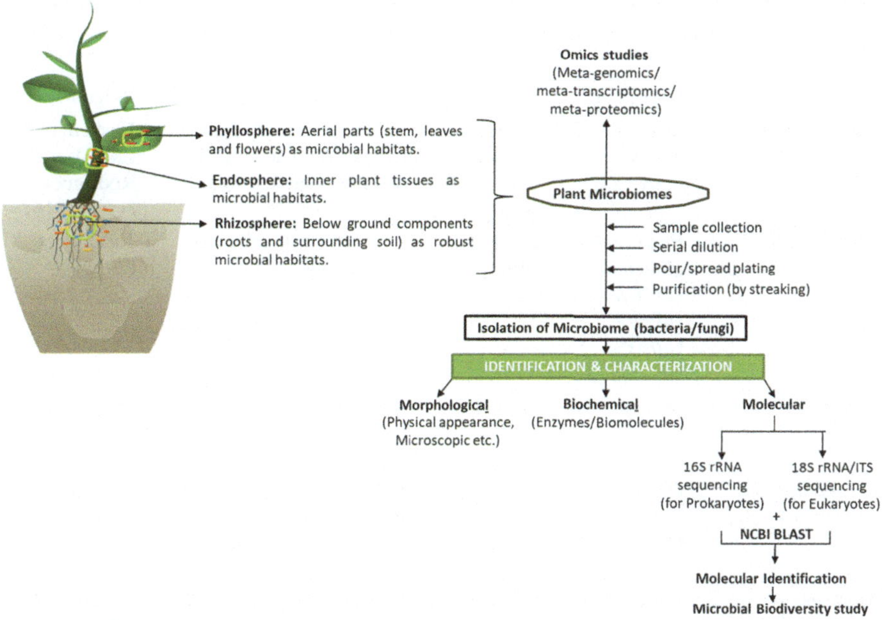

Fig. 8.1 Schematics of morphological, biochemical, and molecular characterization of plant microbiome

system can also alter microbiome diversity. Henceforth, to study the biotic/abiotic factors influencing microbial matrix at phylogenetic and functional level, it is essential to estimate the plant microbiome's heterogeneity individually and in understanding different niches (Santos and Olivares 2021).

Recent studies and advances in the knowledge of plant microbiome have demonstrated that plant-associated heterogenetic microbial consortia influence their growth, development, and productivity (Bhatt et al. 2020; Singh et al. 2019; Trivedi et al. 2020). The microbiome consists of some microorganisms, which can interact with other species/host and influence theorganization microbial community. Such microorganisms are known as "hub-microorganisms" of the microbiome (Agler et al. 2016). Microbiome diversity and composition are majorly affected by soil profile and physiochemical characteristics (texture, moisture proportion, pH, temperature, salinity, organic matter, etc.). Moreover, microbe–microbe/plant–microbe interaction, microbial/host secretome, anthropogenic factors, cultivation exercises, and environmental changes also influence at a certain extent (Bhatt et al. 2020; Li et al. 2020; Oyserman et al. 2021).

Table 8.1 Examples of bacterial strains involved in plant microbiome assembly, their function and relevance in normal and stressed conditions

Strains	Host-plant	Growth parameters	Activity and function	References
Bacillus subtilis	*Zea mays*, *Acacia gerrardii*	Nutrient deficiency and salinity stress	Solubilization of supplements (Mg, Ca, N, P, K) and their delivery to plant	Hashem et al. (2016), Rodrigues et al. (2016)
Pseudomonas putida	*Pennisetum glaucum*, *Hordeum vulgare*, *Citrus macrophylla*	Salt and salinity stress	Tolerance and enhanced phytoremediation	Jodeh et al. (2015)
Novosphingobium sp.	*Glycine max*	Normal	Improved nitrogen digestion and transport of supplement	Hara et al. (2019)
Pseudomonas fluorescens	*Oryza sativa*, *Citrus aurantifolia*	Drought stress	Enhanced phosphate solubilization, growth, and chlorophyll content	Shahsavar et al. (2016)
Enterobacter sp.	*Pennisetum glaucum*, *Brassica juncea*	Metal stress	Enhanced phosphate solubilization, growth, and phytohormones	Durán et al. (2018)
Klebsiella sp.	*Avena sativa*	Salt stress	Enhanced salt tolerance	Sapre et al. (2018)
Acinetobacter calcoaceticus	*Agave americana*	Normal	Improved growth and sugar content	De La Torre-Ruiz et al. (2016)
Streptomyces sp.	*Pisum sativum*	Drought stress	Improved salt tolerance, seed germination, root/shoot length, and ROS gathering	Zahir et al. (2008)
Saccharibacteria, *Verrucomicrobia* *Firmicutes*	*Triticum aestivum*	Normal	Antifungal property against *Microdochium*, *Neoascochyta, and Fusarium* spp.	Kinnunen-Grubb et al. (2020)
Sinorhizobium and Rhizobium	*Hordeum vulgare*	Oils sand mining	Increased plant growth, improved resistance against biotic/abiotic stresses	Mitter et al. (2017)
Acremonium sp. *Mesorhizobium* *Mesorhizobium ciceri*	*Cicer arietinum*	Normal	Improved knob size, yield, and nitrogen content. Antifungal property against *Botrytis*, *Fusarium, and Rhizoctonia solani*	Alok et al. (2020), Qin et al. (2017)
Rhizobium daejeonense	*Agave americana*	Normal	Enhanced growth, sugar, and solubilization of supplements.	De La Torre-Ruiz et al. (2016)

Table 8.2 Examples of fungal strains involved in plant microbiome assembly, their function and relevance in normal and stressed conditions

Strains	Host-plant	Growth parameters	Activity and function	References
Penicillium menonorum	*Cucumis sativus*	Normal	Plant-growth-promotion activity via production of siderophore and indole-3 acetic acid and phosphate solubilization	Babu et al. (2015)
Acaulospora scrobiculata, *Funneliformis mosseae*	*Morus alba*	Normal	Enhanced growth due to increased chlorophyll content and rate of photosynthesis. Improved conductance of stomata.	Shi et al. (2016)
Rhizophagus intraradices, *R. fasciculates*, *Glomus mosseae*, *Scutellospora* sp., and *Gigaspora margarita*	*Solanum lycopersicum*	Drought and salt stress	Enhanced water transport, water pressure, and phosphorous aggregation.	Chitarra et al. (2016), Mohumad Tahat (2012), Padmavathi et al. (2015), Viscardi et al. (2016)
Rhizophagus clarus	*Sorghum bicolor*, *Capsicum annuum*	Normal	Improved plant growth	Lee and Eom (2015)
Glomus etunicatum	*Citrus aurantifolia*	Drought stress	Improved plant growth, chlorophyll and photosynthesis.	Shahsavar et al. (2016)
Glomus species	*Allium cepa*	Normal	Improved plant growth, chlorophyll, and photosynthesis.	Shuab et al. (2014)
Funneliformis mosseae, *Rhizoglomus intraradices*	*Lactuca sativa*	Normal	Enhanced zinc absorption	Konieczny and Kowalska (2016), Zhang et al. (2020)

8.3 Approaches for Studying Plant Microbiome Diversity

Plant microbiome exhibits extensively wide-ranging diversity depending upon habitat and plant organs such as phyllo/endo/rhizospheric regions. Plant microbiome diversity investigation aims to unravel the structural/functional diverseness of a specific host plant-associated microbial community in a definite habitat (Bhatt et al. 2020; Dastogeer et al. 2020; Liu et al. 2019). The colonization of microbiome is facilitated by the host's secondary metabolite secretome and specialized structures such as hairs and trichomes (Gupta et al. 2021).

Conventionally, identification and characterization of plant microbiomes (bacteria and fungi) consists of sequencing of universally conserved sequences and bioinformatic analysis. Briefly, this method consists of different sequential steps as: (1) microbial isolation from phyllo/endo/rhizosphere, (2) morphological characterization (for phenotypic and microscopic characteristics), (3) biochemical characterization (for biomolecules and enzymes), (4) Sanger's sequencing (for 16S rNRA/18S rRNA), and (5) Basic Local Alignment Search Tool (BLAST) analysis using National Centre for Biological Information (NCBI) database.

These conventional methods give the outlook information only, henceforth holistic approaches using modern biotechnological techniques to unwrap complete multi-dimension information to understand the mechanistic component are essential. In this context, during last few years various studies implemented advance techniques to unwrap the microbiome exclusively and understand their interaction mechanism with host (Fitzpatrick et al. 2020). It has been established that along with the host and habitat, microbiomes are also affected by their environment and developmental stages. In recent years, advanced modified Sanger's sequencing techniques including 454-pyrosequencing, Oxford nanopore, Illumina, Ion- torrent, PacBio, and hybrid-platform were used to reveal rare microbial taxa in definite niche (Bhatt et al. 2020; Gupta et al. 2021). These techniques extract complete genomic information leading to comprehensive microbial diversity study in a given population through phylogenetic and comparative genomics studies. Nevertheless, these techniques fail to differentiate dead/live cells, which raises the possibility of errors in estimation or hiked microbiome diversity, which are being addressed by managing PCR error using specifically tagged DNA (Gupta et al. 2021).

Moreover, microbiome also contains enormous diversity in uncultivable microbes, which makes it difficult to select a suitable method for their identification. However, in recent years advanced culture-independent direct-environmental DNA cloning, next generation sequencing, single strand conformation polymorphism, denaturing/temperature gradient gel-electrophoresis, terminal restriction/restriction fragment length polymorphism, and fatty acid methyl esters were reported for plant microbiome studies (Bodor et al. 2020; Gupta et al. 2021; del Orozco-Mosqueda et al. 2018).

Omics technologies (genomics/metagenomics, transcriptomics/meta-transcriptomics, proteomics/meta-proteomics, and metabolomics) give a comprehensive genetic, structural, and functional information of the plant microbiomes (Sharma et al. 2020). This information, coupled with the metabolic engineering at cellular and/or molecular level, enhances the knowledge on dynamics of microbiomes.

8.4 Factors Affecting Plant Microbiome Diversity

Microbiome consists of several microbes; however, various factors including genome organization, microbial interactions, biotic/abiotic components, local agricultural practices, environmental factors, and physiochemical factors greatly affect plant microbiome organization in local habitat. Some important factors are described below:

8.4.1 Impact of Genomic Organization

Plants benefit from capacities of their microbiome, which is the reason they devote some fraction of their carbon sources toward the development and sustenance of the microbiota. Plants with different genotypes perform distinctively toward metabolism of roots, composition of root secretome, recognition system, and native defense mechanism (Santos and Olivares 2021). Plant genotype alters the root metabolism by modulating the metabolic intermediates (pyruvic acid, citric acid, malic acid, lactic acid, etc.), nitrogenase activity, and reactive oxygen species (ROS) generation (Shcherbakova et al. 2017). Genotype also modulates composition of root secretome in terms of organic acids (citric, malic, and succinic), sugars (ribose, glucose, galactose, and fructose), amino acids (glycine, serine, lysine, and histidine), isoflavones, enzymes (chitinases, proteases, phytases), and phytoalexins (Shcherbakova et al. 2017). Genotype mediates the localized/systemic immune responses of plants by transcriptional changes leading to the elaboration of physical barriers, synthesis of antimicrobial compounds, defense phytohormones, jasmonic acid, salicylic acid, and gaseous ethylene (Lebeis et al. 2015).

Studies demonstrate that varieties of this nature can modify the design and action of the plant microbiome, which thus follows up on development, sustenance, and protection from biotic and abiotic stresses (Santos and Olivares 2021). Curlango-Rivera et al. demonstrated the impact of genotype toward composition of root microbiome, wherein cotton cultivars showed structural differences in their border cells (Curlango-Rivera et al. 2013). These cells are shaped from meristematic cells and were at first thought to be "dead" and answerable for the mechanical security of the root tip. Today, it is realized that line cells are connected with colonization of the root by microorganisms. Essentially, plant roots store high and low sub-atomic weight compounds at the root–soil interface, the rhizodeposition, which might shift as per their genotype. Along these lines, plants with various genotypes can deliver compounds with various creations, advancing explicit microbial flagging and colonization. While examining the seed microbiome, various creators distinguished the effect of the plant genotype on the microbial synthesis of the seed. In the tomato phyllosphere, a review uncovered that four of the nine genotypes tried had an alternate microbial arrangement, which demonstrates that the hereditary elements of the host plant might shape the related microbiota (Morella et al. 2020). In the rhizosphere, microorganisms from the families Solibacteraceae, Pseudomonadaceae, Cytophagaceae, and Bacillaceae were more bountiful in Fusarium-safe bean cultivars (Mendes et al. 2019). Fusarium opposition formed the microbial gathering of the rhizosphere and chose bacterial taxa with biocontrol action (Santos and Olivares 2021). In recent study, transplantation of rhizospheric soil from a *Ralstonia solanacearum*-resistant tomato plant to a prone plant genotype stifled the susceptibility (Kwak et al. 2018).

8.4.2 Impact of Agricultural Activities

Agricultural activities alter soil's nutrient and physiochemical properties. Nutrient properties act as selection pressure for selection of plant microbiome and altered nutrition influences plant's microbiome profile in negative way and reduces their productivity. Several studies reported the change in microbiome in different agricultural practices using green manure (Bergottini et al. 2017), organic manure (Lupatini et al. 2017), mulch (Qin et al. 2017), nitrogen fertilizers (Cai et al. 2017; Zhu et al. 2016), and crop rotation (Hong et al. 2020). Therefore, for achieving high agricultural productivity, it is essential to comprehend the impact of feasible practices on the design of the plant's microbiome referring native plants and vegetable/animal residues (Santos and Olivares 2021).

8.4.3 Impact of Bioinoculants

Bioinoculant is a practical strategy to build crop efficiency while lessening the utilization of synthetic chemicals, though information on its effect on the plant microbiome is extremely scant. Subsequently, understanding the connection among inoculated and native microbial networks in plants will add to the improvement of natural products.

Microbial inoculants proficiency relies upon the inoculum characteristics and its capacity to bridge local microbial contest or its layout in the rhizosphere. Rhizospheric fitness (in terms of efficient microbial contents) is an essential factor for promising plant responses (plant-microbe interactions), however, has been explored below par. The cooperation between local miniature life forms of the plant and miniature organic entities (called "transients") can be competitive, parasitic, mutualistic, or predatory (Santos and Olivares 2021), wherein secretomes containing probiotic/antimicrobial compounds selectively stimulate/inhibit the growth of microorganisms (Pieterse et al. 2016). Taken together, the native microbial organization of the host plant is urgent for the achievement of its colonization by transient microbes. Different signalling routes between plants and microorganisms as well as among microorganisms make up the base of the microorganisms in the root area. During plant-microbe interactions the population density is managed by releasing low atomic weight quorum sensing compounds such as acylated-homoserine-lactones (Hartmann et al. 2014).

The impact of the bioinoculant is regarded as indirect when it causes changes in the native microbiome structure and function by producing probiotic or antimicrobial compounds (Santos and Olivares 2021). Several reports stated the alteration in endophytic population's structure while using inoculants (Andreote et al. 2014; Conn and Franco 2004). In another study, changes in the phenotype of the endophytic potato were reported after *Methylobacterium* sp. inoculation to it, wherein relative proportion of biocontrol regulators (*Acinetobacter* sp., *Massilia* sp., *Phoma* sp., and *Entyloma* sp.) were also increased (Ardanov et al. 2016).

The bioinoculants effects are considered as direct if inoculum directly alters/improves the metabolism of plant or root exudates profile, which in turn modulates microbiome's structure and function (Santos and Olivares 2021). Molecular advanced technologies permit portraying the microbiome of various plants or contrasting microbiome of plants treated with bio-inoculants. This information will be helpful in developing novel bioinoculants, or shaping it as per the agricultural need.

8.4.4 Impact of Pathogens

The endophytic plant community can harbor useful and destructive microbes in an inactive state. These microbes can be "reactivated" by extrinsic factors including pathogen's attack (Santos and Olivares 2021). The endophytic community encourages plant resistance either with the presence of resistance inducing microbes or reactivation of endophytic population by entry of a microbial biocontrol strain/abiotic stimulus (Podolich et al. 2015). Nevertheless, reactivation of inactive endophytic microbes after pathogenic attack is more proficient (Santos and Olivares 2021). Agler et al. reported phyllosphere's reformed colonization of endophytic and endophytic bacteria by pathogenic fungi *Albugo* sp. and *Dioszegia* sp. (Agler et al. 2016). Likewise, a decrease in the number of *Bacillus* species in roots, and *Streptomyces & Flavisolibacter* in the rhizosphere of Chinese cabbage was observed with the presence of pathogenic *Plasmodiophora brassicae* (Lebreton et al. 2019). These findings show that the microbiome can protect plants against pathogenic attacks, hence it is conceivable that it can change the microbiota of the host plant.

8.4.5 Impacts of Abiotic Factors

Plants and their microbiome are naturally exposed to different environmental variations including humidity, pH, temperature, salinity, and ultraviolet rays, which straightforwardly or by implication alter the microbial composition. The microbiome profiles of soil from various climate zones shows enormous taxonomic and functional variety, although arid conditions revealed the least functional variation (Tripathi et al. 2017). Lower functional diversity in arid regions is associated with the up-regulation of sporulation/dormancy-associated genes and down-regulation of nutrient (N, P, and S) cycling and stress-associated genes (Tripathi et al. 2017). This study has significant ramifications for understanding the effects of environmental change on various microbiomes. In comparison of soil rhizospheric microbiome of eight different countries, highly significant variation was displayed and credited to difference in pH of different country's soil (Simonin et al. 2020).

Water-stress/drought has maximum impact on plant microbiome (Santos and Olivares 2021). Santos-Medellín et al. studied rice plants exposed to water pressure had their root microbiome rebuilt, which can add to the endurance of plants in this condition (Santos-Medellín et al. 2017). Soil heat (50–80 °C) exposure influences microbiome and disease resistance due to the reduction in antagonistic bacterial loads of Mycobacteriaceae, Micrococcaceae, and Streptomycetaceae (van der Voort et al. 2016). Likely, with routine exposure of plants to UV radiation due to DNA damage phyllospheric microbiome stratification occurs; however, some bacteria acquire UV protection by endospore formation, exo-polysaccharides, and pigments such as melanin, xanthomonadine (Kumar et al. 2019).

Taken together, abiotic factor greatly influences the structure and diversity of plant microbiomes.

8.5 Role of Plant Microbiome in Sustainable Agriculture

Plant microbiome influences agriculture via enhancing soil fertility and promotion of plant growth. Microbes are diverse in nature and constitute of beneficial, harmful, and pathogenic microorganisms.

Plant growth-promoting bacteria (PGPB) synthesize phytohormones (cytokinin, auxin: indole-3-acetic acid, gibberellin, jasmonic acid, and salicylic acid), which act as phytostimulant and enhances plant growth (Kaur 2020; Rastegari 2020; Santos and Olivares 2021). Moreover, ethylene (a stress hormone) is regulated by the enzyme 1-aminocyclopropane-1-carboxylate deaminase (ACC deaminase), which breaks down the molecule ACC and stimulates plant development by lowering ethylene levels (Glick 2014; Santos and Olivares 2021; Shahid et al. 2021). Some common ACC deaminase producers are *Pseudomonas* spp., *Bacillus* spp., *Paraburkholderia* spp., *Arthrobacter* spp., and *Pantoea* spp. (Rastegari 2020; Santos and Olivares 2021). These genera exhibit various properties including phytohormones, phosphate solubilization, and nitrogen fixation which enhances nutrient uptake and stress tolerance. In contrast, some bacteria produce phytoactive compounds, which cause disease symptoms in plants. Some common examples of pathogenic bacteria are *Pseudomonas syringae* (infects tomato, tobacco, olive, and green bean), *Xylella fastidiosa* (infects potato and banana), *Erwinia amylovora* (infects ornamental plants), *Xanthomonas* spp. (infects banana), *Ralstonia solanacearum* (infects banana and potato) (Gupta et al. 2021; Rastegari 2020).

Several bacteria directly facilitate essential nutrients (iron, nitrogen, phosphorus, etc.) acquisition and promote plant development (del Orozco-Mosqueda et al. 2018).

Plant resistance is influenced by microorganisms which are present above and below the soil due to alteration in plant defense system or commensal interactions (Igiehon and Babalola 2018). Various biocontrol mechanism are involved in plant resistance induction such as production of siderophores (chelating agents with potential of insoluble ferric-ions sequestration), antibiotics (2,4-diacetylphloroglucinol), enzymes (proteases, chitinases, phytases), lipopeptides (bacillomycin-D, mycosubtilin, and iturin-A), volatile organic compounds, and bacteriocins (del Orozco-Mosqueda et al. 2018; Santos and Olivares 2021).

Of note, microbiome not only induces plant growth by these biocontrol mechanisms, but also these perform bio-fertilization (by controlled availability and attainment of nutrients for plants) for regulating plant growth. Some typical examples of biofertilization are symbiotic associations between *Rhizobium* (nitrogen-fixing bacteria), arbuscular-mycorrhizal fungi, and phosphate solubilizing bacteria to deliver nitrogen and phosphorus, to plant, respectively (Santos and Olivares 2021). Some examples of phosphate solubilizing bacteria are *Rhizobium*, *Bacillus*, *Microbacterium*, *Azotobacter*, *Erwinia*, *Serratia*, *Burkholderia*, *Enterobacter*, *Beijerinckia*, *Flavobacterium*, and *Pseudomonas* (Chhabra 2019; Chhabra et al. 2013; Chhabra and Dowling 2017; Kumar et al. 2019).

Rhizobium evolves with leguminous plants and has potential of directly fixing atmospheric nitrogen, hence facilitates self-establishment of plants in low nitrogen soils (Santos and Olivares 2021). Notably, along with *Rhizobium*, leguminous plants also contain some nitrogen-fixing endophytic bacteria such as *Azospirillum*, *Azotobacter*, *Gluconacetobacter*, and cyanobacteria (*Calothrix*, *Nostoc*, *Anabena*) (Kumar et al. 2019). Moreover, there are some nitrogen-fixing microorganisms (*Bacillus*, *Beijerinckia* and *Klebsiella*), which can fix nitrogen in non-leguminous plant in their free form (Santos and Olivares 2021).

Taken together, above-mentioned characteristics of microbiome justify their importance toward sustainable agriculture in terms of inducing crop's nutrient intake, disease resistance, harsh environment tolerance, growth, yield, and productivity.

8.6 Current Trends and Future Perspectives

Continuously increasing global demand of food is challenging for the farmers, wherein they have to tackle with various stresses such as changing hostile climate, nutrient, water scarcity, and pollution. In this background, plant growth-promoting potential natural resources such as plant microbiome become a viable alternative tool for supporting plant health and sustainable agriculture. In plant microbiome establishment microbial diversity, their inter/intra generic interactions and environmental factor are critical. Despite extensive microbiome knowledge to scientists, still advances are needed for comprehensive information. In recent years, advanced biotechnological tools, bioinformatics, and meta-omics approaches together have established pipelines for structural, genomic, and functional organization of plant microbiomes and their interaction with plants. With such all-inclusive information, it would be easy to improve native microorganisms at cellular or molecular level leading to the more efficient bioinoculants development for improving crop's yield and productivity.

References

Aamir M, Samal S, Rai A, Kashyap SP, Singh SK, Ahmed M, Upadhyay RS (2021) Plant microbiome: diversity, distribution, and functional relevance in crop improvement and sustainable agriculture. In: Microbiome stimulants for crops. Elsevier, Amsterdam, pp 417–436. https://doi.org/10.1016/b978-0-12-822122-8.00001-7

Agler MT, Ruhe J, Kroll S, Morhenn C, Kim ST, Weigel D, Kemen EM (2016) Microbial hub taxa link host and abiotic factors to plant microbiome variation. PLoS Biol 14:1–31. https://doi.org/10.1371/journal.pbio.1002352

Alok D, Annapragada H, Singh S, Murugesan S, Singh NP (2020) Symbiotic nitrogen fixation and endophytic bacterial community structure in Bt-transgenic chickpea (*Cicer arietinum* L). Sci Rep 10:1–12. https://doi.org/10.1038/s41598-020-62199-1

Andreote FD, Gumiere T, Durrer A (2014) Exploring interactions of plant microbiomes. Sci Agric 71:528–539. https://doi.org/10.1590/0103-9016-2014-0195

Ardanov P, Lyastchenko S, Karppinen K, Häggman H, Kozyrovska N, Pirttilä AM (2016) Effects of *Methylobacterium* sp. on emergence, yield, and disease prevalence in three cultivars of potato (*Solanum tuberosum* L.) were associated with the shift in endophytic microbial community. Plant Soil 405:299–310. https://doi.org/10.1007/s11104-015-2500-y

Babu AG, Kim SW, Yadav DR, Hyum U, Adhikari M, Lee YS (2015) *Penicillium menonorum*: a novel fungus to promote growth and nutrient management in cucumber plants. Mycobiology 43: 49. https://doi.org/10.5941/MYCO.2015.43.1.49

Bergottini VM, Hervé V, Sosa DA, Otegui MB, Zapata PD, Junier P (2017) Exploring the diversity of the root-associated microbiome of *Ilex paraguariensis* St. Hil. (Yerba mate). Appl Soil Ecol 109:23–31. https://doi.org/10.1016/J.APSOIL.2016.09.013

Bhatt P, Verma A, Verma S, Anwar MS, Prasher P, Mudila H, Chen S (2020) Understanding Phytomicrobiome: a potential reservoir for better crop management. Sustainability 12:5446. https://doi.org/10.3390/SU12135446

Bodor A, Bounedjoum N, Vincze GE, Erdeiné Kis Á, Laczi K, Bende G, Szilágyi Á, Kovács T, Perei K, Rákhely G (2020) Challenges of unculturable bacteria: environmental perspectives. Rev Environ Sci Bio/Technol 191(19):1–22. https://doi.org/10.1007/S11157-020-09522-4

Cai F, Pang G, Miao Y, Li R, Li R, Shen Q, Chen W (2017) The nutrient preference of plants influences their rhizosphere microbiome. Appl Soil Ecol 110:146–150. https://doi.org/10.1016/J.APSOIL.2016.11.006

Chhabra S (2019) Phosphorus management in agroecosytems and role and relevance of microbes in environmental sustainability. In: Sustainable green technologies for environmental managment. Springer, Singapore, pp 53–66. https://doi.org/10.1007/978-981-13-2772-8_3

Chhabra S, Dowling DN (2017) Endophyte-promoted nutrient acquisition: phosphorus and iron. In: Functional importance of the plant microbiome. Implications for agriculture, forestry and bioenergy. Springer, Cham, pp 21–42. https://doi.org/10.1007/978-3-319-65897-1_3

Chhabra S, Brazil D, Morrissey J, Burke JI, O'Gara F, Dowling ND (2013) Characterization of mineral phosphate solubilization traits from a barley rhizosphere soil functional metagenome. Microbiology 2:717–724. https://doi.org/10.1002/MBO3.110

Chitarra W, Pagliarani C, Maserti B, Lumini E, Siciliano I, Cascone P, Schubert A, Gambino G, Balestrini R, Guerrieri E (2016) Insights on the impact of arbuscular mycorrhizal symbiosis on tomato tolerance to water stress. Plant Physiol 171:1009–1023. https://doi.org/10.1104/pp.16.00307

Compant S, Samad A, Faist H, Sessitsch A (2019) A review on the plant microbiome: ecology, functions, and emerging trends in microbial application. J Adv Res 19:29–37. https://doi.org/10.1016/j.jare.2019.03.004

Conn VM, Franco CMM (2004) Effect of microbial inoculants on the indigenous actinobacterial endophyte population in the roots of wheat as determined by terminal restriction fragment length polymorphism. Appl Environ Microbiol 70:6407–6413. https://doi.org/10.1128/AEM.70.11.6407-6413.2004

Curlango-Rivera G, Huskey DA, Mostafa A, Kessler JO, Xiong Z, Hawes MC (2013) Intraspecies variation in cotton border cell production: rhizosphere microbiome implications. Am J Bot 100:1706–1712. https://doi.org/10.3732/ajb.1200607

Dastogeer KMG, Tumpa FH, Sultana A, Akter MA, Chakraborty A (2020) Plant microbiome—an account of the factors that shape community composition and diversity. Curr Plant Biol 23:100161. https://doi.org/10.1016/j.cpb.2020.100161

De La Torre-Ruiz N, Ruiz-Valdiviezo VM, Rincón-Molina CI, Rodríguez-Mendiola M, Arias-Castro C, Gutiérrez-Miceli FA, Palomeque-Dominguez H, Rincón-Rosales R (2016) Effect of plant growth-promoting bacteria on the growth and fructan production of *Agave Americana* L. Brazilian J Microbiol 47:587–596. https://doi.org/10.1016/j.bjm.2016.04.010

del Orozco-Mosqueda MC, del Rocha-Granados MC, Glick BR, Santoyo G (2018) Microbiome engineering to improve biocontrol and plant growth-promoting mechanisms. Microbiol Res 208:25–31. https://doi.org/10.1016/J.MICRES.2018.01.005

Durán P, Tortella G, Viscardi S, Barra PJ, Carrión VJ, De La Luz Mora M, Pozo MJ (2018) Microbial community composition in take—all suppressive soils. Front Microbiol 9:1–15. https://doi.org/10.3389/fmicb.2018.02198

Fitzpatrick CR, Salas-González I, Conway JM, Finkel OM, Gilbert S, Russ D, Teixeira PJPL, Dangl JL (2020) The plant microbiome: from ecology to reductionism and beyond. Annu Rev Microbiol 74:81–100. https://doi.org/10.1146/annurev-micro-022620-014327

Glick BR (2014) Bacteria with ACC deaminase can promote plant growth and help to feed the world. Microbiol Res 169:30–39. https://doi.org/10.1016/J.MICRES.2013.09.009

Gupta R, Anand G, Gaur R, Yadav D (2021) Plant–microbiome interactions for sustainable agriculture: a review. Physiol Mol Biol Plants 27:165–179. https://doi.org/10.1007/s12298-021-00927-1

Hara S, Matsuda M, Minamisawa K (2019) Growth stage-dependent bacterial communities in soybean plant tissues: *Methylorubrum* transiently dominated in the flowering stage of the soybean shoot. Microbes Environ 34:446–450. https://doi.org/10.1264/jsme2.ME19067

Hartmann A, Rothballer M, Hense BA, Schröder P (2014) Bacterial quorum sensing compounds are important modulators of microbe-plant interactions. Front Plant Sci 5:131. https://doi.org/10.3389/FPLS.2014.00131/BIBTEX

Hashem A, Abd-Allah EF, Alqarawi AA, Al-Huqail AA, Wirth S, Egamberdieva D (2016) The interaction between arbuscular mycorrhizal fungi and endophytic bacteria enhances plant growth of *Acacia gerrardii* under salt stress. Front Microbiol 7:1–15. https://doi.org/10.3389/fmicb.2016.01089

Hong S, Jv H, Lu M, Wang B, Zhao Y, Ruan Y (2020) Significant decline in banana Fusarium wilt disease is associated with soil microbiome reconstruction under chilli pepper-banana rotation. Eur J Soil Biol 97:103154. https://doi.org/10.1016/J.EJSOBI.2020.103154

Igiehon NO, Babalola OO (2018) Below-ground-above-ground plant-microbial interactions: focusing on soybean, rhizobacteria and mycorrhizal fungi. Open Microbiol J 12:261. https://doi.org/10.2174/1874285801812010261

Jodeh S, Alkowni R, Hamed R, Samhan S (2015) The study of electrolyte leakage from barley (*Hordeum vulgare L*) and pearlmillet using plant growth promotion (PGPR) and reverse osmosis. J Food Nutr Res 3:422–429. https://doi.org/10.12691/jfnr-3-7-3

Kaur T (2020) Fungal endophyte-host plant interactions: role in sustainable agriculture. In: Sustainable crop production. IntechOpen, London. https://doi.org/10.5772/INTECHOPEN.92367

Kinnunen-Grubb M, Sapkota R, Vignola M, Nunes IM, Nicolaisen M (2020) Wheat domestication has dramatically affected the root-associated microbiome 1–23. https://doi.org/10.21203/rs.3.rs-20588/v1

Konieczny A, Kowalska I (2016) The role of arbuscular mycorrhiza in zinc uptake by lettuce grown at two phosphorus levels in the substrate. Agric Food Sci 25:124–137. https://doi.org/10.23986/afsci.55534

Kumar I, Mondal M, Gurusamy R, Balakrishnan S, Natarajan S (2019) Plant-microbiome interaction and the effects of biotic and abiotic components in agroecosystem. In: Microbial interventions in agriculture and environment: volume 2: rhizosphere, microbiome and agro-ecology. Springer, Cham, pp 517–546. https://doi.org/10.1007/978-981-13-8383-0_18

Kwak MJ, Kong HG, Choi K, Kwon SK, Song JY, Lee J, Lee PA, Choi SY, Seo M, Lee HJ, Jung EJ, Park H, Roy N, Kim H, Lee MM, Rubin EM, Lee SW, Kim JF (2018) Rhizosphere microbiome structure alters to enable wilt resistance in tomato. Nat Biotechnol 3611(36): 1100–1109. https://doi.org/10.1038/nbt.4232

Lebeis SL, Paredes SH, Lundberg DS, Breakfield N, Gehring J, McDonald M, Malfatti S, Del Rio TG, Jones CD, Tringe SG, Dangl JL (2015) Salicylic acid modulates colonization of the root microbiome by specific bacterial taxa. Science 349:860–864. https://doi.org/10.1126/SCIENCE.AAA8764/SUPPL_FILE/PAPV2.PDF

Lebreton L, Guillerm-Erckelboudt AY, Gazengel K, Linglin J, Ourry M, Glory P, Sarniguet A, Daval S, Manzanares-Dauleux MJ, Mougel C (2019) Temporal dynamics of bacterial and fungal communities during the infection of *Brassica rapa* roots by the protist *Plasmodiophora brassicae*. PLoS One 14:e0204195. https://doi.org/10.1371/JOURNAL.PONE.0204195

Lee EH, Eom AH (2015) Growth characteristics of *Rhizophagus Clarus* strains and their effects on the growth of host plants. Mycobiology 43:444–449. https://doi.org/10.5941/MYCO.2015.43.4.444

Li R, Feng Y, Chen H, Zhang C, Huang Y, Chen L, Hao Q, Cao D, Yuan S, Zhou X (2020) Whole-genome sequencing of *Bradyrhizobium diazoefficiens* 113-2 and comparative genomic analysis provide molecular insights into species specificity and host specificity. Front Microbiol 11:1–17. https://doi.org/10.3389/fmicb.2020.576800

Liu F, Hewezi T, Lebeis SL, Pantalone V, Grewal PS, Staton ME (2019) Soil indigenous microbiome and plant genotypes cooperatively modify soybean rhizosphere microbiome assembly. BMC Microbiol 19:1–19. https://doi.org/10.1186/s12866-019-1572-x

Lupatini M, Korthals GW, de Hollander M, Janssens TKS, Kuramae EE (2017) Soil microbiome is more heterogeneous in organic than in conventional farming system. Front Microbiol 7:2064. https://doi.org/10.3389/FMICB.2016.02064/BIBTEX

Mendes LW, de Chaves MG, de Fonseca MC, Mendes R, Raaijmakers JM, Tsai SM (2019) Resistance breeding of common bean shapes the physiology of the rhizosphere microbiome. Front Microbiol 10:1–10. https://doi.org/10.3389/fmicb.2019.02252

Mitter EK, de Freitas JR, Germida JJ (2017) Bacterial root microbiome of plants growing in oil sands reclamation covers. Front Microbiol 8. https://doi.org/10.3389/fmicb.2017.00849

Mohumad Tahat M (2012) The potential of endomycorrhizal fungi in controlling tomato bacterial wilt *Ralstonia solanacearum* under glasshouse condition. African J Biotechnol 11:13085–13094. https://doi.org/10.5897/ajb11.3629

Morella NM, Weng FCH, Joubert PM, Jessica C, Lindow S, Koskella B (2020) Successive passaging of a plant-associated microbiome reveals robust habitat and host genotype-dependent selection. Proc Natl Acad Sci U S A 117:1148–1159. https://doi.org/10.1073/pnas.1908600116

Oyserman BO, Cordovez V, Flores SS, Leite MFA, Nijveen H, Medema MH, Raaijmakers JM (2021) Extracting the GEMs: genotype, environment, and microbiome interactions shaping host phenotypes. Front Microbiol 11:1–8. https://doi.org/10.3389/fmicb.2020.574053

Padmavathi T, Dikshit R, Seshagiri S (2015) Effect of *Rhizophagus* spp. and plant growth-promoting Acinetobacter junii on Solanum lycopersicum and Capsicum annuum. Rev Bras Bot 38:273–280. https://doi.org/10.1007/s40415-015-0144-z

Pieterse CMJ, de Jonge R, Berendsen RL (2016) The soil-borne supremacy. Trends Plant Sci 21: 171–173. https://doi.org/10.1016/J.TPLANTS.2016.01.018

Podolich O, Ardanov P, Zaets I, Pirttilä AM, Kozyrovska N (2015) Reviving of the endophytic bacterial community as a putative mechanism of plant resistance. Plant Soil 388:367–377. https://doi.org/10.1007/S11104-014-2235-1/FIGURES/2

Qin S, Yeboah S, Xu X, Liu Y, Yu B (2017) Analysis on fungal diversity in rhizosphere soil of continuous cropping potato subjected to different furrow-ridge mulching managements. Front Microbiol 8:845. https://doi.org/10.3389/fmicb.2017.00845

Rastegari AA (2020) Microbiomes for sustainable. Agriculture. https://doi.org/10.1007/978-3-030-38453-1

Rodrigues AA, Forzani MV, de Soares RS, Sibov ST, Vieira JDG (2016) Isolation and selection of plant growth-promoting bacteria associated with sugarcane. Pesqui Agropecuária Trop 46:149–158. https://doi.org/10.1590/1983-40632016v4639526

Santos LF, Olivares FL (2021) Plant microbiome structure and benefits for sustainable agriculture. Curr Plant Biol 26:100198. https://doi.org/10.1016/j.cpb.2021.100198

Santos-Medellín C, Edwards J, Liechty Z, Nguyen B, Sundaresan V (2017) Drought stress results in a compartment-specific restructuring of the rice root-associated microbiomes. MBio 8: e00764-17. https://doi.org/10.1128/MBIO.00764-17

Sapre S, Gontia-Mishra I, Tiwari S (2018) *Klebsiella* sp. confers enhanced tolerance to salinity and plant growth promotion in oat seedlings (*Avena sativa*). Microbiol Res 206:25–32. https://doi.org/10.1016/j.micres.2017.09.009

Shahid M, Khan MS, Syed A, Marraiki N, Elgorban AM (2021) Mesorhizobium ciceri as biological tool for improving physiological, biochemical and antioxidant state of *Cicer arietinum* (L.) under fungicide stress. Sci Rep 11(1):1–18. https://doi.org/10.1038/s41598-021-89103-9

Shahsavar AR, Refahi A, Zarei M, Aslmoshtaghi E (2016) Analysis of the effects of *Glomus etunicatum* fungi and *Pseudomonas fluorescence* bacteria symbiosis on some morphological and physiological characteristics of Mexican lime (*Citrus aurantifolia* L.) under drought stress conditions. Adv Hortic Sci 30:39–45. https://doi.org/10.13128/ahs-18700

Sharma M, Sudheer S, Usmani Z, Rani R, Gupta P (2020) Deciphering the omics of plant-microbe interaction: perspectives and new insights. Curr Genomics 21:343–362. https://doi.org/10.2174/1389202921999200515140420

Shcherbakova EN, Shcherbakov AV, Andronov EE, Gonchar LN, Kalenskaya SM, Chebotar VK (2017) Combined pre-seed treatment with microbial inoculants and Mo nanoparticles changes composition of root exudates and rhizosphere microbiome structure of chickpea (*Cicer arietinum* L.) plants. Symbiosis 73:57–69. https://doi.org/10.1007/S13199-016-0472-1/TABLES/4

Shi SM, Chen K, Gao Y, Liu B, Yang XH, Huang XZ, Liu GX, Zhu LQ, He XH (2016) Arbuscular mycorrhizal fungus species dependency governs better plant physiological characteristics and leaf quality of mulberry (*Morus alba* L.) seedlings. Front Microbiol 7:1–11. https://doi.org/10.3389/fmicb.2016.01030

Shuab R, Lone R, Naidu J, Sharma V, Imtiyaz S, Koul KK (2014) Benefits of inoculation of arbuscular mycorrhizal fungi on growth and development of onion (*Allium cepa*) plant. Environ Sci 14:527–535. https://doi.org/10.5829/idosi.aejaes.2014.14.06.12347

Simonin M, Dasilva C, Terzi V, Ngonkeu ELM, Diouf D, Kane A, Béna G, Moulin L (2020) Influence of plant genotype and soil on the wheat rhizosphere microbiome: evidences for a core microbiome across eight African and European soils. FEMS Microbiol Ecol 96:1–18. https://doi.org/10.1093/femsec/fiaa067

Singh D, Raina TK, Kumar A, Singh J, Prasad R (2019) Plant microbiome: a reservoir of novel genes and metabolites. Plant Gene. https://doi.org/10.1016/j.plgene.2019.100177

Tripathi BM, Moroenyane I, Sherman C, Lee YK, Adams JM, Steinberger Y (2017) Trends in taxonomic and functional composition of soil microbiome along a precipitation gradient in Israel. Microb Ecol 74:168–176. https://doi.org/10.1007/S00248-017-0931-0/FIGURES/4

Trivedi P, Leach JE, Tringe SG, Sa T, Singh BK (2020) Plant–microbiome interactions: from community assembly to plant health. Nat Rev Microbiol:607–621. https://doi.org/10.1038/s41579-020-0412-1

van der Voort M, Kempenaar M, van Driel M, Raaijmakers JM, Mendes R (2016) Impact of soil heat on reassembly of bacterial communities in the rhizosphere microbiome and plant disease suppression. Ecol Lett 19:375–382. https://doi.org/10.1111/ELE.12567

Viscardi S, Ventorino V, Duran P, Maggio A, De Pascale S, Mora ML, Pepe O (2016) Assessment of plant growth promoting activities and abiotic stress tolerance of *Azotobacter chroococcum* strains for a potential use in sustainable agriculture. J Soil Sci Plant Nutr 16:848–863. https://doi.org/10.4067/s0718-95162016005000060

Zahir ZA, Munir A, Asghar HN, Shaharoona B, Arshad M (2008) Effectiveness of rhizobacteria containing ACC deaminase for growth promotion of peas (*Pisum sativum*) under drought conditions. J Microbiol Biotechnol 18:958–963

Zhang QM, Gong M, Liu K, Chen Y, Yuan J, Chang Q (2020) Rhizoglomus intraradices improves plant growth, root morphology and Phytohormone balance of *Robinia pseudoacacia* in arsenic-contaminated soils. Front Microbiol 11:1428. https://doi.org/10.3389/FMICB.2020.01428/BIBTEX

Zhu S, Vivanco JM, Manter DK (2016) Nitrogen fertilizer rate affects root exudation, the rhizosphere microbiome and nitrogen-use-efficiency of maize. Appl Soil Ecol 107:324–333. https://doi.org/10.1016/J.APSOIL.2016.07.009

Chapter 9
Decoding Beneficial Plant Microbe Association with Latest Techniques for Sustainable Agriculture

Bhawana Jangra, Pooja Bhadrecha, Pankaj Kumar, and Jitendra Kumar

Abstract One of the most precious treasures offered by mother nature to humans is plants and to aid in their outstanding performances, plants maintain allelopathic relationships with microorganisms, especially in roots and rhizosphere, along with providing food, fodder, and employment. But plants have to face various detrimental conditions such as abiotic and biotic stress conditions caused through pollution, climatic situations, pathogenic microorganisms, competing herbs, and weeds. Association of beneficial microorganisms with the plant roots and rhizosphere provide privilege to plant development, potential to resist stress, hence maximize yield even in adverse environment. Therefore, researchers worldwide have concentrated their attention toward exploring the rhizospheric microbiota with the help of conventional methods as well as latest techniques like metabolomic tools and CRISPR/Cas system. Here we discuss various beneficial and mutualistic relationships between plants and associated rhizospheric microbiota and spotlight the latest research techniques and findings aimed at understanding and utilizing microorganisms for sustainable agricultural practices.

Keywords Omics technologies · GC-MS · Maldi TOF · NMR · Sustainable agriculture

B. Jangra · P. Bhadrecha
University Institute of Biotechnology, Chandigarh University, Mohali, Punjab, India

P. Kumar
Department of Microbiology, Dolphin (PG) Institute of Biomedical and Natural Science, Dehradun, Uttarakhand, India

J. Kumar (✉)
Department of Biosciences, UIBT, Chandigarh University, Mohali, Punjab, India

S. Chhabra et al. (eds.), *Plant Microbiome for Plant Productivity and Sustainable Agriculture*, Microorganisms for Sustainability 37,
https://doi.org/10.1007/978-981-19-5029-2_9

9.1 Introduction

The microbiome is native to the microbial population that exists in a particular environment. The plant ecosystem is considered the ecological niches that foster in their rhizosphere, phyllosphere, and endosphere, which carry a varied range of interactions like beneficial, neutral, and harmful. The microorganisms syndicate with plants and assist in the soil nutrients acquisition, stress tolerance, and disease surveillance that promote plant growth and ecological fitness under the agricultural system. In ecosystem various types of interaction exist between plant and microorganisms. However, the interaction between the plant root and soil microorganism is essential, and the soil microbes stimulated toward rhizosphere because of unveiling of root exudates. Microbes exploit the root through varied plant growth-promoting (PGP) aspects. Phyllospheric microbiomes are considered stress-tolerant microbes that have the potential to tolerate UV exposure and torrid heat as compared to other microbiomes (Liu et al. 2020). The endophytic microbiome is considered the most beneficial microbiome due to the invasion of plant microbes in plant tissues like root, stem, flower, fruits, and seeds (Rana et al. 2020; Prasad and Zhang 2022). The most limiting element is nitrogen which is responsible for plant germination. However, microbes (*Rhizobium*, *Azospirillum*, *Gluconoacetobacter*) adjunct with the plants and make available the atmospheric nitrogen for plants. Furthermore, the use of some of these microorganisms as biofertilizer (natural fertilizers that contain living microorganisms (bacteria, fungi, and cyanobacteria) in the soil to provide nutrition to the plant source has emerged as a sustainable technology for nitrogen fixation that promotes crop yield. Phosphorus is another limiting element for plant development. The rhizospheric microbiomes are often considered to have the best ecological approach to transform insoluble inorganic phosphorous into soluble phosphorus, and also help in seed germination and help in plant growth promotion under acidity/alkalinity, salinity, drought, and foster the plants (Verma et al. 2013). Due to the inclusion of systematic resistance, the defense system of plants activates furthermore which prevents plants from initiating pathogen attack through niche competition, production of antimicrobial eventually leading to release of secondary metabolites. The microbial population is harmed by human arbitration like discerning procreation, and pumping of agrochemicals in the agricultural biosphere (White et al. 2019).

9.1.1 Microbiomes and Potential

The microbiome is a mass of microbes (bacteria, fungi, algae, yeast) that provide nutrients to the plants. Various publications have mentioned its importance, and researchers have studied different compartments of the plant to soil continuum for sustainable agriculture (Bertola et al. 2021). Among all the plant organs, the most studied organs are vegetative and reproductive, where the microorganisms can colonize and establish. For last few decades, most focused research is on the

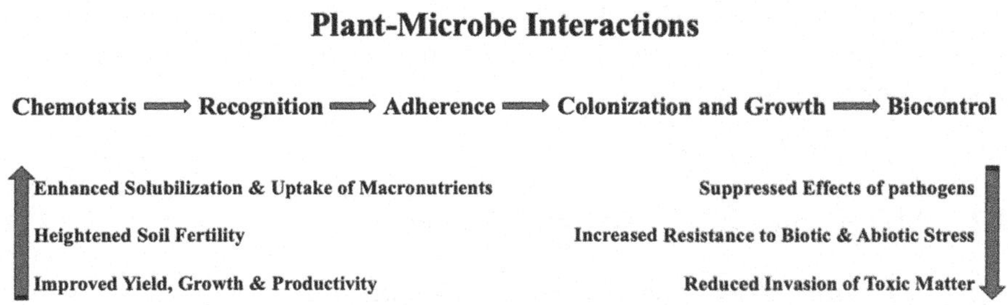

Fig. 9.1 Plant–microbe interactions (Andreote et al. 2010)

rhizosphere and phyllosphere (Nelson 2018). However, habitats like spermosphere (the zone where interaction between soil, microbes, and seed germination occurs) and carposphere (the zone around the fruits where the microbes are present) are less studied (Nelson 2018). The rhizobiome contributes to the plant in nutrient acquisition, phytohormone production, increases plant immunity and improves soil fertility (Nwachukwu and Babalola 2021). Rhizosphere microbiota influences plant growth through detrimental and beneficial microorganisms such as biocontrol microorganisms bacteria and plant pathogenic fungi (Srivastava et al. 2014; Prasad and Varma 2014). Root exudates are the source for nutrient acquisition that produces different inorganic and organic compounds known as rhizodeposits which provide an exclusive environment and attract the microbes with high nutrient utility (Kohli et al. 2012). The multiple microbial colonies make effective soil because of releasing the regulatory substrates that help many to survive in an extreme environment. The rhizoderm cells consider organic acids, siderophores, inorganic acids, sugars, amino acids, vitamins, purines, nucleosides, and polysaccharide mucilage-like compounds that consider hundred times more microorganisms than aggregated soil.

Microorganisms can control nutrient acquisition and promote stress tolerance that influences agricultural productivity. The diversity of microorganisms and species evenness are the impact factors on agricultural productivity which are still not defined. The current focus of research is on specific niches of plants and the regulation of microbial communities. Understanding the phylogenetic and functional perspectives should be heightened on the different parts of plant microbiome individually. Microbiomes from rhizosphere, endosphere, and phyllosphere have diverse microbial communities (Hirsch et al. 2013). Understanding the significance of plant-associated microbiome preventing the plants from harmful pathogens (Fig. 9.1). Mainly focus should be on the appropriate methodology for studying the interaction between plants and microbes (Carvalhais et al. 2013).

The molecular techniques determine the effects incited through stress factors on rhizospheric microbiome and interaction between microbes and plants in changing environments. The main factor is improvement of microbial stress alleviation in crops helps to understand the plant and microbe interaction (Barea 2015). Stress factors such as salt stress, drought, nutritional deficit, impurity, diseases, pests affect the plant and microbe interaction. Scientists discovered a new finding to improve aging factor of water and nutrient for essential crops. The important nutrients and water acquisition occur through the roots. Therefore, understanding the regulation of root branching is essential.

9.2 Abiotic and Biotic Stress Tolerance

9.2.1 Abiotic Stress and Microbial Potential

Regulation of plant–microbe interaction has an essential perspective toward a variety of stress like metal pollution, drought, salt stress, and temperature (Muller et al. 2016).

9.2.2 Salt Stress and Heavy Metal Stress

In semiarid and arid conditions, high salt accumulation is present which is the major abiotic stress for agriculture. In a study, the antioxidative *Trichoderma asperelloides* was recognized with antioxidant properties against salt stress (Brotman et al. 2013). In case of tomato, *Azotobacter chroococcum* was recognized with antioxidant properties (Viscardi et al. 2016). The Plant Growth-Promoting Bacteria (PGPB) and endophytic fungus promote PGP metabolite secretion, IAA production, and ACC deaminase activity in saline conditions (Prasad et al. 2015). Usage of PGPB was found promising for salinity tolerance in crops like wheat (Nadeem et al. 2016) and maize (Sorty et al. 2016). The bacteria found in siderophore are used to reduce the pollution due to heavy metal tolerance in canola, Indian mustard, and tomato cultivation (Burd et al. 2000).

9.2.3 Thermal and Radiation Stress

As we know high temperature is the major issue in today's environment due to climate change or human indulging activities such as increase of CO_2 by using CFC, increment of industries which are affecting crop growth all over the world. Microorganisms are auspicious factor to reduce temperature stress. *Pseudomonas putida*, *Sphingomonas*, *Azospirillum brasilense*, and *Acinetobacter* are stress-tolerant that increase the maize plant biomass (Molina-Romero et al. 2017), whereas high molecular weight proteins produced by using *Pseudomonas* are described in improving metabolites level in sorghum and wheat (Ali et al. 2011). Drought and ultraviolet-B radiations are the strongest stress effector. Some bacteria like *Enterobacter cloacae* in rice and *Clavibacter michiganensis* in groundnut were found to prevent from the UV-B radiations (Rodriguez and Duran 2020).

9.2.4 Drought Stress

Among all stresses drought is known as the major aspect for high loss of crop yield (Barnawal et al. 2013), due to instability of water in plant (Pareek et al. 2009). Various microbes act against abiotic stress. The microbiomes that are commonly used are *Raoultella planticola*, *Azospirillum lipoferum*, *Klebsiella variicola*, *Pseudomonas fluorescens*, and *Pseudomonas putida* which can tolerate the drought stress in maize crops (Ansary et al. 2012; Bano et al. 2013; Gou et al. 2015). Microorganisms perform the essential role in the functions of maize like water potential, stomatal closure, promotes hormonal potential by accumulating abscisic acid, auxin, choline, gibberellin, and proline. Furthermore, the species like *Azospirillum brasilense* reduce the osmotic stress in rice crops and make the better root growth (de Souza et al. 2019). In the study of drought in wheat, some of the microbes are used such as *Rhizobium leguminosarum*, *Mesorhizobium ciceri*, and *Rhizobium phaseoli* (Hussain et al. 2014; Timmusk et al. 2014). Mostly PGPR is approached in the pepper crop due to which rhizobia and the endosphere lead to the ability which can tolerate the shortage of water through increment of roots (Rolli et al. 2015) (Fig. 9.2).

9.3 Biotic Stress and Microbial Potential

Microorganisms play most essential role in reduction of biotic stress, and help in improving plant growth and vegetative vigor and tolerate diseases (De Corato 2020). For example, PGPR releases various significant biomolecules like 1-amino-cyclopropane-1-1-carboxylate (ACC) deaminase, hydrogen cyanide (HCN), phytohormones, siderophores, which prevent the invasion of harmful phytopathogens (Kaur et al. 2018). The species like *Arthrobacternitroguajacolicus*, *B. cereus*, *B. megaterium*, *B. mojavensis*, *P. azotoformans*, and *P. frederisksbergensis* suppress the *Alternaria* and *Fusarium* activity of wilt disease and mortality in tobacco

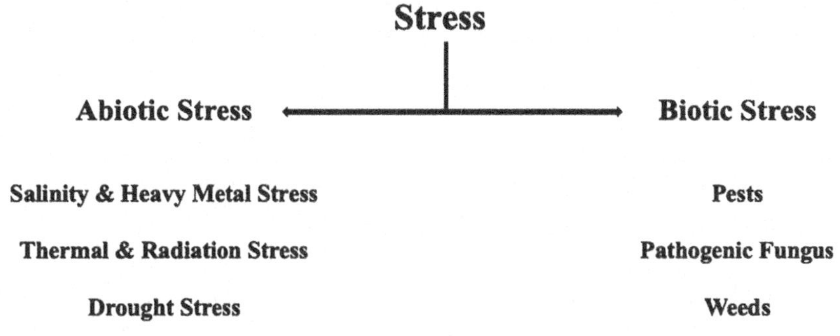

Fig. 9.2 Abiotic and biotic stress

(Santhanam et al. 2015). *Bacillus mojavensis* is an endophytic pathogenic against *Fusarium verticillioides* found in maize kernels and possesses growth-promoting traits (Bacon and Hinton 2011). Along with disease, the pests can also be controlled by using the microbes that contain activity like pathogen. Among all, the soil-dwelling bacteria *Bacillus thuringiensis* (Bt) is most widely exploited for pest suppression. Insects like *Lepidoptera*, *Coleoptera*, and *Diptera* are too degradative for the plants and hence Bt. variants like *Kurstaki*, *Tenebrionis*, and *Israelensis* are popularly employed. *M. robertsii* boosts the root growth of *Panicum virgatum* and *Phaseolus vulgaris* (Sasan and Bidochka 2012). The most widely exploited endophytic microbes reported are *Lecanicillium* and *Akanthomyces* (Nicoletti and Becchimanzi 2020). The endophytic strain *L. lecanii* prevents the cotton plant from aphids and *Aphis gossypii* (Nicoletti and Becchimanzi 2020). *Lecanicillium* species act as anti-fungal against diseases which are caused by powdery mildew and rust fungi (Jackson et al. 2012). Some fungal endophytes produce compounds which act against pests like argot alkaloids, indole-diterpenes, lolines, and paramine (Panaccione et al. 2014) (Fig. 9.2).

9.4 Modern Approaches for Sustainable Agriculture

Implementation of the modern approach in sustainable agriculture can help in understanding the communications among plants and PGPR in terms of biomolecules and signals. Some of the OMIC tools that help in better understanding the plant microbial interactions or the plant microbiome are: metabolome, proteome, genome, transcriptome.

The genetic modification of microbe or plant and microbial consortia in sustainable cultivation can be handled by modern approaches; genomics approach is more accurate compared to other approaches because of precise genomic sequences without the insertion of foreign gene.

Metablomics is an advance technique in micrbiolgical world for decoding the microbes however, its still lacking in accuracy and other various challenges. In present situation, various modern tools and techniques like GC-MS (gas chromatography-mass spectrometry) or LC-MS (liquid chromatography-mass spectrometry) are emerging whereas techniques like nuclear magnetic resonance (NMR) are daunting (Fig. 9.3).

9.4.1 Clustered Regularly Interspaced Short Palindromic Repeats (CRISPR)/Cas System

In the current era of changing climate, the environmental effects impact negatively on economic and agronomic level. To resolve this issue, new technologies are used

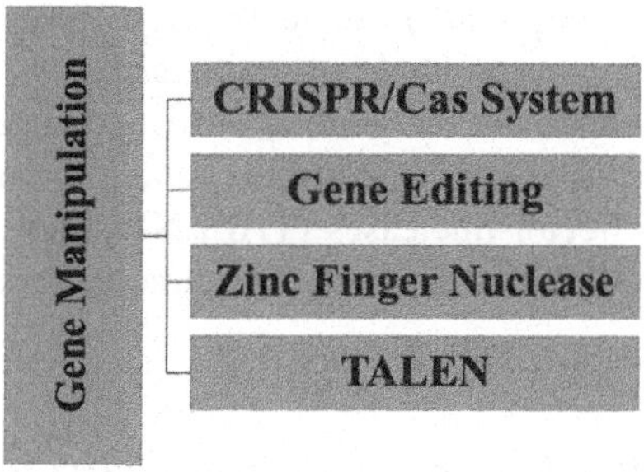

Fig. 9.3 Latest techniques for gene manipulation. *CRISPR* clustered regularly interspaced short palindromic repeats, *TALEN* transcription activator like effector nuclease

in the field of biotechnology. This technique represents a new prospective for genetic engineering and gene editing tools. It is based on the two main components: Cas endonuclease (sgRNA) and single gRNA (Jinek et al. 2012). Its main function is the modification of desired DNA fragment using molecular scissor (Kanchiswamy et al. 2016). About 11 CRISPR/cas systems divided as Type I-III have been reported (Ma and Liu 2016). The most beneficial factor to use this technique is short recognition sequence, i.e., protospacer adjacent motif (PAM) adjacent to gRNA. In this technique, it is easy to use the sgRNA designing for target.

Cas9 is a DNA endonuclease mainly for inhibition of targeted foreign DNA using RNA. Transcription and processing of CRISPR produce mature RISPR (cr) RNAs that introduce the cleavage in the RNA/DNA of the attacker (Krishnakumar and Kraus 2010; Sarma et al. 2021). CRISPR is popular for its specificity which is sustained by gRNA. It permits the certain binding to targeted DNA. Binding of upstream and downstream gRNA prevent the off target editing in the double nickase system. Moreover, it is corrected by using immobilized cas9. Restriction enzyme nuclease activity gets activated when upstream and downstream gRNA is in accessibility (Guilinger et al. 2014). The addition and removal of gene of interest is done by inclusion of DSBs in the target site (To et al. 2014). In the current scenario, diverse computational tools are emerging such as CHOPCHOP, E-CRISPR, and CRISPR designing tool which are used for identifying the probable sequence of cleavage. The conclusion of this technique is that basically it helps to design the gRNA (Hsu et al. 2013; Heigwer et al. 2014; Montague et al. 2014).

9.4.2 Gene Editing

Gene editing is the technique that has the ability to change the organism's DNA. This technique is useful for improve of different variety of crops so that higher yield can be obtained, increased the stress tolerance, resist from pest and various diseases and increased nutritional value. Crops can be improved followed by a few modifications among nucleotides present in the genome of living cells, and/or by editing whole allele, or insertion of new gene at a target site in the genome. However, gene editing is the best method for increasing the crops yield in the harsh climate. Genome editing can be done through three different ways such as alter few nucleotides, replace an allele from already existing one, and insert new genes in predetermined regions in the genome. Most of gene editing tools include the double strand breaks named as CRISPR-Cas, Zinc finger nuclease (ZFN), and transcription activator like effector nucleases (TALEN) which are corrected by non-homologous end joining (NHEJ) or homologous directed repair (HDR) (Szankasi and Smith 1992).

9.4.3 Transcription Activator like Effector Nucleases (TALEN)

TALEN are restriction enzymes that use TAL effector DNA binding domains to break target DNA. An easy "code" that fits the 2-amino acid sequence in the 33–35 amino acid preserved target sequence aids selective targeting. Discovery of diverse methods for facile synthesis and assembly of TALENs, as well as advancements in gene editing tools, enables for productive editing at several sites. TALENs were used to successfully knock out the HIV-resistant CCR5 gene in human cells, destroy the disease related to rice gene (Ayala Garcia et al. 2012; Li et al. 2012), disrupt swine LDL receptor (Carlson et al. 2012), and replace Zebra fish tyrosine hydroxylase gene (Xiao et al. 2013).

9.5 Analytical Tools and Techniques

The techniques like NMR, LC-MS, GC-MS, HPLC, and DFI-MS, which are approached for understanding the plant metabolome, are used for studying the plant metabolome as they speed up metabolic profiling. The metabolome is a group of metabolites in an organism which is used to characterize genetic and environmental variations. Metabolomics plays a significant role in exploring environment–gene interactions, mutant characterization, phenotyping, identification of biomarkers, and drug discovery. Metabolomics is an approach to decipher various metabolic networks that are linked with stress tolerance. Single analytical procedures cannot sense all of the metabolites exist in a sample in plant metabolomics. Alternatively, a combination of approaches is needed to offer highest degree of metabolite coverage (Razzaq et al. 2019). Metabolites research is generally based on the MS techniques, such as MALDI, FTICR-MS, CE-MSIMS, GC-MS, and NMR (Bianchi et al. 2018) (Fig. 9.4).

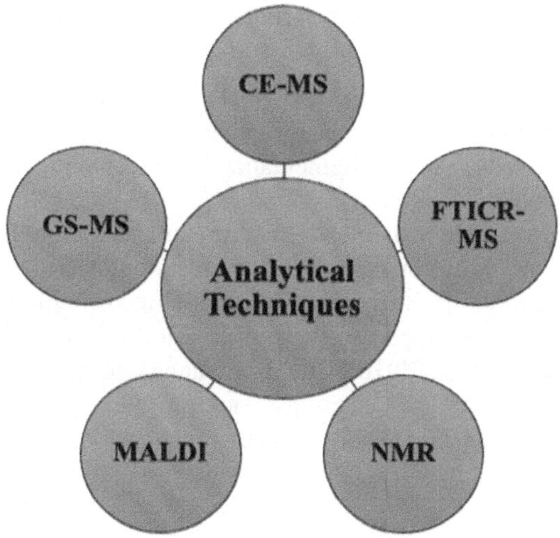

Fig. 9.4 Latest analytical techniques. *GC-MS* gas chromatography-mass spectrometry, *CE-MS* capillary electrophoresis-mass spectrometry, *FTICR-MS* Fourier transform ion cyclotron resonance-mass spectrometry, *NMR* nuclear magnetic resonance, *MALDI* matrix-assisted laser desorption/ionization

9.5.1 Gas Chromatography-Mass Spectrometry (GC-MS)

It is an analytical technique used to identify or quantify the tiny metabolites such as amines, sterols, sugars, alcohols, amino acids, hydroxyl acids, and fatty acids (Guo et al. 2019). Furthermore, different types of methods can also be used such as acylation, trimethylsilylation, silylation, and alkylation. In this technique, there are two major derivatization steps for the isolation and recognition of metabolites. The initial step involves transforming all the carbonyl groups into corresponding oximes through methoxyamine hydrochloride. In next step, the volatility of the derivative metabolites increases through trimethylsilylation reaction using derivatizing reagents such as N, O-bis-(trimethylsilyl) trifluoroacetamide (BSTFA), and N-Methyl-N-(trimethylsilyl) trifluoroacetamide (MSTFA) (Harvey and Vouros 2020). GC-MS is used to identify the compounds with low molecular weight that thermally stable and volatile metabolites through chemical derivatization before research (Hall 2006).

GC-MS contains chemical ionization (CI) and electron ionization (EI). The GC with TOF-MS is popular technique for metabolomic profiling because it contains the ability of faster acquisition time, improved deconvolution, and higher mass accuracy outcomes for compounded mixtures while the GC with EI detector attire with individual quadrupole (Q) mass analyzer contains reproducibility, resolution, robustness, and high sensitivity suffer from stagnant examine agility and with deprived accuracy of mass (Kumar et al. 2016). Concluding all the metabolite techniques the GC-MS is the superior approach due to its better working features such as productivity, standardization, and efficiency (Tsugawa et al. 2011).

9.5.2 Capillary Electrophoresis-Mass Spectrometry

Capillary electrophoresis mass spectrometry used to evaluate diversity of classical metabolites on the basis of charge and size proportion ratio (Obata and Fernie 2012). Based on mass fragmentation, this technique characterizes the metabolites and the outcomes come up with fast and high resolution (Salem et al. 2020). Capillary Electrophoresis-mass spectrometry technique is time saving because capillary mass spectrometry (CE-MS) metabolite analysis imbricates with GC-MS and no need for derivatization. This technique is implemented in fused silica capillary tubes, the ends dipped in the buffer solutions and across which high voltage is applied (Jorge et al. 2016). The disadvantageous features of this technique are poor mitigation time, low sensitivity, lack of reference libraries, and reproducibility (Williams et al. 2007). The advantageous features of this technique separation are done in small amount which is suitable for plant metabolomics (Salem et al. 2020).

9.5.3 Fourier Transform Ion Cyclotron Resonance-Mass Spectrometry (FTICR-MS)

This approach furnishes by means of high resolving power and mass specificity out of all kinds of mass spectrometry (Nikolaev et al. 2016). It becomes the essential technique for proteomics and metabolomics due to its specific analytical features. FTICR-MS is used as a part of metabolomics procedures because of its ability to furnish ultimate elevated resolution and elevated mass specific data (Jorge et al. 2016). This technique is also consistent with multi-stage mass spectrometry analyzers. The high cost is the only barrier to widespread application in plant metabolomics (Ahmad et al. 2018).

9.5.4 Matrix-Assisted Laser Desorption/Ionization (MALDI)

This non-targeted analytical approach is implemented for quantitative and qualitative image of a broad range of metabolites (Baker et al. 2017). The componenets like sugars, metabolites, and lipids from various parts of the plants such as leaves, stem, roots, and flower most of the studies used the MALDI-MS technique (Jung et al. 2010). The plant tissues contain many peptides and proteins that can be analyzed through this technique which involves the coating of a fine film of a matrix comprising either sinapinic acid, alpha-cyano-4-hydroxycinnamic acid (CHCA), and 2, 5-dihydroxy benzoic acid (2,5-DHBA) on the tissue surface. The penetration of laser beam through the matrix-coated tissue generates the mass spectrum. MALDI is the most common method for protein/metabolites imaging, merged with the broad varieties of various mass analyzers such as Fourier ICR transform (FT-ICR), QqToF (quadrupole time of flight), linear and spherical ion trap. Other emerging ionization techniques like laser ablation electrospray ionization (LAESI), desorption electrospray ionization (DESI), and secondary ion mass spectrometry (SIMS) are already studied (Dong et al. 2016).

9.5.5 Nuclear Magnetic Resonance (NMR)

NMR is a technique that furnishes the qualitative or quantitative data from biological extracts used to investigate the different types of plant metabolome (Kim et al. 2011). NMR approach is based on the electromagnetic radiation, occurs when the atomic nuclei is placed in the strong magnetic field. Currently, the NMR is an emerging technique because chromatic separation and sample derivatization needs not much higher requirement to perform the technique (Foroutan et al. 2019). This technique contains various advantages such as quick analysis, non-destructive nature, and excellent repeatability, however, it is less sensitive than MS, rather it is pH-sensitive. Buffered solutions are usually needed at stable pH (Deborde et al. 2017; Kim et al. 2011).

9.6 OMICS Approaches

9.6.1 Genomics

The study of genetic material is known as the genomics (Mulder et al. 2017). In this field, various combinations of recombinant DNA are used such as the DNA sequencing method, bioinformatics to sequence, analyze, and assemble its genomic and functional structure. Using the next-generation sequencing data generates the data that permit faster and cheaper sequencing (Grelewska-Nowotko et al. 2018). The functional and structural genomic approaches are the divisions of genomics, the structural genomics reveals the complete sequence DNA of an organism and it is also the preliminary phase of genome study. Moreover, assessing the function of gene the genome sequence is used in functional genomic (Leister 2004). Genomics provides information about the resistant gene in the plant genome along with its location on the chromosome. For developing the new improved crop varieties, the advance tactics for plant biotechnology allow specified and practicable approaches for molecular plant breeding and marker-assisted selection (Wang et al. 2017). The disease-resistant R gene or nucleotide-binding leucine-rich repeats are detected by the RenSeq tool. In greenhouse conditions, The Rpi-amr3i gene was found that completely prevented phytophthora in the transgenic potatoes (Witek et al. 2016).

9.6.2 Transcriptomics

The transcriptome approach is the complete set of RNA transcripts of an individual's genome under specific conditions/specific cells. The technique like microarray analysis is used in the transcriptomic approach. The microarray quantifies the steady-state mRNA levels which are different from the total functional protein present in the plant tissue. Based on the RNA-sequence, comparison of transcriptomic data allows the recognition and better knowledge of varied expressed gene patterns, splice variants, and uncommon genes in a particular cell type (Gao et al. 2013; Peter et al. 2018). The RNA genome of some plant viruses is hard to analyze through standard molecular techniques. Understanding the plant virus interaction between plant and microorganisms transcriptomic approaches is used such as Degradome sequencing, RNA sequencing and sRNA sequencing is increasing tremendously (Zanardo et al. 2019). The RNA sequence is performed to analyze the differential gene expression in the resistant cotton. Plant science has transcriptomic applications such as the progress of molecular markers, identifying the genes and pathways related to the stress, and searching the genes which are responsible for secondary metabolism and plant growth. Instead of pathologically important genes, transcriptomics analyzes the crops whose sequence is still not discovered. This approach leads to the genes that improve the concerning traits (Agarwal et al. 2014). This technique assists in the examination of genes that are implied in basic growth and stress mechanisms (Zan et al. 2013).

9.6.3 Proteomics

Proteomics is an appropriate approach to understand the molecular mechanism of plant development and its growth. Proteomics is popular because it enables deep study of the molecular mechanism of plant growth and development (Chen and Harmon 2006) through functional and structural proteomics and protein-protein interaction analysis (PPI), presence of the sequence of primary amino acids of proteins, post translocation modification state, estimation of their relative amounts and protein function characterization and structure are identified (Rhee et al. 2006). Determining the plant reactions to peripheral incentives such as ailment and insect infestation helps in unraveling the functions of different proteins by protein expression profile. Furthermore, it helps in understanding the resistance mechanism of plants, mode of action for insecticides and their biodegradation, supporting in the progress of effective, and safe agricultural pests (Van Emon 2016).

9.6.4 Metabolomics

It is an OMIC approach which measures all the metabolites and the molecules whose molecular weight is lower in organisms, cells, and tissues (Castro-Moretti et al. 2020). To generate data in metabolomic approach, the nuclear magnetic resonance and mass spectrometry are used (Winning et al. 2009). Metabolomics is easy to study molecular mechanism of plant development under stress conditions and normal conditions. Along with the improvements in metabolomic techniques, breeding of plants have various applications and futuristic approaches toward the development and resistance against cultivators (Fernie and Schauer 2009). The expeditious testing of plant metabolomic responses to environmental and genetic changes helps in screening individuals which are tolerant and resistant (Castro-Moretti et al. 2020). Identification of several metabolites which are capable to tolerate the stress in plants can be recognized. For instance, resistant and acquiescent cultivation of rice are blighted with *Xanthomonas oryzae* pv. *oryzae*, which causes leaf blight, differs strikingly for different metabolites named lipids, xanthophylls, alkaloids, acetophenone and carbohydrates (Sana et al. 2010). The *Fusarium graminearum*-resistant agmatine coumaroyltransferase (ACT) gene was detected through the analysis of hydrocinnamic acid amide compounds. In conclusion, culturing based on metabolomics helps in developing new plant varieties resisted to diseases (Hong et al. 2016) (Fig. 9.5).

9.7 Conclusion

Manipulating the genome to recruit a coding sequence of interest has made it possible to achieve/enhance desired characteristics and to suppress/delete the unwanted traits, all thanks to latest gene editing techniques. Regular developments and updating of latest analytical tools and techniques are additional benefits to record the effects gene editing by analyzing the production yield and quality of metabolites. Ongoing research and development in techniques like CRISPR/Cas and OMICS will surely be giving us excellency in applicational benefits of these approaches. Hence, utilizing them to decode allelopathic relationships between a plant and rhizospheric microbiome will bring about incredible results in terms of plant growth, yield,

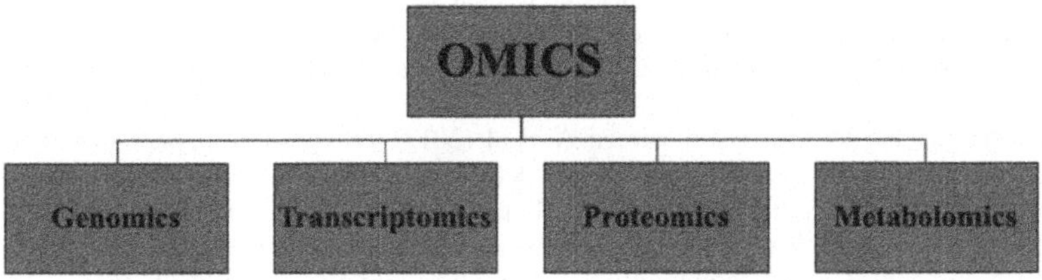

Fig. 9.5 OMICS technologies

production of metabolites beneficial to humans and with great applications in industries, better survival and sustainability of the plants to withstand biotic and abiotic conditions.

References

Agarwal P, Parida SK, Mahto A et al (2014) Expanding frontiers in plant transcriptomics in aid of functional genomics and molecular breeding. Biotechnol J 9:1480–1492. https://doi.org/10.1007/s10123-020-00140-2

Ahmad P, Ahanger MA, Singh VP et al (2018) Plant metabolites and regulation under environmental stress. Academic, Amsterdam

Ali SZ, Sandhya V, Grover M et al (2011) Effect of inoculation with a thermotolerant plant growth promoting pseudomonas putida strains AKMP7 on growth of wheat (*Triticum* spp.) under heat stress. J Plant Interact 6:239–246

Andreote FD, Da Rocha UN, Araujo WL et al (2010) Effect of bacterial inoculation, plant genotype and developmental stage on root-associated and endophytic bacterial communities in potato (*Solanum tuberosum*). Antonie Van Leeuwenhoek 97:389–399. https://doi.org/10.1007/s10482-010-9421-9

Ansary MH, Rahmani HA, Ardakai MR et al (2012) Effect of pseudomonas fluorescent on proline and phytohormonal status of maize (*Zea mays* L.) under water deficit stress. Ann Biol Res 3: 1054–1062

Ayala Garcia MA, Gonzalez Yebra B, Lopez Flores AL, Guani Guerra E (2012) The major histocompatibility complex in transplantation. J Transp Secur 2012:842141. https://doi.org/10.1155/2012/842141

Bacon CW, Hinton DM (2011) Bacillus mojavensis: its endophytic nature, the surfactins, and their role in the plant response to infection by *Fusarium verticillioides*. In: Bacteria in agrobiology: plant growth responses. Springer, New York, pp 21–39

Baker TC, Han J, Brochers CH (2017) Recent advancements in matrix-assisted laser desorption/ionization mass spectrometry imaging. Curr Opin Biotechnol 43:62–69. https://doi.org/10.1007/978-3-642-20332-9_2

Bano Q, Ilyas N, Bano A et al (2013) Effect of *Azospirillum* inoculation on maize (*Zea mays* L.) under drought stress. Pak J Bot 45:13–20

Barea JM (2015) Future challenges and perspectives for applying microbial biotechnology in sustainable agriculture based on a better understanding of plant-microbiome interactions. J Soil Sci Plant Nutr 15:261–282. https://doi.org/10.4067/S0718-95162015005000021

Barnawal D, Maji D, Bharti N et al (2013) ACC deaminase-containing *Bacillus subtilis* reduces stress ethylene-induced damage and improves mycorrhizal colonization and rhizobial nodulation stress, in *Trigonella foenum-graecum* under drought stress. J Plant Growth Regul 32:809–822

Bertola M, Ferrarini A, Visiolo G (2021) Improvement of soil microbial diversity through sustainable agricultural practices, food quality and human safety. Microorganisms 9:1400

Bianchi F, Ilag L, Termopoli V, Mendez L (2018) Advances in MS-based analytical methods: innovations and future trends. J Anal Methods Chem 2018:1. https://doi.org/10.1155/2018/2084567

Brotman Y, Landau U, Cuadros-Inostroza A et al (2013) *Trichoderma*-plant root colonization: escaping early plant defense responses and activation of the antioxidant machinery for saline stress tolerance. PLoS Pathog 9:e1003221. https://doi.org/10.1371/journal.ppat.1003221

Burd GI, Dixon DG, Glick BR (2000) Plant growth-promoting bacteria that decrease heavy metal toxicity in plants. Can J Microbiol 46:237–245. https://doi.org/10.1139/w99-143

Carlson DF, Tan W, Lillico SG et al (2012) Efficient TALEN-mediated gene knockout in livestock. Proc Natl Acad Sci U S A 109:17382–17387. https://doi.org/10.1073/pnas.1211446109

Carvalhais LC, Dennis PG, Tyson GW, Schenk PM (2013) Rhizosphere metatranscriptomics: challenges and opportunities. In: Molecular microbial ecology of the rhizosphere, vol 2. Wiley-Blackwell, Hoboken, pp 1137–1144

Castro-Moretti FR, Gentzel IN, Mackey D, Alonso AP (2020) Metabolomics as an emerging tool for the study of plant-pathogen interactions. Metabolites 10:52. https://doi.org/10.3390/metabo10020052

Chen S, Harmon AC (2006) Advances in plant proteomics. Proteomics 6:5504–5516

De Corato U (2020) Soil microbiota manipulation and its role in suppressing soil-borne plant pathogens in organic farming systems under the light of microbiome-assisted strategies. Chem Biol Technol Agric 7(1):17. https://doi.org/10.1186/s40538-020-00183-7

de Souza EM, Galindo FS, Teixeira M et al (2019) Does the nitrogen application associated with Azospirillum brasilense inoculation influence corn nutrition and yield? Rev Bras Eng Agricola e Ambient 23:53–59

Deborde C, Moing A, Roch L et al (2017) Plant metabolism as studied by NMR spectroscopy. Prog Nucl Magn Reson Spectrosc 102:61–97. https://doi.org/10.1016/j.pnmrs.2017.05.001

Dong Y, Li B, Malitsky S et al (2016) Sample preparation for mass spectrometry imaging of plant tissues: a review. Front Plant Sci 7:60. https://doi.org/10.3389/fpls.2016.00060

Fernie AR, Schauer N (2009) Metabolomics-assisted breeding: a viable option for crop improvement? Trends Genet 25:39–48. https://doi.org/10.1016/j.tig.2008.10.010

Foroutan A, Goldansaz SA, Lipfert M, Wishart DS (2019) Protocols for NMR analysis in livestock metabolomics. Metabol Mol Biol 1996:311–324. https://doi.org/10.1007/978-1-4939-9488-5_23

Gao Y, Xu H, Shen Y, Wang J (2013) Transcriptomic analysis of rice (*Oryza sativa*) endosperm using the RNA-seq technique. Plant Mol Biol 81:363–378. https://doi.org/10.1007/s11103-013-0009-4

Gou W, Tian L, Ruan Z et al (2015) Accumulation of choline and grycinebetaine and drought stress tolerance induced in maize (*Zea mays*) by three plant growth promoting rhizobacteria (PGPR) strains. Pak J Bot 47:581–586

Grelewska-Nowotko K, Zurawaska-Zajfert M, Zmijewska E, Sowa S (2018) Optimization and verification of droplet digital PCR even-specific methods for the quantification of GM maize DAS1507 and NK603. Appl Biochem Biotechnol 185:207–220. https://doi.org/10.1007/s12010-017-2634-x

Guilinger JP, Thompson DB, Liu DR (2014) Fusion of catalytically inactive Cas9 to FokI nuclease improves the specificity of genome modification. Nat Biotechnol 32:577–582. https://doi.org/10.1038/nbt.2909

Guo Y, Tian D, Shen F, Yang G, Long L, He J, Song C, Zhang J, Zhu Y, Huang C, Deng S (2019) Transparent cellulose/technical lignin composite films for advanced packaging. Polymers (Basel) 11(9):1455

Hall RD (2006) Plant metabolomics: from holistic hope, to hype, to hot topic. New Phytol 169:453–468. https://doi.org/10.1111/j.1469-8137.2005.01632.x

Heigwer F, Kerr G, Boutros M (2014) E-CRISP: fast CRISPR target site identification. Nat Methods 11:122–123. https://doi.org/10.1038/nmeth.2812

Hirsch PR, Miller AJ, Dennis PG (2013) Do root exudates exert more influence on rhizosphere bacterial community structure than other rhizodeposits? In: de Bruijn FJ (ed) Molecular microbial ecology of the rhizosphere, vol 1. Wiley, New York, pp 229–242. https://doi.org/10.1002/9781118297674.ch22

Hong J, Yang L, Zhang D, Shi J (2016) Plant metabolomics: an indispensable system biology tool for plant science. Int J Mol Sci 17:767. https://doi.org/10.3390/ijms17060767

Hsu PD, Scott DA, Weinstein JA et al (2013) DNA targeting specificity of RNA-guided Cas9 nucleases. Nat Biotechnol 31:827–832. https://doi.org/10.1080/17429145.2010.545147

Hussain MB, Zahir ZA, Asghar M (2014) Can catalase and exopolysaccharides producing rhizobia ameliorate drought stress in wheat? Int J Agric Biol 16:3–13

Jackson D, Skillman J, Vandermeer J (2012) Indirect biological control of the coffee leaf rust, Hemileia vastatrix, by the entomogenous fungus *Lecanicillium lecanii* in a complex coffee agroecosystem. Biol Control 61:89–97. https://doi.org/10.1016/j.biocontrol.2012.01.004

Jinek M, Chylinski K, Fonfara I et al (2012) A programmable dual-RNA-guided DNA endonuclease in adaptive bacterial immunity. Science 80(337):816–821. https://doi.org/10.1126/science.1225829

Jorge TF, Rodrigues JA, Caldana C et al (2016) Mass spectrometry-based plant metabolomics: metabolite responses to abiotic stress. Mass Spectrom Rev 35:620–649. https://doi.org/10.1002/mas.21449

Jung JH, Song Z, Liu Z et al (2010) High-spatial and high mass resolution imaging of surface metabolites of *Arabidopsis thaliana* by laser desorption-ionization mass spectrometry using colloidal silver. Anal Chem 82:3255–3265. https://doi.org/10.1021/ac902990p

Kanchiswamy CN, Maffei M, Malnoy M et al (2016) Fine-tuning next generation genome editing tools. Trends Biotechnol 34:562–574. https://doi.org/10.1016/j.tibtech.2016.03.007

Kaur A, Devi SR, Vyas P (2018) Stress-tolerant antagonist plant growth-promoting rhizobacteria from *Zea mays*. J Plant Prot Res 58:115–123. https://doi.org/10.24425/119127

Kim HK, Choi YH, Verpoorte R (2011) NMR-based plant metabolomics: where do we stand, where do we go? Trends Biotechnol 29:267–275. https://doi.org/10.1016/j.tibtech.2011.02.001

Kohli A, Narciso JO, Miro B, Raorane M (2012) Root proteases: reinforced links between nitrogen uptake and mobilization and drought tolerance. Physiol Plant 145:165–179

Krishnakumar R, Kraus WL (2010) The PARP side of the nucleus: molecular actions, physiological outcomes, and clinical targets. Mol Cell 39:8–24. https://doi.org/10.1016/j.molcel.2010.06.017

Kumar M, Kuzhiumparambil U, Pernice M et al (2016) Metabolomics: an emerging frontier of systems biology in marine macrophytes. Algal Res 16:76–92. https://doi.org/10.1016/j.algal.2016.02.033

Leister D (2004) Plant functional genomics. CRC Press, Boca Raton

Li T, Liu B, Spalding MH et al (2012) High efficiency TALEN-based gene editing produces disease resistant rice. Nat Biotechnol 30:390–392. https://doi.org/10.1038/nbt.2199

Liu H, Brettell LE, Singh B (2020) Linking the phyllosphere microbiome to plant health. Trends Plant Sci 25:841–844. https://doi.org/10.1016/j.tplants.2020.06.003

Ma X, Liu Y-G (2016) CRISPR/Cas9-based multiplex genome editing in monocot and dicot plants. Curr Protoc Mol Biol 115:31–36. https://doi.org/10.1002/cpmb.10

Molina-Romero D, Baez A, Quintero-Hernandez V et al (2017) Compatible bacterial mixture, tolerant to desiccation, improves maize plant growth. PLoS One 12:e0187913. https://doi.org/10.1371/journal.pone.0187913

Montague TG, Cruz JM, Gagnon JA et al (2014) CHOPCHOP: a CRISPR/Cas9 and TALEN web tool for genome editing. Nucleic Acids Res 42:W401–W407. https://doi.org/10.1093/nar/gku410

Mulder N, Adebamowo CA, Adebamowo SN et al (2017) Genomic research data generation, analysis and sharing—challenges in the African setting. Data Sci J 16:1–15. https://doi.org/10.5334/dsj-2017-049

Muller DB, Vogel C, Bai Y, Vorholt JA (2016) The plant microbiota: systems-level insights and perspectives. Annu Rev Genet 50:211–234. https://doi.org/10.1146/annurev-genet-120215-034952

Nadeem SM, Ahmad M, Naveed M et al (2016) Relationship between in vitro characterization and comparative efficacy of plant growth-promoting rhizobacteria for improving cucumber salt tolerance. Arch Microbiol 198:379–387. https://doi.org/10.1007/s00203-016-1197-5

Nelson EB (2018) The seed microbiome: origins, interactions, and impacts. Plant Soil 422:7–34. https://doi.org/10.1007/s11104-017-3289-7

Nicoletti R, Becchimanzi A (2020) Endophytism of *Lecanicillium* and *Akanthomyces*. Agriculture 10:205. https://doi.org/10.3390/agriculture10060205

Nikolaev EN, Kostyukevich YI, Vladimirov GN (2016) Fourier transform ion cyclotron resonance (FT ICR) mass spectrometry: theory and simulations. Mass Spectrom Rev 35:219–258. https://doi.org/10.1002/mas.21422

Nwachukwu BC, Babalola OO (2021) Perspective for sustainable agriculture from the microbiome in plant rhizosphere. Plant Biotechnol Rep 15:259–278

Obata T, Fernie AR (2012) The use of metabolomics to dissect plant responses to abiotic stresses. Cell Mol Life Sci 69:3225–3243. https://doi.org/10.1007/s00018-012-1091-5

Panaccione DG, Beaulieu WT, Cook D (2014) Bioactive alkaloids in vertically transmitted fungal endophytes. Funct Ecol 28:299–314. https://doi.org/10.1111/1365-2435.12076

Pareek A, Sopory SK, Bohnert HJ et al (2009) Abiotic stress adaptation in plants. Springer, New York. https://doi.org/10.1007/978-90-481-3112-9

Peter SC, Dhanjal JK, Malik V et al (2018) In: Ranganathan S, Grib-skov M, Nakai K, Schonbach C (eds) Encyclopedia of bioinformatics and computational biology. Elsevier, Amsterdam, pp 661–676

Prasad R, Varma A (2014) Biology of underground root engineering. In: Bagde US, Jadhav BL, Jadhav UA (eds) Current trends in life sciences. LAP LAMBERT Academic Publishing, Germany, pp 5–8

Prasad R, Kumar M, Varma A (2015) Role of PGPR in soil fertility and plant health. In: Egamberdieva D, Shrivastava S, Varma A (eds) Plant growth-promoting rhizobacteria (PGPR) and medicinal plants. Springer International Publishing, Cham, pp 247–260

Prasad R, Zhang S-H (2022) Benefical microorganisms in agriculture. Springer, Singapore. https://link.springer.com/book/9789811907326

Rana KL, Kour D, Kaur T et al (2020) Endophytic microbes: biodiversity, plant growth-promoting mechanisms and potential applications for agricultural sustainability. Antonie Van Leeuwenhoek 113:1075–1107. https://doi.org/10.1007/s10482-020-01429-y

Razzaq A, Sadia B, Raza A et al (2019) Metabolomics: a way forward for drop improvement. Metabolites 9:303. https://doi.org/10.3390/metabo9120303

Rhee SY, Dickerson J, Xu D (2006) Bioinformatics and its applications in plant biology. Annu Rev Plant Biol 57:335–360. https://doi.org/10.1146/annurev.arplant.56.032604.144103

Rodriguez R, Duran P (2020) Natural holobiome engineering by using native extreme microbiome to counteract the climate change effects. Front Bioeng Biotechnol 8:568. https://doi.org/10.3389/fbioe.2020.00568

Rolli E, Marasco R, Vigani G et al (2015) Improved plant resistance to drought is promoted by the root associated microbiome as a water stress dependent trait. Environ Microbial 17:316–331. https://doi.org/10.1111/1462-2920.12439

Salem MA, de Souza L, Serag A et al (2020) Metabolomics in the context of plant natural products research: from sample preparation to metabolite analysis. Metabolites 10:37. https://doi.org/10.3390/metabo10010037

Sana TR, Fischer S, Wohlgemuth G et al (2010) Metabolomics and transcriptomic analysis of the rice response to the bacterial blight pathogen *Xanthomonas oryzae*. Metabolomics 6:451–465. https://doi.org/10.1007/s11306-010-0218-7

Santhanam R, Weinhold A, Goldberg J et al (2015) Native root-associated bacteria rescue a plant from a sudden-wilt disease that emerging during continuous cropping. Proc Natl Acad Sci 112: E5013–E5020. https://doi.org/10.1073/pnas.1505765112

Sarma H, Forid N, Prasad R, Prasad MNV, Ma LQ, Rinklebe J (2021) Enhancing phytoremediation of hazardous metal(loid)s using genome engineering CRISPR–Cas9 technology. J Hazard Mater. https://doi.org/10.1016/j.jhazmat.2021.125493

Sasan RK, Bidochka MJ (2012) The insect-pathogenic fungus *Metarhizium robertsii* (Clavicipitaceae) is also an endophyte that stimulated plant root development. Am J Bot 99: 101–107. https://doi.org/10.3732/ajb.1100136

Shrivastava S, Prasad R, Varma A (2014) Anatomy of root from eyes of a microbiologist. In: Morte A, Varma A (eds) Root engineering, vol 40. Springer-Verlag, Berlin, pp 3–22

Sorty AM, Meena KK, Choudhary K et al (2016) Effect of plant growth promotion bacteria associated with halophytic weed (*Psoralea corylifolia* L) on germination and seedling growth of wheat under saline conditions. Appl Biochem Biotechnol 180:872–882. https://doi.org/10.1007/s12010-016-2139-z

Szankasi P, Smith GR (1992) A DNA exonuclease induced during meiosis of *Schizosaccharomyces pombe*. J Biol Chem 267:3014–3023. https://doi.org/10.1016/S0021-9258(19)50688-3

Timmusk S, IA AEI-D, Copolovici L et al (2014) Drought-tolerance of wheat improved by rhizosphere bacteria from harsh environments: enhanced biomass production and reduced emissions of stress volatiles. PLoS One 9:e96086. https://doi.org/10.1371/journal.pone.0096086

To A, Durourek K, De Cian A, Concordet JP, Del Bene F (2014) Highly efficient CRISPR/Cas-mediated knock-in in zebrafish by homology-independent DNA repair. Genome Res 24:142–153. https://doi.org/10.1101/gr.161638.113

Tsugawa H, Tsujimoto Y, Arita M et al (2011) GC/MS based metabolomics: development of a data mining system for metabolite identification by using soft independent modelling of class analogy (SIMCA). BMC Bioinformatics 12:1–13. https://doi.org/10.1186/1471-2105-12-131

Van Emon JM (2016) Theomic evolution in agricultural research. J Agric Food Chem 64:36–44. https://doi.org/10.1021/acs.jafc.5b04515

Verma P, Yadav AN, Kazy SK et al (2013) Elucidating the diversity and plant growth promoting attributes of wheat (*Triticum aestivum*) associated acidotolerant bacteria from southern hills zone of India. Natl J Life Sci 10(2):219–227

Viscardi S, Ventorino V, Duran P et al (2016) Assessment of plant growth promoting activities and abiotic stress tolerance of *Azotobacter chroococcum* strains for a potential use in sustainable agriculture. J Soil Sci Plant Nutr 16:848–863. https://doi.org/10.4067/S0718-95162016005000060

Wang S, Suh JH, Zheng X et al (2017) Identification and quantification of potential anti-inflammatory hydroxycinnamic acid amides from wolfberry. J Agric Food Chem 65(2):364–372. https://doi.org/10.1021/acs.jafc.6b0513665:364-372

White JF, Kingsley KL, Zhang Q et al (2019) Endophytic microbes and their potential applications in crop management. Pest Manag Sci 7:2558–2565. https://doi.org/10.1002/ps.5527

Williams BJ, Cameron CJ, Workman R et al (2007) Amino acid profiling in plant cell cultures: an inter-laboratory comparison of CE-MS and GC-MS. Electrophoresis 28:1371–1379. https://doi.org/10.1002/elps.200600609

Winning H, Roldan-Marin E, Dragsted LO et al (2009) An exploratory NMR nutri-metabolomic investigation reveals dimethyl sulfone as a dietary biomarker for onion intake. Analyst 134:2344–2351. https://doi.org/10.1039/b918259d

Witek K, Jupe F, Witek AI et al (2016) Accelerated cloning of a potato late blight—resistance gene using RenSeq and SMRT sequencing. Nat Biotechnol 34:656–660. https://doi.org/10.1038/nbt.3540

Xiao A, Wang Z, Hu Y et al (2013) Chromosomal deletions and inversions mediated by TALENs and CRISPR/Cas in Zebrafish. Nucleic Acids Res 41:e141. https://doi.org/10.1093/nar/gkt464

Zan Y, Ji Y, Zhang Y et al (2013) Genome-wide identification, characterization and expression analysis of populus leucine-rich repeat receptor-like protein kinase genes. BMC Genomics 14:1–14. https://doi.org/10.1186/1471-2164-14-318

Zanardo LG, de Souza GB, Alves MS (2019) Transcriptomics of plant—virus interactions: a review. Theor Exp Plant Physiol 31:103–125. https://doi.org/10.1007/s40626-019-00143-z

Chapter 10
Phosphate Solubilizing Microorganisms: Multifarious Applications

Mahendra Kumar, Ajay Shankar, Shivani Chaudhary, and Vishal Prasad

Abstract Phosphorus is a key element for plant growth and development. Phosphate sources are available in soil present in both forms organic as well as inorganic. Due to its highly reactive nature, phosphate forms insoluble complex with several metal ions (Fe, Al, and Ca) and becomes unavailable for plant uptake and thus acts as a major limiting factor. Phosphate solubilizing microorganisms (PSMs) such as bacteria, fungi, and actinomycetes possess the ability to solubilize insoluble phosphate and convert into available form as orthophosphate ions thereby helping in plant growth, crop yield, and simultaneously improve soil health. In addition, these PSMs also play major role in various other key activities of environmental significance. A few such activities include ecological restoration, heavy metal decontamination and immobilization, promoting sustainable agricultural practices in saline-alkaline and other unsuitable soils. Overall, these PSMs are evolving as worthy candidates with multifarious application for environmental sustainability. This chapter covers several aspects of phosphate solubilization and mobilization by PSMs in the soil and further describes several other beneficial applications of these PSMs.

Keywords Phosphorus · Phosphatase · Phytase · PSM · Solubilization

10.1 Introduction

Phosphorus (P) is the second most essential macronutrient for plant growth and development after nitrogen and it accounts for around 0.2% of plant dry weight (Lin et al. 2006; Sharma et al. 2013). Phosphorus plays very crucial roles in several of the metabolic processes such as synthesis of biomolecules, energy transfer reactions, photosynthesis, cell division, plasma membrane components, signaling molecules, nucleic acid components, flowers and seeds formation, enzyme activities regulation

M. Kumar · A. Shankar · S. Chaudhary · V. Prasad (✉)
Institute of Environment and Sustainable Development, Banaras Hindu University, Varanasi, UP, India
e-mail: vp.iesd@bhu.ac.in

S. Chhabra et al. (eds.), *Plant Microbiome for Plant Productivity and Sustainable Agriculture*, Microorganisms for Sustainability 37,
https://doi.org/10.1007/978-981-19-5029-2_10

by phosphorylation of serine, histidine, aspartate, threonine, and tyrosine amino acids (Raghothama 1999). Plants absorb P as phosphate anions (HPO_4^{2-} or $H_2PO_4^-$) from soil (Rodriguez and Fraga 1999). However, these phosphate anions are highly reactive in the soil and immediately form complex compound with metals ions like Ca^{2+}, Al^{3+}, and Fe^{3+}; and their precipitation reaction depends on pH of the soil. In acidic soils, phosphate anions react with oxides and hydroxides of Al and Fe and form insoluble complexes whereas in alkaline soils Ca reacts with these phosphate anions and fixes the available P (Igual et al. 2001). Due to high reactivity of these phosphate anions with metal cations, most soils are deficient in available P for the plant uptake thus making it to act as a limiting factor for plant growth and development in tropical and subtropical regions (Richardson 2001). To cope with this P limitation, farmers apply synthetic fertilizer but approximately 70–90% of applied P fertilizers get precipitated rapidly after their application in the soil (Mikanova and Novakova 2002). Consequently, overuse of chemical fertilizer causes nutrient imbalance and puts a burden on arable land which is a serious concern.

Soil microbiomes play essential role in P cycle. The P cycle in the biosphere is considered as open or sedimentary because there are neither gaseous intermediary forms nor any interchange between soil and air. This P cycle takes place by means of oxidation and reduction of phosphorus compounds in which oxidation state of phosphorus ranges from phosphine (−3) to phosphate (+5) (Ohtake et al. 1996; Behera et al. 2014). PSMs eventually increase the availability of soluble phosphate in the soil and boost plant growth and development by enhancing the efficacy of biological nitrogen fixation or enhancing the availability of other trace element such as iron and zinc and by producing plant growth promoting regulators (Ponmurugan and Gopi 2006). In the recent past, these PSMs have been explored for their various other potentials like that of heavy metal removal, ecological restoration, organic pollutant clean-up, and others. In this chapter, we focus on several aspects of phosphate solubilization and mobilization by PSMs in the soil and extend our discussion further to several other beneficial applications of these PSMs.

10.2 Phosphorus in Soil

Soil is a dynamic system and an ecological niche of various biological activities. The inorganic phosphate (Pi) available for biosynthetic purposes depends not only on the total amount of P in the environment but also on its availability as well as solubility. Phosphorus is one of the most important macronutrients for plant growth and development. The concentration of soluble P in soil is generally very low, normally at level of 1 ppm or less (Goldstein 1994). Mineral forms of P are represented in soil primarily by minerals such as apatite, hydroxyapatite, and oxyapatite and their main characteristic is insolubility, though they can be solubilized under suitable conditions and become available for uptake by microorganisms and plants. In soil, P exists in both organic and inorganic form. In cultivated soil around 70–80% of P exist in inorganic form applied in the form of P fertilizers (Foth and Foth 1990). Organic forms of P may constitute 30–50% of the total P in most soil, although its concentration may vary from as low as 5% to as high 95% (Paul and Clark 1988). Organic form of P exists mainly as inositol phosphate (soil phytate) accounting for approximately 50% of the total organic P. It is most stable form of organic P in the soil and is synthesized by microorganisms and plants. The other organic P forms include phosphomonoesters, phosphodiesters, phospholipids, phosphotriesters, and nucleic acids.

10.3 Phosphate Solubilizing Microorganisms

Diverse types of microorganisms are the major players in various processes that are linked with transformation of soil P and thus are an integral part of the soil P cycle. Soil microorganisms play very crucial role in making the availability of P from inorganic and organic pool of the total soil P by solubilization and mineralization (Hilda and Fraga 1999). Evidences of naturally occurring rhizospheric PSMs date back to 1903 (Khan et al. 2007). Microorganisms involved in phosphate solubilization consist of mycorrhizal fungi, bacteria, and actinomycetes, among the microbial population in soil, phosphate solubilizing bacteria (PSB) constitute 1–50%, while phosphate solubilizing fungi (PSF) are only 0.1–0.5% in solubilization potential (Chen et al. 2006). A considerably higher number of PSB are present in the rhizospheric soil than in non-rhizospheric soil (Raghu and Mac Rae 1966). The PSMs found in the plant rhizosphere are reported to be metabolically more active (Vazquez et al. 2000). Population of PSB in soil depends on its physical and chemical properties, organic matter, and P content (Kim et al. 1998). Strains from bacterial genera *Pseudomonas*, *Bacillus*, *Rhizobium*, *Enterobacter* along with *Penicillium* and *Aspergillus* from fungi are the most powerful phosphate solubilizers (White la 2000). Bacteria such as *Bacillus megaterium*, *Bacillus circulans*, *Bacillus subtilis*, *Bacillus polymyxa*, *Bacillus sircalmous*, *Pseudomonas striata*, and *Enterobacter* are referred as most important phosphate-solubilizing strains (Kucey et al. 1989). Among the soil bacterial populations, ectorhizospheric strains from *Pseudomonas* and *Bacillus* and endosymbiotic rhizobia have been observed as effective phosphate solubilizers (Igual et al. 2001). A fungus *Arthrobotrys oligospora* has also been reported for its ability to solubilize the phosphate rocks (Duponnois et al. 2006).

10.4 Need of Phosphate Solubilizing Microorganism

Chemical phosphatic fertilizers are made by a highly energy-intensive process that consumes energy worth US $4 billion per annum in order to fulfill the global needs (Goldstein et al. 1993). Further almost 75–90% of applied phosphatic fertilizers are precipitated by certain metal ions due to complex formation in the soil and it has been estimated that the precipitated phosphates in agricultural soils are sufficient to sustain crop production worldwide for nearly 100 years (Goldstein et al. 1993). Microorganisms are reported to play very crucial role in soil P cycle and relocating P between different soil P pools (Prasad et al. 2018). Hence it is evident to explore sources of phosphate solubilizers. Under different soil and agro-climatic conditions, these PSMs have proved to be an economical alternative to the more expensive chemical phosphatic fertilizers with greater agronomic utility (Ngalimat et al. 2021). The PSMs increase the availability of usable form of phosphate and can improve plant growth and soil health by producing other plant growth promoting substances, increasing biological N-fixation and enhancing the availability of other important trace element like iron and zinc (Ponmurugan and Gopi 2006; Nath et al. 2018).

10.5 Mechanisms of Phosphate Solubilization

There are various mechanisms such as production of low molecular weight organic acids, lowering of pH through H^+ extrusion, production of inorganic acids, and secretion of different enzymes like phosphatases, phytases, and phosphonatases for solubilizing the insoluble phosphate by PSMs. The various ways of phosphate solubilization by PSMs are summarized in Fig. 10.1. Based on the source of insoluble P, the phosphate solubilization mechanism can be categorized into two categories, i.e., inorganic P solubilization and organic P solubilization (Surange et al. 1995; Dutton and Evans 1996; Nahas 1996).

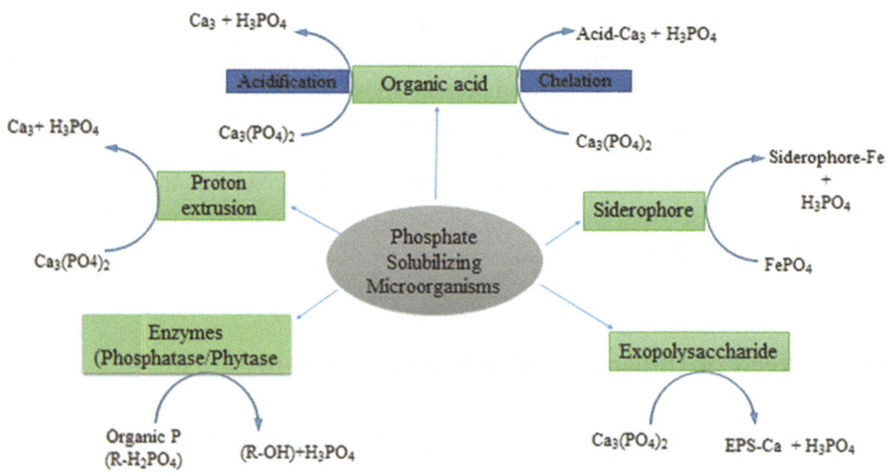

Fig. 10.1 Various mechanisms of phosphate solubilization by phosphate solubilizing microorganisms

10.5.1 Inorganic Phosphate Solubilization

The principal mechanisms of inorganic phosphate solubilization involve production of low molecular weight organic acids (Sperber 1957; Goldstein 1995; Buch et al. 2008). The organic acids produced by PSMs include acetic acid, malic acid, oxalic acid, succinic acid, citric acid, gluconic acid, 2-ketogluconic acid, tartaric acid, and many more (Kalayu 2019). The secretion of these organic acids by the microbial cells brings about acidification of its surroundings. These organic acids produced are consequently changed into ionic forms liberating proton (H^+) which replaces the metal cations like Fe^{3+}, Al^{3+}, and Ca^{2+} from the insoluble phosphate complex and makes available the soluble phosphate for plant uptake; or sometimes carboxylic anions chelate cations and release phosphate anion (Rodriguez and Fraga 1999; Hwangbo et al. 2003; Chen et al. 2006; Lin et al. 2006; Park et al. 2009; Kumar and Rai 2015; Prasad et al. 2018). Among the various organic acids produced by PSMs, gluconic acid and keto-gluconic acid are considered as major ones for P solubilization by the lowering of pH in the rhizosphere. The pH of rhizosphere is also supposed to be lowered through production of proton/bicarbonate release and gaseous (CO_2/O_2) exchange. The bacterial strains which produce the above-mentioned as well as several other organic acids are reported to belong to *Pseudomonas* (Park et al. 2009), *Enterobacter* (Hwangbo et al. 2003; Kumar et al. 2014), and *Burkholderia* (Lin et al. 2006). Based on several studies where cloning and characterization of the genes involved in organic acids production has been carried out, it had been concluded that genes involved directly or indirectly in the synthesis of organic acid or regulation of the expression of genes responsible for organic acid synthesis are also responsible for inorganic phosphate solubilization (Rodriguez et al. 2006; Buch et al. 2010; Chhabra et al. 2013). Some researchers also believe that proton translocation ATPase play an important role in P mineralization as it helps in proton extrusion to the outer surface as well as proton exchange for cation uptake (Illmer and Schiner 1995). The organic acids produced by PSMs in the medium can be identified and measured by using high performance liquid chromatography technique (Park et al. 2009; Kumar and Rai 2015). Furthermore, the other mechanism of inorganic P solubilization takes place as a result of nitrogen assimilation (nitrate formation), evolution of carbon dioxide, and oxidation of sulfur. These processes lead to the formation of nitric acid, carbonic acid, and sulfuric acid (Sperber 1957). However, the efficiency and their impact on release of bound P in soils seem to be less than organic acid production. The concept of organic acid production and phosphate solubilization hardly have any correlation between the concentration of organic acid and amount of solubilized phosphate in the culture medium hence acidification could not be the sole mechanism of inorganic phosphate solubilization (Parks et al. 1990). Solubilization of calcium phosphate has been reported to occur even in the absence of organic acid (Illmer and Schiner 1992). Furthermore, siderophores and exopolysaccharide synthesized by PSMs bring out locked phosphate into soluble form mainly by charge-related interaction (Yi et al. 2008; Sharma et al. 2013). As can be seen from above interpretations the organic

Table 10.1 Principal organic acids by phosphate solubilizing microorganisms for phosphate solubilization

PSMs	Organic acid	References
Bacteria		
Arthrobacter sp.	Malonic acid, oxalic acid	Banik and Dey (1982)
Enterobacter intermedium	2-ketogluconic acid	Zaidi et al. (2009)
Azospirillum sp.	Citric acid, fumaric acid succinic acid, gluconic acid	Kalayu (2019)
Enterobacter ludwigii	Acetic acid, gluconic acid, succinic acid	Tahir et al. (2013)
Pseudomonas cepacia	Gluconic acid, 2-ketogluconic acid	Zaidi et al. (2009)
Bacillus firmus	Oxalic acid	Banik and Dey (1982)
Bacillus megaterium	Gluconic acid	Chen et al. (2006)
Pseudomonas fluorescence	Citric acid, malic acid, tartaric acid, gluconic acid	Zaidi et al. (2009)
Fungus		
Aspergillus flavus	Citric acid, gluconic acid, oxalic acid, succinic acid	Rashid et al. (2004)
Aspergillus foetidus	Citric acid, gluconic acid, oxalic acid, succinic acid, tartaric acid	Zaidi et al. (2009)
Aspergillus japonicus	Citric acid, gluconic acid, oxalic acid, succinic acid, tartaric acid	Zaidi et al. (2009)
Penicillium sp.	Citric acid, gluconic acid, glycolic acid, malic acid, oxalic acid, succinic acid	Sane and Mehta (2015)
Penicillium radicum	Gluconic acid	Fenice et al. (2000)
Penicillium rugulosum	Citric acid, gluconic acid	Reyes et al. (2002)

acid as well as chelating and reducing molecules produced by PSMs are the key factors responsible for inorganic phosphate solubilization and these organic acids are also utilized as an alternate source of energy by PSMs resulting in the improved biomass yield (Buch et al. 2010; Kumar and Rai 2015). Table 10.1 shows various organic acids produced different PSMs.

10.5.2 Organic Phosphate Solubilization

The process of solubilization of organic phosphate is also known as mineralization of organic phosphate. The mineralization of organic phosphate is performed by different types of enzymes mainly phosphatases, phytases, and phosphonatases.

10.5.3 Phosphatase

Phosphatase (Phosphohydrolase) is an enzyme that acts by hydrolyzing phosphoester and phosphoanhydride bonds of organic matter. Phosphatase enzymes are classified into two types on the basis of their optimum pH: alkaline phosphatase ($pH > 7$) and acidic phosphatase ($pH < 6$). Their predominance is determined by pH of soil; in acidic soil acid phosphatases are predominant while in neutral and alkaline soil alkaline phosphatases are predominant (Rodriguez and Fraga 1999; Sharma et al. 2013). Table 10.2 shows some of the PSMs reported to exhibit acid and alkaline phosphatase activities.

10.5.4 Phytase

Phytase is an enzyme which acts on inositol phosphate component of phytate and releases utilizable phosphate. Phytate is the major source of inositol phosphate and accounts for more than 50% of organic phosphate form present in the soil (Rodriguez et al. 2006; Prasad et al. 2018). Phytate is synthesized by microorganism, plant seeds, and pollen grains (Rodriguez and Fraga 1999). Primarily phytases were used to improve animal nutrition; nevertheless, the contemporary approach may be the use of phytase secreting PSMs to improve plant growth and development (Richardson and Simpson 2011). *Arabidopsis* plants genetically engineered with phytase gene from *Aspergillus niger* were capable to procure phosphate from phytate. The growth and P content of the plants were equivalent to those plants supplied with soluble phosphate (Richardson and Simpson 2011). Table 10.3 enlists several of the phytase producing PSMs.

Table 10.2 List showing PSMs with phosphatase activity

Enzyme	Microorganism	References
Acid phosphatase	*Emericella rugulosa*	Yadav and Tarafdar (2007b)
	Serratia marcescens	Hameeda et al. (2006)
	Chaetomium globosum	Hameeda et al. (2006)
	Serratia marcescens	Ryu et al. (2005)
	P. fluorescens	Ryu et al. (2005)
	Burkholderia cepacia	Unno et al. (2005)
	Pseudomonas sp.	Richardson et al. (2001)
	Enterobacter aerogenes	Thaller et al. (1995)
	Enterobacter cloacae	Thaller et al. (1995)
	Citrobacter freundii	Thaller et al. (1995)
Alkaline phosphatase	*Bacillus flexus*	Patel (2016)
	Bacillus megaterium	Priya et al. (2014)
	E. coli	Bhattacharjee et al. (2018)

Table 10.3 List showing PSMs with phytase activity

Enzyme	Microorganism	References
Phytase	*Discosia* sp.	Rahi et al. (2009)
	Rhizobacteria	Hariprasad and Niranjana (2009)
	Rhizobacteria	Patel et al. (2010)
	Serratia marcescens	Hameeda et al. (2006)
	Pseudomonas sp.	Hameeda et al. (2006)
	Emericella rugulosa	Yadav and Tarafdar (2007a)
	Bacillus mucilaginous	Li et al. (2007)

10.5.5 Phosphonatases

Phosphonatases and C-P lyases hydrolyze C-P bond of organo-phosphonates and release phosphate (Rodriguez et al. 2006; Prasad et al. 2018). However, phosphonatases are not the major contributors in soil due to limited availability of their substrates (Rodriguez et al. 2006).

10.6 Application of Phosphate Solubilizing Microorganisms

There are several very important applications of PSMs in the arena of agriculture and allied activities. The PSMs play considerable role in the various ways such as plant growth and development promoters, salinity tolerance, drought tolerance, heavy metal tolerance, soil health repair, and ecological restoration in the management of agriculture (Malla et al. 2004; Prasad et al. 2018; Chhabra 2019). A few such applications of PSMs are discussed below.

10.6.1 Phosphate Solubilizing Microorganisms as Plant Growth Promoters

There are many PSMs present in soil but their numbers are not enough to compete with other microorganisms commonly found in rhizosphere, therefore the amount of phosphate released by these PSMs are generally not in enough quantity which is required by plants for better growth and development. Therefore, inoculation with selected microorganisms at a much higher concentration than that normally found in soil is necessary to harness the benefit of their phosphate solubilizing ability for plant yield increment (Singh and Kapoor 1998; Peix et al. 2001; Bharadwaj et al. 2008; Babu et al. 2015). The effectiveness of PSMs under natural conditions depends on its ability to persist and proliferate in the soil. In general, the density or population of applied PSM decreases rapidly upon introduction into soil due to various environmental factors (Ho and Ko 1985). The various factors that affect the survival of the inoculant PSM include abiotic and biotic factors. Abiotic factors include features such as soil composition, physiological condition, temperature and soil moisture (Bashan et al. 1995; Prasad et al. 2018). The biotic factors such as competition, predation, and root growth are the ones that facilitate substrate availability to the inoculated microorganisms (Mendes et al. 2013). These PSMs enhance plant growth and yield by two mechanisms, i.e., direct and indirect mechanism. The direct mechanism of growth promotion involves making the availability of nutrient by phosphate solubilization, N-fixation, and production of phytohormones (Rodriguez and Fraga 1999; Sharma et al. 2013; Chhabra 2019). Indirect mechanism of growth promotion involves synthesis of antibiotics and siderophores which help in prevention of deleterious effects of pathogenic microorganism (Rodriguez and Fraga 1999; Sharma et al. 2013; Chhabra and Dowling 2017). Several researches have been conducted to evaluate the biofertilization ability of different PSMs to enhance crop productivity in different parts of the world. Growth and crop production of mung bean was increased with the inoculation of *Bacillus circulans* (Singh and Kapoor 1998), common bean with *Burkholderia cepacia* (Peix et al. 2001), potato with *Pseudomonas*, *Stenotrophomonas*, *Arthrobacter*, *Microbacterium*, and *Pantoea* (Bharadwaj et al. 2008; Babu et al. 2015). Enhanced production of peanut (Dey et al. 2004), chickpea (Zaidi et al. 2003), radish (Antoun et al. 1998), maize (Hameeda et al. 2008; Kaur and Reddy 2014), rice (Vasudevan et al. 2002), tomato (Ghosh et al. 2015), and sugarcane (Sundara et al. 2002) has also been demonstrated with the use of different PSMs as biofertilizers.

10.6.2 Phosphate Solubilizing Microorganisms in Ecological Restoration and Phosphorus Cycling

Degraded ecosystems are characterized by extremely low levels of soil nutrients including P (Li 2006). Restoring these ecosystems requires the recovery of soil P cycling (Huang et al. 2012). Microbes play an integral role in soil P cycling, as they mediate bioavailable soil P (Rodriguez and Fraga 1999; Richardson and Simpson 2011). A global meta-analysis of 173 terrestrial studies revealed that plant responses in terrestrial ecosystems to P addition were not significantly different from those to N addition (Elser et al. 2007). Despite such an observation, substantial variations in plant responses to P and/or N addition were found between sub-habitats (e.g., forest, grassland, tundra, and wetland) within terrestrial environments (Elser et al. 2007), indicating that whether the soil is more limited to P or N is dependent on the specific ecosystem considered. Further, another global meta-analysis of 50 terrestrial studies showed that the plant responses in terrestrial ecosystems to P addition were more pronounced under elevated than under ambient N, indicating that P limitation in terrestrial ecosystems will become more pronounced under increasing atmospheric N deposition in the future (Li et al. 2016). Therefore, mitigating terrestrial P

Table 10.4 List of PSMs with active role in ecological restoration

Microorganism	Process	References
Arthrobacter sp.	P cycling	Banik and Dey (1982)
Enterobacter intermedium	P cycling	Zaidi et al. (2009)
Azospirillum sp.	P cycling	Kalayu (2019)
Enterobacter ludwigii	P cycling	Tahir et al. (2013)
Pseudomonas cepacia	P cycling	Zaidi et al. (2009)
Bacillus firmus	P cycling	Banik and Dey (1982)
Bacillus megaterium	P cycling	Chen et al. (2006)
Pseudomonas fluorescence	P cycling	Zaidi et al. (2009)

limitation is increasingly recognized as a major priority in ecosystem management and restoration (Penuelas et al. 2013). A set of PSMs-derived enzymes, such as acid phosphatase, alkaline phosphatase, phytase, phosphonatase, and C-P lyase are able to release free orthophosphate ions from recalcitrant organic P forms (Rodriguez and Fraga 1999; Richardson and Simpson 2011) and a variety of organic acids, including citric acid, formic acid, gluconic acid, malic acid, oxalic acid, are involved in the microbial solubilization of recalcitrant inorganic P forms (Rodriguez and Fraga 1999; Richardson and Simpson 2011). Several of the PSMs involved in ecological restoration are listed in Table 10.4.

10.6.3 Phosphate Solubilizing Microorganisms in Sustainable Agriculture

Mostly the challenge of P deficiency in agriculture is addressed by the application of P fertilizers. However, the majority of the applied phosphatic fertilizer is not available to plants and the addition of inorganic fertilizers in excess of the amount that is commonly employed to overcome this effect can lead to environmental problems such as groundwater contamination and waterway eutrophication (Kang et al. 2011). It is therefore of great interest to investigate management strategies that are capable of improving phosphate fertilization efficiency, increase crop yields, and reduce environmental pollution caused by phosphate drainage from the soil. Soil microorganisms enhance plant nutrient acquisition. They are involved in a wide range of biological processes including the transformation of insoluble soil nutrients (Babalola and Glick 2012). Several PSMs are capable of solubilizing and mineralizing insoluble soil phosphorus for the growth of plants. In the natural environment, numerous PSMs in the soil and rhizosphere are effective at releasing phosphate from bound soil phosphate through solubilization and mineralization (Bhattacharyya and Jha 2012). Several salt-tolerant or halophilic soil microorganisms which also exhibit the ability to solubilize insoluble phosphate holds promises for facilitation and development of saline-alkali soil-based agriculture (Zhu et al. 2011). The inoculation of soil or crop with PSMs is therefore a promising strategy for the improvement of

Table 10.5 List of PSMs involved in promotion of sustainable agriculture

PSM	Test crop	Result	References
Aspergillus niger	Wheat	Improved growth	Xiao et al. (2013)
Serratia sp.	Wheat	Increased growth	Swarnalakshmi et al. (2013)
Aspergillus awamoriS29	Mung bean	Increased plant growth, total P content, and plant biomass	Jain et al. (2012)
Azotobacter chroococcum and *Bacillus subtilis*	Wheat	Enhanced productivity of wheat	Kumar et al. (2014)
P. favisporus TG1R2	Soybeans	Increased dry biomass	Fernandez et al. (2011)

plant absorption of phosphate and thereby reducing the use of chemical fertilizers that have a negative impact on the environment (Alori et al. 2012). Mobilization of soil inorganic phosphate and increasing its bioavailability for plant use by harnessing soil PSM promotes sustainable agriculture, improves the fertility of the soil, and hence increases crop productivity. Various PSMs which imparted positive effects on crop production and promoted sustainable agriculture are listed in Table 10.5.

10.6.4 Phosphate Solubilizing Microorganisms in Immobilization of Heavy Metals

Soil contaminated with heavy metals has become a severe problem in many parts of the world (Li et al. 2014). Heavy metals are naturally occurring ingredient of the earth's crust (Pan et al. 2016); however, there are various anthropogenic activities, like ore mining, e-waste recycling, and sewage irrigation that had greatly increased the concentrations of heavy metals in the soil. Exposure of human to soil-heavy metals mainly includes the leaching of heavy metals from soil into water and the consumption of edible plants grown in the contaminated soil (Cao et al. 2009). Addition of different kinds of phosphate-containing compound into contaminated soil to immobilize heavy metals (like Pb, Zn, Cu, and Cd) has been well documented because of the formation of highly insoluble metal–phosphate precipitates (Liang et al. 2014), especially Pb–phosphate minerals, pyromorphites [$Pb_5(PO_4)_3X$, where $X = Cl^-, OH^-, F^-$], which are the most thermodynamically stable and insoluble Pb minerals over a broad pH and EC range. The efficiency of phosphate addition-induced heavy metals immobilization depends on the solubility of both the heavy metals and the added phosphate (Park et al. 2011a). Although soluble phosphate compounds like sodium-, potassium-, and ammonium phosphates having high water solubility had been widely applied to remediate heavy metals contamination and had achieved high immobilization efficiency, they are relatively more expensive than insoluble phosphate compounds and are more prone to cause eutrophication (Park et al. 2011a, b). In soils, PSMs could produce organic acids and phosphate enzymes

Table 10.6 Heavy metals immobilizing PSMs

Microorganism	Heavy metals	References
Achromobacter xylosoxidans Ax10	Cu	Ma et al. (2009)
Azotobacter chroococcum HKN5, *Bacillus megaterium* HKP-1, *Bacillus mucilaginosus HKK-1*	Pb and Zn	Wu et al. (2006)
Bacillus sp. PSB10	Cr	Wani and khan (2010)
Bacillus subtilis SJ-101	Ni	Zaidi et al. (2006)
Pseudomonas sp. M6, *Pseudomonas jessenii* M15	Ni, Cu, and Zn	Rajkumar and Freitas (2008)

to enhance the solubilization of insoluble phosphate compounds (Chen et al. 2006), and hence, PSMs have been widely used as inoculants to increase soil available phosphate contents and thus act as a good heavy metal immobilizer (Rodriguez and Fraga 1999). Microbial-immobilized remediation technology refers to using soil microorganisms to immobilize heavy metals, causing heavy metals to precipitate or be adsorbed and fixed in the soil, reducing their absorption by plants (Han et al. 2018). PSMs are capable of producing siderophore which is a metal-binding ligand molecule and chelates with several heavy metals such as cadmium, nickel, lead, arsenic, and many others that can help in adsorbing or absorbing these heavy metals from the soil and minimize the toxic effect of heavy metal accumulation by immobilizing them (Ma et al. 2009). Heavy metal-immobilizing bacteria have been widely studied and applied as excellent heavy metal passivators. Heavy metal-immobilizing PSMs not only immobilize heavy metals, alter the existing state of heavy metals in soil, and reduce the absorption of heavy metals by crops, but also promote the growth of crops and improve the quality of crops (Zhao et al. 2019). PSMs increased the growth and heavy metal resistance of vegetables by producing indole acetic acid (IAA), siderophores, 1-aminocyclopropane-1-carboxylate deaminase, and arginine decarboxylase (Teng et al. 2019). Table 10.6 enlists some of the heavy metal immobilizing microorganisms. To cope with stress caused by heavy metals, microorganisms have evolved mechanisms to overcome toxicity, including metal reduction, cell permeability reduction, and extracellular isolation (Noisangiam et al. 2011).

10.7 Conclusion

The current overuse of synthetic phosphatic fertilizers poses greater threats to the environment and also creates soil nutrient imbalance. Therefore, the application of PSMs is an eco-friendly and economically viable and efficient approach for the utilization of fixed phosphate present in the soil. Application of PSMs on one hand reduces the agricultural input cost by curtailing the use of highly priced synthetic fertilizers and on the other gives a more organic and natural crop yield. These PSMs with their additional qualities like that of heavy metal and salinity tolerance hold great potential for future of environmental sustainability.

References

Alori E, Fawole O, Afolayan A (2012) Characterization of arbuscular mycorrhizal spores isolated from Southern Guinea Savanna of Nigeria. J Agric Sci 4:13–19

Antoun H, Beauchamp C, Goussard N et al (1998) Potential of Rhizobium and Brady rhizobium species as plant growth promoting rhizobacteria on non-legumes. Plant Soil 204:57–67

Babalola OO, Glick BR (2012) Indigenous African agriculture and plant associated microbes: current practice and future transgenic prospects. Sci Res Essays 7:2431–2439

Babu AN, Jogaiah S, Ito S et al (2015) Improvement of growth, fruit weight and early blight disease protection of tomato plants by rhizosphere bacteria is correlated with their beneficial traits and induced biosynthesis of antioxidant peroxidase and polyphenol oxidase. Plant Sci 231:62–73

Banik S, Dey BK (1982) Available phosphate content of an alluvial soil as influenced by inoculation of some isolated phosphate solubilizing microorganisms. Plant Soil 69:353–364

Bashan Y, Puente ME, Rodriquez MN et al (1995) Survival of *Azorhizobium brasilense* in the bulk soil and rhizosphere of 23 soil types. Appl Environ 61(5):1938–1945

Behera BC, Singdevsachan SK, Mishra RR et al (2014) Diversity, mechanism and biotechnology of phosphate solubilising microorganism in mangrove—a review. Biocatal Agric Biotechnol 3:97–110

Bharadwaj DP, Lundquist PO, Alstrom S (2008) Arbuscular mycorrhizal fungal spore associated bacteria affect mycorrhizal colonization, plant growth and potato pathogens. Soil Biol Biochem 40:2494–2501

Bhattacharjee M, Banerjee M, Mitra P (2018) Partial purification and characterization of periplasmic alkaline phosphatase from *E. coli* isolated from water sample. Biosci Biotechnol 9(1): 200–208

Bhattacharyya PN, Jha DK (2012) Plant growth-promoting rhizobacteria (PGPR): emergence in agriculture. World J Microbiol Biotechnol 28:1327–1350. https://doi.org/10.1007/s1127401109799

Buch A, Archana G, Kumar GN (2008) Metabolic channelling of glucose towards gluconate in phosphate-solubilizing *Pseudomonas aeruginosa* P4 under phosphorus deficiency. Res Microbiol 159:635–642

Buch A, Archana G, Kumar GN (2010) Heterologous expression of phosphoenol pyruvate carboxylase enhances the phosphate solubilizing ability of fluorescent pseudomonads by altering the glucose catabolism to improve biomass yield. Bioresour Technol 101:679–687

Cao XD, Wahbi A, Ma LN et al (2009) Immobilization of Zn, Cu, and Pb in contaminated soils using phosphate rock and phosphoric acid. J Hazard Mater 164:555–564

Chen YP, Rekha PD, Arun AB et al (2006) Phosphate solubilizing bacteria from subtropical soil and their tricalcium phosphate solubilizing abilities. Appl Soil Ecol 34:33–41

Chhabra S (2019) Phosphorous management in agroecosytems and role and relevance of microbes in environmental sustainability. In: Shah S, Ramanan VV, Prasad R (eds) Sustainable green technologies for environmental management. Springer, New York, pp 53–66. https://doi.org/10.1007/978-981-13-2772-8

Chhabra S, Dowling DN (2017) Endophyte promoted nutrient acquisition: phosphorous and iron. In: Doty S (ed) Functional importance of the plant microbiome: implications for agriculture, forestry and bioenergy. Springer, Cham. https://doi.org/10.1007/978331965897

Chhabra S, Brazil D, Morrissey J et al (2013) Characterization of mineral phosphate solubilization traits from a barley rhizosphere soil functional metagenome. Microbiol Open 2:717–724. https://doi.org/10.1002/mbo3.110

Dey R, Pal KK, Bhatt DM et al (2004) Growth promotion and yield enhancement of peanut (*Arachis hypogaea* L.) by application of plant growth promoting rhizobacteria. Microbiol Res 159:371–394

Duponnois R, Kisa M, Olenchette C (2006) Phosphate solubilizing potential of the nemato fungus Arthrobotrys oligospore. J Plant Nutr Soil Sci 169:280–282

Dutton VM, Evans CS (1996) Oxalate production by fungi; its role in pathogenicity and ecology in the soil environment. Can J Microbiol 42:881–895

Elser JJ, Bracken ME, Cleland EE et al (2007) Global analysis of nitrogen and phosphorus limitation of primary producers in freshwater, marine and terrestrial ecosystems. Ecol Lett 10: 1135–1142

Fenice M, Selbman L, Federici F et al (2000) Application of encapsulated Penicillium variabile P16 in solubilisation of rock phosphate. Bioresour Technol 73:157–162

Fernandez BL, Silvani V, Colombo R et al (2011) Pre-symbiotic and symbiotic interactions between Glomus intraradices and two Paenibacillus species isolated from AM propagules: in vitro and in vivo assays with soybean (AG043RG) as plant host. Soil Biol Biochem 43: 1866–1872

Foth H, Foth HD (1990) Soil chemistry. In: Foth HD (ed) Fundamentals of soil science, 8th edn. Wiley, New York, pp 164–185

Ghosh P, Rathinasabapathi B, Ma LQ (2015) Phosphorus solubilization and plant growth enhancement by arsenic-resistant bacteria. Chemosphere 134:1–6

Goldstein AH (1994) Involvement of the quino protein glucose dehydrogenase in the solubilization of exogenous phosphates by Gram-negative bacteria. In: Torriani-Gorini A, Yagil E, Silver S (eds) Phosphate in microorganisms: cellular and molecular biology. ASM Press, Washington, DC, pp 197–203

Goldstein AH (1995) Recent progress in understanding the molecular genetics and biochemistry of calcium phosphate solubilization by Gram-negative bacteria. Biol Agric Hortic 12:185–193

Goldstein AH, Rogers RD, Mead G (1993) Mining by microbe. Nat BioTechnol 11:1250–1212

Hameeda B, Rupela OP, Reddy G et al (2006) Application of plant growth-promoting bacteria associated with composts and macro fauna for growth promotion of Pearl millet (Pennisetum glaucum L.). Biol Fertil Soils 43:221–227

Hameeda B, Harini G, Rupela OP et al (2008) Growth promotion of maize by phosphate–solubilizing bacteria isolated from composts and macro fauna. Microbiol Res 163:234–242

Han H, Sheng XF, Hu JW et al (2018) Metal-immobilizing *Serratia liquefaciens* CL-1 and *Bacillus thuringiensis* X30 increase biomass and reduce heavy metal accumulation of radish under field conditions. Ecotoxicol Environ Saf 161:526–533

Hariprasad P, Niranjana SR (2009) Isolation and characterization of phosphate solubilizing rhizobacteria to improve plant health of tomato. Plant Soil 316:13–24

Hilda R, Fraga R (1999) Phosphate solubilizing bacteria and their role in plant growth promotion. Biotechnol Adv 17:319–359

Ho WC, Ko WH (1985) Effect of environmental edaphic factors. Soil Biol Biochem 17:167–170

Huang L, Baumgartl T, Mulligan D (2012) Is rhizosphere remediation sufficient for sustainable revegetation of mine tailings? Ann Bot 110:223–238

Hwangbo H, Park RD, Kim YW et al (2003) 2-Ketogluconic acid production and phosphate solubilization by Enterobacter intermedium. Curr Microbiol 47:87–92

Igual JM, Valverde A, Cervantes E et al (2001) Phosphate-solubilizing bacteria as inoculants for agriculture: use of updated molecular techniques in their study. Agronomie 21:561–568

Illmer P, Schiner F (1992) Solubilisation of inorganic phosphates by microorganisms isolated from forest soils. Soil Biol Biochem 24:257–263

Illmer P, Schiner F (1995) Solubilisation of inorganic calcium phosphates-solubilization mechanisms. Soil Biol Biochem 27(3):257–263

Jain R, Saxena J, Sharma V (2012) Effect of phosphate-solubilizing fungi *Aspergillus awamori* S29 on mung bean (*Vigna radiata* cv. RMG 492) growth. Folia Microbiol 57:533–541. https://doi.org/10.1007/s1222301201679

Kalayu G (2019) Phosphate solubilizing microorganisms: promising approach as bio fertilizers. Int J Agron 10:1–7

Kang J, Amoozegar A, Hesterberg D et al (2011) Phosphorus leaching in a sandy soil as affected by organic and in composted cattle manure. Geoderma 161:194–201. https://doi.org/10.1016/j.geoderma.201012019

Kaur G, Reddy MS (2014) Influence of P-solubilizing bacteria on crop yield and soil fertility at multi locational sites. Eur J Soil Biol 61:35–40

Khan MS, Zaidi A, Wani PA (2007) Role of phosphate-solubilizing microorganisms in sustainable agriculture—a review. Agron Sustain Dev 27:29–43

Kim KY, Jordan D, Mc Donald GA (1998) Effect of phosphate solubilizing bacteria and vesicular-arbuscular mycorrhizae on tomato growth and soil microbial activity. Biol Fertil Soils 26:79–87

Kucey RMN, Janzen H, Leggett M (1989) Microbially mediated increases in plant available phosphorus. Adv Agron 42:199–228

Kumar A, Rai LC (2015) Proteomic and biochemical basis for enhanced growth yield of *Enterobacter* sp. LCR1 on insoluble phosphate medium. Microbiol Res 170:195–204

Kumar S, Bauddh K, Barman SC et al (2014) Amendments of microbial bio fertilizers and organic substances reduces requirement of urea and DAP with enhanced nutrient availability and productivity of wheat (*Triticum aestivum* L.). Ecol Eng 71:432–437. https://doi.org/10.1016/j.ecoleng.201407007

Li MS (2006) Ecological restoration of mineland with particular reference to the metalliferous mine wasteland in China: a review of research and practice. Sci Total Environ 357:38–53

Li J, Xu L, Yang F (2007) Expression and characterization of recombinant thermostable alkaline phosphatase from a novel thermophilic bacterium *Thermus thermophilus* XM. Acta Biochim Biophys Sin 39:844–850

Li ZY, Ma ZW, van der Kuijp TJ et al (2014) A review of soil heavy metal pollution from mines in China: pollution and health risk assessment. Sci Total Environ 468-469:843–853

Li Y, Niu S, Yu G (2016) Aggravated phosphorus limitation on biomass production under increasing nitrogen loading. A meta-analysis. Glob Chang Biol 22:934–943

Liang Y, Cao XD, Zhao L, Arellano E (2014) Biochar and phosphate induced immobilization of heavy metals in contaminated soil and water: implication on simultaneous remediation of contaminated soil and groundwater. Environ Sci Pollut Res 21:4665–4674

Lin TF, Huang HI, Shen FT et al (2006) The protons of gluconic acid are the major factor responsible for the dissolution of tricalcium phosphate by *Burkholderia cepacia* CC-A174. Bioresour Technol 97:957–960

Ma Y, Rajkumar M, Freitas H (2009) Improvement of plant growth and nickel uptake by nickel resistant-plant-growth promoting bacteria. J Hazard Mater 166(2):1154–1161

Malla R, Prasad R, Kumari R, Giang PH, Pokharel U, Oelmueller R, Varma A (2004) Phosphorus solubilizing symbiotic fungus *Piriformospora indica*. Endocytobiosis Cell Res 15(2):579–600

Mendes R, Garbeva P, Raaijmakers JM (2013) The rhizosphere microbiome: significance of plant beneficial, plant pathogenic, and human pathogenic microorganisms. FEMS Microbiol Rev 37(5):634–663. https://doi.org/10.1111/1574697612028

Mikanova O, Novakova J (2002) Evaluation of the P-solubilizing activity of soil microorganisms and its sensitivity to soluble phosphate. Rostl Vyroba 48:397–400

Nahas E (1996) Factors determining rock phosphate solubilization by microorganism isolated from soil. World J Microbiol Biotechnol 12:567–572

Nath M, Bhatt D, Bhatt MD, Prasad R, Tuteja N (2018) Microbe-mediated enhancement of nitrogen and phosphorus content for crop improvement. In: Prasad R, Gill SS, Tuteja N (eds) Crop improvement through microbial biotechnology. Elsevier, pp 291–301

Ngalimat MS, Hata EM, Zulperi D et al (2021) Plant growth-promoting bacteria as an emerging tool to manage bacterial rice pathogens. Microorganisms 9(682):1–23. https://doi.org/10.3390/microorganisms9040682

Noisangiam R, Nuntagij A, Pongsilp N et al (2011) Heavy metal tolerant Metalliresistens boonkerdii gen. nov, sp. nov, a new genus in the family Brady rhizobiaceae isolated from soil in Thailand. Syst Appl Microbiol 34:166–168

Ohtake H, Kato J, Kuroda A et al (1996) Chemo lactic signal transduction in *Pseudomonas aeruginosa*. In: Nakazawa T, Furukawa K, Hass D, Silver S (eds) Molecular biology and biotechnology. American Society of Microbiology, Washington DC, pp 188–194

Pan LB, Ma J, Wang XL et al (2016) Heavy metals in soils from a typical county in Shanxi Province, China: levels, sources and spatial distribution. Chemosphere 148:248–254

Park KH, Lee CY, Son HJ (2009) Mechanism of insoluble phosphate solubilization by Pseudomonas fluorescens RAF15 isolated from ginseng rhizosphere and its plant growth-promoting activities. Lett Appl Microbiol 49:222–228

Park JH, Bolan N, Megharaj M et al (2011a) Concomitant rock phosphate dissolution and lead immobilization by phosphate solubilizing bacteria (*Enterobacter* sp.). J Environ Manag 92: 1115–1120

Park JH, Bolan N, Megharaj M, Naidu R (2011b) Isolation of phosphate solubilizing bacteria and their potential for lead immobilization in soil. J Hazard Mater 185:829–836

Parks EJ, Olson GJ, Brickman FE (1990) Characterization of high-performance liquid chromatography (HPLC) of the solubilization of phosphorus in iron one by a fungus. Indian J Microbiol 5: 183–190

Patel FR (2016) Purification and characterization of alkaline phosphatase from a halotolerant facultative alkaliphile Bacillus flexus FPB17. Int J Pharma Sci Res 7(6):2641–2647

Patel KJ, Singh AK et al (2010) Organic acid producing phytate mineralizing rhizobacteria and their effect on growth of pigeon pea (*Cajanus cajan*). Appl Soil Ecol 44(3):252–261

Paul EA, Clark FE (1988) Soil microbiology and biochemistry. Academic, San Diego, CA

Peix A, Mateos PF, Rodriguez-Barrueco C et al (2001) Growth promotion of common bean (Phaseolus vulgaris L.) by a strain of *Burkholderia cepacia* under growth chamber conditions. Soil Biol Biochem 33:1927–1935

Penuelas J, Poulter B, Sardans J et al (2013) Human-induced nitrogen-phosphorus imbalances alter natural and managed ecosystems across the globe. Nat Commun 4:2934

Ponmurugan P, Gopi C (2006) In vitro production of growth regulators and phosphatase activity by phosphate solubilizing bacteria. Afr J Biotechnol 5:348–350

Prasad V, Chaudhary S, Singh A (2018) Improving phosphorus fertility in soil through microbial mediators. Int J Plant Env 4(2):74–80

Priya D, Mahesh Kumar DJ, Kalaichelvan PT (2014) Optimization and production of extracellular alkaline phosphatase from Bacillus megaterium. Int J Chem Tech Res 6(9):4251–4258

Raghothama KG (1999) Phosphate acquisition. Annu Rev Plant Physiol Plant Mol Biol 50:665–693

Raghu K, Mac Rae IC (1966) Occurrence of phosphate-dissolving microorganisms in the rhizosphere of rice plants and in submerged soils. J Appl Microbiol 29:582–586

Rahi P, Vyas P, Sharma S et al (2009) Plant growth promoting potential of the fungus Discosia sp. FIHB571 from tea rhizosphere tested on chickpea, maize and pea. Indian. J Microbiol 49(2): 128–133

Rajkumar M, Freitas H (2008) Influence of metal resistant-plant growth-promoting bacteria on the growth of Ricinus communis in soil contaminated with heavy metals. Chemosphere 71(5): 834–842. https://doi.org/10.1016/j.chemosphere.200711038

Rashid M, Khalil S, Ayud N (2004) Organic acids production and phosphate solubilization by phosphate solubilizing microorganisms (PSM) under in vitro conditions. Pakistan J Biol Sci 7: 187–196

Reyes I, Bernier L, Antoun H (2002) Rock phosphate solubilisation and colonization of maize rhizosphere by wild and genetically modified strains of *Penicillium rugulosum*. Microb Ecol 44: 39–48

Richardson AE (2001) Prospects for using soil microorganisms to improve the acquisition of phosphorus by plants. Aust J Plant Physiol 28:897–906

Richardson AE, Simpson RJ (2011) Soil microorganisms mediating phosphorus availability. Plant Physiol 156:989–996

Richardson AE, Hadobas PA, Hayes JE et al (2001) Utilization of phosphorus by pasture plants supplied with myo-inositol hexaphosphate is enhanced by the presence of soil micro-organisms. Plant Soil 229:47–56

Rodriguez H, Fraga R (1999) Phosphate solubilizing bacteria and their role in plant growth promotion. Biotechnol Adv 17:319–339

Rodriguez H, Fraga R, Gonzalez T et al (2006) Genetics of phosphate solubilisation and its potential applications for improving plant growth-promoting bacteria. Plant Soil 287:15–21

Ryu CM, Hu CH, Locy RD et al (2005) Study of mechanisms for plant growth promotion elicited by rhizobacteria in Arabidopsis thaliana. Plant Soil 268:285–292

Sane SA, Mehta SK (2015) Isolation and evaluation of rock phosphate solubilizing fungi as potential biofertilizer. J Fertil Pestic 6:1–6

Sharma SB, Sayyed RZ, Trivedi MH (2013) Phosphate solubilizing microbes: sustainable approach for managing phosphorus deficiency in agricultural soils, vol 2. Springer, Cham, pp 587–600. https://doi.org/10.1186/2193-1801-2-587

Singh S, Kapoor KK (1998) Effects of inoculation of phosphate-solubilizing microorganisms and an arbuscular mycorrhizal fungus on mung bean grown under natural soil conditions. Mycorrhiza 7:249–253

Sperber JI (1957) Solution of mineral phosphates by soil bacteria. Nature 180:994–995

Sundara B, Natarajan V, Hari K (2002) Influence of phosphorus solubilizing bacteria on the changes in soil available phosphorus and sugarcane and sugar yields. Field Crops Res 77:43–49

Surange S, Wollum AG, Kumar N et al (1995) Characterization of Rhizobium from root nodules of leguminous trees growing alkaline soils. Can J Microbiol 43:891–894

Swarnalakshmi K, Prasanna R, Kumar A et al (2013) Evaluating the influence of novel cyanobacterial biofilmed biofertilizers on soil fertility and plant nutrition in wheat. Eur J Soil Biol 169:123–133. https://doi.org/10.1007/s120100129967218661872

Tahir M, Mirza MS, Zaheer A et al (2013) Isolation and identification of phosphate solubilizer Azospirillum, Bacillus and Enterobacter strains by 16S rRNA sequence analysis and their effect on growth of wheat (Triticum aestivum L.). Aust J Crop Sci 7(9):1284–1292

Teng ZD, Shao W, Zhang KY et al (2019) Characterization of phosphate solubilizing bacteria isolated from heavy metal contaminated soils and their potential for lead immobilization. J Environ Manag 231:189–197

Thaller MC, Berlutti F, Schippa S et al (1995) Heterogeneous patterns of acid phosphatases containing low-molecular-mass Poli peptides in members of the family Enterobacteriaceae. Int J Syst Evol Microbiol 4:255–261

Unno Y, Okubo K, Wasaki J et al (2005) Plant growth promotion abilities and micro scale bacterial dynamics in the rhizosphere of Lupin analysed by phytate utilization ability. Environ Microbiol 7(3):396–404

Vasudevan P, Reddy MS, Kavitha S et al (2002) Role of biological preparations in enhancement of rice seedling growth and grain yield. Curr Sci 83:1140–1143

Vazquez P, Holguin G, Puente ME et al (2000) Phosphate solubilizing microorganism associated with the rhizosphere of mangroves in a semiarid coastal lagoon. Biol Fertil Soils 30:460–468

Wani PA, Khan MS (2010) Bacillus species enhance growth parameters of chickpea (Cicer arietinum L.) in chromium stressed soils. Food Chem Toxicol 48(11):3262–3267

White la MA (2000) Growth promotion of plants inoculated with phosphate solubilizing fungi. Adv Agron 69:99–151

Wu CH, Wood TK, Mulchandani A et al (2006) Engineering plant-microbe symbiosis for rhizoremediation of heavy metals. Appl Environ Microbiol 72(2):1129–1134

Xiao C, Zhang H, Fang Y et al (2013) Evaluation for rock phosphate solubilization in fermentation and soil–plant system using a stress-tolerant phosphate-solubilizing *Aspergillus niger* WHAK1. Appl Microbiol Biotechnol 169(1):123–133. https://doi.org/10.1007/s1201001299672

Yadav BK, Tarafdar JC (2007a) Availability of unavailable phosphate compounds as a phosphorus source for cluster bean (*Cyamopsis tetragonoloba* (L.) Taub.) through the activity of phosphatase and phytase produced by actinomycetes. J Ari Legu 4:110–116

Yadav BK, Tarafdar JC (2007b) Ability of *Emericella rugulosa* to mobilize unavailable P compounds during Pearl millet (*Pennisetum glaucum* (L.) R.Br.) crop under arid condition. Indian J Microbiol 47:57–63

Yi Y, Huang W, Ge Y (2008) Exopolysaccharide: a novel important factor in the microbial dissolution of tricalcium phosphate. World J Microbiol Biotechnol 24:1059–1065

Zaidi A, Khan MS, Amil M (2003) Effect of rhizospheric microorganisms on yield and nutrient uptake of chickpea (*Cicer arietinum* L.). Eur J Agron 19:15–21

Zaidi S, Usmani S, Singh BR et al (2006) Significance of Bacillus subtilis strain SJ-101 as a bioinoculant for concurrent plant growth promotion and nickel accumulation in Brassica juncea. Chemosphere 64(6):991–997

Zaidi A, Khan MS, Ahemad M et al (2009) Plant growth promotion by phosphate solubilizing bacteria. Acta Microbiol Immunol Hung 56:263–284

Zhao XQ, Wang M, Wang H et al (2019) Study on the remediation of Cd pollution by the biomineralization of urease-producing bacteria. Int J Environ Res Public Health 16:268–282

Zhu F, Qu L, Hong X et al (2011, 2011) Isolation and characterization of a phosphate solubilizing halophilic bacterium *Kushneria* sp. YCWA18 from Daqiao Saltern on the coast of yellow sea of China. Evid Based Complement Alternat Med:615032. https://doi.org/10.1155/2011/615032

Chapter 11
Bacillus and *Streptomyces* for Management of Biotic Stresses in Plants for Sustainable Agriculture

A. Manikandan, R. Anandham, I. Johnson, R. Krishnamoorthy, M. Senthilkumar, R. Raghu, N. O. Gopal, and P. K. Mukherjee

Abstract Plant pathogens cause various crop plant diseases and are considered one of the biotic stresses, accounting for 20–40% of economic losses. Chemical-based treatments are currently believed to be an effective and reliable agricultural management technique for disease control. Agrochemicals are highly effective and easy to use, yet they pose a danger to the ecosystem. The use of microbial agents to combat plant diseases is a good alternative to the use of chemicals, which are harmful to the environment and human health. Members of the *Bacillus* and *Streptomyces* genera are two of the most frequently used biocontrol agents to suppress plant diseases. Plant growth-promoting (PGP) characteristics are shared by both bacterial species. This chapter comprehensively reports the mechanisms used by *Bacillus* and *Streptomyces* in their behavior as biocontrol and PGP agents. Also discussed are the various commercial formulations made from these genera. The application of biocontrol agents made from viable microbial strains to the field bodes well for successful management of the disease for sustainable agriculture.

A. Manikandan · R. Anandham (✉) · R. Raghu · N. O. Gopal
Department of Agricultural Microbiology, Tamil Nadu Agricultural University (TNAU), Coimbatore, Tamil Nadu, India

I. Johnson
Department of Plant Pathology, Tamil Nadu Agricultural University (TNAU), Coimbatore, Tamil Nadu, India

R. Krishnamoorthy
Department of Crop Management, Vanavarayar Institute of Agriculture, Pollachi, Tamil Nadu, India

M. Senthilkumar
Agricultural College and Research Institute, Tamil Nadu Agricultural University (TNAU), Eachangkottai, Tamil Nadu, India

P. K. Mukherjee
Nuclear Agriculture and Biotechnology Division, Bhabha Atomic Research Centre, Trombay, Mumbai, India

S. Chhabra et al. (eds.), *Plant Microbiome for Plant Productivity and Sustainable Agriculture*, Microorganisms for Sustainability 37,
https://doi.org/10.1007/978-981-19-5029-2_11

Keywords *Bacillus* · *Streptomyces* · Gamma irradiation · PGPR · Plant defense enzyme · Disease control · Biotic stress · Sustainable agriculture

11.1 Introduction

11.1.1 General

Numerous biotic stresses that plants frequently experience are having an impact on crop growth and yields. The existence of numerous pathogens, pests, and parasites is what causes biotic stresses. Around 20–40% of the yield losses occur worldwide only through the diseases caused by several plant pathogenic fungi, bacteria, viruses, and various pests. Microorganisms could play an important role in adaptation strategies and increase of tolerance to biotic stresses in agricultural plants (Sobiczewski et al. 2017). Plant growth-promoting rhizobacteria (PGPR) mitigate most effectively the impact of biotic stresses on plants through the production of exopolysaccharides, biofilm formation, hydrolytic enzymes, lipopeptides, siderophore production of volatile organic compounds, degradation of the ethylene precursor ACC by bacterial ACC deaminase, and induction of systemic resistance in plants (Prasad et al. 2015). *Bacillus* and *Streptomyces* genera are widely used as the biological control agents due to their prevalence in the agro-ecological region's natural environment and a wide range of biocontrol mechanisms (Manikandan et al. 2022). In this chapter, the mode of actions, production of metabolites, and improvement of biocontrol mechanisms by gamma-induced mutations and effective formulations of these biocontrol agents are discussed.

11.2 Biotic Stress

Numerous biotic stresses that plants frequently experience are having an impact on crop development and yields. Plant infections and biotic stresses are caused by the abundance of pathogens, pests, and parasites. The major plant fungus is necrotrophic (killing the host by producing toxins, for example, *Sclerotinia sclerotiorum* and *Fusarium* spp.) and biotrophic (having a long-term relationship with the host for feeding, for example, *Basidiomycota* and *Ascomycota* groups). They can cause leaf spots, cankers, and vascular wilts in plants (Laluk and Mengiste 2010; Doughari 2015; Sobiczewski et al. 2017). The plant pathogenic bacteria also cause many diseases and exhibit various types of symptoms which include galls, overgrowths, soft rots, wilts, blights, leaf spots, and cankers. Nematodes also cause major soil-borne diseases. They feed plant parts and cause stunted growth, wilting, and nutrient deficiency (Lambert and Bekal 2002; Bernard et al. 2017; Osman et al. 2020). Likewise, viruses also cause chlorosis and stunted growth in plants (Pallas and García 2011). On the other hand, insects and mites can cause damage to plants by feeding and also act as a vector for various bacteria and viruses (Schumann and

D'arcy 2006). Plants have developed various immune systems to overcome biotic stress (Rejeb et al. 2014). All the plants are having first-line defense mechanisms to avoid pathogen and insect attacks, which also consist of physical barriers such as wax, trichomes, and cuticle layer. Plants are also able to produce metabolite compounds to protect from pathogens (Taiz and Zeiger 2006; Hammond-Kosack 2000). Furthermore, stress reduction can be improved by having a detailed understanding of the molecular mechanisms underpinning plant defense systems (Cheng et al. 2012; Wang et al. 2019; Daayf et al. 2012).

Around 20–40% of the yield losses occur worldwide only through the diseases caused by several plant pathogenic fungi, bacteria, viruses, and various pests (Cabral-Pinto et al. 2020). To overcome these pests and diseases, farmers are following various approaches to suppress the plant diseases. For control of plant pathogens, they majorly depend on chemicals which include fungicides, pesticides, and nematicides that arrest pests and plant pathogens. These chemicals effectively control plant diseases; however, the excessive use of these chemicals could lead the pathogen resistance against these chemicals and also harm the environment and human health (Tilman et al. 2002). For example, captan, a fungicide which is banned in Mexico because of its carcinogenic effect (González et al. 2018). On the other hand, DTT (dichlorodiphenyltrichloroethane), which is a synthetic pesticide effectively used to control plant pathogens, was found to cause endocrine disorders and genotoxicity in humans (Cohn et al. 2007). Hence, there is a need for the development of alternative plant pathogen control strategies. Currently, the plant pathologists focus their research on developing a new alternative that could eventually replace the usage of chemicals in plant disease control. Apart from this, cultural methods such as crop rotation and the system of planting in different plots might reduce the occurrence of pests and diseases; however, these are not completely effective (Sainju et al. 2016). Biocontrol agents as a promising solution to reduce the adverse effects on agricultural yield and quality caused by various diseases minimize the resistance of plant pathogens and reduce contamination of environments. This helps produce safe foods while also lowering agricultural production costs (Tena et al. 2015).

The first biological control agent was used in the early 1800s, to control different crop diseases by using living organisms or their byproducts (Badii and Abreu 2006). The significance of a broad group of rhizosphere microorganisms in the management of plant pathogenic organisms has been explored in recent years. These rhizosphere organisms act as the plant's front line of defense against plant pathogens, preventing their development in the roots (Tejera-Hernández et al. 2011). Various direct and indirect biocontrol methods have been observed in various biological control agents. *Bacillus* and *Streptomyces* genera are widely used for biocontrol purposes because of their abundance in nature in the agroecological region and owing to various biocontrol mechanisms shown in Fig. 11.1 (Jog et al. 2012; Tejera-Hernández et al. 2011).

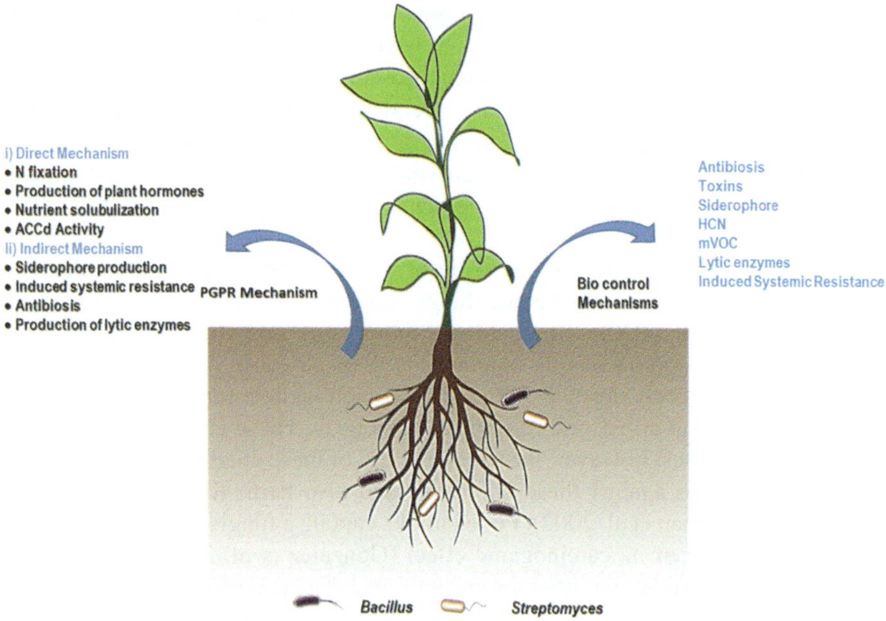

Fig. 11.1 PGPR and biocontrol mechanisms of *Bacillus* and *Streptomyces*

11.3 *Bacillus* and *Streptomyces*

Bacillus is a genus of the Gram-positive group, rod-shaped, and spore-forming bacterium belonging to the phylum *Firmicutes* which is found in mostly all types of environments including soil rhizosphere region. Cultivation-dependent method was conducted by Vargas-Ayala et al. (2000) and the results showed the occurrence of several isolates that were phenotypically and phylogenetically related species of *Bacillus subtilis* and *Bacillus cereus* with the range of 10^3 to 10^6 (CFU)/g of soil. *Bacillus* is considered one of the most important and widely used biocontrol agents against various groups of plant pathogens. It also produces numerous types of plant growth-promoting characters and various hydrolytic enzymes and lipopeptides which help to inhibit the plant pathogenic organisms and also improve the plant growth and health. It forms endospore under the adverse climatic condition to remain viable in all climatic conditions (Teixeira et al. 2010; Fira et al. 2018; Shafi et al. 2017; Verschuere et al. 2000; Cavaglieri et al. 2005).

Streptomyces is a group of Gram-positive, filamentous, aerobic, and non-motile bacterium placed under the phyla of *Actinobacteria*, which are usually persistent in soil; it accounts the ~10% of the total microbial load in the soil (Samac et al. 2003; Hayakawa et al. 2004; Law et al. 2017; Schrey and Tarkka 2008). *Streptomyces* is known for the production of a plethora of secondary metabolites including antibiotics; however, in recent years attention has been paid towards plant growth-promoting promoting (PGP) characteristics such as IAA production, phosphate solubilization, siderophore production, volatile organic compounds (VOC), and induced systematic responses (ISR) (Jog et al. 2012; Lehr et al. 2007, 2008). Many *Streptomyces* strains have been used for the control of various plant pathogens (Hanif et al. 2019; Hue et al. 2001; Inbar et al. 2005; Gopalakrishnan et al. 2014).

11.4 Antibiotics from *Bacillus* and *Streptomyces*

Bacillus species can produce numerous types of antimicrobial compounds. In their total genomes, around 4–5% is employed only for the synthesis of antimicrobial compounds (Stein 2005). Based on the biosynthesis nature, antimicrobial substances are classified into two categories. One is lipopeptides, which are produced non-ribosomally. Another one is bacteriocin, which is produced ribosomally (Arnison et al. 2013; Ongena and Jacques 2008). The antimicrobial compounds produced by the *Bacillus* sp. include bacillomycin, subtilin, bacilysin, mycobacillin, iturin, surfactin, and fengycin, which have both antifungal and antibacterial activities. Recently lipopeptides are receiving more attention due to their nature of wide antimicrobial ability, less toxic, easily degradable, and stable even in high temperatures (Meena and Kanwar 2015). Lipopeptides are cyclic, low molecular weight compounds, which are divided into three major groups, namely, fengycin, surfactin, and iturin (Kakinuma et al. 1969; Vanittanakom et al. 1986).

Fengycin is majorly produced by the *B. subtilis* and *B. amyloliquefaciens* (Steller and Vater 2000; Hanif et al. 2019). Fengycin helps in plant growth (Deleu et al. 2005) and inhibits various plant pathogens including *Fusarium oxysporum*, *F. oxysporum* Forma specialis *spinaciae*, *F. solani*, *F. solani* f. sp. *radicicola*, *F. verticillioides*, *F. graminearum*, and *Plasmodiophora*. Surfactin is produced by most of the *Bacillus* spp. and it is having antibacterial and antifungal properties (Hanif et al. 2019; Hue et al. 2001). Surfactin has arrested the growth of various fungal pathogens such as *F. verticillioides*, *F. verticillioides*, *F. oxysporum*, and *Colletotrichum gloeosporiodes* (Fan et al. 2005; Dunlap et al. 2011; Snook et al. 2009; Geissler et al. 2017). Iturin is also mostly produced by the *B. subtilis* and *B. amyloliquefaciens* species (Ongena and Jacques 2008; Nasir and Besson 2012). It also shows antifungal activity against the *Colletotrichum*, *Botrytis*, *F. graminearum*, *F. oxysporum*, *Rhizoctonia solani*, and *Penicillium* (Meena et al. 2014; Calvo et al. 2019; Zalila-Kolsi et al. 2016; Fujita and Yokota 2019). Surfactin and fengycin lipopeptides also induce the systemic resistance of plants (Cawoy et al. 2014). A dual plate assay was carried on *Bacillus* spp. against the *Macrophomina phaseolina* pathogen. Results showed the stunted hyphal growth in the contact point of the pathogen near the antagonistic organisms (Fig. 11.2).

Streptomyces is known for the production of versatile types of antibiotic compounds, including antibacterial, antifungal, antiviral, insecticides, and herbicides (Sharma et al. 2014). Oxytetracycline is produced from *Streptomyces rimosus* and showed antagonistic activity against fire blight (*Erwinia amylovora*). Streptomycin from *Streptomyces griseus* prevents plant diseases caused by *Pseudomonas tabaci*,

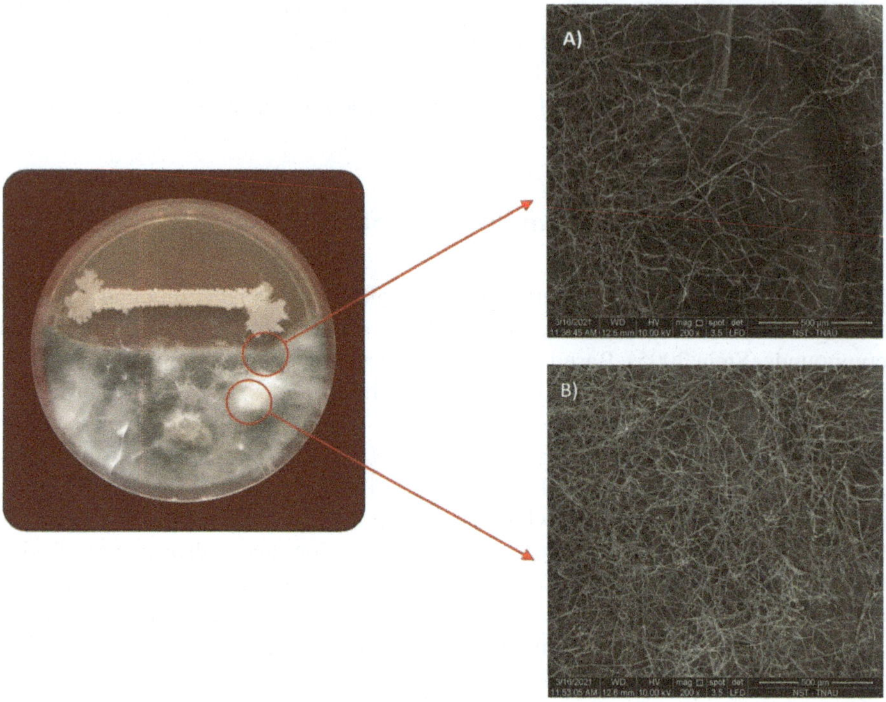

Fig. 11.2 Dual plate assay of *Bacillus* spp. against the *Macrophomina phaseolina* pathogen. (**a**) SEM image was taken in the normal hyphal region of pathogen. (**b**) SEM image was taken in the pathogen and antagonistic contact region

Xanthomonas oryzae, and *Xanthomonas citri*. Blasticidin-S of *Streptomyces griseochromogenes* inhibits the rice blast (*Pyricularia oryzae*). Kasugamycin produced by *Streptomyces kasugaensis* arrests the leaf spot in sugar beet and mycostop produced by *Streptomyces* sp. K61 inhibits the damping-off caused by *Rhizoctonia solani, Fusarium*, and *Phytophthora*. Actinovate produced by *S. lydicus* WYEC 108 inhibits the soil-borne diseases, namely, *Pythium*, *Fusarium*, *Phytophthora*, *Rhizoctonia*, and *Verticillium* (Aggarwal et al. 2016). Lipopeptides and antibiotics produced from different *Bacillus* spp. and *Streptomyces* spp. are listed in Table 11.1.

11.5 Volatile Organic Compounds (VOCs)

Volatile organic compounds (VOCs) are a group of odorous, low molecular weight (<300 Da) secondary metabolites produced by soil microorganisms. Many VOCs can control the plant pathogens and induce the systemic resistance in the plants (Kanchiswamy et al. 2015; Raza et al. 2016; Tahir Ha et al. 2017). Major chemical

Table 11.1 Lipopeptides and antibiotics produced from *Bacillus* spp. and *Streptomyces* spp.

Species	Stains name	Lipopeptide/ antibiotic	Pathogens	References
Bacillus amyloliquefaciens	YN201732	Bacillomycin D	*Fusarium solani*	Jiao et al. (2021)
B. velezensis	S499, FZB42 QST713	Fengycins Iturins	*Rhizoctonia variabilis*	Zihalirwa Kulimushi et al. (2017)
B. subtilis	98S	Fengycins	*Botrytis cinerea*, *F. oxysporum* *Pythium aphanidermatum*	Cawoy et al. (2015)
B. velezensis	SQR9	Fengycins Iturins Surfactins Bacillibactin	*Verticillium dahliae* *F. oxysporum*, *Phytophthora parasitica var. nicotianae* *Sclerotinia sclerotiorum*	Li et al. (2014)
B. subtilis	B9-5	Fengycins Surfactins	*R. stolonifer* *F. sambucinum* *Verticillium dahliae*	DeFilippi et al. (2018)
B. velezensis	QST713	Surfactins	*T. aggressivum* f. sp. *europaeum*	Pandin et al. (2019)
B. amyloliquefaciens	SB-1	Surfactin, Iturin A1, Bacillibactin, Bacillaene, Fengycin A, Macrolactin E	*F. oxysporum* *F. moniliforme*, *F. solani*, *Aspergillus flavus*, *A. niger*, and *Curvularia* sp.	Shahid et al. (2021)
B. subtilis	A-2	Surfactin Iturin A1 Bacillibactin Bacillaene Bacillomycin D Fengycin A Macrolactin E	*F. oxysporum* *F. moniliforme*, *F. solani*, *A. flavus*, *A. niger*, and *Curvularia* sp.	Shahid et al. (2021)
B. tequilensis	A-3	Surfactin, Iturin A1 Bacillibactin, Bacillaene	*F. oxysporum*, *F. moniliforme*, *F. solani*, *A. flavus*, *A. niger*, and *Curvularia* sp.	Shahid et al. (2021)
B. amyloliquefaciens	A3	Surfactins, iturins, and fengyins	*Dickeya dadantii*	Hossain et al. (2020)
B. coagulans	ATCC 7050	Lactosporin	*Micrococcus luteus* and *Listeria monocytogenes*	Riazi et al. (2009)
B. tequilensis	*A-3*	Iturins, fengycins, macrolactins, bacillomycin-D	*F. moniliforme*, *F. solani*, *A. flavus*, *A. niger*, and *Curvularia* sp.	Shahid et al. (2021)

(continued)

Table 11.1 (continued)

Species	Stains name	Lipopeptide/ antibiotic	Pathogens	References
Streptomyces griseochromogenes	–	Blasticidin-S	*Pyricularia oryzae*	Misato et al. (1959)
S. kasugaensis	–	Kasugamycin	*Venturia* spp., *Cercospora* spp., *M. grisea*	Hamada et al. (1965)
S. natalensis	–	Natamycin	*F. oxysporum*	Copping and Duke (2007)
S. rimosus	–	Oxytetracycline	*Erwinia amylovora*, *Xanthomonas* sp.	Copping and Duke (2007)
S. cacoai var. asoensis	–	Polyoxin B	*Sphaerotheca* spp. *Botrytis cinerea* *Sclerotinia sclerotiorum* *Corynespora melonis* *Alternaria alternata*	Copping and Duke (2007)
S. griseus	–	Streptomycin	*Xanthomonas oryzae* *X. citri* *Pseudomonas tabaci* *P. lachrymans*	Copping and Duke (2007)
S. hygroscopicus	–	Validamycin	*R. solani* and *Rhizoctonia*	Copping and Duke (2007)
S. scabies	–	Macrolactin A	*F. oxysporum*	Copping and Duke (2007)

classes of microbial volatile organic compounds (mVOCs) are alcohols, ketones, aromatic compounds, terpenes, organic acids, esters, aldehydes, sulphur compounds, alkanes, and nitrogen compounds (Schenkel et al. 2015; Schmidt et al. 2017). Approximately 2000 different types of VOCs with various synonyms have been reported from the 1000 bacterial species (Lemfack et al. 2018).

Numerous volatile organic compounds with antifungal and antimicrobial properties are produced by *Bacillus* species (Leelasuphakul et al. 2008; Chen et al. 2008). The first reported mVOCs 2,3-butanediol and acetoin produced by *B. subtilis* enhance the growth and systemic resistance in *Arabidopsis* (Ryu et al. 2003). Wu et al. (2019) reported the volatile compounds 2-nonanone and 2-heptanone produced by *Bacillus amyloliquefaciens*, which effectively control the *Fusarium* wilt of watermelon caused by *F. oxysporum* f. sp. *niveum*. The compounds acetoin and 2,3-butanediol help plant growth. *Bacillus subtilis* CF-3 produced 62 potential VOCs and among these benzothiazoles, benzoic acid, 3-methylbutanal, 2,4-di-*tert*-butylphenol, and 1-octanol showed the antagonism towards the *Macrophomina fructicola* and *Clostridium gloeosporioides* (Gao et al. 2018). Benzenoids class of

VOCs such as 1-butanol, 3-methyl-, hexanol, 2-ethyl decanal, 2-ethyl nonanal are produced by the *B. subtilis* and effective against the *F. oxysporum* f. sp. *lactucae*, *F. oxysporum* f. sp. *cubense*, *R. solani*, *S. sclerotiorum*. Similarly, sulphur-containing volatiles such as carbon disulphide, dimethyl trisulfide, and thiophene showed effectiveness against *Colletotrichum gloeosporioide*, *Alternaria brassicae*, *A. solani*, *Ascochyta citrullina*, *B. cinerea*, *Cercospora kikuchii*, *F. graminerum*, *F. oxysporum*.

Streptomyces genus produced nearly 10,000 secondary metabolites including volatile compounds (VOCs) (Berdy 2005; Hopwood 2007; van Wezel et al. 2009). Various groups of VOCs include alkanes, alkenes, alcohols, esters, ketones, genomics, and sulphur compounds produced by the *Actinomyces* (Palaniyandi et al. 2014). *Streptomyces albulus* NJZJSA2 produced over 13 VOCs, and among these 4-methoxystyrene, 2-pentylfuran, and anisole showed the antifungal activity against the *Sclerotinia sclerotiorum* and *Fusarium oxysporum* (Wu et al. 2015). The VOCs produced from *S. platensis* F-1 effectively control the fungal pathogens including *S. sclerotiorum*, *R. solani*, and *Botrytis cinerea* (Wan et al. 2008). Volatile compounds produced by *S. philanthi* RM-1-138 effectively control *R. solani* (Boukaew et al. 2013). *Streptomyces alboflavus* TD-1 produced nearly 35 compounds and among these benzenamine and dimethyl trisulfide exhibited the control of mycelial growth of *Aspergillus ochraceus* (Yang et al. 2019). Wang et al. (2013) reported VOCs from *Streptomyces alboflavus* and *Streptomyces philanthi* which inhibited the growth of *Fusarium moniliforme*, *Fusarium fujikuroi*, *Aspergillus flavus*, *Aspergillus ochraceus*, *Aspergillus niger*, and *Penicillium citrinum*.

11.6 Plant Growth-Promoting (PGP) Traits

Plant rhizosphere regions are populated with a wide range of plant growth-promoting rhizobacteria (PGPR) and are involved in the various direct and indirect mechanisms to help the plant development and also help in phytopathogen control. PGPR includes the genera of *Azospirillum*, *Bacillus*, *Pseudomonas*, *Rhizobium*, and *Burkholderia* (Glick et al. 1998; Solanki et al. 2017; Kalita et al. 2015). Furthermore, actinobacterial strains including *Streptomyces* spp., *Streptosporangium* sp., and *Micromonospora* sp. show the various PGPR and biocontrol activities (Gomes et al. 2000; Sousa et al. 2008; Franco-Correa et al. 2010). PGPR and biocontrol mechanisms of *Bacillus* and *Streptomyces* are shown in Fig. 11.1. Similarly, different PGPR and biocontrol traits of *Bacillus* and *Streptomyces* were given in Table 11.2.

Table 11.2 PGPR and biocontrol traits of *Bacillus* and *Streptomyces*

Species	Strain	Host plant	PGP character	Biocontrol of disease/pathogen	References
B. vallismortis	RHFS10	–	Siderophore production, IAA, Phosphate solubilization and hydrolytic activities	*M. phaseolina*	Castaldi et al. (2021)
B. subtilis	MF497446	Maize	Siderophore and an increased amount of catalase (CAT), peroxidase (PO), and polyphenol oxidase (PPO)	*Cephalosporium maydis*	Ghazy and El-Nahrawy (2021)
B. subtilis	PM32	Potato	Zinc and potassium solubilization, biological nitrogen fixation, ammonia, and siderophore production, and extracellular enzyme (cellulase, catalase, amylase, protease, pectinase, and chitinase)	*R. solani*	Mehmood et al. (2021)
B. velezensis	AK-0	Apple	Antifungal activity and lipopeptides genes such as *ituD* and *bacD* were expressed	*C. gloeosporioides*, *F. solani*, *A. niger*, and *C. musae*	Kim et al. (2021)
Bacillus sp.	PM31	Potato	Zinc and potassium solubilization, nitrogen fixation, siderophore, exopolysaccharides production, and extracellular enzyme activities	*F. solani*	Mehmood et al. (2021)
B. amyloliquefaciens	SB-1	–	IAA, protease, lipase, and cellulase	*F. moniliforme, F. solani*, *A. flavus*, *A. niger*, and *Curvularia* sp.	Shahid et al. (2021)
Bacillus sp.	BA97, BN17, BN20, and BR20	Common bean	Phosphate solubilization Siderophore, IAA	*M. phaseolina*	Bojórquez-Armenta et al. (2021)
B. siamensis	LZ88	Tobacco	Peroxidase (PO) and polyphenol oxidase (PPO)	*Alternaria alternata*	Xie et al. (2021)
B. amyloliquefaciens	MBI600	–	Siderophore	*P. syringae* pv. *tomato*	Dimopoulou et al. (2021)
B. ginsengihumi	M2.11	–	Siderophore and IAA	*Fusarium* sp.	Itkina et al. (2021)

B. xiamenensis PM14		Sugarcane	Induced systemic resistance	Red rot	Mehmood et al. (2021)
B. subtilis	MML2490	Turmeric	Production of lytic enzymes IAA, P solubilization, ammonia production	*R. solani*, *M. phaseolina*, *F. graminearum*	Chenniappan et al. (2019)
B. velezensis	XT1	Tomato and strawberry	Induced systemic resistance	*Botrytis cinerea*	Toral et al. (2020)
B. pumilus	HR10	Vegetables	Induced systemic resistance and extracellular polysaccharides (EPS)	*R. solani*	Zhu et al. (2020)
B. licheniformis	POT1	Potato	Activation of phenylpropanoid flavonoid biosynthesis	*Alfalfa mosaic virus*	Abdelkhalek et al. (2020)
Streptomyces pactum	A12	Cucurbits	Induced systemic resistance	*Didymella bryoniae*	Zhao et al. (2012)
S. albospinus	CT205	Cucumber	Chitinase, β-glucanase	*Fusarium*	Wang et al. (2016)
S. enissocaesilis	BYC	Sugar beet	K-solubilizing, auxin-producing, and inhibitory activities	*Fusarium* spp.	Aallam et al. (2021)
S. bellus	AYD	Sugar beet	K-solubilizing, auxin-producing, and inhibitory activities	*Fusarium* spp.	Aallam et al. (2021)
S. saprophyticus	DE2	Sugar beet	K-solubilizing, auxin-producing, and inhibitory activities	*Fusarium* spp.	Aallam et al. (2021)
S. tricolor	HM10	–	Siderophores, IAA, and P solubilization	*C. gloeosporioides*	Rehan et al. (2021)
Streptomyces spp.	KS62	Chickpea	Siderophore, protease and chitinase production	*F. oxysporum* f. sp. *ciceris*	Amini et al. (2016)
Streptomyces sp.	MR14	Tomato	IAA, siderophore, and ammonia production	*F. moniliforme*	Kaur et al. (2019)
S. griseorubens	E44G	Maize	Protease activity	*R. solani*	Abdelkhalek et al. (2020)

(continued)

Table 11.2 (continued)

Species	Strain	Host plant	PGP character	Biocontrol of disease/pathogen	References
Streptomyces spp.	A20	Rice	Siderophores, IAA, extracellular enzymes and solubilizing phosphate	*Burkholderia glumae*	Suárez-Moreno et al. (2019)
Streptomyces spp.	SNL2	Tomato	Hydrogen cyanide, IAA and siderophores, chitinase and β-1,3-glucanase activities, P solubilization	*F. oxysporum* f. sp. *radicis*	Goudjal et al. (2016)
Streptomyces spp.	PM3	Tomato	Induced system resistance, siderophore production, P solubilization	Soft rot in tomato	Dias et al. (2017)

11.6.1 Indole Acetic Acid (IAA)

IAA is one of the most studied phytohormones which belongs to classes of auxin. It helps in plant growth and root development. 80% of the soil rhizosphere microorganisms are able to produce IAA (Neubauer et al. 2000). Microbial IAA is the effective signaling molecule for the plant–microbe interaction and plant growth (Matsuda et al. 2018). *B. subtilis* showed the IAA activity in various crops which improved plant growth in rice, maize, and onion (Park et al. 2013; Ahmad et al. 2017; Reetha et al. 2014). Similarly, *B. amyloliquefaciens* displays the IAA production in various crops such as banana, cucumber, rice, potato, and brassica (Wang et al. 2016; Blom et al. 2012; Shao et al. 2015; Niazi et al. 2014). Manulis et al. (1994) reported the IAA synthesis and its pathways in *Streptomyces* sp. including *S. griseus*, *S. violaceus*, *S. exfoliates*, *S. coelicolor*, and *S. lividans*. El-Shanshoury (1991) reported that *Streptomyces* sp. induces the IAA production in plants under greenhouse conditions.

11.6.2 1-Aminocyclopropane-1-Carboxylic Acid Deaminase (ACCd) Activity

According to Brandl and Lindow (1997), the plant hormone ethylene aids in the growth and development of plants, but high levels can also cause defoliation and lower plant metabolisms. PGPR has the enzyme 1-aminocyclopropane-1-carboxylate (ACC deaminase) which controls ethylene production. It converts ACC into ammonia and α-ketobutyrate (Zahir et al. 2008). *Bacillus subtilis* Rhizo SF 48 exhibits the ACCd activity in tomatoes, significantly improving the plant growth even under different stress conditions (Gowtham et al. 2020). Similarly, Misra and Chauhan (2020) recorded ACCd activity in three *Bacillus* strains, namely, *B. subtilis* NBRI 28B, *B. subtilis* NBRI33N, and *B. safensis* NBRI 12M, which mitigate the salt stress in maize. Similarly, ACCd activity and exopolysaccharide-producing *B. cereus* effectively reduced the heat stress in tomatoes (Mukhtar et al. 2020). *Streptomyces* spp. show the various PGP activities including ACCd, improving the growth of maize plants (Nozari et al. 2021). In a study done by Yoolong et al. (2019), the *acdS* gene which is responsible for ACC deaminase was taken from *S. venezuelae* ATCC 10712 and cloned into *S. venezuelae*. The mutants show a significant improvement in salt stress alleviation in rice compared to wild strains. *Streptomyces* sp. strain PGPA39 exhibits the ACCd activity and improved the growth in *Arabidopsis* under in vitro condition (Palaniyandi et al. 2014).

11.6.3 Siderophore Production

Siderophore is a low molecular weight compound that chelates with iron (Fe^{3+}) from the environment (Jha and Saraf 2015). In the soil, iron is mostly existing in the ferric iron form and it remains unavailable to support microbial growth and development (Das et al. 2007). In order to prevent other microbes, notably harmful pathogens, from obtaining the iron in the rhizosphere, PGPR Produces siderophores (Freitas et al. 2015). *Bacillus subtilis* CAS15 can control the *Fusarium* wilt in pepper crops and also exhibits siderophore production (Yu et al. 2011). Antagonistic bacterium *Bacillus licheniformis* K11 that effectively inhibits *Phytophthora capsici* also produces the siderophore and cellulose (Jung et al. 2007). Shobha and Kumudini (2012) revealed that the *Bacillus* spp. produced siderophore also shows antagonistic activity against *Fusarium oxysporum*. Jog et al. (2012) reported that *Streptomyces rochei*, *S. carpinensis*, and *S. thermolilacinus* effectively improve the plant growth and produce the siderophore. Similarly, siderophore-producing *Streptomyces* was tested in the saline soil condition that improved the plant growth (Akram and Anjum 2011).

11.6.4 Induced Systematic Resistance (ISR)

PGPR can also control plant diseases through the indirect mechanism, the so-called induced systematic resistance (ISR). ISR improves defense-related enzymes or proteins against plant pathogens. ISR activates the plant defense response after the pathogen attack (Pieterse et al. 2014). Niu et al. (2011) reported that PGPR *Bacillus cereus* AR156 activates the ISR through the salicylic acid (SA) pathway in *Arabidopsis*. *B. subtilis* increased the plant defense enzymes such as PAL, PPO, and PO and chitinase activity against the various plant pathogens such as *F. oxysporum* (Akram and Anjum 2011), *Colletotrichum acutatum* (Wang et al. 2014), *Alternaria solani* (Latha et al. 2009), and *Xanthomonas oryzae* pv. *oryzae* (Udayashankar et al. 2011). *B. pumilus* activates the phenolic compounds in pea against the *F. oxysporum* f. sp. *pisi* (Benhamou et al. 1996). *Streptomyces* AcH 505 activates the ISR through SA and jasmonic acid pathway in oak trees against the powdery mildew (Kurth et al. 2014). Similarly, *Streptomyces* strains activate the SA/JA pathway in *Arabidopsis* to protect against fungal pathogens (Conn et al. 2008).

11.7 Mutation

Mutation is a commonly used tool for the enhancement of secondary metabolite and biocontrol ability in various beneficial microorganisms (Spadaro and Gullino 2005). Two types of mutation have been employed to create mutants: (1) physical and

(2) chemical methods. The chemicals such as methyl methanesulfonate (MMS), ethyl methanesulfonate (EMS), hydroxylamine (HA), and *N*-methyl-*N*′-nitro-*N*-nitrosoguanidine (MNNG) are widely used to create mutants. The physical mutation was achieved by ultraviolet and gamma rays (Baltz 2001). Through gamma, irradiation mutation was achieved in *Bacillus thuringiensis* NM101-19 to enhance the chitinase production. Mutants showed a 2.6-fold higher chitinase activity compared to wild-type strains. Also, mutants showed higher inhibitory activity against *Verticillium* sp., *Pythium* sp., and *Rhizopus stolonifer* (Gomaa and El-Mahdy 2018). Khaliq et al. (2009) studied to improve the tylosin production through UV and gamma irradiation in *Streptomyces fradiae* NRRL-2702. Similarly, gamma irradiation was performed in *Streptomyces rimosus* strain CN08 to improve the oxytetracycline antibiotic production, and the mutant *S. rimosus* γ-45 showed 19-fold increased antibiotic production (Lazim et al. 2010). Numerous studies have been conducted in *Trichoderma* to improve the biocontrol ability through gamma mutations against various phytopathogens including *F. oxysporum* f. sp. *radicis-cucumerinum* (Sahampoor et al. 2020), *Fusarium graminearum*, *Sclerotinia sclerotiorum*, *Rhizoctonia* (Soufi et al. 2021), *Macrophomina phaseolina*, and *Rhizoctonia solani* (Abbasi et al. 2016). Afsharmanesh et al. (2013) reported that enhanced biofilm production was achieved in *Bacillus subtilis* UTB1 by gamma irradiation and also exhibited improved biocontrol ability towards *Aspergillus flavus*. Similarly, Manikandan et al. (2022) reported that the enhanced biocontrol ability in mutants of *Bacillus subtilis* BRBac4, *Bacillus siamensis* BRBac21, and *Streptomyces cavourensis* BRAcB10 was achieved through gamma irradiation against the *Macrophomina phaseolina* and *Fusarium oxysporum* f. sp. *udum*.

Repetitive element sequence-based polymerase chain reaction (rep-PCR) is a genomic fingerprinting technique which is used for the identification of genetic variants in the strains. The DNA primers used in this technique are corresponding to naturally interspaced sequences in bacteria such as ERIC, BOX, and REP sites (Versalovic et al. 1991). Moreover, it has been shown that random mutagenesis such as gamma irradiation induces the rearrangement of chromosomal genome (Najafi et al. 2011). In this case, rep-PCR is an effective tool for the identification of diversity of mutants in comparison with wild and mutant strains (Kousar and Babu 2010). The rep-PCR fingerprinting technique is used to determine the genome diversity and polymorphic bands between the mutants and wild-type strains *Bacillus subtilis* UTB1 (Afsharmanesh et al. 2014), *Trichoderma harzianum* (Abbasi et al. 2016) (Sahampoor et al. 2020), *Bacillus thuringiensis* NM101-19 (Gomaa and El-Mahdy 2018), *Trichoderma aureoviride* (Soufi et al. 2021), *Bacillus subtilis* BRBac4, *Bacillus siamensis* BRBac21, and *Streptomyces cavourensis* BRAcB10 (Manikandan et al. 2022).

11.8 Formulations

The formulation is a commercially available product made from experimentally competent biocontrol strains (Prasad et al. 2019). It is a major challenge for all agro-based industries to make their product a success or failure. There are four types of formulations that are commercially available in the market: liquid, powder, granules, and slurry (Bashan et al. 2014; Bashan 1998). Furthermore, formulations should be made from low-cost materials such as peat, charcoal, vermiculite, lignite, and vermicompost (Catroux et al. 2001). All the biocontrol strains may not give the same results as that in the laboratory conditions when comes to the field conditions, which is a crucial factor in the development of bioinoculants for industrial purposes. The biocontrol agent efficiency is affected by the soil pH, nutrient availability, moisture level, and organic matter content of the soil. Differences in environmental conditions also influence the efficiency of biofertilizer strain under greenhouse/field conditions that are performed well in the laboratory conditions (Suprapta 2012).

Carrier is a transporter of bioinoculants from industry to field level (Bashan 1998). It has to maintain the right number of cell load and also give temporary protection in the field conditions both physically and nutritionally (Arora et al. 2010). The carriers are classified into four types based on their origins: (1) Solid-based inoculum includes soils (inorganic soil, peat, and coal), (2) plant-based materials (farmyard manure, compost), (3) inert materials (vermiculite, perlite, alginate beads, and ground rock phosphate), and (4) lyophilized bioinoculant cells, liquid carriers, and pellets containing spore or cells (Bashan 1998). Dry inoculants (powders) are produced from the soil and organic and inorganic carriers. Mostly dry inoculants are formed from the peat. Granules are made from calcite or silica grain, which is moistened with the adhesive material and mixed with powdered bioinoculants. Granules are coated with the targeted bioinoculant/s (Stephens and Rask 2000). The liquid formulation is based on broth culture, polymer-based substance, and mineral/organic oil (Malusá et al. 2012; Xavier et al. 2004). Liquid-based formulations are an easier way to handle and apply directly onto soil or seeds (Stephens and Rask 2000; Xavier et al. 2004). Dried synthetic carriers (lyophilized freeze-dried cells) stay dormant for a longer amount of time, although they may induce cell death during dehydration (Bashan 1998). A list of commercially available biocontrol agents of *Bacillus* and *Actinobacteria* is given in Table 11.3.

11.9 Prospects and Conclusions

The use of biocontrol agents as a key to modern agriculture is fundamental, based on its renewable, low-cost, and eco-friendly potential in ensuring sustainable agriculture. Importantly, the application of biocontrol as an integral component of agricultural practice in promoting plant yield has gained more attraction recently in meeting

Table 11.3 List of commercially available *Bacillus* and *Streptomyces* biofertilizers

Strains	Formulations	Company	Pathogen control
Bacillus subtilis	WP	STING T. Stanes and Company Limited	*Rhizoctonia, Botrytis, Oidiopsis, Leveillula, Pythium, Alternaria, Xanthomonas, Phakopsora, Sclerotium, Phytophthora, Peronospora, Sclerotinia*
Bacillus megaterium var. phosphaticum	WP	SYMBION-P	P solubilizers
Bacillus subtilis	Liquid	Samridhi™ Jaipur Bio Fertilizers	It acts against deuteromycetes, oomycetes, ascomycetes, and bacterial plant pathogens
Streptomyces K61	WP	Mycostop® AgBio, Inc., USA	*Fusarium, Phytophthora, Alternaria*, and *Pythium*
Streptomyces lydicus	Liquid	Bio-Nata Farmers Bio-fertilizers and Organics	*Fusarium, Rhizoctonia, Pythium, Phytophthora, Phytomatotricum, Aphanomyces, Botrytis, Sclerotinia, Postia, Verticillium* and *Geotrichum, Monosporascus, Armillaria, Alternaria solani, Pyricularia, Monilinia, Colletotrichum, Cladosporium*
Streptomyces griseus	Liquid	Green Marvel Greenlife Bio-tech Lab	Citrus canker, pomegranate bacterial blight

the demand for food production. Employing *Bacillus* and *Streptomyces* in the production of biocontrol agents for disease management has recorded success. Moreover, the new technology which involves enhancing the shelf life of these biocontrol agents with various organic and organic amendments in their growth medium and the development of newer formulations such as lyophilized cells, spore suspension, and metabolites are yet to be explored. However, an appropriate SCAR (sequence characterized amplified regions) marker has to be developed to track the introduced biocontrol strain formulation at the field level.

In conclusion, overdependence on the use of chemical fertilizers has encouraged industries to produce chemicals that are toxic to human health, thus causing ecological imbalances. These drawbacks are combined with a high cost of production which is beyond the purchasing capacity of many farmers in the developing world. The application of biocontrol agents is eco-friendly, relatively inexpensive, nontoxic, and possesses the significant potential to increase plant yield. Thus, the application of biocontrol agents made from viable microbial strains to the field bodes well for the successful management of the disease for sustainable agriculture.

Acknowledgement The authors are thankful to the Government of India, Board of Research in Nuclear Sciences (GoI-BRNS), Department of Atomic energy (DAE) for funding.

Funding Information This study was supported by the Government of India, the Board of Research in Nuclear Science (GoI-BRNS), and the Department of Atomic Energy (DAE. 35/14/17/2018-BRNS/10395).

Competing Interests The authors declare that they have no conflict of interest.

References

Aallam Y, Maliki BE, Dhiba D et al (2021) Multiple potential plant growth promotion activities of endemic Streptomyces spp. from Moroccan sugar beet fields with their inhibitory activities against *Fusarium* spp. Microorganisms 9:1429

Abbasi S, Safaie N, Shams-Bakhsh M, Shahbazi S (2016) Biocontrol activities of gamma induced mutants of *Trichoderma harzianum* against some soilborne fungal pathogens and their DNA fingerprinting. Iran J Biotechnol 14:260

Abdelkhalek A, Al-Askar AA, Behiry SI (2020) *Bacillus licheniformis* strain POT1 mediated polyphenol biosynthetic pathways genes activation and systemic resistance in potato plants against *Alfalfa* mosaic virus. Sci Rep 10:1–16

Afsharmanesh H, Ahmadzadeh M, Javan-Nikkhah M, Behboudi K (2014) Improvement in biocontrol activity of *Bacillus subtilis* UTB1 against Aspergillus flavus using gamma-irradiation. Crop Prot 60:83–92

Afsharmanesh H, Ahmadzadeh M, Majdabadi A, Motamedi F, Behboudi K, Javan-Nikkhah M (2013) Enhancement of biosurfactants and biofilm production after gamma irradiation-induced mutagenesis of Bacillus subtilis UTB1, a biocontrol agent of Aspergillus flavus. Arch Phytopathol PFL 46:1874–1884

Aggarwal N, Thind SK, Sharma S (2016) Role of secondary metabolites of actinomycetes in crop protection. In: Plant growth promoting actinobacteria. Springer, New York, pp 99–121

Ahmad Z, Wu J, Chen L, Dong W (2017) Isolated *Bacillus subtilis* strain 330-2 and its antagonistic genes identified by the removing PCR. Sci Rep 7:1–13

Akram W, Anjum T (2011) Quantitative changes in defense system of tomato induced by two strains of *Bacillus* against *Fusarium* wilt. Indian J Fund Appl Life Sci 1:7–13

Amini J, Agapoor Z, Ashengroph M (2016) Evaluation of *Streptomyces* spp. against *Fusarium oxysporum* f. sp. ciceris for the management of chickpea wilt. J Plant Protection Res 56(3): 257–264

Arnison PG, Bibb MJ, Bierbaum G et al (2013) Ribosomally synthesized and post-translationally modified peptide natural products: overview and recommendations for a universal nomenclature. Nat Prod Rep 30:108–160

Arora NK, Khare E, Maheshwari DK (2010) Plant growth promoting rhizobacteria: constraints in bioformulation, commercialization, and future strategies. In: Plant growth and health promoting bacteria. Springer, New York, pp 97–116

Badii MH, Abreu JL (2006) Control biológico una forma sustentable de control de plagas (biological control a sustainable way of pest control). Daena 1:82–89

Baltz RH (2001) Genetic methods and strategies for secondary metabolite yield improvement in actinomycetes. Antonie Van Leeuwenhoek 79:251–259

Bashan Y (1998) Inoculants of plant growth-promoting bacteria for use in agriculture. Plant Soil 378:1–33

Benhamou N, Kloepper JW, Quadt-Hallmann A, Tuzun S (1996) Induction of defenserelated ultrastructural modifications in pea root tissues inoculated with endophytic bacteria. Plant Physiol 112:919–929

Bashan Y, De-Bashan LE, Prabhu SR, Hernandez J-P (2014) Advances in plant growth-promoting bacterial inoculant technology: formulations and practical perspectives, pp 1998–1201

Berdy J (2005) Bioactive microbial metabolites. J Antibiot 58:1–26

Bernard GC, Egnin M, Bonsi C (2017) The impact of plant-parasitic nematodes on agriculture and methods of control. In: Nematology-concepts, diagnosis and control, vol 10. IntechOpen, London, p 192

Blom J, Rueckert C, Niu B, Wang Q, Borriss R (2012) The complete genome of *Bacillus amyloliquefaciens* subsp. *plantarum* CAU B946 contains a gene cluster for nonribosomal synthesis of iturin A. J Bacteriol 194(7):1845–1846

Bojórquez-Armenta YDJ, Mora-Romero GA, López-Meyer M et al (2021) Evaluation of *Bacillus* spp. isolates as potential biocontrol agents against charcoal rot caused by Macrophomina phaseolina on common bean. J Gen Plant Pathol 87:377–386

Boukaew S, Plubrukam A, Prasertsan P (2013) Effect of volatile substances from Streptomyces philanthi RM-1-138 on growth of Rhizoctonia solani on rice leaf. BioControl 58:471–482

Brandl MT, Lindow SE (1997) Environmental signals modulate the expression of an indole-3-acetic acid biosynthetic gene in Erwinia herbicola. Mol Plant-Microbe Interact 10:499–505

Cabral-Pinto MMS, Inácio M, Neves O et al (2020) Human health risk assessment due to agricultural activities and crop consumption in the surroundings of an industrial area. Exposure Health 12:629–640

Calvo H, Mendiara I, Arias E, Blanco D, Venturini ME (2019) The role of iturin A from *B. amyloliquefaciens* BUZ-14 in the inhibition of the most common postharvest fruit rots. Food Microbiol 82:62–69

Castaldi S, Petrillo C, Donadio G et al (2021) Plant growth promotion function of *Bacillus* sp strains isolated from salt-pan rhizosphere and their biocontrol potential against *Macrophomina phaseolina*. Int J Mol Sci 22:3324

Catroux G, Hartmann A, Revellin C (2001) Trends in rhizobial inoculant production and use. Plant Soil 230:21–30

Cavaglieri L, Orlando J, Etcheverry M (2005) In vitro influence of bacterial mixtures on *Fusarium verticillioides* growth and fumonisin B1 production: effect of seeds treatment on maize root colonization. Lett Appl Microbiol 41:390–396

Cawoy H, Debois D, Franzil L, De Pauw E, Thonart P, Ongena M (2015) Lipopeptides as main ingredients for inhibition of fungal phytopathogens by *Bacillus subtilis/amyloliquefaciens*. Microb Biotechnol 8:281–295

Cawoy H, Mariutto M, Henry G et al (2014) Plant defense stimulation by natural isolates of Bacillus depends on efficient surfactin production. Mol Plant-Microbe Interact 27:87–100

Chen H, Xiao X, Wang J, Wu L, Zheng Z, Yu Z (2008) Antagonistic effects of volatiles generated by *Bacillus subtilis* on spore germination and hyphal growth of the plant pathogen, Botrytis cinerea. Biotechnol Lett 30:919–923

Cheng X, Tian C-J, Li A-N, Qiu J-L (2012) Advances on molecular mechanisms of plant-pathogen interactions. Yi Chuan Hereditas 34:134–144

Chenniappan C, Narayanasamy M, Daniel GM et al (2019) Biocontrol efficiency of native plant growth promoting rhizobacteria against rhizome rot disease of turmeric. Biol Control 129:55–64

Cohn BA, Wolff MS, Cirillo PM, Sholtz RI (2007) DDT and breast cancer in young women: new data on the significance of age at exposure. Environ Health Perspect 115:1406–1414

Conn VM, Walker AR, Franco CMM (2008) Endophytic actinobacteria induce defense pathways in Arabidopsis thaliana. Mol Plant-Microbe Interact 21:208–218

Copping LG, Duke SO (2007) Natural products that have been used commercially as crop protection agents. Pest Manag Sci 63:524–554

Daayf F, El Hadrami A, El-Bebany AF et al (2012) Phenolic compounds in plant defense and pathogen counter-defense mechanisms. In: Recent advances in polyphenol research, vol 3, pp 191–208

Das A, Prasad R, Srivastava A, Giang PH, Bhatnagar K, Varma A (2007) Fungal siderophores: structure, functions and regulations. In: Varma A, Chincholkar SB (eds) Microbial siderophores, vol 12. Springer-Verlag, Berlin, pp 1–42

Defilippi S, Groulx E, Megalla M, Mohamed R, Avis TJ (2018) Fungal competitors affect production of antimicrobial lipopeptides in *Bacillus subtilis* strain B9–5. J Chem Ecol 44: 374–383

Deleu M, Paquot M, Nylander T (2005) Fengycin interaction with lipid monolayers at the air–aqueous interface implications for the effect of fengycin on biological membranes. J Colloid Interface Sci 283:358–365

Dias MP, Bastos MS, Xavier VB, Cassel E, Astarita LV, Santarém ER (2017) Plant growth and resistance promoted by *Streptomyces* spp. in tomato. Plant Physiol Biochem 118:479–493

Dimopoulou A, Theologidis I, Benaki D et al (2021) Direct antibiotic activity of Bacillibactin broadens the biocontrol range of *Bacillus amyloliquefaciens* MBI600. Msphere 6:e00376–e00321

Dunlap CA, Schisler DA, Price NP, Vaughn SF (2011) Cyclic lipopeptide profile of three *Bacillus subtilis* strains; antagonists of *Fusarium* head blight. J Microbiol 49:603

El-Shanshoury AR (1991) Biosynthesis of indole-3-acetic acid in Streptomyces atroolivacetls and its changes during spore germination and mycelial growth. Microbios 67:159–164

Fan H, Leve EW, Scullin C et al (2005) Surfactant-assisted synthesis of water-soluble and biocompatible semiconductor quantum dot micelles. Nano Lett 5:645–648

Fira D, Dimkić I, Berić T, Lozo J, Stanković S (2018) Biological control of plant pathogens by *Bacillus* species. J Biotechnol 285:44–55

Franco-Correa M, Quintana A, Duque C, Suarez C, Rodríguez MX, Barea J-M (2010) Evaluation of actinomycete strains for key traits related with plant growth promotion and mycorrhiza helping activities. Appl Soil Ecol 45:209–217

Freitas MA, Medeiros FHV, Carvalho SP et al (2015) Augmenting iron accumulation in cassava by the beneficial soil bacterium *Bacillus subtilis* (GBO3). Front Plant Sci 6:596

Fujita S, Yokota K (2019) Disease suppression by the cyclic lipopeptides iturin A and surfactin from *Bacillus* spp. against *Fusarium* wilt of lettuce. J Gen Plant Pathol 85:44–48

Gao H, Li P, Xu X, Zeng Q, Guan W (2018) Research on volatile organic compounds from *Bacillus subtilis* CF-3: biocontrol effects on fruit fungal pathogens and dynamic changes during fermentation. Front Microbiol 9:456

Geissler M, Oellig C, Moss K, Schwack W, Henkel M, Hausmann R (2017) High-performance thin-layer chromatography (HPTLC) for the simultaneous quantification of the cyclic lipopeptides Surfactin, Iturin A and Fengycin in culture samples of Bacillus species. J Chromatogr B 1044:214–224

Ghazy N, El-Nahrawy S (2021) Siderophore production by *Bacillus subtilis* MF497446 and Pseudomonas koreensis MG209738 and their efficacy in controlling Cephalosporium maydis in maize plant. Arch Microbiol 203:1195–1209

Glick BR, Penrose DM, Li J (1998) A model for the lowering of plant ethylene concentrations by plant growth-promoting bacteria. J Theor Biol 190:63–68

Gomaa EZ, El-Mahdy OM (2018) Improvement of chitinase production by *Bacillus thuringiensis* NM101-19 for antifungal biocontrol through physical mutation. Microbiology 87:472–485

Gomes RC, Semedo L, Soares RMA, Alviano CS, Linhares LF, RRR C (2000) Chitinolytic activity of actinomycetes from a cerrado soil and their potential in biocontrol. Lett Appl Microbiol 30: 146–150

González FB, Beltan Camacho V, Del Carmen Colin Olmos M, Vandame R (2018) Highly hazardous pesticides in Mexico. RAMPAM, México

Gopalakrishnan S, Vadlamudi S, Bandikinda P et al (2014) Evaluation of *Streptomyces* strains isolated from herbal vermicompost for their plant growth-promotion traits in rice. Microbiol Res 169:40–48

Goudjal Y, Zamoum M, Sabaou N, Mathieu F, Zitouni A (2016) Potential of endophytic *Streptomyces* spp. for biocontrol of Fusarium root rot disease and growth promotion of tomato seedlings. Biocontrol Sci Tech 26:1691–1705

Gowtham HG, Singh B, Murali M et al (2020) Induction of drought tolerance in tomato upon the application of ACC deaminase producing plant growth promoting rhizobacterium *Bacillus subtilis* Rhizo SF 48. Microbiol Res 234:126422

Hamada M, Hashimoto T, Takahashi T et al (1965) Antimicrobial activity of kasugamycin. J Antibiot A 18:104–106

Hammond-Kosack K, 2000. Responses to plant pathogens. In Buchanan B, Gruissem W, Jones R Biochemistry and molecular biology of plants, Wiley New York 1102–1109

Hanif A, Zhang F, Li P et al (2019) Fengycin produced by *Bacillus amyloliquefaciens* FZB42 inhibits Fusarium graminearum growth and mycotoxins biosynthesis. Toxins 11:295

Hayakawa M, Yoshida Y, Iimura Y (2004) Selective isolation of bioactive soil actinomycetes belonging to the *Streptomyces violaceusniger* phenotypic cluster. J Appl Microbiol 96:973–981

Hopwood DA (2007) *Streptomyces* in nature and medicine: the antibiotic makers. Oxford University Press

Hossain A, Islam Masum MM, Wu X et al (2020) Screening of *Bacillus* strains in biocontrol of pathogen Dickeya dadantii causing stem and root rot disease of sweet potato. Biocontrol Sci Tech 30:1180–1198

Hue N, Serani L, Laprévote O (2001) Structural investigation of cyclic peptidolipids from *Bacillus subtilis* by high-energy tandem mass spectrometry. Rapid Commun Mass Spectrom 15:203–209

Inbar E, Green SJ, Hadar Y, Minz D (2005) Competing factors of compost concentration and proximity to root affect the distribution of streptomycetes. Microb Ecol 50:73–81

Itkina DL, Suleimanova AD, Sharipova MR (2021) *Pantoea brenneri* AS3 and *Bacillus ginsengihumi* M2. 11 as potential biocontrol and plant growth-promoting agents. Microbiology 90:210–218

Doughari J (2015) An overview of plant immunity. J Plant Pathol Microbiol 6:11

Jha CK, Saraf M (2015) Plant growth promoting rhizobacteria (PGPR). J Agricult Res Dev 5:0108–0119

Jiao R, Cai Y, He P et al (2021) *Bacillus amyloliquefaciens* YN201732 produces lipopeptides with promising biocontrol activity against fungal pathogen Erysiphe cichoracearum. Front Cell Infect Microbiol 11:598999

Jog R, Nareshkumar G, Rajkumar S (2012) Plant growth promoting potential and soil enzyme production of the most abundant *Streptomyces* spp. from wheat rhizosphere. J Appl Microbiol 113:1154–1164

Jung H-K, Kim J-R, Woo S-M, Kim S-D (2007) Selection of the auxin, siderophore, and cellulase-producing PGPR, *Bacillus licheniformis* K11 and its plant growth promoting mechanisms. Appl Biol Chem 50:23–28

Kakinuma A, Hori M, Isono M, Tamura G, Arima K (1969) Determination of amino acid sequence in surfactin, a crystalline peptide lipid surfactant produced by Bacillus subtilis. Agric Biol Chem 33:971–972

Kalita M, Bharadwaz M, Dey T, *et al.*, 2015. Developing novel bacterial based bioformulation having PGPR properties for enhanced production of agricultural crops

Kanchiswamy CN, Malnoy M, Maffei ME (2015) Chemical diversity of microbial volatiles and their potential for plant growth and productivity. Front Plant Sci 6:151

Kaur T, Rani R, Manhas RK (2019) Biocontrol and plant growth promoting potential of phylogenetically new *Streptomyces* sp. MR14 of rhizospheric origin. AMB Express 9:1–14

Khaliq S, Akhtar K, Ghauri MA, Iqbal R, Khalid AM, Muddassar M (2009) Change in colony morphology and kinetics of tylosin production after UV and gamma irradiation mutagenesis of *Streptomyces fradiae* NRRL-2702. Microbiol Res 164:469–477

Kim YS, Lee Y, Cheon W et al (2021) Characterization of *bacillus velezensis* AK-0 as a biocontrol agent against apple bitter rot caused by *Colletotrichum gloeosporioides*. Sci Rep 11:1–14

Kousar M, Babu GS (2010) Mutagenic effectiveness and efficiency of gamma rays, sodium azide and their synergistic effects in urd bean (*Vigna mungo* L.). World J Agricult Sci 6:234–237

Kurth F, Mailänder S, Bönn M et al (2014) Streptomyces-induced resistance against oak powdery mildew involves host plant responses in defense, photosynthesis, and secondary metabolism pathways. Mol Plant-Microbe Interact 27:891–900

Laluk K, Mengiste T (2010) Necrotroph attacks on plants: wanton destruction or covert extortion? Arabidopsis Book 8:e0136

Lambert K, Bekal S (2002) Introduction to plant-parasitic nematodes. Plant Health Instr 10:1094–1218

Latha P, Anand T, Ragupathi N, Prakasam V, Samiyappan R (2009) Antimicrobial activity of plant extracts and induction of systemic resistance in tomato plants by mixtures of PGPR strains and Zimmu leaf extract against *Alternaria solani*. Biol Control 50:85–93

Law JW-F, Ser H-L, Khan TM et al (2017) The potential of *Streptomyces* as biocontrol agents against the rice blast fungus, Magnaporthe oryzae (Pyricularia oryzae). Front Microbiol 8:3

Lazim H, Slama N, Mankai H, Barkallah I, Limam F (2010) Enhancement of oxytetracycline production after gamma irradiation-induced mutagenesis of *Streptomyces rimosus* CN08 strain. World J Microbiol Biotechnol 26:1317–1322

Leelasuphakul W, Hemmanee P, Chuenchitt S (2008) Growth inhibitory properties of *Bacillus subtilis* strains and their metabolites against the green mold pathogen (*Penicillium digitatum* Sacc.) of citrus fruit. Postharvest Biol Technol 48:113–121

Lehr NA, Schrey SD, Bauer R, Hampp R, Tarkka MT (2007) Suppression of plant defence response by a mycorrhiza helper bacterium. New Phytol 174:892–903

Lehr NA, Schrey SD, Hampp R, Tarkka MT (2008) Root inoculation with a forest soil streptomycete leads to locally and systemically increased resistance against phytopathogens in Norway spruce. New Phytol 177:965–976

Lemfack MC, Gohlke B-O, Toguem SMT, Preissner S, Piechulla B, Preissner R (2018) mVOC 2.0: a database of microbial volatiles. Nucleic Acids Res 46:D1261–D12D5

Li B, Li Q, Xu Z, Zhang N, Shen Q, Zhang R (2014) Responses of beneficial *Bacillus amyloliquefaciens* SQR9 to different soilborne fungal pathogens through the alteration of antifungal compounds production. Front Microbiol 5:636

Malusá E, Sas-Paszt L, Ciesielska J (2012) Technologies for beneficial microorganisms inocula used as biofertilizers. Sci World J 2012:491206

Manulis S, Shafrir H, Epstein E, Lichter A, Barash I (1994) Biosynthesis of indole-3-acetic acid via the indole-3-acetamide pathway in *Streptomyces* spp. Microbiology 140:1045–1050

Matsuda R, Handayani ML, Sasaki H, Takechi K, Takano H, Takio S (2018) Production of indoleacetic acid by strains of the epiphytic bacteria *Neptunomonas* spp. isolated from the red alga Pyropia yezoensis and the seagrass Zostera marina. Arch Microbiol 200:255–265

Meena KR, Kanwar SS (2015) Lipopeptides as the antifungal and antibacterial agents: applications in food safety and therapeutics. Biomed Res Int 2015:473050

Meena KR, Saha D, Ramashish K (2014) Isolation and partial characterization of iturin like lipopeptides (a bio-control agent) from a *Bacillus subtilis* strain. Int J Curr Microbiol App Sci 3:121–126

Manikandan A, Johnson I, Jaivel N, Krishnamoorthy R, Kumar S, Murugaiyan R, Rajasekaran G, Olaganathan N, Mukherjee PK, Anandham R (2022) Gamma-induced mutants of *Bacillus* and *Streptomyces* display enhanced antagonistic activities and suppression of the root rot and wilt diseases in pulses. Biomolecular Concepts 13:103–118

Mehmood S, Khatoon Z, Ahmad AI et al (2021) *Bacillus* sp. PM31 harboring various plant growth-promoting activities regulates Fusarium dry rot and wilt tolerance in potato. Archiv Agron Soil Sci:1–15

Misato T, Ishii I, Asakawa M, Okimoto Y, Fukunaga K (1959) Antibiotics as protectant fungicides against rice blast (2) the therapeutic action of Blasticidin S. Ann Phytopathol Soc Japan 24:302–306

Misra S, Chauhan PS (2020) ACC deaminase-producing rhizosphere competent *Bacillus* spp. mitigate salt stress and promote Zea mays growth by modulating ethylene metabolism. 3Biotech 10:119

Mukhtar T, Smith D, Sultan T et al (2020) Mitigation of heat stress in *Solanum lycopersicum* L. by ACC-deaminase and exopolysaccharide producing *Bacillus cereus*: effects on biochemical profiling. Sustainability 12:2159

Najafi N, Hosseini R, Ahmadi AR (2011) Impact of gamma rays on the *Phaffia rhodozyma* genome revealed by RAPD-PCR. Iran J Microbiol 3:216

Nasir MN, Besson F (2012) Interactions of the antifungal mycosubtilin with ergosterol-containing interfacial monolayers. Biochim Biophys Acta 1818:1302–1308

Neubauer U, Nowack B, Furrer G, Schulin R (2000) Heavy metal sorption on clay minerals affected by the siderophore desferrioxamine B. Environ Sci Technol 34:2749–2755

Niazi A, Manzoor S, Asari S, Bejai S, Meijer J, Bongcam-Rudloff E (2014) Genome analysis of *Bacillus amyloliquefaciens* subsp. *plantarum* UCMB5113: a rhizobacterium that improves plant growth and stress management. PLoS One 9:e104651

Niu D-D, Liu H-X, Jiang C-H et al (2011) The plant growth–promoting rhizobacterium *Bacillus cereus* AR156 induces systemic resistance in *Arabidopsis thaliana* by simultaneously activating salicylate-and jasmonate/ethylene-dependent signaling pathways. Mol Plant-Microbe Interact 24:533–542

Nozari RM, Ortolan F, Astarita LV, Santarém ER (2021) Streptomyces spp. enhance vegetative growth of maize plants under saline stress. Braz J Microbiol 52(3):1371–1383

Ongena M, Jacques P (2008) *Bacillus* lipopeptides: versatile weapons for plant disease biocontrol. Trends Microbiol 16:115–125

Osman HA, Ameen HH, Mohamed M, Elkelany US (2020) Efficacy of integrated microorganisms in controlling root-knot nematode *Meloidogyne javanica* infecting peanut plants under field conditions. Bullet Natl Res Centre 44:1–10

Palaniyandi SA, Damodharan K, Yang SH, Suh JW (2014) *Streptomyces* sp. strain PGPA39 alleviates salt stress and promotes growth of ‘Micro Tom’ tomato plants. J Appl Microbiol 117:766–773

Pallas V, García JA (2011) How do plant viruses induce disease? Interactions and interference with host components. J Gen Virol 92:2691–2705

Pandin C, Darsonval M, Mayeur C, Le Coq D, Aymerich S, Briandet R (2019) Biofilm formation and synthesis of antimicrobial compounds by the biocontrol agent *Bacillus velezensis* QST713 in an Agaricus bisporus compost micromodel. Appl Environ Microbiol 85:e00327–e00319

Park K, Park J-W, Lee S-W, Balaraju K (2013) Disease suppression and growth promotion in cucumbers induced by integrating PGPR agent *Bacillus subtilis* strain B4 and chemical elicitor ASM. Crop Prot 54:199–205

Pieterse CMJ, Zamioudis C, Berendsen RL, Weller DM, Van Wees SCM, PaHM B (2014) Induced systemic resistance by beneficial microbes. Annu Rev Phytopathol 52:347–375

Prasad R, Kumar M, Varma A (2015) Role of PGPR in soil fertility and plant health. In: Egamberdieva D, Shrivastava S, Varma A (eds) Plant growth-promoting rhizobacteria (PGPR) and medicinal plants. Springer International Publishing, Cham, pp 247–260

Prasad R, Kumar V, Kumar M, Choudhary D (2019) Nanobiotechnology in bioformulations. Springer International Publishing. ISBN 978-3-030-17061-5. https://www.springer.com/gp/book/9783030170608

Raza W, Ling N, Yang L, Huang Q, Shen Q (2016) Response of tomato wilt pathogen *Ralstonia solanacearum* to the volatile organic compounds produced by a biocontrol strain Bacillus amyloliquefaciens SQR-9. Sci Rep 6:1–13

Reetha S, Bhuvaneswari G, Thamizhiniyan P, Mycin TR (2014) Isolation of indole acetic acid (IAA) producing rhizobacteria of pseudomonas fluorescens and *Bacillus subtilis* and enhance growth of onion (*Allium cepa* L.). Int J Curr Microbiol App Sci 3:568–574

Rehan M, Alsohim AS, Abidou H, Rasheed Z, Al Abdulmonem W (2021) Isolation, identification, biocontrol activity, and plant growth promoting capability of a superior *Streptomyces tricolor* strain HM10. Pol J Microbiol 70:245

Rejeb IB, Pastor V, Mauch-Mani B (2014) Plant responses to simultaneous biotic and abiotic stress: molecular mechanisms. Plan Theory 3:458–475

Riazi S, Wirawan RE, Badmaev V, Chikindas ML (2009) Characterization of lactosporin, a novel antimicrobial protein produced by *Bacillus coagulans* ATCC 7050. J Appl Microbiol 106: 1370–1377

Ryu C-M, Farag MA, Hu C-H et al (2003) Bacterial volatiles promote growth in Arabidopsis. Proc Natl Acad Sci 100:4927–4932

Sahampoor L, Zaker Tavallaie F, Fani SR, Shahbazi S (2020) In vitro efficiency of *Trichoderma harzianum* mutants in biocontrol of *Fusarium oxysporum* f. sp. radicis-cucumerinum. J Crop Protect 9:285–300

Sainju UM, Lenssen AW, Allen BL, Stevens WB, Jabro JD (2016) Nitrogen balance in response to dryland crop rotations and cultural practices. Agric Ecosyst Environ 233:25–32

Samac DA, Willert AM, Mcbride MJ, Kinkel LL (2003) Effects of antibiotic-producing *Streptomyces* on nodulation and leaf spot in alfalfa. Appl Soil Ecol 22:55–66

Schenkel D, Lemfack MC, Piechulla B, Splivallo R (2015) A meta-analysis approach for assessing the diversity and specificity of belowground root and microbial volatiles. Front Plant Sci 6:707

Schmidt R, De Jager V, Zühlke D et al (2017) Fungal volatile compounds induce production of the secondary metabolite Sodorifen in Serratia plymuthica PRI-2C. Sci Rep 7:1–14

Schrey SD, Tarkka MT (2008) Friends and foes: *Streptomycetes* as modulators of plant disease and symbiosis. Antonie Van Leeuwenhoek 94:11–19

Schumann GL, D'arcy CJ (2006) Essential plant pathology. American Phytopathological Society (APS Press), St Paul

Shafi J, Tian H, Ji M (2017) *Bacillus* species as versatile weapons for plant pathogens: a review. Biotechnol Biotechnol Equip 31:446–459

Shahid I, Han J, Hanooq S, Malik KA, Borchers CH, Mehnaz S (2021) Profiling of metabolites of *Bacillus* spp. and their application in sustainable plant growth promotion and biocontrol. Front Sustain Food Syst 5:37

Shao J, Li S, Zhang N et al (2015) Analysis and cloning of the synthetic pathway of the phytohormone indole-3-acetic acid in the plant-beneficial *Bacillus amyloliquefaciens* SQR9. Microb Cell Factories 14:1–13

Sharma M, Dangi P, Choudhary M (2014) *Actinomycetes*: source, identification, and their applications. Int J Curr Microbiol App Sci 3:801–832

Shobha G, Kumudini BS (2012) Antagonistic effect of the newly isolated PGPR *Bacillus* spp. on *Fusarium oxysporum*. Int J Appl Sci Eng Res 1:463–474

Snook ME, Mitchell T, Hinton DM, Bacon CW (2009) Isolation and characterization of Leu7-surfactin from the endophytic bacterium *Bacillus mojavensis* RRC 101, a biocontrol agent for *Fusarium verticillioides*. J Agric Food Chem 57:4287–4292

Sobiczewski P, Iakimova ET, Mikiciński A, Węgrzynowicz-Lesiak E, Dyki B (2017) Necrotrophic behaviour of *Erwinia amylovora* in apple and tobacco leaf tissue. Plant Pathol 66:842–855

Solanki MK, Wang Z, Wang F-Y et al (2017) Intercropping in sugarcane cultivation influenced the soil properties and enhanced the diversity of vital diazotrophic bacteria. Sugar Tech 19:136–147

Soufi E, Safaie N, Shahbazi S, Mojerlou S (2021) Gamma irradiation induces genetic variation and boosting antagonism in Trichoderma aureoviride. Arch Phytopathol Plant Protect 54:1649–1674

Sousa CDS, Soares ACF, Garrido MDS (2008) Characterization of *streptomycetes* with potential to promote plant growth and biocontrol. Sci Agric 65:50–55

Spadaro D, Gullino ML (2005) Improving the efficacy of biocontrol agents against soilborne pathogens. Crop Prot 24:601–613

Stein T (2005) Bacillus subtilis antibiotics: structures, syntheses and specific functions. Mol Microbiol 56:845–857

Steller S, Vater J (2000) Purification of the fengycin synthetase multienzyme system from *Bacillus subtilis* b213. J Chromatogr B Biomed Sci Appl 737:267–275

Stephens JHG, Rask HM (2000) Inoculant production and formulation. Field Crop Res 65:249–258

Suárez-Moreno ZR, Vinchira-Villarraga DM, Vergara-Morales DI et al (2019) Plant-growth promotion and biocontrol properties of three *Streptomyces* spp. isolates to control bacterial rice pathogens. Front Microbiol 10:290

Suprapta DN (2012) Potential of microbial antagonists as biocontrol agents against plant fungal pathogens. J ISSAAS 18:1–8

Tahir Ha S, Gu Q, Wu H et al (2017) Effect of volatile compounds produced by *Ralstonia solanacearum* on plant growth promoting and systemic resistance inducing potential of *Bacillus* volatiles. BMC Plant Biol 17:1–16

Taiz L, Zeiger E (2006) Plant physiology/by Lincoln Taiz and Eduardo Zeiger. Sinauer Associates, Sunderland

Teixeira LCRS, Peixoto RS, Cury JC et al (2010) Bacterial diversity in rhizosphere soil from Antarctic vascular plants of Admiralty Bay, maritime Antarctica. ISME J 4:989–1001

Tejera-Hernández B, Rojas-Badía MM, Heydrich-Pérez M (2011) Potentialities of the genus *Bacillus* in the promotion of plant growth and the biological control of phytopathogenic fungi. CENIC Magaz Biol Sci 42(3):131–138

Tena AR, Enríquez GR, Pérez LL, Martínez ZE, Aguilar EEQ (2015) Lucha entre microbios: una herramienta para el control de enfermedades de plantas 16(11). ISSN 1607 –6079

Tilman D, Knops J, Wedin D, Reich P (2002) Plant diversity and composition: effects on productivity and nutrient dynamics of experimental grasslands. In: Biodiversity and ecosystem functioning: synthesis and perspectives. Oxford University Press, Oxford, pp 21–35

Toral L, Rodríguez M, Béjar V, Sampedro I (2020) Crop protection against *Botrytis cinerea* by rhizosphere biological control agent *Bacillus velezensis* XT1. Microorganisms 8:992

Udayashankar AC, Nayaka SC, Reddy MS, Srinivas C (2011) Plant growth-promoting rhizobacteria mediate induced systemic resistance in rice against bacterial leaf blight caused by *Xanthomonas oryzae* pv. *oryzae*. Biol Control 59:114–122

Van Wezel GP, Mckenzie NL, Nodwell JR (2009) Applying the genetics of secondary metabolism in model actinomycetes to the discovery of new antibiotics. Methods Enzymol 458:117–141

Vanittanakom N, Loeffler W, Koch U, Jung G (1986) Fengycin-a novel antifungal lipopeptide antibiotic produced by *Bacillus subtilis* F-29-3. J Antibiot 39:888–901

Vargas-Ayala R, RodrÍGuez-Kábana R, Morgan-Jones G, Mcinroy JA, Kloepper JW (2000) Shifts in soil microflora induced by velvetbean (*Mucuna deeringiana*) in cropping systems to control root-knot nematodes. Biol Control 17:11–22

Versalovic J, Koeuth T, Lupski R (1991) Distribution of repetitive DNA sequences in eubacteria and application to fingerpriting of bacterial genomes. Nucleic Acids Res 19:6823–6831

Verschuere L, Rombaut G, Sorgeloos P, Verstraete W (2000) Probiotic bacteria as biological control agents in aquaculture. Microbiol Mol Biol Rev 64:655–671

Wan M, Li G, Zhang J, Jiang D, Huang H-C (2008) Effect of volatile substances of Streptomyces platensis F-1 on control of plant fungal diseases. Biol Control 46:552–559

Wang C, Wang Z, Qiao X et al (2013) Antifungal activity of volatile organic compounds from *Streptomyces alboflavus* TD-1. FEMS Microbiol Lett 341:45–51

Wang S, Liang Y, Shen T, Yang H, Shen B (2016) Biological characteristics of *Streptomyces albospinus* CT205 and its biocontrol potential against cucumber *Fusarium* wilt. Biocontrol Sci Tech 26:951–963

Wang X, Wang L, Wang J, Jin P, Liu H, Zheng Y (2014) *Bacillus cereus* AR156-induced resistance to *Colletotrichum acutatum* is associated with priming of defense responses in loquat fruit. PLoS One 9:e112494

Wang Z, Ma L-Y, Cao J et al (2019) Recent advances in mechanisms of plant defense to *Sclerotinia sclerotiorum*. Front Plant Sci 10:1314

Wu Y, Yuan J, Yaoyao E, Raza W, Shen Q, Huang Q (2015) Effects of volatile organic compounds from *Streptomyces albulus* NJZJSA2 on growth of two fungal pathogens. J Basic Microbiol 55: 1104–1117

Wu Y, Zhou J, Li C, Ma Y (2019) Antifungal and plant growth promotion activity of volatile organic compounds produced by *Bacillus amyloliquefaciens*. Microbiol Open 8:e00813

Xavier IJ, Holloway G, Leggett M (2004) Development of rhizobial inoculant formulations. Crop Manag 3:1–6

Xie Z, Li M, Wang D et al (2021) Biocontrol efficacy of *Bacillus siamensis* LZ88 against brown spot disease of tobacco caused by *Alternaria alternata*. Biol Control 154:104508

Yang M, Lu L, Pang J et al (2019) Biocontrol activity of volatile organic compounds from *Streptomyces alboflavus* TD-1 against *Aspergillus flavus* growth and aflatoxin production. J Microbiol 57:396–404

Yoolong S, Kruasuwan W, Thanh Phạm HT, Jaemsaeng R, Jantasuriyarat C, Thamchaipenet A (2019) Modulation of salt tolerance in Thai jasmine rice (*Oryza sativa* L. cv. KDML105) by *Streptomyces venezuelae* ATCC 10712 expressing ACC deaminase. Sci Rep 9:1275

Yu X, Ai C, Xin L, Zhou G (2011) The siderophore-producing bacterium, *Bacillus subtilis* CAS15, has a biocontrol effect on Fusarium wilt and promotes the growth of pepper. Eur J Soil Biol 47: 138–145

Zahir ZA, Munir A, Asghar HN, Shaharoona B, Arshad M (2008) Effectiveness of rhizobacteria containing ACC deaminase for growth promotion of peas (*Pisum sativum*) under drought conditions. J Microbiol Biotechnol 18:958–963

Zalila-Kolsi I, Mahmoud AB, Ali H et al (2016) Antagonist effects of *Bacillus* spp. strains against *Fusarium graminearum* for protection of durum wheat (*Triticum turgidum* L. subsp. durum). Microbiol Res 192:148–158

Zhao J, Xue Q-H, Shen G-H, Xue L, Duan J-L, Wang D-S (2012) Evaluation of *Streptomyces* spp. for biocontrol of gummy stem blight (*Didymella bryoniae*) and growth promotion of *Cucumis melo* L. Biocontrol Sci Tech 22:23–37

Zhu M-L, Wu X-Q, Wang Y-H, Dai Y (2020) Role of biofilm formation by *Bacillus pumilus* HR10 in biocontrol against pine seedling damping-off disease caused by *Rhizoctonia solani*. Forests 11:652

Zihalirwa Kulimushi P, Argüelles Arias A, Franzil L, Steels S, Ongena M (2017) Stimulation of fengycin-type antifungal lipopeptides in *Bacillus amyloliquefaciens* in the presence of the maize fungal pathogen *Rhizomucor variabilis*. Front Microbiol 8:850

Chapter 12
Omic Route to Utilize Endophytes and Their Functional Potentials in Plant Growth Advancement

Charu Gupta, Dhan Prakash, Mohd Fadzelly Abu Bakar, and Worapong Kitdamrongtham

Abstract Plant–bacteria associations have been studied for several decades. Conversely, a complete understanding of the mechanisms employed by plant growth-promoting bacteria remained somewhat elusive. It is therefore difficult to take full advantage of such complex relationships to reproducibly improve the growth of plants in a pragmatic background. Endophytic microorganisms such as bacteria and fungi are better adapted and protected by their host plants and are considered superior to their rhizospheric counterparts. The current scenario is that bacteria can positively impact plant growth and health while plants can select their microbiome in order to have beneficial bacterial colonizers, including those living within the plant tissues. They function by increasing the bioavailability of essential nutrients and modulating hormone levels. It is assumed that endophytes have the potential to replace or augment several chemicals currently used in agricultural practice including fertilizers, pesticides and chemical remediation agents for a number of environmental hazards. Therefore, the prime focus of this chapter is to describe the role of endophytes in the production of phytohormones and the utilization of some of the functional traits in plant growth promotion. Further, the emphasis would also be to cover the 'omic' approach to unravel plant–endophyte communications under abiotic stress.

Keywords Endophytes · Plant growth promoting bacteria (PGPB) · Phytohormones · Abiotic stress

C. Gupta (✉) · D. Prakash
Amity Institute of Herbal Research and Studies, Amity University Uttar Pradesh, Noida, India

M. F. A. Bakar
Faculty of Applied Sciences and Technology, Universiti Tun Hussein Onn Malaysia (UTHM), Johor, Malaysia

W. Kitdamrongtham
Faculty of Engineering and Technology, North-Chiang Mai University, Chiang Mai, Thailand

S. Chhabra et al. (eds.), *Plant Microbiome for Plant Productivity and Sustainable Agriculture*, Microorganisms for Sustainability 37,
https://doi.org/10.1007/978-981-19-5029-2_12

12.1 Introduction

Endophytes are a group of symbiotic microorganisms that vastly spread through all plants and they colonize the intra- and intercellular spaces of the host without causing any substantial morphological modification or visual infection (Fouda et al. 2019). They are mainly located in roots inside stems and in distinct compartments of the plant (Compant et al. 2010). The structure and composition of endophytes are determined by environment and plant-associated factors, such as the plant genotype, developmental stage, phenology and edaphic properties (da Silva et al. 2016; Goulart et al. 2019).

The coexistence of microbes in the plant affects the health and growth performance of the plant, which effectively improves the agricultural properties such as root length and fresh and dry weight of shoot and root and improves yield, quality of soils, and nutrient cycling (Habib et al. 2020; Soliman et al. 2019; Elkelish et al. 2020). A symbiotic linkage is formed between microorganisms and plants, which is useful for both partners. The endophytic microbes produce secondary active metabolites that protect plants from phytopathogens, in addition to exo-enzyme production, which could be supporting in plant colonization (Gill et al. 2016). Endophytes may enhance plant growth by phytohormone production and support plant growth under adverse biotic and abiotic stress (Eid et al. 2019). Endophytes can typically interact with their hosts more effectively than their plant growth-promoting rhizospheric counterparts (Rashid et al. 2012). A variety of endophytes have been isolated from different tissue types in numerous species of plants, and often multiple species of endophytes are found within a single plant (Kobayashi and Palumbo 2000). Recently, bacterial endophytes have been used in different biotechnological sectors, such as biofertilizers to improve crop production and significantly reduce the chemical input into the environment (ALKahtani et al. 2020), as well as in nanotechnology for the fabrication of various nanoparticles incorporated in different applications (Fouda et al. 2020; Eid et al. 2020; Prasad 2017).

The genetic and biochemical factors that contribute towards successful endophytic colonization and maintenance are not well understood. Nevertheless, there have been a large number of studies that described the use of endophytic bacteria as components of various phytoremediation schemes (Doty 2008). As a starting point, it should be emphasized that all available evidence indicates that endophytic bacteria employ similar plant growth promotion mechanisms to those used by rhizospheric bacteria (Santoyo et al. 2016). The existing evidence suggests that endophytic bacteria are more effective than similar non-endophytic bacterial strains in promoting plant growth under a wide range of environmental conditions (Ali et al. 2012, 2014). In order to ensure their efficiency, the endophytes should ameliorate multiple abiotic stresses and should be good plant colonizers with broad host ranges. They should be good soil and plant competitors to compete with native soil and plant microbes for entry into plant tissues, should not be pathogenic to plants or animals when exogenously applied, and should not interfere with the functions of the plant microbiome. Thus, the present chapter discusses the mechanism employed by growth-promoting bacteria to improve plant growth, the role of endophytes in the production of phytohormones, and the utilization of functional traits in plant growth promotion and furthermore emphasizes the 'omic' approach of plant–endophyte communications under abiotic stress.

12.2 Role of Endophytes in Combating Abiotic Stress

Plant stress implies some adverse effects on the physiology of a plant induced upon a sudden transition from some optimal environmental condition where homeostasis is maintained to some suboptimal condition which disrupts this initial homeostatic state.

Reactive oxygen species (ROS) may be considered endogenously produced signal molecules or regulators produced by several plant organelles, including mitochondria, chloroplast, or peroxisomes under stresses. ROS consists of a group of chemically reactive oxygen molecules such as hydrogen peroxide (H_2O_2), superoxide radical ($O^{2\bullet-}$), hydroxyl radical ($OH^{\bullet}$) and singlet oxygen ($_1O^2$) and are produced in plants under stress conditions (Peleg and Blumwald 2011). Abiotic stress leads to the overproduction of ROS that must be managed in a homeostatic pool; however, excess concentrations of ROS cause oxidative stress, which results in denaturation of protein structure, lipid peroxidation and nucleotide disruption and may affect plant physiology which ultimately leads to the death of plants (Demidchik 2015; Nath et al. 2017).

Many plants have the capacity to tolerate a particular stress and hence are considered to be stress resistant. Stress resistance requires that the organism exhibit the capacity to adjust or to acclimate to the stress. Plants have various mechanisms that allow them to survive and often prosper in the complex environments in which they live.

In the plant system, mitigation of ROS excess concentrations generally leads to activation of enzymatic or non-enzymatic antioxidant systems. Plants secrete several enzymes, including catalase (CAT), ascorbate peroxidase (APX), superoxide dismutase (SOD), glutathione reductase (GR), dehydroascorbate reductases (DHAR) and monodehydroascorbate reductases (MDHAR); the non-enzymatic system involves quenching of ROS via synthesis of ascorbic acid (AsA), glutathione (GSH) and carotenoids, which quench free radicals and protect the plant cell from oxidative stress (Hasanuzzaman et al. 2020; Kumari et al. 2020).

The endophytic microbiome shows mutualistic relations with the host plant in maintaining health or vigour (Hardoim et al. 2015). Moreover, it is essentially involved directly or indirectly in the growth and development of host plants via secreting various growth-promoting attributes like phytohormone synthesis, nutrient acquisition and siderophore production, antibiotic production, phosphate solubilization and mitigation of various biotic and abiotic conditions (Shade et al. 2017). However, this has been described in earlier studies on impacts of drought and salinity stress on the effect of growth, productivity, or survivability of plants (White and Torres 2010).

Zhou et al. (2021) reported improved seedling growth of *Pinus tabulaeformis* after inoculation of endophytic strain *Phoma* sp. under drought condition. Wu et al. (2008) described decreased leaf area, photosynthetic pigments and photosynthetic efficiency under drought stress. Higher salinity in soil affects the survivability of plants by altering chemical, morphological and physiological processes (Otlewska et al. 2020). Naveed et al. (2014) reported improved growth, water availability, and photosynthetic activity in maize cultivars under drought after inoculation of endophytic bacterial strains *Burkholderia phytofirmans* strain PsJN and *Enterobacter* sp. FD17. The endophytes inoculation improved seedling growth, shoot and root biomass and photochemical efficiency of PSII.

Yandigeri et al. (2012) demonstrated the potential of endophytic bacterial strains *Streptomyces coelicolor* DE07, *S. olivaceus* DE10 and *S. geysiriensis* DE27, isolated from arid and drought affected regions, to increase tolerance of plants to intrinsic water stress and showed plant growth promotion after application to wheat seedlings. Additionally, the combined application of *S. olivaceus* DE10 + *S. geysiriensis* DE27 strains showed synergistic effects and improved response in terms of stress mitigation and growth promotion. Morsy et al. (2020) reported that the endophytic fungal strains *Ampelomyces* sp. and *Penicillium* sp. isolated from stress inducing soil (drought and high salinity) enhanced drought tolerance (*Ampelomyces* sp.) and salinity tolerance (*Penicillium* sp.) in tomato.

Endophytic microorganisms present inside the host plant play an essential role in host fitness, nutrient supply and stress tolerance. Endophytes are often used in sustainable agriculture as biofertilizers, biopesticides and inoculants to mitigate abiotic stresses including salinity, drought, cold and pH variation in the soil. Plants experience stress conditions that involve endogenous boosting of their immune system or the overexpression of their defensive redox regulatory systems with increased reactive oxygen species (ROS). However, rising stress factors overwhelm the natural redox protection systems of plants, which leads to massive internal oxidative damage and death. Endophytes are an integral internal partner of hosts and have been shown to mitigate abiotic stresses via modulating local or systemic mechanisms and producing antioxidants to counteract ROS in plants. Endophytes also provide support in acclimatizing crop plants under abiotic stress conditions, growth promotion and management of phytopathogens and help in activating stress responsive/induced genes of plants that are not usually activated under stress conditions (Verma et al. 2021).

Thus microbial endophytes appear to be a suitable alternative for drought and salinity stress management. In the recent past, various microbial strains have been successfully utilized to increase drought tolerance. Inoculation of microbial endophytes or exogenous supply of phytohormones significantly enhanced adaptive behaviour of plants via improving photosynthetic activity, chlorophyll contents, root growth, water status, antioxidant enzymes, phytohormone signalling and nutrient uptake under drought conditions (Khan et al. 2016; Singh et al. 2018). Recent advances of research on positive interaction between microbial endophytes and their plant host under drought and salinity stress conditions are described and compiled in Table 12.1.

Table 12.1 Example of positive interaction amongst microbial endophytes and their plant host under stress conditions adapted from Verma et al. (2021)

Endophyte: ameliorate stress and beneficial effect	Mode of action	Plant host
Phoma species (fungi)	Increased proline peroxidase (POD), catalase (CAT), superoxide dismutase (SOD)	*Pinus tabulaeformis*
Glomus mosseae, *G. versiforme*, *G. diaphanum* (fungi)	Increase of peroxidase activity and beneficial effects on soil structure	*Poncirus trifoliata*
Endophyte consortia (*Rhodotorula graminis*, *Burkholderia vietnamiensis*, *Rhizobium tropici*, *Acinetobacter calcoaceticus*, *Rahnella* sp., *Burkholderia* sp., *Enterobacter asburiae*, *Sphingomonas yanoikuyae*, *Pseudomonas* sp., *Curtobacterium* sp.) (fungi + bacteria)	Reduced damage by reactive oxygen species (ROS), increment of IAA	*Populus* sp.
Bacillus, *Achromobacter*, *Klebsiella*, *Citrobacter* (bacteria)	Production of 1-aminocyclopropane-1-carboxylate (ACC) deaminase	*Capsicum annuum*
Burkholderia phytofirmans, *Enterobacter* sp. (bacteria)	Reduced H_2O_2 induced damage	*Zea mays*
Bacillus subtilis (bacteria)	Upregulation of the drought-response genes, such as DREB2B-like, DHN3-like and LEA-14-A-like and modulation of the DNA methylation genes, such as MET1B-like, CMT3-like and DRM2-like that regulate the process	*Brachypodium distachyon*
Bacillus subtilis (bacteria)	Enhanced level of ROS scavenging antioxidant enzymes (superoxide dismutase, peroxidase, catalase)	*Cicer arietinum* seedling
Curvularia sp. (fungi)	Elevates antioxidant enzymes (SOD and APX)	Poplar plant
Piriformospora indica (fungi)	Enhanced plant growth and attenuated the NaCl-induced lipid peroxidation, metabolic heat efflux and fatty acid desaturation in leaves. In addition, significantly elevated the amount of ascorbic acid and increased the activities of antioxidant enzymes catalase, ascorbate peroxidase, dehydroascorbate reductase, monodehydroascorbate reductase and glutathione reductase	*Hordeum vulgare* seedling

12.3 Phytohormones Production by Endophytes

12.3.1 Auxin

Auxins are plant hormones that play vital roles in almost every step of life and daily growth of a plant (Grossmann 2010). The most common naturally occurring auxin is indole-3-acetic acid (IAA), which is produced by plants, bacteria and fungi through at least three diverse tryptophan-dependent IAA production pathways (Duca et al. 2014). The type of pathway a bacterium uses (genetically and/or environmentally dependent) to produce IAA within plants or in their close vicinity can determine the nature of the resulting plant–microbe interactions. It is important to note that not all the IAA producing bacteria are beneficial to plants. Interestingly, many plant-beneficial bacteria produce IAA via the indole-3-pyruvate (IPyA) pathway, whereas many pathogenic bacteria mainly synthesize IAA via indole-3-acetamide (IAM) pathway (Hardoim et al. 2008).

The areas of activity of IAA mainly include, but are not limited to, cell division and elongation, initiation of root systems, leaves, flowers, fruit development and senescence (Phillips et al. 2011). Numerous PGPB, including both Gram-positive and Gram-negative bacteria, have been reported to produce IAA (Rashid et al. 2012). The prominent IAA producing endophytic bacterial genera include *Pseudomonas*, *Rhizobium*, *Azospirillum*, *Enterobacter*, *Azotobacter*, *Klebsiella*, *Alcaligenes*, *Pantoea*, *Acetobacter*, *Herbaspirillum*, *Burkholderia*, *Bacillus*, *Rhodococcus* and *Streptomyces* (Duca et al. 2014; Rashid et al. 2012). Generally, plants are very sensitive to the amount of IAA present in plant tissues at any particular time. Since plants also produce IAA, in order to regulate plant growth, an IAA producing PGPB must provide the appropriate amount of IAA when combined with the amount of hormone produced by plant. In fact, phytopathogens are often characterized by their ability to produce IAA at high concentrations (Kunkel and Chen 2006).

Numerous bacterial endophytes have been reported to promote plant growth by their ability to biosynthesize IAA. For example, one study reported IAA production as a common growth promotion trait in bacterial endophytes isolated from apple tree buds, in which 8 of 18 isolates exhibited IAA production of 1.2–2.4 μg/mL (Miliūtė and Buzaitė 2011). In another report, Vendan et al. (2010) investigated the various plant growth-promoting capabilities, including IAA production, of bacterial endophytes from ginseng (*Panax ginseng* C.A. Meyer). A total of 51 bacterial endophytes were isolated from ginseng stem and clustered in four groups, namely, *Firmicutes*, *Actinobacteria*, *α-Proteobacteria* and *γ-Proteobacteria*, with *Firmicutes* being the most prominent group. Some 18 representatives of all groups were further characterized, and 14 of these 18 endophytic isolates produced significant amounts of IAA when supplemented with tryptophan as a precursor. The highest amount of IAA (13.93 μg/mL) was produced by isolate E-I-4 (*Micrococcus luteus*) and was followed by the isolates E-I-20 (*Lysinibacillus fusiformis*) and E-I-8 (*Bacillus cereus*), which produced 7.23 and 4.61 μg/mL, respectively. The population diversity and plant growth promotion effects of IAA produced by endophytic and epiphytic bacteria isolated from soybeans (cultivars: Foscarin and Cristalina) have also been investigated. Isolates that presented plant growth promotion capabilities were identified as belonging to genera *Pseudomonas*, *Ralstonia*, *Enterobacter*, *Pantoea* and *Acinetobacter*. Moreover, endophytic soybean cultivars exhibited more abundant IAA producing abilities (34%) than that of epiphytic population (21%) (Kuklinsky-Sobral et al. 2004). Further seed endophytes were isolated from 12 different cultivars of soybeans and identified by amplified ribosomal DNA restriction analysis (ARDRA) grouping and by partially sequencing of their 16S rRNA gene. These endophytes were classified as *Acinetobacter*, *Bacillus*, *Brevibacterium*, *Chryseobacterium*, *Citrobacter*, *Curtobacterium*, *Enterobacter*, *Methylobacterium*, *Microbacterium*, *Micromonospora*, *Pantoea*, *Paenibacillus*, *Pseudomonas*, *Ochrobactrum*, *Streptomyces* and *Tsukamurella*. They all produced IAA in vitro at significant levels but only one strain (*Enterobacter* sp.) significantly increased the root dry biomass when soybean seeds were pre-treated with this strain (Assumpção et al. 2009). Moreover, endophytic bacteria were isolated from crops like berseem clover or canola in rotation with rice using IAA production as a primary screening trait. This study demonstrated plant growth-promoting features of seven isolates and rice seedlings inoculated with any one of these isolates exhibited higher shoot biomass, root length, and the number of colonizing bacteria than did control plants inoculated with endophytic strains that did not produce IAA (Etesami et al. 2015).

12.3.2 Gibberellin

Gibberellins (GAs) stimulate a number of plant metabolic functions, which are essential for plant growth and development (Khan et al. 2015) including seed germination, stem elongation, flowering, fruit formation and senescence. To date, there have been ~136 GAs identified (Bömke and Tudzynski 2009). Different GAs are named according to their order of discovery (Khan et al. 2015). The full bacterial gibberellin biosynthesis pathway has been described (Nett et al. 2017) but very little is known about GA production by bacterial endophytes. Only a few studies have described this potential plant growth-promoting trait of bacterial endophytes (Khan et al. 2014a, b). Two bacterial endophytes, namely, *Acetobacter diazotrophicus* and *Herbaspirillum seropedicae*, have been reported to produce gibberellins (GA1, GA3 and IAA). These bacteria establish an endophytic relationship with graminaceae species where they promote growth and overall yield (Bastián et al. 1998). In addition, *Azospirillum lipoferum* strain op 33 has been recognized as an endophyte of grasses and a GA producer in vitro. The gibberellins produced by these bacterial endophytes were measured in chemically defined media by capillary gas chromatography-mass spectrometry (Bastián et al. 1998). In one study, spent culture from 10 wild-type and mutant strains (including *nod*$^{-}$ and *fix*$^{-}$) of *Rhizobium phaseoli* were screened for the presence of GAs and IAA by a high-performance liquid chromatography (HPLC) immunoassay. The presence of GA1, GA4, GA9 and GA20-like molecules was confirmed by combined gas chromatography mass spectrometry; however, the GA20-like molecule was present only in some cultures and not in all, and GA9 was detected only in small amounts compared to GA1 and GA4 (Atzorn et al. 1988). The study of *nod*$^{-}$ and *fix*$^{-}$ mutants indicated that the production of GAs was independent of genes involved in nodulation and nitrogen fixation in *Rhizobium phaseoli*. In another study, a bacterial endophyte *Sphingomonas* sp. LK11, originally isolated from the leaves of *Tephrosia apollinea* (a legume native to parts of Asia and Africa), has shown the ability, by advanced chromatographic and spectroscopic techniques, to synthesize physiologically active GA4 and inactive GA9 and GA20 in culture media. Tomato plants inoculated with *Sphingomonas* sp. LK11 displayed a significant increase in plant shoot length, chlorophyll contents and shoot and root dry weights compared to control plants (Khan et al. 2014a, b).

The other large group of endophytic microorganisms that produces GAs is fungi. A number of fungi have been identified with the capacity to synthesize physiologically active GAs (Leitão and Enguita 2016). Hamayun et al. (2009a) recognized two fungal endophytes *Aspergillus fumigatus* and *Scolecobasidium tshawytschae*, which were isolated from a drought stressed cultivar (Hwangkeumkong) and a salt stressed cultivar (Daewonkong) of soybean, respectively. Both strains were identified by morphological characteristics and phylogenetic analysis of 18S ribosomal RNA gene sequences. *Scolecobasidium tshawytschae* produced physiologically active GA1, GA3, GA4 and GA7 and inactive GA15 and GA24, whereas *Aspergillus fumigatus* synthesized active GA3, GA4 and GA7 along with physiologically inactive GA5, GA19 and GA24.

Subsequently, rice and soybean plants that were treated with these fungal endophytes showed a significant increase in plant length and plant fresh and dry weight compared to plants treated with *Gibberella fujikuroi*, which is also a non-endophytic gibberellin producing fungus and was used as control for these studies (Hamayun et al. 2009a, b). In addition, the fungal endophyte *Aspergillus fumigatus* sp. LH02 facilitated soybean plant growth under salt stress (70 and 140 mM). The soybean plants pre-treated with this fungus exhibit significant increase compared to control in shoot length, shoot fresh, dry biomass, leaf area, chlorophyll contents and photosynthetic rate (Khan et al. 2011). It was argued that the treatment of plants with the fungal endophyte, *Aspergillus fumigatus* sp. LH02, increased the plant's levels of proline, salicylic acid (SA) and jasmonic acid (JA) and lowered the abscisic acid (ABA) concentration compared to control (Khan et al. 2011). Moreover, the treated plants had a higher level of isoflavones, which was independent of the level of salt stress, notwithstanding the fact that isoflavones are considered to be a factor in helping soybean plants to cope with salt stress (Khan et al. 2011).

Similarly, a group of 11 fungal endophytes was isolated from the sand dune plant *Elymus mollis* and screened for the production of GAs and their plant growth-promoting capacities on Waito-C rice (GAs deficient rice) and *Atriplex gmelinii* (saltbush). Altogether, 7 of 11 fungal endophytes promoted the growth of both plants and one isolate, EM-7-1, showed significantly higher plant growth compared to control plants (Afzal Khan et al. 2009). Screening of the culture filtrate of isolate

EM-7-1 revealed the presence of GA1, GA3, GA4 and GA7 as well as physiologically inactive GA5, GA9, GA20 and GA24. Subsequently, this isolate was identified as *Gliomastix murorum* (Afzal Khan et al. 2009).

A similar study was reported by Khan et al. (2014a), who isolated and characterized two fungal endophytes from the bark of *Moringa peregrina* (a tree indigenous to the Horn of Africa), *Aspergillus caespitosus* LK12 and *Phoma* sp. LK13, and showed that these fungi could produce a variety of GAs in culture filtrate. Both fungal strains promoted the growth of rice that lacks gibberellins biosynthesis (Khan et al. 2014a). Furthermore, two GAs and IAA producing fungal endophytes *Phoma glomerata* LWL2 and *Penicillium* sp. LWL3 have been shown to provide protection to cucumber plants under salinity and drought stresses. The treated plants exhibited significantly higher contents of a range of nutrients than the untreated control plants. On the other hand, under salinity stress, treated cucumber plants upregulated SA levels, altered JA levels, and downregulated ABA levels and glutathione, catalase, peroxidase and polyphenol oxidase activities. This change in the metabolome of the treated cucumber plants compared to uninoculated control plants is ascribed to the endophytic fungus, which apparently ameliorated the detrimental effects of the stress (Waqas et al. 2012). The effects of GAs producing strains of *Penicillium* sp. on plants under salt stress was evaluated and the role of the fungus in overcoming the stress incurred by salinity was evaluated (Leitão and Enguita 2016).

12.3.3 Cytokinin

Cytokinins are a group of hormones that promote cell division in plant roots, shoots and growing buds. These hormones have been found in all complex plants as well as mosses, fungi and bacteria. There are about 200 different natural and synthetic cytokinins known to botanists today. Most cytokinins are produced in the meristem of the roots and transported to the other parts of the plant through the xylem (vascular system). Cytokinins have been reported to be present in the culture filtrate of a number of bacteria including *Azotobacter* sp., *Rhizobium* sp., *Pantoea agglomerans*, *Rhodospirillum rubrum*, *Pseudomonas fluorescens*, *Bacillus subtilis* and *Paenibacillus polymyxa* (Glick 2014). Although these bacteria have been documented as endophytes of different plants, little evidence has been found to definitively link bacterial cytokinin production with plant growth promotion. One study described the isolation, identification and characterization of bacterial endophytes that produce cytokinin-like molecules (Bhore et al. 2010). In this study, three bacterial endophytes were isolated from Sambung Nyawa [*Gynura procumbens* (Lour.) Merr.] and identified as *Pseudomonas resinovorans*, *Paenibacillus polymyxa* and *Acinetobacter calcoaceticus*. The ethyl acetate extract of bacterial culture media was used to inoculate cucumber cotyledons in a greening bioassay. The assay indicated positive results only for the strains *Pseudomonas resinovorans* and *Paenibacillus polymyxa* with the suggestion that these bacterial endophytes might be used as plant growth-promoting agents for Sambung Nyawa (Bhore et al. 2010).

In a similar study, crude bacterial suspensions of 115 bacterial isolates from 72 different plant species were screened using the cucumber cotyledon greening bioassay to investigate if these endophytes could produce cytokinins, but the study found that none yielded better results than the control (Bhore and Sathisha 2010). Recently, workers engineered a strain of *Sinorhizobium meliloti* to overproduce cytokinin by expressing an *Agrobacterium ipt* gene under the control of the *E. coli trp* promoter (Xu et al. 2012). Following a period of severe drought stress, alfalfa plants inoculated with the engineered *S. meliloti* strain were significantly larger than plants inoculated with the parental strain. This experiment indicates, despite the fact that cytokinin-producing plant growth-promoting bacteria appear to be relatively uncommon, that rhizobial strains synthesizing higher than normal levels of cytokinin may improve plant tolerance to severe drought stress.

12.3.4 Endophytic ACC Deaminase Production

The microbial enzyme 1-aminocyclopropane 1-carboxylate (ACC) deaminase (E.C. 4.1.99.4) is used as a dynamic enzyme in sustainable agriculture. ACC deaminase is a multimeric enzyme that requires pyridoxal 5′-phosphate as an essential co-factor for enzymatic activity (Glick 2014) and cleaves ACC to α-ketobutyrate and ammonia, where ACC is the immediate precursor of phytohormone ethylene. Thus, ACC deaminase lowers the levels of deleterious ethylene in higher plants (Sun et al. 2009). Ethylene, like other phytohormones, is crucial for plant growth, development and stress signalling (Glick 2004; Rashid et al. 2012). Plant growth-promoting endophytes expressing the enzyme ACC deaminase have been revealed to help protect plants from a number of different biotic and abiotic stresses and to promote the growth of plants in the absence of stressful conditions. For example, a group of 25 endophytes, originally isolated from tomato plants, that contained ACC deaminase all demonstrated the ability to significantly promote canola seedling growth compared to un-inoculated canola seedlings (Rashid et al. 2012). Several studies have defined the potential of ACC deaminase containing endophytes in promoting plant growth in tomato (Abbamondi et al. 2016), rice (Raweekul et al. 2016) and ginger (Jasim et al. 2014).

Further, the potential of the bacterial endophyte *Burkholderia phytofirmans* PsJN was evaluated in a field study of switchgrass at two different soil sites over 2 years. The inoculated switchgrass displayed enhanced biomass production, increased root growth, tillering and greater early season plant growth vigor than that of untreated control plants. Moreover, the plants grown on a low fertility soil site performed better with bacterial endophyte treatment. These researchers suggested that the mechanism of this plant growth promotion, especially at the poor soil site, might include the possession of ACC deaminase by this endophyte and the interaction of auxins and ethylene in response to the action of ACC deaminase (Lowman et al. 2015). In addition, it has been previously shown that a mutant of *Burkholderia phytofirmans* PsJN that lacks the ability to produce ACC deaminase (i.e. *acdS*$^{-}$)

could not promote canola seedling growth in a growth pouch root elongation assay (Sun et al. 2009). Complementing the mutant with exogenous DNA carrying the ACC deaminase gene from wild-type strain restored the ability of the mutant to promote canola root elongation, therefore proving the importance of this gene for the observed growth promotion (Sun et al. 2009). Further, it was also found that *B. phytofirmans* PsJN specifically promoted the growth of a certain genotype of switchgrass, which led to the isolation and characterization of another bacterial endophyte, *Pantoea agglomerans* strain PaKM, from the surface of sterilized seeds of switchgrass (Kim-Dura et al. 2016). Strain PaKM was able to promote the growth of at least eight varieties of switchgrass in in vitro conditions; subsequently, two of these varieties were screened more extensively in greenhouse and field environments, and a significant difference in the biomass of endophytic treated plants was observed (Kim-Dura et al. 2016). *P. agglomerans* strain PaKM was able to protect switchgrass under salt and drought stress in in vitro conditions; however, this endophyte does not contain ACC deaminase (Kim-Dura et al. 2016).

When the bacterial endophyte *Pseudomonas migulae* 8R6, an ACC deaminase containing bacterium, was utilized as a biocontrol agent against yellow disease of grapevines caused by phytoplasma, it significantly protected periwinkle, a model plant hosting phytoplasma. The results have shown that the density of the phytoplasma inside the leaf tissue was unaffected by this bacterial endophyte; however, the symptoms of the disease were significantly reduced in plants treated with wild-type bacterium compared with either untreated or treated plants with an ACC deaminase minus mutant (*acdS*$^{-}$) of strain 8R6 (Gamalero et al. 2016). These trials suggest that ACC deaminase played a key role in protecting the plant from the biotic stress of phytoplasma infection. Moreover, ACC deaminase containing endophytes have also been found to protect plants from salinity and other abiotic stress. The rice endophyte *Pseudomonas stutzeri* A1501 has demonstrated rice seedling growth promotion in moderate (0.12 M) and high (2 M) salt (i.e. NaCl) and in the presence of 0.3 mM heavy metals (Cu, Co, Ni and Zn) (Han et al. 2015). In these experiments, the bacterial treatment was given to surface-sterilized rice seeds and plant biometrics were collected after 7 days. In order to validate that the ACC deaminase activity is the main driving force in protecting and facilitating rice seedling growth in presence of these abiotic stresses, a mutant of the *acdS* gene was constructed. It was observed that the seeds treated with wild-type *P. stutzeri* A1501 displayed significantly longer roots and higher fresh and dry weights compared to the plants either untreated or treated with the mutant (Han et al. 2015). In another study, endophytic bacteria isolated from date palm were assayed for growth promotion of canola roots in the presence and absence of 100 mM salt. The majority of endophytes tested exhibited canola root elongation under salt stress compared to un-inoculated control plants in gnotobiotic conditions; however, the researchers pointed out that these endophytes, in addition to ACC deaminase, could also produce IAA and increase the uptake of nutrients that enable them to benefit the host under stress (Yaish et al. 2015). *Brachybacterium paraconglomeratum* is an ACC deaminase producing salt-tolerant bacterial endophyte, which was isolated from the surface-sterilized roots of the medicinal plant *Chlorophytum borivilianum* (Barnawal et al. 2016). This bacterium promoted host plant growth by reducing oxidative and osmotic damages caused by salinity (150 mM). Moreover, biochemical analysis of bacterially treated and untreated plants, both grown in the presence of salt, revealed that there were high amounts of ACC, proline, malondialdehyde (MDA) and abscisic acid (ABA) in untreated controls. Increased levels of proline and MDA indicate osmotic and oxidative stress, respectively, whereas increased ABA and ACC levels are thought to be the consequence of osmo-oxidative damage (Barnawal et al. 2016). However, plants treated with the wild-type endophyte (containing ACC deaminase) show reduced levels of proline, MDA, ABA and ACC and increased total chlorophyll contents, IAA levels and plant biomass compared to untreated controls (Barnawal et al. 2016).

Additionally, the expression of five genes involved in the stress response (*CaACCO*, *CaLTPI*, *CaSAR82A* and putative *P5CR* and *P5CS*) in pepper plants (*Capsicum annuum* L.) was investigated. These plants were given a mild osmotic stress in the presence and absence of two plant growth-promoting bacterial endophytes, *Arthrobacter* spp. EZB4 and *Bacillus* spp. EZB8 (Sziderics et al. 2007). The pepper plant gene *CaACCO*, which encodes the enzyme ACC oxidase that catalyses the final step in the biosynthesis of ethylene, was strongly upregulated in non-inoculated stressed plant root and leaf tissue. This gene was significantly less upregulated in leaf tissue, unaffected in plant roots treated with strain EZB4, and unaffected in the leaf and root tissue of plants treated with strain EZB8. The pepper plant gene *CaLTPI* encodes a lipid transfer protein that may be induced by ethylene in addition to a number of other stress factors (Jung et al. 2003). The gene expression of *CaLTPI* was significantly upregulated under osmotic stress in non-inoculated plants, unaltered in plants (leaves and roots) treated with strain EZB4 and the leaf tissue of plants treated with strain EZB8, and significantly downregulated in the roots treated with strain EZB8. The putative *P5CR* and *P5CS* are involved in proline biosynthesis and were significantly downregulated in the leaves of plants treated with both endophytic strains under stress conditions and remained unaffected in the roots of stressed plants. The expression of pepper plant gene *CaSAR82A*, which is also a stress-inducible gene, was not consistent under stress conditions in either leaf or root tissues with bacterial treatments but was significantly upregulated in non-inoculated stressed plants (Sziderics et al. 2007). Altogether, it was speculated that because ethylene can act as signalling molecule and subsequently could regulate the gene expression under stress conditions, the addition of endophytes that can lower stress ethylene levels by functioning of ACC deaminase could ameliorate the damage caused by such stress. Nonetheless, independent of the levels of altered gene expression of above-mentioned genes, all of the pepper plants treated with bacterial endophytes *Arthrobacter* spp. EZB4 and *Bacillus* spp. EZB8 showed significant increases in biomass compared to the non-inoculated control plants under mild osmotic stress (Sziderics et al. 2007).

In addition to plant growth facilitation under stressful environmental conditions, ACC deaminase has also been documented to help endophytic colonization within plants (Hardoim et al. 2008). Ethylene levels in plant tissues modulate plant colonization by endophytes (Iniguez et al. 2005). The bacterial endophyte *Klebsiella pneumoniae* strain 342 can establish endophytic relationships with *Medicago truncatula*; however, the colonization of plants by this endophyte was found to be under the control of ethylene. In an ethylene-insensitive mutant of *Medicago truncatula*, this endophyte hypercolonizes the plant compared to the wild-type (Iniguez et al. 2005) *Medicago truncatula*, which displayed a low level of endophytic colonization in the presence of ACC, and an increase in endophytic colonization was observed when the ethylene inhibitor 1-methylcyclopropene was introduced to the plant (Iniguez et al. 2005). Since ACC deaminase is able to lower the levels of ethylene by cleaving ACC, it may help some endophytes to efficiently colonize plant tissues, thereby giving those endophytes an additional advantage in their interaction with plants.

12.4 Plant Growth Promotion and Stress Management by Endophytes

Endophytes provide support in acclimatizing crop plants under abiotic stress conditions, growth promotion and management of phytopathogens and help in activating stress responsive/induced genes of plants that are not usually activated under stress conditions. The endophytic microbiome shows mutualistic relations with the host plant in maintaining health or vigour (Hardoim et al. 2015). Moreover, they are also essentially involved directly or indirectly in the growth and development of host plants via secreting various growth-promoting attributes, namely, phytohormone synthesis, nutrient acquisition and siderophore production, antibiotic and phosphate solubilization and by mitigating various biotic and abiotic conditions (Shade et al. 2017). Various microbial strains have been successfully utilized to increase drought tolerance. Inoculation of microbial endophytes or exogenous supply of phytohormones significantly enhanced the adaptive behaviour of plants via improving photosynthetic activity, chlorophyll contents, root growth, water status, antioxidant enzymes, phytohormone signalling and nutrient uptake under drought conditions (Singh et al. 2018).

Naveed et al. (2014) reported improved growth, water availability, and photosynthetic activity in maize cultivars under drought after inoculation of endophytic bacterial strains *Burkholderia phytofirmans* strain PsJN and *Enterobacter* sp. FD17. The endophytes inoculation improved seedling growth, shoot and root biomass and photochemical efficiency of PSII. Yandigeri et al. (2012) revealed the potential of endophytic bacterial strains *Streptomyces coelicolor* DE07, *S. olivaceus* DE10 and *S. geysiriensis* DE27, isolated from arid and drought affected regions, to increase tolerance of plants to intrinsic water stress and showed plant growth promotion after application to wheat seedlings. Additionally, the combined application of *S. olivaceus* DE10 + *S. geysiriensis* DE27 strains showed synergistic effects and improved response in terms of stress mitigation and growth promotion. Jayakumar et al. (2020) reported that several endophytic bacterial strains, including *Bacillus* sp.,

Providencia sp. and *Staphylococcus* spp., isolated from *Ananas comosus*, enhanced drought tolerance and promoted growth and pathogen resistance. Likewise, Sandhya et al. (2017) reported that several endophytic bacterial strains isolated from various crops in which most of the strains conferred drought tolerance up to (−1.02) matric potential also had growth promotion potential. Chen et al. (2017) stated that endophytic strain *Pantoea alhagi* isolated from *Alhagi sparsifolia*, after inoculation, enhanced the growth of wheat seedlings under drought conditions. Additionally, the endophyte-treated plant showed enhanced accumulation of soluble sugars and decreased concentrations of malondialdehyde. In the grass *Brachypodium distachyon*, drought stress was mitigated with the help of an endophytic bacterium *Bacillus subtilis* B26, which also upregulated the stress responsive genes (Gagné-Bourque et al. 2015). Morsy et al. (2020) described that endophytic fungal strains *Ampelomyces* sp. and *Penicillium* sp., isolated from stress inducing soil (drought and high salinity), enhanced drought tolerance (*Ampelomyces* sp.) and salinity tolerance (*Penicillium* sp.) in tomato.

12.5 'Ome' Approach of Plant-Endophyte Communications Under Abiotic Strain

Endophytic microbes are known to modulate the genome, epigenome, proteome and metabolome of their hosts after inoculation to cope with abiotic stress. Plants with their modulated 'ome' after inoculation with endophytes bear better potential to ameliorate various abiotic stresses including drought and salinity. The molecular basis of endophytes in mitigating abiotic stress in crops is poorly understood. The recent developments in high-throughput technologies of sequencing and mass-spectroscopy based omics techniques have generated hopes for a detailed gene and protein study of molecular insights into the interaction of plant-endophytes during abiotic stress conditions.

Culture-independent sequencing approaches including metagenomics, meta-transcriptomics and meta-proteomics have emerged as new tools for studying the unexplored wealth of endophytes for conferring abiotic stress tolerance in plants. Shotgun metagenome analysis of uncultured microbe communities of endophytic bacteria revealed the population of Proteobacteria and Actinobacteria which can play a role in plant growth promotion and abiotic stress tolerance (Hong et al. 2019). The change in endophytic bacterial communities of wheat, as assessed by 16S rRNA sequencing, was associated with the change in drought stress conditions (Žiarovská et al. 2020). Not only the endophytes, but the 'ome' of plants is also modulated during their interactions with endophytes while coping with abiotic stress. The 'omics' of endophytes also may be modulated by 'horizontal gene transfer' and synergism while interacting with their host crop (Tiwari and Bae 2020). Coutinho et al. (2015) reported that the influence of host crop *Oryza sativa* on gene expression of endophytic *Burkholderia kururiensis* M130 was related to biofilm regulation and iron transport. Some of the endophytic *Rhizobium* and *Xanthomonas* sp. associated with crops have shown the transfer of genes responsible for plant adaptation and survival (Van Elsas et al. 2003). Comparative transcriptomics and proteomics studies associated with *Atractylodes lancea* in response to endophytic fungus *Gilmaniella* sp. AL12 revealed regulated plant metabolites, with upregulation in terpene skeleton biosynthesis and upregulated genes annotated as β-farnesene synthase and β-caryophyllene synthase (Yuan et al. 2019). Similarly, to understand the interaction of endophytic *Piriformospora indica* and host *Brassica napus*, an LC-MS/MS-based label-free quantitative proteome technique was used, revealing the change in metabolic pathways and stress response and increase in stress adapting metabolites after endophytic interactions (Shrivastava et al. 2018). Understanding the roles of endophytes–plant interactions at a molecular level is crucial to understanding crop coping mechanisms to abiotic stress and may lead to more sustainable agriculture. The uncultured microbiome of endophytes can also be exploited for coping with the abiotic stress using the next generation of sequencing technologies.

12.5.1 Metaproteogenomics

Metaproteogenomics links proteome and genome of the environmental samples and allows identification of more proteins (functions) than proteomics alone. It involves combinatorial study of metagenome and metaproteome of the same sample. Knief et al. (2012) have used metaproteogenomic approach to study microbial communities in the phyllosphere and rhizosphere of rice. The results showed that despite the presence of nifH genes in both microenvironments, expression was found in rhizosphere only. If such an approach could be applied to study the endosphere, more significant data regarding endophyte functionality can be collected. Characterization of the metaproteogenome is expected to provide data linking the genetic and functional diversity of microbial communities. Proteins involved in plant endophyte interactions that could not be studied in cultivated isolates are new targets for functional studies. Plant associated bacterial protein secretion system can be successfully used for determining plant bacterial interactions (Downie 2010). Delmotte et al. (2009) have successfully used community proteogenomics to identify the unique traits of phyllosphere bacteria. Bacterial proteogenomic pipeline and other tools are available for proteogenomic analysis studies (Uszkoreit et al. 2014). The technique offers insights into possible strategies adopted for an endophytic lifestyle. The combined metagenome and metaproteome analysis would allow overcoming the limitations of protein identifications as in the metaproteomic approach due to the non-availability of closely related reference genomes (Kaul et al. 2016).

12.5.2 Microarray-Based Techniques

Microarray technique has equipped the modern genome-based studies with the tools for genome-specific gene expression, endophyte gene profiling, exploration of host plant–symbiont interactions and many other transcriptome analysis (Felitti et al. 2006). Barnett et al. (2004) used the dual genome Symbiosis Chip based tool to study symbiotic interactions. Symbiosis chip allow simultaneous analysis of gene expression in both partners of the association and can easily be used to study the endophyte–host interaction. Barnett et al. (2004) studied the coordinate differentiation and response generated from signal exchange between two symbiotic partners simultaneously, namely, α-proteobacterium *Sinorhizobium meliloti* and its legume partner *Medicago truncatula* during nodule development. They designed a custom Affymetrix Gene Chip with the complete *S. meliloti* genome and ≈10,000 probe sets for *M. truncatula*.

Genomic interspecies microarray hybridization technique has proved to be useful in the characterization of previously untouched genomes, provided that the genome of a close relative has already been fully sequenced (Dong et al. 2001). Microarray technique allows the identification of a number of genes in an uncharacterized genome without the need for genome sequencing. However, reference selection is a critical step in microarray studies as non-specific references may generate ambiguous results. Non-availability or limited access to the specific gene expression/ profiling databases has restricted such studies (Kaul et al. 2016) (Fig. 12.1).

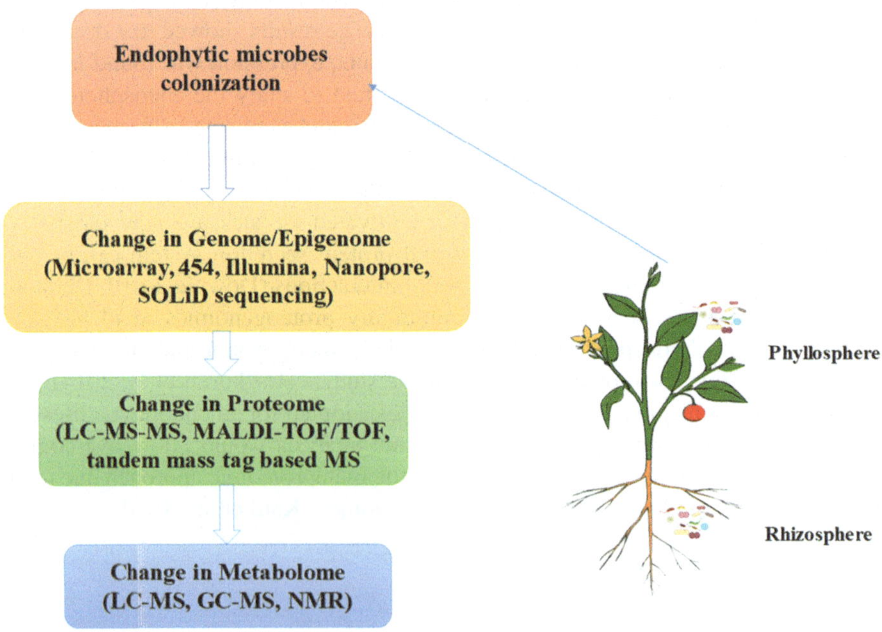

Fig. 12.1 Summary of the 'ome' of the plant-endophyte interactions under abiotic stress

12.6 Conclusions and Future Perspectives

Plants harbour and interact with a variety of microbial populations at various stages of their lifetime; these dynamic interactions may be harmful, benign, or beneficial to plants. Endophytism is a mutualistic plant–microbe interaction where plants provide a safe home and secure supply of food to microbes and microbes, in return, benefit the plants enormously. Microbial endophyte biology is a growing field of research and a number of researchers have planned these beneficial interactions where endophytes provide nutrients, act as growth regulators, fix nitrogen, antibiotics, secondary metabolites, and protect plants from both abiotic and biotic stresses. Moreover, endophytes are thought to be superior to their rhizospheric counterparts in facilitating plant growth as they have the ability to colonize the interior of plants where they can rapidly sense any environmental change and quickly respond to their needs. Since endophytes are generally not host specific, they can be readily introduced into plants other than their natural host.

The increasing research output over the past three decades shows that there has been an increasingly growing interest in the study of endophytic microbes. A significant knowledge has been accumulated on endophytic microbes and their effects on plants. It is known that microbes colonize in shoots and roots of plants. In many cases, microbes actually enter into plant cells and involve in the rhizophagy cycle. This suggests that interaction between endophytic microbe and plant may be to the extent of a direct protoplast interaction within plant cells. A deep understanding of endophyte–host interactions is the need of the hour in order to realize the use of endophytes as plant probiotics. The major areas of future research are the knowledge of the intimate microbe cell to plant cell interactions or 'cross talk' that results in beneficial effects in plants. The 'signals' expressed between endophytes and host plant cells result in oxidative stress tolerance in plants. The signal sent to host cells that triggers the oxidative response may be the key to understanding the endophyte–host interaction. It is that crosstalk between endophyte and host which determines if the plant recognizes the microbes as friendly endophyte or pathogen. The complementary information generated through modern "omics" studies in association with other system biological techniques is inevitable to build up models for prediction and explain endophyte-mediated processes. Advanced techniques can be used with accuracy for bacterial and fungal endophytes to reveal their genetic and metabolic potential, ecology and evolution. The prime focus is on microbial endophytes to improve plant/crop productivity and sustainable agriculture with minimum or nil environmental degradation. In light of these considerations, the practice of endophytes in agriculture can offer an economic means to achieve sustainable high crop productivity.

Acknowledgements The authors are grateful to Dr. Ashok K. Chauhan, Founder President, Dr. Atul Chauhan, Chancellor, Prof. Dr. Balvinder Shukla Vice Chancellor and Amity University, UP, Noida, for the motivation, research facilities and ASEAN-India Science & Technology Development Fund, Department of Science and Technology (DST-SERB), New Delhi India for financial support.

Conflict of Interest The author(s) confirm that this article content has no conflict of interest.

Ethical Issues There is none to be declared.

References

Abbamondi GR, Tommonaro G, Weyens N, Thijs S, Sillen W, Gkorezis P, Iodine C, Rangel WM, Nicolaus B, Vangronsveld J (2016) Plant growth-promoting effects of rhizospheric and endophytic bacteria associated with different tomato cultivars and new tomato hybrids. Chem Biol Technol Agric 3:1

Afzal Khan S, Hamayun M, Kim H, Yoon H, Lee I, Kim J (2009) Gibberellin production and plant growth promotion by a newly isolated strain of *Gliomastix murorum*. World J Microbiol Biotechnol 25(5):829–833

Ali S, Charles TC, Glick BR (2012) Delay of flower senescence by bacterial endophytes expressing 1-aminocyclopropane-1-carboxylate deaminase. J Appl Microbiol 113(5):1139–1144

Ali S, Charles TC, Glick BR (2014) Amelioration of high salinity stress damage by plant growth-promoting bacterial endophytes that contain ACC deaminase. Plant Physiol Biochem 80:160–167

ALKahtani MD, Fouda A, Attia KA, Al-Otaibi F, Eid AM, Ewais EE-D, Hijri M, St-Arnaud M, Hassan SE-D, Khan N (2020) Isolation and characterization of plant growth promoting Endophytic bacteria from desert plants and their application as bioinoculants for sustainable agriculture. Agronomy 10:1325

Assumpção LDC, Lacava PT, Dias AC, de Azevedo JL, Menten JOM (2009) Diversity and biotechnological potential of endophytic bacterial community of soybean seeds. Pesqui Agropecu Bras 44(5):503–510

Atzorn R, Crozier A, Wheeler CT, Sandberg G (1988) Production of gibberellins and indole-3-acetic acid by *rhizobium phaseoli* in relation to nodulation of *Phaseolus vulgaris* roots. Planta 175(4):532–538

Barnawal D, Bharti N, Tripathi A, Pandey SS, Chanotiya CS, Kalra A (2016) ACC-deaminase-producing endophyte *Brachybacterium paraconglomeratum* strain SMR20 ameliorates *Chlorophytum* salinity stress via altering Phytohormone generation. J Plant Growth Regul 35(2):553–564

Barnett MJ, Toman CJ, Fisher RF, Long SR (2004) A dual-genome Symbiosis Chip for coordinate study of signal exchange and development in a prokaryote–host interaction. Proc Natl Acad Sci U S A 101:16636–16641

Bastián F, Cohen A, Piccoli P, Luna V, Baraldi R, Bottini R (1998) Production of indole-3-acetic acid and gibberellins A1 and A3 by *Acetobacter diazotrophicus* and *Herbaspirillum seropedicae* in chemically-defined culture media. Plant Growth Regul 24(1):7–11

Bhore SJ, Sathisha G (2010) Screening of endophytic colonizing bacteria for cytokinin-like compounds: crude cell-free broth of endophytic colonizing bacteria is unsuitable in cucumber cotyledon bioassay. World J Agri Sci 6(4):345–352

Bhore SJ, Nithaya R, Loh CY (2010) Screening of endophytic bacteria isolated from leaves of Sambung Nyawa [*Gynura procumbens* (Lour.) Merr.] for cytokinin-like compounds. Bioinformation 5(5):191–197

Bömke C, Tudzynski B (2009) Diversity, regulation, and evolution of the gibberellin biosynthetic pathway in fungi compared to plants and bacteria. Phytochemistry 70(15–16):1876–1893

Chen C, Xin K, Liu H, Cheng J, Shen X, Wang Y, Zhang L (2017) *Pantoea alhagi*, a novel endophytic bacterium with ability to improve growth and drought tolerance in wheat. Sci Rep 7: 1–14

Compant S, Clément C, Sessitsch A (2010) Plant growth-promoting bacteria in the rhizo- and endosphere of plants: their role, colonization, mechanisms involved and prospects for utilization. Soil Biol Biochem 42:669–678

Coutinho BG, Licastro D, Mendonça-Previato L, Cámara M, Venturi V (2015) Plant-influenced gene expression in the rice endophyte *Burkholderia kururiensis* M130. Mol Plant-Microbe Interact 28:10–21

da Silva KJ, de Armas RD, Soares CR, Ogliari JB (2016) Communities of endophytic microorganisms in different developmental stages from a local variety as well as transgenic and conventional isogenic hybrids of maize. World J Microbiol Biotechnol 32:189

Delmotte NL, Kniefa C, Chaffronb S, Innerabnera G, Roschitzkic B, Schlapbachc R et al (2009) Community proteogenomics reveals insights into the physiology of phyllosphere bacteria. Proc Natl Acad Sci U S A 106:16428–16433

Demidchik V (2015) Mechanisms of oxidative stress in plants: from classical chemistry to cell biology. Environ Exp Bot 109:212–228

Dong Y, Glasner JD, Blattner FR, Triplett EW (2001) Genomic interspecies microarray hybridization: rapid discovery of three thousand genes in the maize endophyte, *Klebsiella pneumoniae* 342, by microarray hybridization with *Escherichia coli* K-12 open reading frames. Appl Environ Microbio 67:1911–1921

Doty SL (2008) Enhancing phytoremediation through the use of transgenics and endophytes. New Phytol 179(2):318–333

Downie JA (2010) The roles of extracellular proteins, polysaccharides and signals in the interactions of rhizobia with legume roots. FEMS Microbiol Rev 34:150–170

Duca D, Lorv J, Patten CL, Rose D, Glick BR (2014) Indole-3-acetic acid in plant-microbe interactions. Antonie Van Leeuwenhoek 106(1):85–125

Eid AM, Salim SS, Hassan SE-D, Ismail MA, Fouda A (2019) Role of endophytes in plant health and abiotic stress management. In: Kumar V, Prasad R, Kumar M, Choudhary DK (eds) Microbiome in plant health and disease: challenges and opportunities. Springer, Singapore, pp 119–144

Eid AM, Fouda A, Niedbała G, Hassan SE-D, Salem SS, Abdo AM, Hetta H, Shaheen TI (2020) Endophytic *Streptomyces laurentii* mediated green synthesis of Ag-NPs with antibacterial and anticancer properties for developing functional textile fabric properties. Antibiotics 9:641

Elkelish A, Qari SH, Mazrou YSA, Abdelaal KAA, Hafez YM, Abu-Elsaoud AM, Batiha GE-S, El-Esawi MA, El Nahhas N (2020) Exogenous ascorbic acid induced chilling tolerance in tomato plants through modulating metabolism, osmolytes, antioxidants, and transcriptional regulation of catalase and heat shock proteins. Plan Theory 9:431

Etesami H, Alikhani HA, Hosseini HM (2015) Indole-3-acetic acid (IAA) production trait, a useful screening to select endophytic and rhizosphere competent bacteria for rice growth promoting agents. MethodsX 2:72–78

Felitti S, Shields K, Ramsperger M, Tian P, Sawbridge T, Webster T et al (2006) Transcriptome analysis of *Neotyphodium* and *Epichloe* grass endophytes. Fungal Genet Biol 43:465–475

Fouda A, Hassan S, Eid AM, El-Din Ewais E (2019) The interaction between plants and bacterial endophytes under salinity stress. In: Endophytes and secondary metabolites. Springer, Cham, pp 1–18

Fouda A, Hassan SED, Abdo AM, El-Gamal MS (2020) Antimicrobial, antioxidant and larvicidal activities of spherical silver nanoparticles synthesized by endophytic *Streptomyces* spp. Biol Trace Elem Res 195:707–724

Gagné-Bourque F, Mayer BF, Charron JB, Vali H, Bertrand A, Jabaji S (2015) Accelerated growth rate and increased drought stress resilience of the model grass *Brachypodium distachyon* colonized by *Bacillus subtilis* B26. PLoS One 10:e0130456

Gamalero E, Marzachì C, Galetto L, Veratti F, Massa N, Bona E, Novello G, Glick BR, Ali S, Cantamessa S, D'Agostino G, Berta G (2016) An 1-Aminocyclopropane-1-carboxylate (ACC) deaminase-expressing endophyte increases plant resistance to flavescence dorée phytoplasma infection. Plant Biosyst:1–10

Gill SS, Gill R, Trivedi DK, Anjum NA, Sharma KK, Ansari MW, Ansari AA, Johri AK, Prasad R, Pereira E, Varma A, Tuteja N (2016) *Piriformospora indica*: potential and significance in plant stress tolerance. Front Microbiol 7:332. https://doi.org/10.3389/fmicb.2016.00332

Glick BR (2004) Bacterial ACC deaminase and the alleviation of plant stress. Adv Appl Microbiol 56:291–312

Glick BR (2014) Bacteria with ACC deaminase can promote plant growth and help to feed the world. Microbiol Res 169(1):30–39

Goulart MC, Cueva-Yesquén LG, Hidalgo Martinez KJ, Attili-Angelis D, Fantinatti-Garboggini F (2019) Comparison of specific endophytic bacterial communities in different developmental stages of *Passiflora incarnata* using culture-dependent and culture-independent analysis. Microbiol Open 8:e896

Grossmann K (2010) Auxin herbicides: current status of mechanism and mode of action. Pest Manag Sci 66(2):113–120

Habib N, Ali Q, Ali S, Javed MT, Zulqurnain Haider M, Perveen R, Shahid MR, Rizwan M, Abdel-Daim MM, Elkelish A et al (2020) Use of nitric oxide and hydrogen peroxide for better yield of wheat (*Triticum aestivum* L.) under water deficit conditions: growth, osmoregulation and anti-oxidative defense mechanism. Plan Theory 9:285

Hamayun M, Khan SA, Khan MA, Khan AL, Kang S, Kim S, Joo G, Lee I (2009a) Gibberellin production by pure cultures of a new strain of *Aspergillus fumigatus*. World J Microbiol Biotechnol 25(10):1785–1792

Hamayun M, Khan SA, Kim H, Chaudhary MF, Hwang Y, Shin D, Kim I, Lee B, Lee I (2009b) Gibberellin production and plant growth enhancement by newly isolated strain of *Scolecobasidium tshawytschae*. J Microbiol Biotechnol 19(6):560–565

Han Y, Wang R, Yang Z, Zhan Y, Ma Y, Ping S, Zhang L, Lin M, Yan Y (2015) 1-Aminocyclopropane-1-carboxylate deaminase from *pseudomonas stutzeri* A1501 facilitates the growth of rice in the presence of salt or heavy metals. J Microbiol Biotechnol 25(7): 1119–1128

Hardoim PR, van Overbeek LS, Elsas JD (2008) Properties of bacterial endophytes and their proposed role in plant growth. Trends Microbiol 16(10):463–471

Hardoim PR, van Overbeek LS, Berg G, Pirttilä AM, Compant S, Campisano A, Döring M, Sessitsch A (2015) The hidden world within plants: ecological and evolutionary considerations for defining functioning of microbial endophytes. Microbiol Mol Biol Rev 79:293–320

Hasanuzzaman M, Bhuyan MHMB, Zulfiqar F, Raza A, Mohsin SM, Al Mahmud J, Fujita M, Fotopoulos V (2020) Reactive oxygen species and antioxidant defense in plants under abiotic stress: revisiting the crucial role of a universal defense regulator. Antioxidants 9:681

Hong CE, Kim JU, Lee JW, Bang KH, Jo IH (2019) Metagenomic analysis of bacterial endophyte community structure and functions in *Panax ginseng* at different ages. Biotech 9:1–8

Iniguez AL, Dong Y, Carter HD, Ahmer BMM, Stone JM, Triplett EW (2005) Regulation of enteric endophytic bacterial colonization by plant defenses. Mol Plant-Microbe Interact 18(2):169–178

Jasim B, Joseph AA, John CM, Mathew J, Radhakrishnan EK (2014) Isolation and characterization of plant growth promoting endophytic bacteria from the rhizome of *Zingiber officinale*. 3 Biotech 4:197–204

Jayakumar A, Padmakumar P, Nair IC, Radhakrishnan EK (2020) Drought tolerant bacterial endophytes with potential plant probiotic effects from *Ananas comosus*. Biologia 75:1769–1778

Jung HW, Kim W, Hwang BK (2003) Three pathogen-inducible genes encoding lipid transfer protein from pepper are differentially activated by pathogens, abiotic, and environmental stresses. Plant Cell Environ 26(6):915–928

Kaul S, Sharma T, Dhar MK (2016) "Omics" tools for better understanding the plant–endophyte interactions. Front Plant Sci 7:955

Khan AL, Hamayun M, Kim Y, Kang S, Lee J, Lee I (2011) Gibberellins producing endophytic *Aspergillus fumigatus* sp. LH02 influenced endogenous phytohormonal levels, isoflavonoids production and plant growth in salinity stress. Process Biochem 46(2):440–447

Khan AL, Waqas M, Hussain J, Al-Harrasi A, Al-Rawahi A, Al-Hosni K, Kim M, Adnan M, Lee I (2014a) Endophytes *Aspergillus caespitosus* LK12 and *Phoma* sp. LK13 of *Moringa peregrina* produce gibberellins and improve rice plant growth. J Plant Interact 9(1):731–737

Khan AL, Waqas M, Kang S, Al-Harrasi A, Hussain J, Al-Rawahi A, Al-Khiziri S, Ullah I, Ali L, Jung H, Lee I (2014b) Bacterial endophyte *Sphingomonas sp.* LK11 produces gibberellins and IAA and promotes tomato plant growth. J Microbiol 52(8):689–695

Khan AL, Hussain J, Al-Harrasi A, Al-Rawahi A, Lee I (2015) Endophytic fungi: resource for gibberellins and crop abiotic stress resistance. Crit Rev Biotechnol 35(1):62–74

Khan Z, Rho H, Firrincieli A, Hung SH, Luna V, Masciarelli O, Kim SH, Doty SL (2016) Growth enhancement and drought tolerance of hybrid poplar upon inoculation with endophyte consortia. Curr Plant Biol 6:38–47

Kim-Dura S, Lowman S, Zhang S, Mei C (2016) Growth promotion of switch grass by bacterial endophyte *Pantoea agglomerans* strain PaKM isolated from seeds. J Pathol Microbiol 1(2): 1007

Knief C, Delmotte N, Chaffron S, Stark M, Innerebner G, Wassmann R et al (2012) Metaproteogenomic analysis of microbial communities in the phyllosphere and rhizosphere of rice. ISME J 6:1378–1390

Kobayashi DY, Palumbo JD (2000) Bacterial endophytes and their effects on plants and uses in agriculture. In: Bacon CW, White JF (eds) Microbial endophytes. Marcel Dekker, Inc, New York, pp 199–233

Kuklinsky-Sobral J, Araújo WL, Mendes R, Geraldi IO, Pizzirani-Kleiner AA, Azevedo JL (2004) Isolation and characterization of soybean-associated bacteria and their potential for plant growth promotion. Environ Microbiol 6(12):1244–1251

Kumari M, Pandey S, Mishra SK, Giri VP, Agarwal L, Dwivedi S, Pandey AK, Nautiyal CS, Mishra A (2020) Omics-based mechanistic insight into the role of bioengineered nanoparticles for biotic stress amelioration by modulating plant metabolic pathways. Front Bioengg Biotechnol 17:242

Kunkel BN, Chen Z (2006) Virulence strategies of plant pathogenic bacteria. In: Dworkin M (ed) The prokaryotes. A handbook on biology of bacteria, ecophysiology and biochemistry. Springer, New York, pp 421–440

Leitão AL, Enguita FJ (2016) Gibberellins in *Penicillium* strains: challenges for endophyte-plant host interactions under salinity stress. Microbiol Res 183:8–18

Lowman JS, Lava-Chavez A, Kim-Dura S, Flinn B, Nowak J, Mei C (2015) Switch grass field performance on two soils as affected by Bacterization of seedlings with *Burkholderia phytofirmans* strain PsJN. Bioenergy Res 8(1):440–449

Miliūtė I, Buzaitė O (2011) IAA production and other plant growth promoting traits of endophytic bacteria from apple tree. Biologija 57(2):98–102

Morsy M, Cleckler B, Armuelles-Millican H (2020) Fungal endophytes promote tomato growth and enhance drought and salt tolerance. Plan Theory 9:877

Nath M, Bhatt D, Prasad R, Tuteja N (2017) Reactive oxygen species (ROS) metabolism and signaling in plant-mycorrhizal association under biotic and abiotic stress conditions. In: Varma A, Prasad R, Tuteja N (eds) Mycorrhiza. Springer International Publishing AG, pp 223–232

Naveed M, Mitter B, Reichenauer TG, Wieczorek K, Sessitsch A (2014) Increased drought stress resilience of maize through endophytic colonization by *Burkholderia phytofirmans* PsJN and *Enterobacter sp.* FD17. Environ Exp Bot 97:30–39

Nett RS, Montanares M, Marcassa A, Lu X, Nagel R, Charles TC, Hedden P, Rojas MC, Peters RJ (2017) Elucidation of gibberellin biosynthesis in bacteria reveals convergent evolution. Nat Chem Biol 13(1):69–74

Otlewska A, Migliore M, Dybka-Stepien K, Manfredini A, Struszczyk-Swita K, Napoli R, Białkowska A, Canfora L, Pinzari F (2020) When salt meddles between plant, soil, and microorganisms. Front Plant Sci 11:1429

Peleg Z, Blumwald E (2011) Hormone balance and abiotic stress tolerance in crop plants. Curr Opin Plant Biol 14:290–295

Phillips KA, Skirpan AL, Liu X, Christensen A, Slewinski TL, Hudson C, Barazesh S, Cohen JD, Malcomber S, McSteen P (2011) Vanishing tassel2 encodes a grass-specific tryptophan aminotransferase required for vegetative and reproductive development in maize. Plant Cell 23(2): 550–566

Prasad R (2017) Fungal nanotechnology: applications in agriculture, industry, and medicine. Springer International Publishing. ISBN 978-3-319-68423-9. https://www.springer.com/gb/book/9783319684239

Rashid S, Charles TC, Glick BR (2012) Isolation and characterization of new plant growth-promoting bacterial endophytes. Appl Soil Ecol 61:217–224

Raweekul W, Wuttitummaporn S, Sodchuen W, Kittiwongwattana C (2016) Plant growth promotion by Endophytic bacteria isolated from Rice (*Oryza sativa*). Thammasat Inter J Sci Technol 21(1):6–17

Sandhya V, Shrivastava M, Ali SZ, Sai Shiva Krishna Prasad V (2017) Endophytes from maize with plant growth promotion and biocontrol activity under drought stress. Russ Agric Sci 43: 22–34

Santoyo G, Moreno-Hagelsieb G, del Carmen Orozco-Mosqueda M, Glick BR (2016) Plant growth-promoting bacterial endophytes. Microbiol Res 183:92–99

Shade A, Jacques MA, Barret M (2017) Ecological patterns of seed microbiome diversity, transmission, and assembly. Curr Opin Microbiol 37:15–22

Shrivastava N, Jiang L, Li P, Sharma AK, Luo X, Wu S, Pandey R, Gao Q, Lou B (2018) Proteomic approach to understand the molecular physiology of symbiotic interaction between *Piriformospora indica* and *Brassica napus*. Sci Rep 8:5773

Singh VK, Singh AK, Singh PP, Kumar A (2018) Interaction of plant growth promoting bacteria with tomato under abiotic stress: a review. Agric Ecosyst Environ 267:129–140

Soliman M, Alhaithloul HA, Hakeem KR, Alharbi BM, El-Esawi M, Elkelish A (2019) Exogenous nitric oxide mitigates nickel-induced oxidative damage in eggplant by upregulating antioxidants, osmolyte metabolism, and glyoxalase systems. Plan Theory 8:562

Sun Y, Cheng Z, Glick BR (2009) The presence of a 1-aminocyclopropane-1-carboxylate (ACC) deaminase deletion mutation alters the physiology of the endophytic plant growth-promoting bacterium *Burkholderia phytofirmans* PsJN. FEMS Microbiol Lett 296(1):131–136

Sziderics AH, Rasche F, Trognitz F, Sessitsch A, Wilhelm E (2007) Bacterial endophytes contribute to abiotic stress adaptation in pepper plants (*Capsicum annuum* L.). Can J Microbiol 53(11): 1195–1202

Tiwari P, Bae H (2020) Horizontal gene transfer and endophytes: an implication for the acquisition of novel traits. Plan Theory 9:305

Uszkoreit J, Plohnke N, Rexroth S, Marcus K, Eisenacher M (2014) The bacterial proteogenomic pipeline. BMC Genomics 15:S19

Van Elsas JD, Turner S, Bailey MJ (2003) Horizontal gene transfer in the phytosphere. New Phytol 157:525–537

Vendan RT, Yu YJ, Lee SH, Rhee YH (2010) Diversity of endophytic bacteria in ginseng and their potential for plant growth promotion. J Microbiol 48(5):559–565

Verma H, Kumar D, Kumar V, Kumari M, Singh SK, Sharma VK, Droby S, Santoyo G, White JF, Kumar A (2021) The potential application of endophytes in management of stress from drought and salinity in crop plants. Microorganisms 9:1729

Waqas M, Khan AL, Kamran M, Hamayun M, Kang S, Kim Y, Lee I (2012) Endophytic fungi produce gibberellins and indole acetic acid and promotes host-plant growth during stress. Molecules 17(9):10754–10773

White JF, Torres MS (2010) Is plant endophyte-mediated defensive mutualism the result of oxidative stress protection? Physiol Plant 138:440–446

Wu QS, Xia RX, Zou YN (2008) Improved soil structure and citrus growth after inoculation with three arbuscular mycorrhizal fungi under drought stress. Eur J Soil Biol 44:122–128

Xu J, Li X, Luo L (2012) Effects of engineered *Sinorhizobium meliloti* on cytokinin synthesis and tolerance of alfalfa to extreme drought stress. Appl Environ Microbiol 78(22):8056–8061

Yaish MW, Antony I, Glick BR (2015) Isolation and characterization of endophytic plant growth-promoting bacteria from date palm tree (*Phoenix dactylifera* L.) and their potential role in salinity tolerance. Antonie Van Leeuwenhoek 107(6):1519–1532

Yandigeri MS, Meena KK, Singh D, Malviya N, Singh DP, Solanki MK, Yadav AK, Arora DK (2012) Drought-tolerant endophytic actinobacteria promote growth of wheat (*Triticum aestivum*) under water stress conditions. Plant Growth Regul 68:411–420

Yuan J, Zhang W, Sun K, Tang MJ, Chen PX, Li X, Dai CC (2019) Comparative transcriptomics and proteomics of *Atractylodes lancea* in response to endophytic fungus *Gilmaniella sp.* AL12 reveals regulation in plant metabolism. Front Microbiol 10:1208

Zhou XR, Dai L, Xu GF, Wang HS (2021) A strain of *Phoma species* improves drought tolerance of *Pinus tabulaeformis*. Sci Rep 11:7637

Žiarovská J, Medo J, Kyseľ M, Zamiešková L, Kačániová M (2020) Endophytic bacterial microbiome diversity in early developmental stage plant tissues of wheat varieties. Plan Theory 9:266

Chapter 13
Siderophore Production in Iron Uptake and Plant Biofortification

Neerja Srivastava

Abstract Although iron (Fe) is present abundantly on the earth, its bioavailability to plants is low and is dependent on the soil's redox potential and soil pH. In aerobic or alkaline soil, Fe is readily oxidized and usually forms insoluble complexes as ferric oxides which make it unavailable for plant uptake. Beneficial microorganisms, such as mycorrhiza and root endophytic fungi, dark septate fungi, and plant growth-promoting rhizobacteria, have several functions and can increase mineral Fe uptake. They have also been considered to have long-term efficiency once they are used in the field. Under iron-limited surroundings, some iron chelating bacteria produce iron chelating molecules siderophores to increase iron uptake as well as transport it to their host plants. This chapter covers and describes the role of siderophore in iron uptake and biofortification in plants.

Keywords Siderophores · Iron · Iron deficiency · Biofortification

13.1 Introduction

Micronutrient malnutrition is a serious threat to human health around the world, mainly in developing countries (Kennedy et al. 2003). Iron deficiency is an extremely predominant nutritional ailment affecting 2.5–5 billion people around the world (Yip 2002), where underprivileged households, as well as preschool children, are seriously affected because of the greater demand for iron (Benoist et al. 2008). Iron is a co-factor for many enzymes executing primary functions in the human body. An insufficient amount of iron causes disability and anaemia as well as retards mental growth (Sheftela et al. 2011). The malnourishment of iron can be reduced by enhancing the bio-available iron content through supplementation by fortification of food (Rana et al. 2012a). Biofortification is a method for producing

N. Srivastava (✉)
Department of Life Sciences and Biotechnology, School of Sciences, CSJM University, Kanpur, India
e-mail: neerjasrivastava@csjmu.ac.in

S. Chhabra et al. (eds.), *Plant Microbiome for Plant Productivity and Sustainable Agriculture*, Microorganisms for Sustainability 37,
https://doi.org/10.1007/978-981-19-5029-2_13

micronutrient-laden staple food through traditional breeding as well as transgenic strategies (Murgia et al. 2012). Rhizobacteria promote plant growth, fortify food crops for iron, and also improve the fertility of soil as well as crop yield via siderophore production (Rana et al. 2012a). Siderophores are low molecular weight organic substances that possess a high affinity for iron (Das et al. 2007). They form very stable iron complexes as well as increase the dissolution of minerals that have iron through coordination with an iron atom at the mineral surface. Rhizobacteria produce siderophores and release them into the surroundings to dissolve the iron by making iron chelate complex as well as translocate them to the plant through growing roots. Transporter proteins present on the plasma membrane of plant roots help in Fe uptake by forming iron–siderophore complexes (Boukhalfa and Crumbliss 2002). Moreover, these rhizobacteria can also improve plant growth by providing various advantageous features such as enhanced nitrogen fixation, solubilization of phosphorus, phytohormone synthesis, and organic acid production with decreased vulnerability for disease (Ahemad and Kibret 2014). According to Sharma et al. (2013), the amount of iron in rice and grain was enhanced through the use of plant growth-promoting bacteria (PGPR). Thus, plant biofortification by PGPR is regarded as a safe method to increase iron content in various edible plant parts as well as to remove malnutrition (Khalid et al. 2015).

13.2 Micronutrient Deficiency and Strategies for Control of Micronutrient Malnutrition

To counter the deficiency of micronutrients in food crops, the biofortification approach is a reasonable answer to maintaining an appropriate amount of vital nutrients in the various edible portions of plants. Unfavourable environmental stress conditions impede the micronutrient uptake from the soil into plants, and thus the insufficient level of micronutrients is going to restrict plant growth and development, causing malnutrition amongst the human population of the world. The modern agriculture system aims to yield nutritiously safe food crops free from chemical residues and with enhanced micronutrient level in the edible portion of plants. Most of the human population is highly dependent on foods that are based upon crops for a basic diet and if they take a diet with insufficient amounts of vital micronutrients, they face severe health issues such as anaemia. Implementing of approach such as biofortification is a better option for producing crops that possess an increased level of required micronutrients. According to the available literature, biofortification of food crops can be achieved mainly through three different ways: agronomic biofortification, breeding approach and genetic modifications.

Although these strategies of agronomic as well as genetic biofortification to supplement micronutrients in crops for removing micronutrient malnutrition are regarded as profitable, they are unsuitable in developing countries where the rural population is large (Mayer et al. 2008). Greater use of chemical fertilizers poses several problems such as disruption of soil microbial ecology and loss of soil fertility, which eventually can lead to drastic effects on both crop production and human health. Thus, there is a clear demand for discovering other cheap alternate approaches in place of genetic and agronomic strategies for achieving the goal of fortification. The application of microorganisms, particularly those which promote plant growth, is another alternate strategy for biofortification and is a better option in place of chemical fertilizers (Singh et al. 2017a). Furthermore, the use of plant growth-promoting rhizobacteria as bioinoculants can provide a cheaper and manageable option for increasing the micronutrient level in plants (Shaikh and Saraf 2017). Soil inhabiting microbiome promotes plant growth, preserves soil health, and enables the conversion of complex micronutrients into a simple form that is eventually taken up by plants (Singh and Singh 2017; Shah et al. 2018; Prasad et al. 2016). Plant-associated bacteria induce the host plant growth by various means such as increased mobility, uptake, and enhancement of nutrients in plants (Prasanna et al. 2016; Singh and Prasad 2014; Chhabra et al. 2010, 2013). Plant growth-promoting bacteria (PGPB) increases crop yield through several processes such as biological nitrogen fixation, solubilization of insoluble minerals such as phosphorus, zinc, and calcium, phytohormone production, and biocontrol of insect pests and plant pathogens (Glick et al. 1999; Singh et al. 2010a, b, c, 2011; Singh and Goel 2015; Prasad et al. 2015). Microbes can also affect the availability of nutrients by several properties such as chelation, solubilization, and oxidation or reduction activities (Khan 2005; Singh et al. 2013, 2018c). Soil, as well as rhizosphere-associated microbes, assists in the enhanced nutrient acquisition of vital micronutrients from the soil system to different plant parts. Therefore microbial-assisted precise nutrient acquisition strategy makes microbes ideal candidates for crop biofortification with essential nutrient elements such as Zn, Fe, and Se. Different microbes such as bacteria, cyanobacteria, and fungi can be used for growing plants effectively with enhanced nutrients level, which are achieved through genetic and agronomic-based biofortification.

13.3 Need for Biofortification

Biofortification targets either enhanced storage of micronutrients in edible portions of plants or increase in their bioavailability for plants. It is regarded as an inexpensive approach that emphasizes removing malnutrition in developing nations. Iron is a naturally present metalloid, which is needed in small quantities for humans and other animals. It is a vital micronutrient for every microbe, plant, and animal. It is crucial for plants as it plays a very important role in transporting oxygen, oxidative metabolism, and cell proliferation as well as required in some physiological mechanisms such as N_2 fixation and photosynthesis (Nair and Iyengar 2009). Iron is an important micronutrient in biological systems and getting attention worldwide due to growing reports on iron deficiencies in human populations and crops. Mineral and vitamin deficiencies are responsible for about 7.3% of the global disease burden (WHO 2002). Iron deficiency is amongst the most prominent micronutrient deficiencies and is affecting almost two billion people (McLean et al. 2009) It is one of the major health issues for 39% of children below the age of 5 years, 48% of 5–14 years old children, and 42% of total women while 52% of expecting women in developing countries are anaemic (Zimmermann and Hurrell 2007; Bersamin et al. 2008). In developing countries, legumes, vegetables, and cereal grains are the key sources of nutrition. Today, plant growth-promoting rhizobacteria (PGPR) are applied as substitutes for increased micronutrient uptake in plants. These bacteria produce plant growth-promoting substances, which may have a significant role in increasing enhancing with increasing plant growth through morphological changes such as greater root surface area for nutrient uptake in the soil and also for protecting crops against diseases (Rana et al. 2012b; Patel et al. 2020; Prasad et al. 2015)

13.4 Strategies for Biofortification

Biofortification approaches such as agricultural, plant breeding, genetic manipulation and the use of microbial inoculants are the main processes for increasing the concentration of nutrients in plants. 'Agronomic biofortification' is the process of applying nutrient-laden fertilizers into the soil or on foliage to enhance the micronutrient level in the edible portion of the crops and ultimately improve the use of required micronutrients by users (Carvalho and Vasconcelos 2013). A significant strategy for biofortification 'plant breeding' is being used by farmers for several years whereas traditional plant breeding is the crossing of plants to yield offspring having properties of both the parents (Garcia-Casal et al. 2016), and thus this technique can be employed to produce progenies with a considerable amount of required nutrients. Rice varieties having a greater amount of iron and zinc are crossed with high-yielding rice species to get offspring having both properties such as greater yield and a higher amount of micronutrients (Khush 2003). Biofortification via 'genetic modification' takes a shorter time to produce crops that express the trait of interest like nutritional content in a very refined manner and permits the transfer of specific genes or genes of interest (Garcia-Casal et al. 2016). The utilization of microbial inoculants for biofortification is an inexpensive strategy and provides a sustainable answer for increased micronutrient levels in plants in an eco-friendly way. Useful soil microbes make possible nutrients available to plants through several processes such as atmospheric N_2 fixation, solubilization of the nutrients fixed into the soil, and through the production of phytohormones (Yao et al. 2008). These microbes facilitate further increase in the level of micronutrients in plants because they have a prime role in the mineralization of organic material as well as converting inorganic nutrients (Rana et al. 2012a; Upadhayay et al. 2018). Each biofortification strategy is described in detail in the following section.

13.4.1 Agronomic Interventions

The application of mineral fertilizers to the soil for maintaining soil health and enhancing plant quality is a very old practice (Rengel et al. 1999). It was seen that the used nutrient transferred more noticeably when a particular element is deficient in the soil or the properties of the specific element permit its quick utilization. Supplementation of micronutrients by using chemical fertilizers is an efficient method applied by farmers to get the highest crop production. Although, efficiency of micronutrients utilization is very low in crops and only 2-5% of total applied dose of fertilizer is used (Tian et al. 2008).

13.4.2 Genetic Approaches

Genetic biofortification includes traditional breeding methods for characterization and exploitation of genetic differences such as by using novel strategies including gene discovery and marker-assisted breeding (Grusak 2002). Hindu et al. (2018) utilized genome-wide association studies (GWAS) for the recognition of various genomic regions in maize for kernel zinc and iron biofortification. Velu et al. (2016) proposed that genomic selection (GS) may be a possible breeding technique for iron and zinc biofortification in wheat. There are various benefits of nutrition-oriented breeding of crop plants based on sustainability. Current developments enable us to modify signalling pathways. DNA markers and marker-assisted selection (MAS) techniques are speeding up the growth of nutrient-rich genotypes. Kumar et al. (2018) described quantitative trait loci (QTLs) for iron and zinc biofortification in pearl millet via diversity array technology (DArT) and simple sequence repeat (SSR) markers. There are various reports on Fe and/or Zn biofortified varieties of rice, wheat, and maize which are released globally including in India to lessen malnutrition. In addition to the genetic or plant breeding strategies, various transgenic approaches have also been used for fruitfully biofortifying food crops. Transgenic methods allow the substitution of genes between completely unrelated species or carry new genes into food or cash crops. Ramesh et al. (2004) devised a new strategy for enhancing Zn and Fe content in seeds by overexpressing a zinc transporter in Hordeum vulgare cv. Golden Promise assisted by a ubiquitin promoter. Goto et al. (1999) succeeded in enhancing the Fe concentration three times more in rice grains by *Agrobacterium*-mediated transfer of the whole coding sequence of the ferritin gene of soybean plants. Lucca et al. (2002) produced transgenic rice plants, with more Fe content, rich in phytase and cysteine peptides which facilitated improved iron uptake and bioavailability. Vasconcelos et al. (2003) manipulated the soybean ferritin gene expression regulated by glutelin promoter in a selected Indica rice line with highly needed agronomic and field performance characters. More amount of grain nutrition was reported in brown grains and polished grains. Liu et al. (2004) through genetic engineering produced ferritin-containing rice varieties with 64% more iron amount in the milling where this ferritin gene could be expressed at a higher level, particularly in the transgenic rice endosperm. Genetic strategies might be tough work for breeders in soils typically having fewer iron and zinc micronutrients. To get the full capacity of biofortified species, simultaneous attention is required for additional aspects such as soil pH and organic matter which affect root exudation as well as enzyme activities in the rhizosphere and thus micronutrient intake and storage (Cakmak 2008).

13.4.3 Microorganisms and Plant-Based Strategies

The process of taking iron by higher plants in iron deficiency is classified into two groups (Kobayashi and Nishizawa 2012; Römheld and Marschner 1986): strategy I for non-graminaceous plants and strategy II for graminaceous plants. The two key mechanisms in the strategy I response are (1) ferric chelate reduction at the root surface by the assistance of the ferric reduction oxidase gene (*FRO2*) and (2) the absorption of the produced ferrous ions through the root plasma membrane via iron-regulated transporter gene (*IRT1*). Another mechanism included in strategy I is the removal of proton and phenolic compounds by the roots to the rhizosphere, which improves the solubility of ferric ions or supports the reduction capacity of ferric on the root surface. Strategy II plants uptake iron in iron deficiency through the excretion of phytosiderophores (PSs), which are low molecular weight Fe chelating substances such as mugineic acids (MA) and nicotianamine (NA) having high affinity for ferric and produce an iron–phytosiderophore soluble complex. The iron–phytosiderophore complex is then transferred into root cells by a strong affinity intake scheme. Suzuki et al. (2006) reported that barley plants secreted mugineic acid (MA) phytosiderophore under Zn deficiency and formed Zn(II)–mugineic acid complex and absorbed more Zn^{2+} by the roots of a Zn-deficient plant. The amounts and kinds of phytosiderophores secreted by plants into the rhizosphere vary from species to species (Mori 1999). A large concentration of iron and zinc is found in the earth's crust but not available to plants because they exist in the form of insoluble salts. Plant-dependent intrinsic approaches such as the production of phytosiderophore or organic acid and secretions of chelators are not every time adequate for making micronutrients accessible in micronutrient-deficient soils. With our enhanced knowledge of crosstalk between soils, plants, and microorganisms, higher awareness of the rhizosphere environment has been achieved (De Santiago et al. 2011; Mishra et al. 2011; Pii et al. 2015; Zaidi et al. 2003). Plant growth-promoting microbes play a very important part in the fortification of macronutrients in food crops through several processes such as siderophore production, transformations, nitrogen fixation, and phosphorus mobilization (Khan et al. 2019; Singh et al. 2018b). Microbes play key roles in the zinc and iron biofortification in cereal grains (Gosal et al. 2010; Rana et al. 2012a; Sharma et al. 2012). Both rhizospheric and endophytic microbes prominently affect micronutrient bioavailability in plants. Endophytic microorganisms are regarded as more promising agents to enhance iron and zinc uptake and translocation because endophytic microbes can indirectly influence the regulation of metal transporters (Reiter et al. 2002; Weyens et al. 2013). Bacterial and fungal endophytes have been implicated in the biofortification of grains of wheat and rice with iron and zinc (Abaid-Ullah et al. 2015; Ramesh et al. 2014). Gosal et al. (2010) described an endophytic fungus *Piriformospora indica* as having an important influence on plant growth, biomass, and micronutrient uptake. *Arthrobacter sulfonivorans* (DS-68) and *Enterococcus hirae* (DS-163) improved iron concentration and iron availability in wheat grains (Singh et al. 2017b, 2018b). Rana et al. (2012a) reported that the iron amount in wheat grains improved considerably because of inoculation of *Providencia* sp. PW5. There are several processes by which microorganisms enhance the zinc and iron availability in soil and increase their movement in plant parts or improve the bioavailability of iron and zinc in food grains. These processes are: i) production of siderophores and other chelating substances, ii) secretion of organic acid and proton extrusion, iii) alteration in morphology and anatomy of root, iv) overexpression of zinc and iron transporters, v) decrease in phytic acids or anti-nutritional factors in food grains, vi) phenolics, and its related reducing moieties secretion and vii) phytohormones such as signalling molecules secretion.

13.5 Production of Siderophore and Other Chelating Substances

Low molecular weight iron chelating compounds which possess a strong affinity for Fe(III) (Ganz 2018) are called siderophores. It has been reported by several workers that siderophores are secreted by various microorganisms to counter the iron scarcity in soil (Schalk et al. 2011). Fe(III) is not soluble in soil, but siderophores make siderophore–Fe(III) complexes that can improve the availability of iron in the environment (Saha et al. 2012). Owing to their solubilizing influence on iron hydroxides, the siderophores are produced in the rhizosphere, which is the crucial microbial action for assisting plants in gaining iron (Desai and Archana 2011; Hayat et al. 2012). Khalid et al. (2015) described that through inoculating siderophore-generating fluorescent *Pseudomonas*, iron amount in chickpea grains can be enhanced. In another previous field experiment, inoculation with siderophore generating endophytes *Arthrobacter sulfonivorans* DS-68 and *Enterococcus hirae* DS-163 increased the iron content in grains of low and high iron storing wheat genotypes by 67% and 46%, respectively. Singh et al. (2018b) observed a positive connection between the production of siderophores and the iron storage in wheat grains through endophytes, which shows the significant role of siderophores in gaining iron and movement into the plant. However, its movement within the plant, particularly to grains, involves various steps as well as processes such as transport, remobilization, and accumulation mechanisms which are facilitated through membrane transporters, chelators, and regulatory proteins (Singh and Prasanna 2020).

13.5.1 Siderophores Secreted by Mycorrhizal Fungi

Siderophore ferricrocin is produced by ectomycorrhizal fungus (EMF) *Cenococcum geophilum*. When the fungus was grown in a less iron culture medium, certain siderophores were produced in a culture medium where the key siderophore was ferricrocin, which was confirmed by HPLC, FAB-mass spectrometry, and ^{1}H- and ^{13}C-NMR spectra (Haselwandter and Winkelmann 2002). Two members in the genus *Suillus*, *S. granulatus* and *S. luteus*, are generally EMF species, and they produced cyclic and linear sideramine fusigen, ferrichrome, coprogen together with triacetylfusarinine C in the culture medium (Haselwandter et al. 2011). Two fungi *Wilcoxina mikolae* and *W. rehmii* are new fungi that can have a symbiosis with certain members of the genus *Pinus* to create ectomycorrhizae. When three fungal strains, *W. mikolae* CSY-14 and RMD-947 and *W. rehmii* CSY-85, were grown in a pure culture medium with little iron, they secreted ferricrocin (Prabhu et al. 1996). Several members of the genus *Glomus* can have a symbiosis with various food crops, such as *G. intraradices* with wheat (Mohammad et al. 2004), *G. mossae* with lettuce (*Lactuca sativa*) and onion (*Allium cepa*) (Garcia-Garrido et al. 1992), and *G. versiform* with wheat (Luo et al. 2019). A new siderophore basidiochrome is produced in the orchidaceous mycorrhizal fungi *Ceratobasidium* and *Rhizoctonia* spp. (Haselwandter et al. 2006). When AMF infected *Poncirus trifoliata* seedlings in loam and lime condition, there were no prominent changes in total root length as well as specific root length in comparison to non-AMF seedlings, but shoots iron level was noticeably greater than that of non-AMF seedlings (Treeby 1992). Additionally, siderophores produced through mycorrhizal fungi are vital for iron solubilization. Watteau and Berthelin (1994) observed that ectomycorrhizal fungus *Suillus granulatus* could solubilize Fe form goethite (ferric oxyhydroxide) and solubility can be because of siderophores produced by the fungus. The widespread existence of AMF causes the acidification of rhizosphere area due to the production of organic acids and phenolic substances, with the enhancement of iron availability in soil (White and Broadley 2009). Siderophores are produced by mycorrhizal varieties such as ect-, ectendo-, and endomycorrhizal fungi with orchid mycorrhizal fungi. If these mycorrhizal fungi can have a symbiosis with food crops, they are valuable for iron biofortification.

13.5.2 Siderophores Produced by Dark Septate Fungi and Root Endophytic Fungi

Dark septate fungi (DSFs) are usually studied for their role in plant growth promotion, uptake of nutrition, and tolerance to abiotic and biotic stresses. Hydroxamate siderophores are produced in DSF *Phialocephala fortinii*, which is an anamorphic ascomycete belonging to the order Helotiales (Bartholdy et al. 2001). In the two isolates of the fungus, the highest siderophore production was found at a pH in the range of pH 4.0–4.5, and the optimal concentration of ferric iron was determined to be between 20 and 40 μg/L iron(III) under the experimental conditions. In addition to hydroxamate, ferrirubin and ferrichrome C are also produced in the two isolates. Plant endophytic fungi are those fungi that can complete their life cycle partially or entirely in an inter- and/or intracellular way in the host plant tissues and cause no symptoms of infection in the host plants. Root endophytic fungi comprise a wide variety of fungi having symbiosis with plants. Members of the genus *Trichoderma* are usually employed in biofilm biofertilizers for plant growth and biocontrol. These members produce and discharge several siderophores. For example, *T. asperellum* (Segarra et al. 2010), *T. longibrachiatum*, and *T. pseudokoningii* synthesize coprogen, ferricrocin with a coprogen derivative (Anke et al. 1991), *T. virens* synthesizes ferricrocin (Mukherjee et al. 2018), and *T. harzianum* produces a new siderophore harzianic acid (Vinale et al. 2013). *Beauveria bassiana* and *Metarhizium robertsii* (*M. anisopliae*) are well-known entomopathogenic fungi, but occasionally these two fungi develop into root endophytic fungi to increase plant growth (Greenfield et al. 2016).

13.5.3 Siderophores Produced from PGPR

Plant growth-promoting rhizobacteria (PGPR) are very common in agriculture as well in forestry production which increases plant growthas well as yield and is also used as biocontrol agents (Adesemoye and Egamberdieva 2013; Durairaj et al. 2017; Ferreira et al. 2019). Siderophores secreted by PGPR can work as plant growth promoters, which has been even proven by evidence (Pahari et al. 2017). Members of the genus *Pseudomonas* act as PGPR and they are popular for their functions. These bacteria also produce siderophores, such as pyoverdine with pyochelin as in the case of *P. aeruginosa* (Lehoux et al. 2000; Rossbach et al. 2000; Li et al. 2009) and pyoverdine in *P. fluorescens* (Baysse et al. 2001; Visca et al. 2007). *P. aeruginosa* also works like an opportunistic pathogen. In the bacterium, siderophore-mediated signalling controls the production of virulence factors (Lamont et al. 2002). In *P. fluorescence*, mutations in *hemA* and *hemH* genes affect the biosynthesis of pyoverdine (Baysse et al. 2001), and the Fe-regulated *PbrA* sigma factor controls the biosynthesis of the siderophore pseudobactin. Members of the genus *Bacillus* are also prominent PGPR *Bacillus subtilis* CAS15 synthesized the catecholic siderophore bacillibactin and Fe inhibited siderophore productivity (May et al. 2001; Yu et al. 2011). This bacterium also synthesizes siderophore AH18 (Woo and Kim 2008), while *B. licheniformis* synthesizes a catecholic siderophore (Temirov et al. 2003) and ferrisiderophores are produced by *B. megaterium* (Arceneaux and Byers 1980). *B. stratosphericus* too synthesizes siderophores (Durairaj et al. 2017). *Serratia marcescens* is also among the one of PGPR, and one siderophore-deficient mutant of *S. marcescens* was identified in which one gene *entA* homologue is inactivated (Hofte and Bakker 2007). The gene *entA* codes for 2,3-dihydro-2,3-dihydroxybenzoate dehydrogenase enzyme which is involved in the biosynthetic pathway of enterobactin siderophore. The role of these siderophores synthesized by PGPR is still unidentified for Fe uptake in host plants.

13.6 Effects of Siderophores Secreted by Beneficial Microorganisms on Cellular Structures and Iron Distribution in Host Plants

Siderophores produced by beneficial microorganisms influence cellular structures as well as iron distribution in the roots of host plants. Recently, a study was done on *Pinus sylvestris* using transmission electron microscopy together with energy-dispersive X-ray spectroscopy. Mucha et al. (2019) described that metabolites derived from the ectomycorrhizal fungus *Hebeloma crustuliniforme* entered and interacted with *P. sylvestris* root cells and strongly altered the ultrastructure of root cells. The triacetylfusarinine C and ferricrocin siderophore treatments noticeably altered root cells' ultrastructure. Meanwhile, substances produced through *H. crustuliniforme* enhanced the iron level in the cell wall, cytoplasm, mitochondrion, nucleus, and the vacuole of host cells. There was almost 23 and 15 times enhancement in iron in the cell wall and cytoplasm, respectively. When PGPRs *Arthrobacter sulfonivorans* DS-68 and *Arthrobacter* sp. DS-179 infected wheat seedlings, clear alterations in root anatomy were observed (Singh et al. 2018a). The endodermis, cortical region, root hair extension, xylem and xylem vessels, pericycle, and vascular bundles were more distinct and thicker in comparison to control. These results oncluded that mycorrhizal fungi, as well as PGPR, increased iron uptake of host plants through the production of siderophores with ultrastructural alterations in host root cells. The same type of study is essential for food crops such as maize, rice, and barley to understand in a better way the processes by which mycorrhizal fungi, as well as PGPR, enhance iron uptake together with the biofortification of food crops. Taken together, beneficial microorganisms increase iron uptake through their host plants. Siderophore-secreting systems in beneficial microorganisms have shown their capability in increasing iron bioavailability to plants and thereby reducing iron fertilizer application. Moreover, beneficial microorganisms can also be employed for increased phytoremediation of iron polluted soils (Guo et al. 2013; Abbaszadeh-Dahaji et al. 2016; Mishra et al. 2016; Sut et al. 2016; Liu et al. 2021).

13.7 Conclusion

It can be concluded that food-dependent crops with increased micronutrients level are needed worldwide for removing hidden hunger. Plant-associated microorganisms exist in the soil and root interface and increase uptake of macro- and micronutrients for plant life. Several microbes proficient in the production of siderophore were considered for improving iron in food crops. Microbial inoculants can play a useful role in biofortification and help in decreasing dependency on expensive strategies such as agronomic involvement and genetic modification for increasing micronutrient level in the edible portions of crops. Microbes possess several plant growth stimulating characters as well as various nutrient acquiring mechanisms that can be utilized for the formulation of the biofortification approach. Fe-biofortified plants can be a good source of iron to fulfill the daily needs of human populations and can prevent the diseases caused by iron defciencies.

References

Abaid-Ullah M, Hassan MN, Jamil M, Brader G, Shah MKN, Sessitsch A, Hafeez FY (2015) Plant growth promoting rhizobacteria: an alternate way to improve yield and quality of wheat (Triticum aestivum). Int J Agric Biol 17:51–60

Abbaszadeh-Dahaji P, Omidvari M, Ghorbanpour M (2016) Increasing phytoremediation efficiency of heavy metal-contaminated soil using PGPR for sustainable agriculture. In: Plant-microbe interaction: an approach to sustainable agriculture. Springer, Singapore, pp 187–204

Adesemoye AO, Egamberdieva D (2013) Beneficial effects of plant growth-promoting rhizobacteria on improved crop production: prospects for developing economies. In: Maheshwari D, Saraf M, Aeron A (eds) Bacteria in agrobiology: crop productivity. Springer, Berlin, pp 45–63

Ahemad M, Kibret M (2014) Mechanisms and applications of plant growth-promoting rhizobacteria: current perspective. J King Saud Univ Sci 26(1):1–20

Anke H, Kinn J, Bergquist KE, Sterner O (1991) Production of siderophores by strains of the genus *Trichoderma*. Biol Met 4(3):176–180

Arceneaux JE, Byers BR (1980) Ferrisiderophore reductase activity in *Bacillus megaterium*. J Bact 141:715–721

Bartholdy B, Berreck M, Haselwandter K (2001) Hydroxamate siderophore synthesis by *Phialocephala fortinii*, a typical dark septate fungal root endophyte. Biometals 14:33–42

Baysse C, Matthijs S, Pattery T, Cornelis P (2001) Impact of mutations in hemA and hemH genes on pyoverdine production by *Pseudomonas fluorescens* ATCC17400. FEMS Micro Lett 205(1): 57–63

Benoist BD, McLean E, Andersson M, Rogers L (2008) Iodine deficiency in 2007: global progress since 2003. Food Nutr Bull 29(3):195–202

Bersamin A (2008) Nutrition and health info sheet: iron and iron deficiency anemia. UCANR Publications, pp 1–6

Boukhalfa H, Crumbliss AL (2002) Chemical aspects of siderophore mediated iron transport. Biometals 15(4):325–339

Cakmak I (2008) Enrichment of cereal grains with zinc: agronomic or genetic biofortification? Plant Soil 302(1–2):1–17

Carvalho SM, Vasconcelos MW (2013) Producing more with less: strategies and novel technologies for plant-based food biofortification. Food Res Int 54(1):961–971

Chhabra S, Hogan J, Ryan D, Burke J, Brazil D, Dowling D (2010) Analysis of microbial communities under long-term conventional and reduced-input management of tillage soil. In: Vilas AM (ed) Microorganisms in industry and environment, from scientific and industrial research to consumer products. World Scientific Publications, Singapore, pp 8–11. https://doi.org/10.1142/9789814322119_0002. ISBN: 978-981-4322-10-2

Chhabra S, Brazil D, Morrissey J, Burke J, O'Gara F, Dowling DN (2013) Fertilization management affects the alkaline phosphatase bacterial community in barley rhizosphere soil. Biol Fert Soils 49:31–39. https://doi.org/10.1007/s00374-012-0693-2

Das A, Prasad R, Srivastava A, Giang PH, Bhatnagar K, Varma A (2007) Fungal siderophores: structure, functions and regulations. In: Varma A, Chincholkar SB (eds) Microbial siderophores, vol 12. Springer-Verlag, Berlin, pp 1–42

de Santiago A, Quintero JM, Avilés M, Delgado A (2011) Effect of Trichoderma asperellum strain T34 on iron, copper, manganese, and zinc uptake by wheat grown on a calcareous medium. Plant Soil 342(1–2):97–104

Desai A, Archana G (2011) Role of siderophores in crop improvement. In: Bacteria in agrobiology: plant nutrient management. Springer, New York, pp 109–139

Durairaj K, Velmurugan P, Park JH, Chang WS, Park YJ, Senthilkumar P, Choi K-M, Lee J-H, Oh BT (2017) Potential for plant biocontrol activity of isolated *Pseudomonas aeruginosa* and *Bacillus stratosphericus* strains against bacterial pathogens acting through both induced plant resistance and direct antagonism. FEMS Micro Lett 364(23):fnx225

Ferreira CM, Soares HM, Soares EV (2019) Promising bacterial genera for agricultural practices: an insight on plant growth-promoting properties and microbial safety aspects. Sci Total Env 682: 779–799

Ganz T (2018) Iron and infection. Int J Hematol 107(1):7–15. https://doi.org/10.1007/s12185-017-2366-2

García-Garrido JM, García-Romera I, Ocampo JA (1992) Cellulase production by the vesicular–arbuscular mycorrhizal fungus Glomus mosseae (Nicol. & Gerd.) Gerd. and Trappe. New Phytol 121(2):221–226

Garcia-Casal MN, Peña-Rosas JP, Pachón H, De-Regil LM, Tablante EC, Flores-Urrutia MC (2016) Staple crops biofortified with increased micronutrient content: effects on vitamin and mineral status, as well as health and cognitive function in the general population. Cochrane Database Syst Rev 8

Glick BR, Holguin G, Patten CL, Penrose DM (1999) Biochemical and genetic mechanisms used by plant growth promoting bacteria. Imperial College Press, London

Gosal S, Karlupia A, Gosal S, Chhibba I, Varma A (2010) Biotization with Piriformospora indica and Pseudomonas fluorescens improves survival rate, nutrient acquisition, field performance and saponin content of micropropagated Chlorophytum sp. Indian J Biotechnol 9:289–297

Goto F, Yoshihara T, Shigemoto N, Toki S, Takaiwa F (1999) Iron fortification of rice seed by the soybean ferritin gene. Nat Biotechnol 17:282–286

Greenfield M, Gómez-Jiménez MI, Ortiz V, Vega FE, Kramer M, Parsa S (2016) *Beauveria bassiana* and *Metarhizium anisopliae* endophytically colonize cassava roots following soil drench inoculation. Biol Control 95:40–48

Grusak MA (2002) Enhancing mineral content in plant food products. J Am Coll Nutr 21(suppl 3): 178S–183S

Guo W, Zhao R, Yang H, Zhao J, Zhang J (2013) Using native plants to evaluate the effect of arbuscular mycorrhizal fungi on revegetation of iron tailings in grasslands. Biol Fert Soils 49(6): 617–626

Haselwandter K, Winkelmann G (2002) Ferricrocin—an ectomycorrhizal siderophore of *Cenococcum geophilum*. Biometals 15:73–77

Haselwandter K, Passler V, Reiter S, Schmid DG, Nicholson G, Hentschel P et al (2006) Basidiochrome—a novel siderophore of the orchidaceous mycorrhizal fungi *Ceratobasidium* and *Rhizoctonia* spp. Biometals 19(3):335–343

Haselwandter K, Haninger G, Ganzera M (2011) Hydroxamate siderophores of the ectomycorrhizal fungi *Suillus granulatus* and *S. luteus*. Biometals 24:153–157

Hayat R, Ahmed I, Sheirdil RA (2012) An overview of plant growth promoting rhizobacteria (PGPR) for sustainable agriculture. In: Crop production for agricultural improvement. Springer, New York, pp 557–579

Hindu V, Palacios-Rojas N, Babu R, Suwarno WB, Rashid Z, Usha R, Saykhedkar GR, Nair SK (2018) Identification and validation of genomic regions influencing kernel zinc and iron in maize. Theor Appl Genet 131:1443–1457

Hofte M, Bakker PAHM (2007) Competition for iron and induced systemic resistance by siderophores of plant growth promoting rhizobacteria. In: Varma A, Chincholkar SB (eds) Microbial siderophores. Springer-Verlag, Berlin, pp 121–133

Kennedy G, Nantel G, Shetty P (2003) The scourge of "hidden hunger": global dimensions of micronutrient deficiencies. Food Nutr Agric 32:8–16

Khalid S, Asghar HN, Akhtar MJ, Aslam A, Zahir ZA (2015) Biofortification of iron in chickpea by plant growth-promoting rhizobacteria. Pak J Bot 47:1191–1194

Khan AG (2005) Role of soil microbes in the rhizospheres of plants growing on trace metal contaminated soils in phytoremediation. J Trace Elem Med Biol 18:355–364

Khan A, Singh J, Upadhayay VK, Singh AV, Shah S (2019) Microbial biofortification: a green technology through plant growth promoting microorganisms. In: Sustainable green technologies for environmental management. Springer, Singapore, pp 255–269

Khush G (2003) Productivity improvements in rice. Nut Rev 61(6 Pt 2):S114–S116. https://doi.org/10.1301/nr.2003.jun.S114

Kobayashi T, Nishizawa NK (2012) Iron uptake, translocation, and regulation in higher plants. Annu Rev Plant Biol 63:131–152

Kumar S, Hash C, Nepolean T, Mahendrakar M, Satyavathi C, Singh G, Rathore A, Yadav R, Gupta R, Srivastava R (2018) Mapping grain iron and zinc content quantitative trait loci in an Iniadi-derived immortal population of pearl millet. Genes 9(5):24

Lamont IL, Beare PA, Ochsner U, Vasil AI, Vasil ML (2002) Siderophore-mediated signaling regulates virulence factor production in *Pseudomonas aeruginosa*. PNAS 99(10):7072–7077

Lehoux DE, Sanschagrin F, Levesque RC (2000) Genomics of the 35-kb pvd locus and analysis of novel pvdIJK genes implicated in pyoverdine biosynthesis in *Pseudomonas aeruginosa*. FEMS Micro Lett 190(1):141–146

Li W, Lyte M, Freestone PP, Ajmal A, Colmer-Hamood JA, Hamood AN (2009) Norepinephrine represses the expression of toxA and the siderophore genes in *Pseudomonas aeruginosa*. FEMS Micro Lett 299(1):100–109

Liu QQ, Yao QH, Wang HM, Gu MH (2004) Endosperm-specific expression of the ferritin gene in transgenic rice (Oryza sativa L.) results in increased iron content of milling rice. Yi Chuan Xue Bao 31:518–524

Liu C, Ye Y, Liu J, Pu Y, Wu C (2021) Iron biofortification of crop food by symbiosis with beneficial microorganisms. J Plant Nut 44(18):2793–2810

Lucca P, Hurrel R, Potrykus I (2002) Fighting iron deficiency anemia with iron-rich rice. J Am Coll Nutr 3:184S–190S

Luo W, Li J, Ma X, Niu H, Hou S, Wu F (2019) Effect of arbuscular mycorrhizal fungi on uptake of selenate, selenite, and selenomethionine by roots of winter wheat. Plant Soil 438(1):71–83

May JJ, Wendrich TM, Marahiel MA (2001) The dhb operon of *bacillus subtilis* encodes the biosynthetic template for the catecholic siderophore 2,3-dihydroxybenzoate-glycine-threonine trimeric ester bacillibactin. J Biol Chem 276(10):7209–7217

Mayer JE, Pfeiffer WH, Beyer P (2008) Biofortified crops to alleviate micronutrient malnutrition. Cur Opin Plant Biol 11(2):166–170

McLean E, Cogswell M, Egli I, Wojdyla D, de Benoist B (2009) Worldwide prevalence of anaemia, WHO vitamin and mineral nutrition information system, 1993-2005. Public Health Nutr 12: 444–454

Mishra PK, Bisht SC, Ruwari P, Joshi GK, Singh G, Bisht JK, Bhatt JC (2011) Bioassociative effect of cold tolerant Pseudomonas spp. and Rhizobium leguminosarum-PR1 on iron acquisition, nutrient uptake and growth of lentil (Lens culinaris L.). Eur J Soil Biol 47:35–43

Mishra V, Gupta A, Kaur P, Singh S, Singh N, Gehlot P, Singh J (2016) Synergistic effects of Arbuscular mycorrhizal fungi and plant growth promoting rhizobacteria in bioremediation of iron contaminated soils. Int J Phytoremediation 18(7):697–703

Mohammad A, Mitra B, Khan AG (2004) Effects of sheared-root inoculum of *Glomus intraradices* on wheat grown at different phosphorus levels in the field. Agri Eco Env 103(1):245–249

Mori S (1999) Iron acquisition by plants. Curr Opin Plant Biol 2(3):250–253

Mucha J, Gabała E, Zadworny M (2019) The effects of structurally different siderophores on the organelles of *Pinus sylvestris* root cells. Planta 249:1747–1760

Mukherjee PK, Hurley JF, Taylo JT, Puckhaber L, Lehner S, Druzhinina I, Schumacher R, Kenerley CM (2018) Ferricrocin, the intracellular siderophore of Trichoderma virens, is involved in growth, conidiation, gliotoxin biosynthesis and induction of systemic resistance in maize. Biochem Biophy Res Comm 505(2):606–661

Murgia I, Arosio P, Tarantino D, Soave C (2012) Biofortification for combating 'hidden hunger' for iron. Trends Plant Sci 17(1):47–55

Nair KM, Iyengar V (2009) Iron content, bioavailability and factors affecting iron status of Indians. Ind J Med Res 130(5):634–645

Pahari A, Pradhan A, Nayak SK, Mishra BB (2017) Bacterial siderophore as a plant growth promoter. In: Patra J, Vishnuprasad C, Das G (eds) Micro Biotech. Springer, Singapore, pp 163–180

Patel P, Bhatt S, Patel H, Saraf M (2020) Iron chelating bacteria: a carrier for biofortification and plant growth promotion. J Biol Studies 3(3):111–120

Pii Y, Mimmo T, Tomasi N, Terzano R, Cesco S, Crecchio C (2015) Microbial interactions in the rhizosphere: beneficial influences of plant growth-promoting rhizobacteria on nutrient acquisition process. A review. Biol Fertil Soils 51(4):403–415

Prabhu V, Biolchini PF, Boyer GL (1996) Detection and identification of ferricrocin produced by ectendomycorrhizal fungi in the genus *Wilcoxina*. Biometals 9(3):229–234

Prasad R, Kumar M, Varma A (2015) Role of PGPR in soil fertility and plant health. In: Egamberdieva D, Shrivastava S, Varma A (eds) Plant growth-promoting rhizobacteria (PGPR) and medicinal plants. Springer International Publishing, Cham, pp 247–260

Prasad B, Kumar A, Singh AV, Kumar A (2016) Plant growth and seed yield attributes as influenced by bacterial isolates under glass house. Prog Res 11(IV):2573–2576

Prasanna R, Nain L, Rana A, Shivay YS (2016) Biofortification with microorganisms: present status and future challenges. In: Singh U, Praharaj C, Singh S, Singh N (eds) Biofortification of food crops. Springer, New Delhi. https://doi.org/10.1007/978-81-322-2716-8_19

Ramesh SA, Choimes S, Schachtman DP (2004) Over-expression of an Arabidopsis zinc transporter in Hordeum vulgare increases short term zinc uptake after zinc deprivation and seed zinc content. Plant Mol Biol 54:373–385. https://doi.org/10.1023/B:PLAN.0000036370.70912.34

Ramesh A, Sharma SK, Sharma MP, Yadav N, Joshi OP (2014) Inoculation of zinc solubilizing Bacillus aryabhattai strains for improved grand wheat cultivated in Vertisols of Central India. Appl Soil Ecol 73:87–96

Rana A, Joshi M, Prasanna R, Shivay YS, Nain L (2012a) Biofortification of wheat through inoculation of plant growth promoting rhizobacteria and cyanobacteria. J Soil Biol 50:118–126

Rana A, Saharan B, Nain L, Prasanna R, Shivay YS (2012b) Enhancing micronutrient uptake and yield of wheat through bacterial PGPR consortia. Soil Sci Plant Nutr 58(5):573–582

Reiter B, Pfeifer U, Schwab H, Sessitsch A (2002) Response of endophytic bacterial communities in potato plants to infection with Erwinia carotovora subsp. atroseptica. Appl Environ Microbiol 68:2261–2268

Rengel Z, Batten GD, Crowley DE (1999) Agronomic approaches for improving the micronutrient density in edible portions of field crops. Field Crop Res 60:27–40

Römheld V, Marschner H (1986) Evidence for a specific uptake system for iron phytosiderophores in roots of grasses. Plant Physiol 80(1):175–180

Rossbach S, Wilson TL, Kukuk ML, Carty HA (2000) Elevated zinc induces siderophore biosynthesis genes and a zntA-like gene in *Pseudomonas fluorescens*. FEMS Micro Lett 191(1):61–70

Saha R, Saha N, Donofrio RS, Bestervelt LL (2012) Microbial siderophores: a mini review. J Basic Microbiol 52:1–15

Schalk IJ, Hannauer M, Braud A (2011) New roles for bacterial siderophores in metal transport and tolerance. Environ Microbiol 13:2844–2854

Segarra G, Casanova E, Aviles M, Trillas I (2010) Trichoderma asperellum strain T34 controls Fusarium wilt disease in tomato plants in soilless culture through competition for iron. Micro Eco 59:141–149

Shah S, Ramanan VV, Singh AV, Singh AK (2018) Potential and prospect of plant growth promoting rhizobacteria in lentil. In: Singh AK et al. (eds) Scientific lentil production. Satish Serial Publishing House, Delhi

Shaikh S, Saraf M (2017) Biofortification of Triticum aestivum through the inoculation of zinc solubilizing plant growth promoting rhizobacteria in field experiment. Biocat Agri Biotech 9: 120–126. https://doi.org/10.1016/j.bcab.2016.12.008

Sharma SK, Sharma MP, Ramesh A, Joshi OP (2012) Characterization of zinc-solubilizing Bacillus isolates and their potential to influence zinc assimilation in soybean seeds. J Microbiol Biotechnol 22:352–359. https://doi.org/10.4014/jmb.1106.05063

Sharma A, Shankhdhar D, Shankhdhar SC (2013) Enhancing grain iron content of rice by the application of plant growth promoting rhizobacteria. Plant Soil Environ 59(2):89–94

Sheftela AD, Mason AB, Ponka P (2011) The long history of iron in the universe and in health and disease. Biochim Biophys Acta Gen Subj 1820(3):161–187

Singh AV, Goel R (2015) Plant growth promoting efficiency of *Chryseobacterium* sp. PSR10 on finger millet (*Eleusine coracana*). J Global Biosci 4(6):2569–2575

Singh AV, Chandra R, Reeta G (2013) Phosphate solubilization by Chryseobacterium sp. and their combined effect with N and P fertilizers on plant growth promotion. Arch Agro Soil Sci 59 (5):641–651

Singh AV, Prasad B (2014) Enhancement of plant growth, nodulation and seed yield through plant growth promoting *Rhizobacteria* in lentil (*Lens culinaris* Medik cv. VL125). Int J Curr Microbiol Appl Sci 3(6):614–622

Singh D, Prasanna R (2020) Potential of microbes in the biofortification of Zn and Fe in dietary food grains. A review. Agr Sustain Dev 40:15

Singh J, Singh AV (2017) Microbial strategies for enhanced phytoremediation of heavy metals contaminated soils. In: Bharagava RN (ed) Environmental pollutants and their bioremediation approaches. Taylor & Francis, London, pp 249–264

Singh AV, Agarwal A, Goel R (2010a) Comparative phosphate solubilization efficiency of two bacterial isolates and their effect on *Cicer arietinum* seeds in indigenous and alternative soil system. Environ Ecol 28:1979–1983

Singh AV, Prasad B, Shah S (2010b) Screening plant growth promotory rhizobacteria for improving seed germination and seedling vigor of lentil (*Lens culinaris* Medik). Environ Ecol 28: 2055–2058

Singh AV, Shah S, Prasad B (2010c) Effect of phosphate solubilizing bacteria on plant growth promotion and nodulation in soybean (Glycine max (L.) Merr). J Hill Agri 1(1):35–39

Singh AV, Prasad B, Shah S (2011) Influence of phosphate solubilizing bacteria for enhancement of plant growth and seed yield in lentil. J Crop Weed 7(1):1–4

Singh J, Singh AV, Prasad B, Shah S (2017a) Sustainable agriculture strategies of wheat biofortification through microorganisms. In: Kumar A, Kumar A, Prasad B (eds) Wheat a premier food crop. Kalyani Publishers, New Delhi

Singh D, Rajawat MVS, Kaushik R, Prasanna R, Saxena AK (2017b) Beneficial role of endophytes in biofortification of Zn in wheat genotypes varying in nutrient use efficiency grown in soils sufficient and deficient in Zn. Plant Soil 416(1–2):107–116

Singh D, Geat N, Rajawat MVS, Mahajan MM, Prasanna R, Singh S, Kaushik R, Singh RN, Kumar K, Saxena AK (2018a) Deciphering the mechanisms of endophyte-mediated biofortification of Fe and Zn in wheat. J Plant Growth Regul 37(1):174–182

Singh D, Geat N, Rajawat MVS, Prasanna R, Kar A, Singh AM, Saxena AK (2018b) Prospecting endophytes from different Fe or Zn accumulating wheat genotypes for their influence as inoculants on plant growth, yield, and micronutrient content. Ann Microbiol 68(12):815–833

Singh AV, Prasad B, Goel R (2018c) Plant growth promoting efficiency of phosphate solubilizing *Chryseobacterium sp.* PSR 10 with different doses of N and P fertilizers on lentil (*Lens culinaris var.* PL-5) growth and yield. Int J Curr Microbiol App Sci 7(05):2280–2289

Sut M, Boldt-Burisch K, Raab T (2016) Possible evidence for contribution of arbuscular mycorrhizal fungi (AMF) in phytoremediation of iron-cyanide (Fe-CN) complexes. Ecotoxicology 25: 1260–1269

Suzuki M, Takahashi M, Tsukamoto T, Watanabe S, Matsuhashi S, Yazaki J, Kishimoto N, Kikuchi S, Nakanishi H, Mori S, Nishizawa NK (2006) Biosynthesis and secretion of mugineic acid family phytosiderophores in zinc-deficient barley. Plant J 48(1):85–97

Temirov YV, Esikova TZ, Kashparov IA, Balashova TA, Vinokurov LM, Alakhov YB (2003) A catecholic siderophore produced by the thermoresistant Bacillus licheniformis VK21 strain. Rus J Bio Chem 29(6):542–549

Tian X, Lu X, Mai W, Yang X, Li S (2008) Effect of calcium carbonate content on availability of zinc in soil and zinc and iron uptake by wheat plants. Soil 40(3):425–431

Treeby MT (1992) The role of mycorrhizal fungi and non-mycorrhizal micro-organisms in iron nutrition of citrus. Soil Biol Biochem 24(9):857–864

Upadhayay VK, Singh AV, Pareek N (2018) An insight in decoding the multifarious and splendid role of microorganisms in crop biofortification. Int J Curr Microbiol App Sci 7(06):2407–2418

Vasconcelos M, Datta K, Oliva N, Khalekuzzaman M, Torrizo L, Krishnan S, Oliveira M, Goto F, Datta SK (2003) Enhanced iron and zinc accumulation in transgenic rice with the ferritin gene. Plant Sci 164:71–378

Velu G, Crossa J, Singh RP, Hao Y, Dreisigacker S, Perez-Rodriguez P, Joshi AK, Chatrath R, Gupta V, Balasubramaniam A, Tiwari C (2016) Genomic prediction for grain zinc and iron concentrations in spring wheat. Theor Appl Genet 129(8):1595–1605

Vinale F, Nigro M, Sivasithamparam K, Flematti G, Ghisalberti EL, Ruocco M, Varlese R, Marra R, Lanzuise S, Eid A, Lorito M (2013) Harzianic acid: a novel siderophore from *Trichoderma harzianum*. FEMS Microbiol Lett 347(2):123–129

Visca P, Imperi F, Lamont IL (2007) Pyoverdine synthesis and its regulation in fluorescent Pseudomonads. In: Varma A, Chincholkar SB (eds) Microbial siderophores. Springer-Verlag, Berlin, pp 135–163

Watteau F, Berthelin J (1994) Mineral dissolution of iron and aluminium from soil minerals: efficiency and specificity of hydroxamate siderophores compared to aliphatic acids. Eur J Soil Biol 30:1–9

Weyens N, Beckers B, Schellingen K, Ceulemans R, Croes S, Janssen J, Haenen S, Witters N, Vangronsveld J (2013) Plant associated bacteria and their role in the success or failure of metal phytoextraction projects: first observations of a field related experiment. Microbial Biotech 6: 288–299

White PJ, Broadley MR (2009) Biofortiication of crops with seven mineral elements often lacking in human diets—iron, zinc, copper, calcium, magnesium, selenium and iodine. New Phytol 182: 49–84

Woo S-M, Kim S-D (2008) Structural identification of siderophore (AH18) from *Bacillus subtilis* AH18, a biocontrol agent of Phytophthora blight disease in red-pepper. Korean J Microbiol 36(4):326–335

World Health Organization (2002) The world health report 2002—reducing risks, promoting healthy life. World Health Organization, Geneva

Yao T, Yasmin S, Malik KA, Hafeez FY (2008) Potential role of Rhizobacteria isolated from Northwestern China for enhancing wheat and oat yield. J Agri Sci 146(01):49–56. https://doi.org/10.1017/S0021859607007356

Yip R (2002) Prevention and control of iron deficiency: policy and strategy issues. J Nutr 132(4): 802–805

Yu X, Ai C, Xin L, Zhou G (2011) The siderophore-producing bacterium, *Bacillus subtilis* CAS15, has a biocontrol effect on Fusarium wilt and promotes the growth of pepper. Eur J Soil Biol 47(2):138–145

Zaidi A, Khan MS, Amil MD (2003) Interactive effect of rhizotrophic microorganisms on yield and nutrient uptake of chickpea (Cicer arietinum L.). Eur J Agron 19(1):15–21

Zimmermann MB, Hurrell RF (2007) Nutritional iron deficiency. Lancet 370(9586):511–520

Chapter 14
Plant Microbiome Diversity and Potential for Crops and Sustainable Agriculture

Nazia Rifat Zaman, Farhana Tasnim Chowdhury, Haseena Khan, and Mohammad Riazul Islam

Abstract Since the last few decades plants are not treated as single entities, rather they are considered with association of different microbial communities known as plant microbiome. These microbiomes are now known an extension of a plant genome. Diversity of this microbial population is quite large including different classes of bacteria, archaea, and fungi. They always preserve symbiotic relation with a host plant by providing different growth promoting factors. It is well established that plant microbiota are structured mainly by different plant compartments (e.g., different plant organs, and diverse between rhizosphere, phyllosphere, and endosphere). These plant microbial groups show potential functions related to probiotics and plant protection. They can help a plant with germination and growth, as well as disease prevention, biotic, and abiotic stress resistance. Microbial diversity has been discovered as an important role in disease prevention and could be used as a biomarker in plant protection measures. The study of plant–microbiome interaction has resulted in a paradigm shift in our knowledge of the microbiome's involvement in plant health and disease, with significant implications for biocontrol and growth promotion.

Keywords Plant microbiome · Endophyte · Sustainable agriculture · Biofertilizers · Bioinoculant

N. R. Zaman
Department of Biochemsitry, Primeasia University, Dhaka, Bangladesh

F. T. Chowdhury · H. Khan · M. R. Islam (✉)
Department of Biochemistry and Molecular Biology, Faculty of Biological Sciences, University of Dhaka, Dhaka, Bangladesh
e-mail: mriazulislam@du.ac.bd

S. Chhabra et al. (eds.), *Plant Microbiome for Plant Productivity and Sustainable Agriculture*, Microorganisms for Sustainability 37,
https://doi.org/10.1007/978-981-19-5029-2_14

14.1 Introduction

Increase of global food production has been greatly disrupted due to rapid industrialization, anthropogenic interventions, climate change events, overgrazing, and extensive uses of agrochemicals/fertilizers/pesticides (Raza et al. 2019). These problems are getting worse with shrinking agricultural land, deterioration of soil quality and nutrition profile, and recurring challenges due to several biotic and abiotic stresses. The UN Food and Agriculture Organization (FAO) has projected to increase the food production 70% more by 2050 to meet the global needs of the predicted population of nine billion people on earth (www.fao.org). To meet this challenge a low-cost sustainable agro-ecosystem technology should be adopted.

For the last few decades, lots of approaches have been proposed to improve crop production. Among them, the use of microorganisms to improve soil health, plant growth has got much attention compared to the use of conventional chemical stimulants and pesticides. These microorganisms are plant-associated microbes residing in soil or plant tissues and commonly known as plant microbiome (Singh et al. 2019). Different studies have shown that, use of plant microbiome has a great potentiality in mitigating the challenges currently faced in agriculture. The microbial components of a plant in the rhizosphere, phyllosphere, and endosphere have important functions supporting plant growth and health (Vorholt 2012). Plant microbiota consist of different types of organisms including fungi, archaea, and bacteria. A deeper knowledge of the plant as a meta-organism and how plants can benefit from their microbial partners can be gained by revealing the functioning of plant–microbe interactions and elements involved in community assembly (Hardoim et al. 2015; Hacquard 2016). Multiple factors influence the community structure and dynamics of a host-associated microbial system, including host genotypes, surrounding environment, host–microbe, and microbe–microbe interactions. It has been reported that the existence of diverse and well-balanced microbiome plays a crucial role in crop improvement program by mitigating abiotic and biotic stress response and management of nutrients supporting plant growth and productivity, and therefore reduces the need for chemical fertilizers and pesticides (Compant et al. 2019).

In this chapter, we have discussed the diversity of microbial populations residing in plants, their secreted compounds used for host–microbe interactions, how plant gets benefited through this relation, and their potential uses to increase crop productions.

14.2 Plant Microbiome Diversity

There have been a number of studies identifying individual factors that drive the microbiome composition of plants and their associated organisms and environments. Plant microbiome is known to be mostly structured by plant compartments (e.g., different plant organs, and diverse between endosphere and ectosphere) (Trivedi

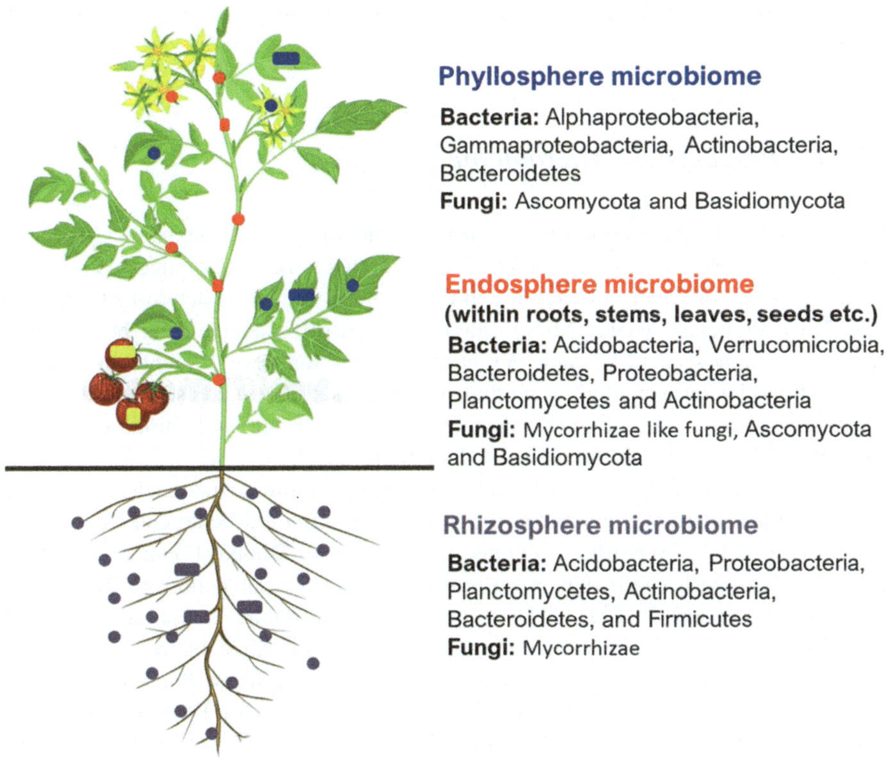

Fig. 14.1 Plant microbiome diversity

et al. 2020). The makeup of both plant and soil microbiomes has been demonstrated to be influenced by plant genotype and developmental phases (Wagner et al. 2016). Plant microbiome composition has been demonstrated to be influenced by soil microbiome composition (Cordovez et al. 2019), and plant pathogens and herbivores produce compositional shifts in plant-associated microbial communities (Lareen et al. 2016). However, we still know little about the relative strength of these factors in shaping plant microbiomes.

Despite the plant microbiota's taxonomic and functional overlap (Bai et al. 2015), distinct microbiomes have been found for each plant organ and species (Vorholt 2012; Philippot et al. 2013; Hardoim et al. 2015) and plant species (Berg and Smalla 2009) (Fig. 14.1). The plant's enrichment of microbes is not a random, but rather deliberate. The function of both chemo-attractants and repellents is influenced by differences in plant root exudates and plant components (Badri and Vivanco 2009). Plant defense signaling plays a part in these processes as well (Doornbos et al. 2012). Multi-omics methods provide significantly more detailed insights into the organization of plant-associated microbial communities, extending and supplementing existing information (Berg et al. 2016; Jansson and Baker 2016). In addition, the tools revealed novel roles of the plant microbiome as well as ecological interconnections.

14.2.1 Rhizosphere Microbiome

The diversified habitat created by soil texture encourages the coexistence of a vast variety of microorganisms, including bacteria, archaea, fungus, oomycetes, viruses, and protists, all of which interact in complex trophic exchange networks (Compant et al. 2019; Wei et al. 2019b). Microorganisms in the rhizosphere can be beneficial or harmful to the host plant's health (Yu et al. 2019). Plant development can be aided by beneficial bacteria (including mutualistic microbes) that increase food availability, produce plant hormones, and improve tolerance to biotic and abiotic challenges (Haney et al. 2015; Jacoby et al. 2017; Yin et al. 2021).

Mycorrhizae, rhizobium bacteria, plant growth promoting rhizobacteria (PGPR), and biocontrol microorganisms are the most often explored beneficial rhizosphere species (Fig. 14.1). According to Gans, Wolinsky, and Dunbar (Gans et al. 2005), 1 g of soil might contain over a million different bacterial genomes. The potato rhizosphere yielded 55,121 OTUs (operational taxonomic units), according to İnceoğlu et al. (2011). The most common bacteria in the rhizosphere belong to the *Acidobacteria*, *Proteobacteria*, *Planctomycetes*, *Actinobacteria*, *Bacteroidetes*, and *Firmicutes* families (Bulgarelli et al. 2015; Uroz et al. 2010). The microbial community composition of bulk soil (soil not linked to the plant root) and rhizosphere soil did not differ significantly as found in several investigations (Lundberg et al. 2012; Schlaeppi and Bulgarelli 2015).

Mycorrhizal fungi are widespread in the rhizosphere ecosystem, having been discovered in over 200,000 plant species and believed to be associated with over 80% of all plants (van der Heijden et al. 2015). By controlling nitrogen and carbon cycles, these mycorrhizae root connections have a significant impact in land ecosystems. Mycorrhizae are vital to plant health because they provide up to 80% of the N (nitrogen) and P (phosphorus) that plants require.

14.2.2 Phyllosphere Microbiome

The phyllosphere (the above-ground surface of plants) is home to a diverse range of microorganisms, and this phyllosphere microbiome interacts with the host plant, influencing its health and function. The phyllosphere's environment is more dynamic than the rhizosphere's and endosphere's. Microbial colonists are exposed to diurnal and seasonal changes in temperature, humidity, and radiation. Furthermore, these environmental factors influence plant physiology (photosynthesis, respiration, water absorption, and so on) as well as microbiome composition indirectly. Microorganisms in the phyllosphere, mostly bacteria and fungus, can function as mutualists, boosting plant growth and stress tolerance, commensals, utilizing the leaf habitat for their own development and reproduction, or antagonistic pathogens (Gong and Xin 2021).

Bacterial assemblages in the phyllosphere are often less species rich than those in the rhizosphere or soil (Knief et al. 2012). On the leaf surface, alpha-proteobacteria are extremely abundant, and these bacteria serve a variety of ecological activities (Ruinen 1965; Innerebner et al. 2011) (Fig. 14.1). Gamma-proteobacteria have also been found in phyllosphere bacterial community composition surveys (Redford et al. 2010; Vorholt 2012). Proteobacteria have a wide range of metabolic capabilities, and bacteria that do methyltrophy, nitrification, nitrogen fixation, or anoxygenic photosynthesis are frequently found in the phyllosphere (Fürnkranz et al. 2008; Atamna-Ismaeel et al. 2012; Watanabe et al. 2016). *Bacteroidetes* and *Actinobacteria* are the next most prevalent bacterial lineages, and both of these phyla are also substantially represented in the rhizosphere (Lauber et al. 2009; Philippot et al. 2013).

Fungi have a vital role in the microbiota of the phyllosphere. The fungal community is made up of organisms that play a wide range of ecological tasks, and whose population levels vary according to the growth season and, eventually, leaf senescence (Morris and Kinkel 2002). Before senescence, *Ascomycota* molds are frequently the dominating fungus on the leaf surface (Bashir et al. 2022; Abdelfattah et al. 2015). Yeasts from the *Ascomycota* and *Basidiomycota* families are also notable fungi (Dickinson 2012). Following leaf senescence, filamentous fungi dominate over the fungal microbiome (Voříšková and Baldrian 2013).

14.2.3 Endosphere Microbiome

The plant endosphere is inhabited by complex microbial communities and microorganisms known as endophytes, which live in the plant interior for at least part of their lives. They host a diversified microbiota in several compartments and tissues, including vegetative organs such as roots, stems, and leaves, as well as reproductive/disseminating organs (flowers, fruits/seeds). Bacteria and, to a lesser extent, Archaea, play a key role in endosphere communities. The plant immune system may be able to regulate endophyte abundance and maintain the most "plant-friendly" bacterial density in various organs (Liu et al. 2017). The root bacterial microbiome has the greatest diversity of microorganisms (although not necessarily) (Amend et al. 2019). *Acidobacteria*, *Verrucomicrobia*, *Bacteroidetes*, *Proteobacteria*, *Planctomycetes*, and *Actinobacteria* are the most commonly found taxa, and the majority of them may also be found in the rhizosphere (Hardoim et al. 2015) (Fig. 14.1). Other microorganisms, such as fungi or mycorrhizae-like fungi, as well as microbial interactions, might affect the diversity of the root endophytic microbiota (Deveau et al. 2018). Several bacterial taxa, including *Azoarcus*, *Gluconacetobacter*, *Herbaspirillum*, and *Klebsiella* spp., have been shown to occupy predominantly intercellular but also intracellular regions inside stems (Turner et al. 2013). The epidermis, xylem vessels, ovary, ovules, and stigma, as well as, other organs including the floral receptacle, petal, sepal, and anthers with their filaments, and pollen, have all been discovered to be colonized by bacteria (Compant et al. 2011). Strains of *Firmicutes*, *Actinobacteria*, and *Proteobacteria*, for example, have been isolated from these organs (Shi et al. 2010; Compant et al. 2011; Fürnkranz et al. 2012). Within the plant, the identity and diversity of the endophytic microbiome differs between above- and below-ground tissues (Dastogeer et al. 2020).

14.3 Plant and Soil Microbiome Interaction

Soil is an ecosystem which has the capacity to produce resources required for the development of plants. The diverse groups of microorganisms (bacteria and fungi) found in the soil are responsible for the decomposition of biomass and circulation of biogenic elements. These allow for the availability of nutrients to plants making the microbial communities in soils work as facilitators of plant processes (Pang et al. 2021). For example, they secrete hormones that change root growth; some microbes help to uptake iron from the soil or solubilize phosphorus and make these nutrients available to the plant. Non-pathogenic microorganisms are known to alter plant immune responses, thereby giving protection against pathogens. Then there are the nitrogen fixers which provide N for plant growth by converting atmospheric N_2 to NH_3. Thus, co-evolution of plants with soil microorganisms serves a variety of useful purposes. A large number of research findings have shown that, in almost all cases, not a single microbe but rather a conglomerate of microorganisms is responsible for the positive effects on plants.

The rhizosphere is a boundary between the plant roots and the soil and is characterized by a dynamic community of microorganisms. The exudates together with mucilages and cells which are shed from the plants influence strongly the soil and its microbiome around the plant root (Turner et al. 2013). These microbial populations form structured and interconnected networks. This leads to an active communication not only between the microbes but between the plants and microbes as well, enabled by molecular signals especially the secondary metabolites that the plants produce (Pang et al. 2021). This cross-talk acts as the primary screen for the microbes that inhabit the root as endophytes.

In such ecosystems, the soil health can be altered by cropping practices and intensive land-use management, which can further impact soil functions. Earlier studies on soil health in relation to agriculture were based on soil eco-functions that looked into non-biological properties like the nutrient and structure of the soil. Only in recent years have biological properties such as soil microorganism garnered emphasis as an essential composition in soil health as well. Agricultural practices also impact the plant microbiome composition and functions (Hartman et al. 2018) with both negative and positive consequences on plant growth. Different cropping systems, which include intercropping and tilling free farming, have been found to increase the diversity of microbial population, yields in crop, and soil levels of organic carbon (Sergaki et al. 2018). Cultivation of a single crop or short rotations of crops rapidly exhaust soil nutrients and increase plant species-specific soil pathogens and root herbivores. This leads to a drastic decrease in crop yields. Following microbial supplementation, some fungi were found to help plants augment resistance against drought. Soil treatment with bacterial inoculants have helped to increase plant tolerance to heat stress by reducing reactive oxygen species, enhanced plant immunity and through assistance with drought stress.

Since relation between plants and soil microbes is significant for plant growth and disease resistance, in recent years a lot of research has focused on understanding the dynamics, assembly and functions of these microbial communities, the mechanism of interactions and the forces that modulate the plant microbiome association. However, the mode of action of such communications and the ways these modulations are driven remain evasive. Exploiting the diverse benefits afforded by the plant microbiome, more knowledge of this system is a viable strategy to promote sustainable agriculture. Understanding the key elements that determine the rhizosphere microbiome assembly, as well as the mechanisms of reciprocal adaptation between microbes and plants in response to changing environmental circumstances, may aid in the identification of prospective breeding and management targets for future crops (Sandrini et al. 2022).

Recent agricultural production strategies have given rise to the concept of "smart agriculture." It is based on intelligent agricultural management. Manipulation of rhizospheric microbiota and its management strategies are now considered "smart farming." This involves the integration of beneficial plant microbiome traits, traits that improve plant growth, resistance to diseases and stress tolerance leading to sustainable agricultural production (Sandrini et al. 2022). How and to which extent the different cropping practices permit the manipulation of soil and root microbiota are the foci of many recent research groups. Land management and tillage intensities influence the dominance or the connectivity between bacteria and fungi in both the soil and the roots. Focus is now on how to define the microbiome functions that can be manipulated through cropping practices.

Now researchers are suggesting that plant–soil feedbacks are shaped by microbial bequests that plants leave in the soil. In bacterial communities, legacies fade away rapidly and bacterial communities are influenced strongly by the current plant (Hannula et al. 2021). However, both fungal and bacterial legacies are found conserved inside the roots of the current plant species and their composition significantly correlate with plant growth. Microbial soil legacies present at the time of plant establishment play a vital role in shaping plant growth even when these legacies had faded away in the soil due to the growth of the current plant species. The soil microbiome legacies have been found to be reversible and versatile.

Mendes and Raaijmakers (2015) have recently shown cross-kingdom similarities in microbiome functions. This is truer for the root and gut microbiome, since the two microbial communities exist under comparable conditions which are basically open systems characterized by gradients of oxygen, water, and pH that create a whole array of diverse niches. The microbial members of the two groups come from the environment: food in humans and soil in plants, respectively. Both the systems are inhabited by a host of similar bacterial phyla (*Firmicutes*, *Bacteroidetes*, *Proteobacteria*, and *Actinobacteria*). Research has shown that raw plant materials are the source of some members of the gut microbiome. The microbial community of both the gut and the root are known to synthesize essential amino acids, vitamins, and many other secondary metabolites that modulate their host immune system: as such, the plant and gut microbiomes can be considered as meta-organs with paramount importance for the health of their hosts. Furthermore, because of the food connection, humans would benefit from consuming unprocessed organic food, which contains beneficial bacteria as well as secondary metabolites. Research on the integral role of microbiomes on their host's metabolism and health should therefore not stop at the human gut microbiome but expand to the microbiota of plants and their function in plant growth and development.

As a result, it's critical to better understand the activities and responsibilities of the hundreds of distinct microbial species that interact with their hosts in a complex network. The subject of how to create and maintain a healthy microbiome is also crucial. Agriculture has consequences for studying microbiomes holistically rather than one bacterium at a time (for example, utilizing tracking devices). However, in spite of its importance, the complex, heterogenic and subterranean setting makes it difficult to study soil microbiome. Fortuitously, emerging technologies, artificial intelligence, advances in modeling biological systems, data interpretation are now allowing us to understand the complex interlink between plants and microbial community.

14.4 Microbiomes and Secreted Metabolites in Plant Growth Promotion

Plant growth-promoting bacteria (PGPB) enhance plant development through a variety of ways. For example, they secrete phytohormones, such as auxins that change root growth. Some bacteria absorb iron and phosphorus, making these minerals accessible to the plant. They can indirectly promote growth by forming a biofilm that acts as a protective layer against pathogens or as a surface for improved nutrient acquisition from soil (Weselowski et al. 2016). Particular non-pathogenic microorganisms may even alter plant immune responses, thereby giving protection against pathogens. Additionally, PGPR increase tolerance to abiotic stress in crops such as salinity and drought (Lucke et al. 2020).

These mechanisms have been grouped into three clusters according to the plant growth promoting effects on plant physiology. These groups are: (1) biofertilization including biological fixation of atmospheric nitrogen, phosphate solubilization, siderophores production, and exopolysaccharides production; (2) phytostimulation including production of indole acetic acid, gibberellins, cytokinins, and ethylene; and (3) biocontrol including induction of systemic resistance, competition for iron,

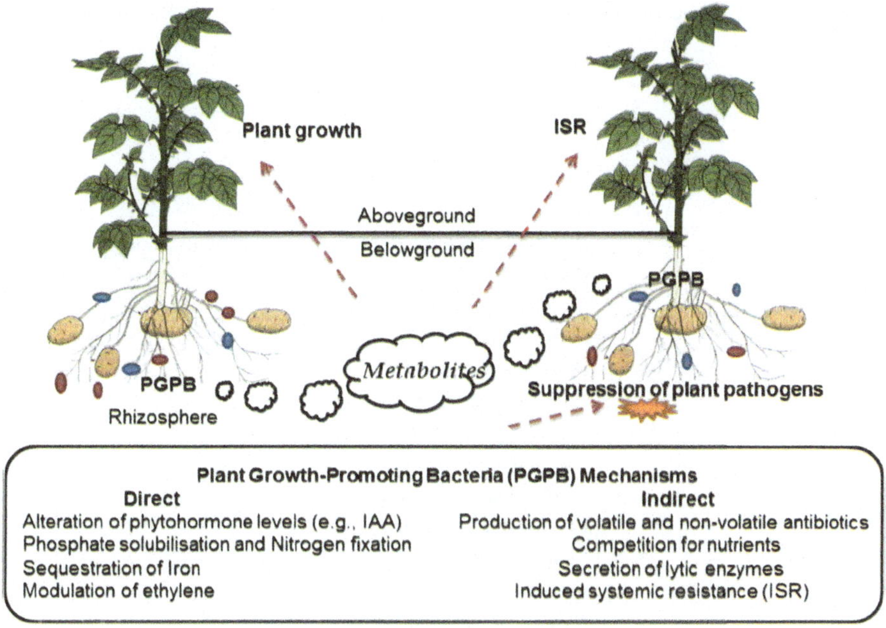

Fig. 14.2 The role of microbial metabolites in plant growth promotion (Velivelli et al. 2014)

nutrient and space, production of antibiotics, lytic enzymes, hydrogen cyanide, and volatile compounds (Mitter et al. 2013; Tsegay et al. 2016) (Fig. 14.2).

14.4.1 Phosphate Solubilization and Mobilization

Phosphorus is one of the vital plant nutrients essential for the proper functioning of plants. Though soil is rich in insoluble phosphates of metal ion like iron, aluminum, and calcium, the soluble form orthophosphate which can be absorbed by plants is deficient which can severely restrict plant growth, development, and yield.

Applying phosphorus fertilizers in soil is the main input of inorganic P. About 75–90% of these added chemical P fertilizer is becomes fixed in soils as metal-cation precipitation and has long-term impacts on the environment in terms of soil fertility depletion, eutrophication, and carbon footprint (Sharma et al. 2013).

Phosphate solubilizing microbes (PSMs) are helpful bacteria that can hydrolyze organic and inorganic insoluble phosphorus compounds into soluble P that plants may easily absorb. PSM provides an eco-friendly and economically alternative to phosphorus fertilizers to overcome the P scarcity (Kalayu 2019).

Several bacterial, fungal, and actinomycetes strains have been identified as PSM for example strains of *Pseudomonas*, *Bacillus*, *Rhizobium*, *Aspergillus*, and *Penicillium* are some known phosphate solubilizers (Table 14.1). In addition to supplying

Table 14.1 List of potential phosphate solubilizing microorganisms (PSMs) (Kalayu 2019)

Types	PSMs
Bacteria	
Bacillus	*Bacillus circulans*
	Bacillus megaterium
	Bacillus polymyxa; *B. subtilis*
	Bacillus pulvifaciens
	Bacillus coagulans; *B. fusiformis*; *B. pumilus*; *B. chitinolyticus*
	Bacillus sircalmous
Pseudomonas	*Pseudomonas canescens*
	Pseudomonas putida
	Pseudomonas calcis
	Pseudomonas fluorescens
	Pseudomonas koreensis
	Pseudomonas striata
Others	*Thiobacillus ferrooxidans*
	Enterobacter cloacae
	Pantoea agglomerans
	Rhizobium meliloti
	Rhizobium leguminosarum
	Mesorhizobium mediterraneum
	Burkholderia cepacia
Fungi	
Aspergillus	*Aspergillus niger*
	Aspergillus clavatus
	Aspergillus awamori
	Aspergillus candidus; *A. parasiticus*; *A. fumigatus*; *A. rugulosus*
	Aspergillus flavus
	Aspergillus foetidus; *A. nidulans*; *A. wentii*
	Aspergillus terreus
	Aspergillus tubingensis
	Aspergillus sydowii; *A. ochraceus*; *A. versicolor*
Penicillium	*Penicillium bilaii*
	Penicillium citrinum
	Penicillium digitatum; *P. lilacinum*; *P. balaji*; *P. funiculosum*
	Penicillium oxalicum
	Penicillium simplicissimum; *P. rubrum*
Others	*Trichoderma viride*
	Arthrobotrys oligospora
Actinomycetes	
	Acinetobacter rhizosphaerae
	Streptomyces albus; *S. cyaneus*; *Streptoverticillium album*
Cyanobacteria	*Calothrix braunii*

soluble P to plants, these microorganisms also promote plant growth by improving the uptake of other nutrients, stimulating the production of some phytohormones and helping the plant to manage biotic and abiotic stresses (Shrivastava et al. 2018; Mitra et al. 2020).

PSMs apply different approach to make phosphorus accessible for plants to absorb like secretion of organic acids thus lowering soil pH, enzyme production, mineralization, and excretion of siderophores that can chelate the metal ions and form complexes, making phosphates available for plant uptake (Zheng et al. 2017).

Organic acid production leading to lowering of soil pH and production of chelating substances are the principal mechanisms adopted by PSB which result in inorganic P solubilization. Alkaline soil with high pH restrict P acquisition by plants (Zheng et al. 2017). Most Gram-negative bacteria show glucose dehydrogenase activity that can solubilize phosphate by extracellular oxidation of glucose to gluconic acid, when glucose is used as carbon source. The PSMs may release several organic acids that can solubilize phosphates like citric, lactic, gluconic, 2-ketogluconic, oxalic, glyconic, acetic, malic, fumaric, succinic, tartaric, malonic, glutaric, propionic, butyric, glyoxalic, and adipic acids; however, gluconic acid is reported as the principal organic acid produced by phosphate solubilizing bacteria (Kalayu 2019). Thus the PSMs have the ability to release such organic acids, through which their hydroxyl and carboxyl groups chelate the cation bound to phosphate, converting it to soluble forms.

Mineralization is another process for solubilizing soil phosphate (P). PSMs mineralize soil organic P by producing phosphatases, which hydrolyze organic forms of phosphate compounds, releasing inorganic phosphorus that will be immobilized by plants.

14.4.2 Nitrogen Fixation

After water, nitrogen is the major factor limiting plant growth under most conditions. Nitrogen is used to synthesize plant proteins and nucleic acids. Although, atmospheric nitrogen is abundant, plants cannot use the available form (N_2). Plants can utilize if nitrogen are combined chemically with oxygen or hydrogen to form different nitrogenous compounds. The commercially available fertilizers are nitrogenous compounds added to the soil in the form of ammonium (NH_4^+) and nitrate (NO_3^+) like ammonia, nitric acid, ammonium nitrate, and urea (Ji et al. 2014). Nitrogen is essential for plant growth and development during vegetative and reproductive phases (Cheremisinoff 2010). For this reason common agricultural practices make use of chemical nitrogenous fertilizers to maintain high crop yields.

PGPB retain more soil organic nitrogen and other nutrients in the plant–soil system, thus reducing the need for fertilizers. Since N_2 fixation is a prime requisite for plant growth, N_2 fixer bacteria, also called “diazotrophs” play a critical role in the plant growth by reducing dinitrogen (N_2) to ammonia (NH_3) by the enzyme nitrogenase.

Many PGPB are capable of biological nitrogen fixation (BNF). Several bacteria such as *Azospirillum*, *Klebsiella*, *Burkholderia*, *Bacillus*, and *Pseudomonas* have been identified as PGPR through BNF, phosphate solubilization, phytohormone production, and biological control of soil pathogens (Kuan et al. 2016; Fukami et al. 2018a, b). Nitrogen-fixing rhizobacteria can promote plant growth; however, the role of BNF in the promotion of plant growth has not been well documented still (Pankievicz et al. 2015).

Nitrogen fixing organisms are generally categorized as (a) symbiotic N_2 fixing bacteria which form symbiosis with leguminous plants and non-leguminous trees and (b) non-symbiotic (free living, associative, and endophytes) nitrogen fixing forms. However, non-symbiotic nitrogen fixing bacteria provide only a small amount of the fixed nitrogen to the host plant (Ahemad and Kibret 2014).

In legumes, its symbiotic soil rhizobial bacteria can provide N through biological nitrogen fixation (BNF). However, most agricultural plants, especially grasses, lack this ability (Charpentier and Oldroyd 2010). Endophytic nitrogen-fixing bacteria are believed to contribute significant amounts of N to some gramineous crops such as rice, maize, wheat, and barley (Ladha et al. 1997). Few examples of endophytic diazotrophic bacteria are *Lysinibacillus sphaericus* (L1), *Klebsiella pneumoniae* (S2), and *Bacillus cereus* (R2) reported in rice (Shabanamol et al. 2018), *Pseudomonas aeruginosa* PM389 in wheat (Amanullah and Fahad 2018), *Bacillus* spp. in corn and banana (Amanullah and Fahad 2018). Endophytic diazotrophic bacteria were also identified belonging to genera: *Bacillus*, *Caballeronia*, *Paenibacillus*, and *Pseudomonas* in pine and spruce trees (Puri et al. 2018).

In order to reduce the excessive use of chemical nitrogenous fertilizer, agricultural crops can be colonized by bacterial inoculants for a significant enhancement of growth through BNF. BNF by PGPR has been reported to contribute up to 12–70% of total N uptake in field crops like in maize, sugarcane, wheat, rice, and oil palm. Generally, it has been estimated that up to 65% of N used in agriculture is contributed by biological nitrogen fixation, and it will be an increasingly key element for more-sustainable agricultural practices in future plant-N management (Kuan et al. 2016; Fukami et al. 2018a, b); Montañez et al. 2009).

14.4.3 Potassium Solubilization and Mobilization

Potassium together with nitrogen and phosphorus is one of the most important essential nutrients for plants and is the third key element in the standard commercial fertilizers NPK (nitrogen, phosphorus, and potassium) (Kour et al. 2020). It plays an important role in the activation of several metabolic processes such as photosynthesis, synthesis of proteins, and enzymes, as well as in resistance to diseases, pests, and abiotic stresses (Teotia et al. 2016; Kumar et al. 2020). In addition, K is required to activate over 80 different enzymes responsible for plant and animal processes (Etesami et al. 2017). Since most of the soil potassium is not readily available for uptake by the plants, potassium fertilizers majorly in the form of the muriate of potash (potassium chloride) or sulfate of potash are used to fulfill the needs of the plants (Scherer 2005).

A wide range of potassium-solubilizing microbes (KSM) have been reported, which include bacteria and fungi; the prominent are *Bacillus* spp. (*B. megaterium, B. mucilaginosus, B. edaphicus, B. circulans, Acidithiobacillus ferrooxidans, Pseudomonas putida, Arthrobacter* sp., and *Paenibacillus* sp.), *Aspergillus* spp., and *Aspergillus terreus, Burkholderia, Enterobacter, Pantoea, Flectobacillus, Klebsiella, Microbacterium, Myroides*, and *Stenotrophomonas* (Teotia et al. 2016; Kour et al. 2020). *Pseudomonas, Bacillus*, and *Aspergillus*, among other rhizosphere K-solubilizing microorganisms, produce organic acids that solubilize the inaccessible potassium and make it available to plant roots (Teotia et al. 2016). Currently there is little information available on the strategy by which KSM can solubilize K-bearing minerals and release K for improving the growth and yield of plant. Similar to the mechanism of P solubilization, the major mechanism of K solubilization is inorganic and organic acids production and production of protons (acidolysis mechanism). Organic acid such as tartaric acid, citric acid, succinic acid, α-ketogluconic acid, and oxalic acid can directly increase dissolution by a proton- or ligand-mediated mechanism. Released protons can directly dissolve the mineral K converting the insoluble K to soluble forms. Organic acids can also indirectly increase dissolution by the formation of complexes in solution with reaction products (Etesami et al. 2017).

14.4.4 Microbial ACC Deaminase

Ethylene is a plant hormone involved in the regulation of various physiological processes of plants like germination, fruit ripening, growth, and senescence. Abiotic stress such as flooding, drought, salinity, wounding, and pathogen attack can induce the increment in ethylene level secreted by plants and has been found in almost all plants growing under stress conditions. Environmental stress-induced ethylene production in plants cause a significant reduction in plant growth, root elongation and lateral root emergence, and even result in plant death (Gupta and Pandey 2019).

1-Aminocyclopropane-1-carboxylic acid (ACC) is the immediate precursor of the hormone ethylene in plants. PGPB possess the ability to produce ACC deaminase that can hydrolyze ACC to ammonia and α-ketobutyrate, thus reducing the indigenous ethylene level inside the plants, thereby increasing the root length and growth and curbing the environmental stress induced ethylene production and its associated harmful effect on plants (Raghuwanshi and Prasad 2018). PGPR containing ACC deaminase thus make the plants more resistant to various abiotic stresses.

ACC deaminase has been widely reported in various species of plant growth promoting endophyte bacteria like *Agrobacterium genomovars, Azospirillum lipoferum, Alcaligenes, Bacillus, Burkholderia, Enterobacter, Methylobacterium, Pseudomonas, Ralstonia solanacearum, Rhizobium, Rhodococcus, Sinorhizobium meliloti, Variovorax paradoxus* (Saleem et al. 2007).

Table 14.2 Inoculation with PGPR containing ACC deaminase and subsequent physiological changes in plants (Saleem et al. 2007)

Plant species	PGPR	Comments
Brassica campestris	*Methylobacterium fujisawaense*	Bacterium promoted root elongation in canola
Brassica campestris	*Bacillus circulans* DUC1, *Bacillus firmus* DUC2, *Bacillus globisporus* DUC3	Bacterial inoculation enhanced root and shoot elongation
Brassica napus	*Alcaligenes* spp., *Bacillus pumilus, Pseudomonas* spp., *Variovorax paradoxus*	Inoculated plant demonstrated more vigorous growth than the control (uninoculated)
Brassica napus	*Enterobacter cloacae*	A significant increase in the root and shoot lengths was observed
Dianthus caryophyllus L.	*Azospirillum brasilense* Cd1843	Inoculated cuttings produced longest roots
Glycine max	*Pseudomonas cepacia*	Rhizobacterium caused an early soybean growth
Pisum sativum L.	*Rhizobium leguminosarum bv. viciae* 128C53K	Bacterium enhanced nodulation in plants
Vigna radiata L.	*Pseudomonas* spp., *Bradyrhizobium* sp.	Bacterium promoted nodulation in mung bean
Vigna radiata L.	*Pseudomonas putida*	The ethylene production was inhibited in inoculated cuttings
Zea mays L.	*Enterobacter sakazakii* 8MR5, *Pseudomonas* spp. 4MKS8, *Klebsiella oxytoca* 10MKR7	Inoculation increased agronomic parameters of maize
Zea mays L.	*Pseudomonas* spp.	Bacterium caused root elongation in maize

A number of studies show, a plant inoculated with bacteria containing ACC deaminase exhibits more root growth (Saleem et al. 2007). Table 14.2 contains some examples of inoculation with PGPR containing ACC deaminase that alter the endogenous levels of ethylene and subsequently brings changes in plant growth.

14.4.5 Siderophores

Siderophores are iron-chelating peptide molecules with high affinity side chains and functional groups for ferric iron (Das et al. 2007). They are low molecular weight molecules and their molecular weights range from 400 to 1500 Da (Beneduzi et al. 2012). These macromolecules are one of the most effective binding agents for soluble Fe^{3+} and they are released by microbes in response to iron deficiency so they may gather iron from the surroundings and provide the required iron mineral to the microbial cell (Kannahi and Senbagam 2014). Microbial siderophores can be used by plants to obtain iron as well (Sessitsch et al. 2004). Several PGP (plant growth promoting) bacteria have been reported for growth promoting potentiality through siderophore production in sugarcane (*Saccharum* spp.), rye grass (*Lolium perenne*), Thai jasmine rice plant, and jute (Haidar et al. 2018). Iron deficient tomato plants supplemented with microbial siderophores are also shown to produce higher crop yields, and had increased chlorophyll and iron content in the leaves (Radzki et al. 2013).

Siderophore-producing PGPR are mostly found closely to the surface of plant roots and rhizosphere area. They not only provide iron nutrition to the plant, but also act as the first defense against harmful parasites that mostly attack plant roots (Sayyed et al. 2013). Siderophore-producing bacteria act as an antagonist against phytopathogens by scavenging the necessary iron from the environment which is needed for growth, nucleic acid synthesis, sporulation, and maintaining cell morphology of the pathogen (Campbell et al. 1986; Pandya and Saraf 2014). Generally the fungi are mostly affected by this iron deprivation as they cannot absorb the iron–siderophore complex (Goswami et al. 2016). Thus PGPR secret the siderophores as the weapons to win the battle to acquire maximum iron from the environment (Balhara et al. 2016).

PGPR usually produce four types of siderophores: hydroxamate, catecholate, salicylate, and carboxylate (Kannahi and Senbagam 2014). Among all the bacteria studied for siderophore activity, *Pseudomonas* species produce the most potent siderophore pyoverdines containing both hydroxamate and catecholate functional groups acting as a growth inhibitor of bacteria and fungi in iron depleted media in vitro (Beneduzi et al. 2012).

The most reported siderophore-producing bacteria are commonly a member of the genus *Pseudomonas*, among them the most known are *Pseudomonas fluorescens* and *Pseudomonas aeruginosa* which secret the siderophores pyochelin and pyoverdine (Goswami et al. 2016). Some other important siderophore producing bacteria are *Bacillus* spp., *Escherichia coli*, *Azotobacter*, *Burkholderia* sp., *Rhizobium radiobacter*, *Pantoea allii*, and *Mycobacterium* species (Ferreira et al. 2019; Beneduzi et al. 2012).

14.4.6 Phytohormones

Phytohormones are small organic signal molecules produced by plants in very low concentration that regulates all the physiological processes in plants like growth, development, pathogen defense, stress tolerance, and reproductive processes.

PGPR directly or indirectly influence plant growth by secreting secondary metabolites that in turn play role in plant hormone synthesis and obtaining nutrient from soil. Almost all the PGPB have the ability to produce phytohormones like auxin, i.e., indole acetic acid (IAA), abscisic acid (ABA), gibberellic acid (GA), and cytokinins. These phytohormones greatly play roles in plant root growth resulting a better uptake of moisture and nutrients from soil (Fukami et al. 2018a, b).

14.4.7 Volatile Compounds

The volatile compounds produced by the plant associated microbiome inhibit the growth of plant pathogens and provide defense against them. They may also increase the immunity of the plants by inducing systemic resistance. *Bacillus amyloliquefaciens* YN201732, an endophytic bacterium found to be antagonistic against 12 pathogenic fungi, was reported to reduce the effects of powdery mild dew in tobacco, possibly by inducing the jasmonic acid/ethylene signaling pathway (Jiao et al. 2020). Volatile organic compounds (VOCs) produced by *Bacillus amyloliquefaciens* FZB42 and *Bacillus atrophaeus* LSSC22 showed inhibitory activity against *Ralstonia solanacearum*, responsible for wilt disease. Benzaldehyde, 1,2-benzisothiazol-3(2 H)-one and 1,3-butadiene in particular exhibited prominent negating effect, manifested by the significant inhibition in the colony size, cell viability, and motility of the pathogens. Moreover, VOCs even genetically altered the virulence and pathogenicity related genes of the pathogen and elicited the upregulation of wilt resistance and pathogen defense related genes in the plant, thus decreasing the wilt disease in tobacco (Tahir et al. 2017). *Streptomyces* sp. CEN26, a root node endophyte of *Centella asiatica* (L.), was found to produce 2,5-bis(hydroxymethyl)furan monoacetate that impede conidial germination and appressorium formation of the plant pathogenic fungi *Alternaria brassicicola* that causes black spot disease on a broad range of hosts (Phuakjaiphaeo et al. 2016). The *Muscodor* endophytic fungi are already considered to be promising biocontrol agents owing to the VOCs they produce (Pena et al. 2019). Blue mold and gray mold caused by *Penicillium expansum* and *B. cinerea* respectively were completely diminished in *M. albus* inoculated fruits. 2-methyl-1-butanol and isobutyric acid abundantly released by the fungi were suggested to be responsible for the disease suppression (Kumar et al. 2021). The VOC emitted by *Flavobacterium johnsoniae* strain GSE09, 2,4-di-tert-butylphenol, hindered the development of the pathogen *Phytophthora capsici* (Sang and Kim 2012). In addition to increasing plant biomass *Trichoderma asperellum* T1 VOCs impeded the growth of two leaf spot fungal pathogens *Corynespora cassiicola* and *Curvularia aeria*. These VOCs also elicited the host defense response as plant increases the activity of chitinase and β-1,3-glucanase (Wonglom et al. 2020). Microbial VOCs also have the potential as insecticides to control insect induced diseases from the roots as seen in case of *Paenibacillus polymyxa* strain BMP-11 (Zhao et al. 2011).

14.4.8 Other Secondary Metabolites

Antibiotics are yet another strategy for microbes to communicate with one another. Antibiotics are most commonly associated with medicine, but soil microbes need them to live in a tough competitive environment (Cornforth and Foster 2015).

PGPR strains' biocontrol capabilities are mostly reliant on extensive root colonization, the generation of systemic defense response in the host plant, and the synthesis of antimicrobial chemicals. Antibiotics are a diverse collection of low-molecular-weight, organic chemicals that prevent other microbes growth or metabolic functions. Antibiotics work in different ways, like by blocking pathogen cell wall synthesis, altering or deforming cellular membrane structures, slowing the formation of initiation complexes on the ribosome, thereby preventing protein biosynthesis, etc. (Beneduzi et al. 2012). Antibiotics also have an important role in the induced systemic resistance (ISR) mechanism in plants, in addition to their direct inhibitory activity (Kenawy et al. 2019).

Actinomycetes, ubiquitous bacteria that play important roles in soil ecology, are sources of 70% of clinical antibiotics—a widely used example is streptomycin (Tkacz and Poole 2021). PGPR belonging to *Bacillus* and *Pseudomonas* species play an active role in the suppression of pathogenic microorganisms producing antibiotics. These PGPR secrete this extracellular metabolites that are inhibitory to plant pathogen even at low concentration (Goswami et al. 2016).

Phenazine, 2,4-diacetylphloroglucinol (DAPG), pyrrolnitrin, surfactin, iturin, fengycin, viscosinamide, kanosamine, zwittermicin-A, polymyxin, circulin, pantocin, subtilin, and subtilosin are some of the antibiotics synthesized by PGPR, having potential biocontrol abilities (Kenawy et al. 2019).

Pseudomonads synthesize an antifungal drug phenazine, which has redox action potential and can reduce plant infections by inhibiting plant pathogens like *Fusarium oxysporum* and *Gaeumannomyces graminis*. The antibiotic DAPG, generated by *Pseudomonas fluorescens*, has been found to be efficient in destroying the cellular membrane of phytopathogen *Fusarium oxysporum* sp. *niveum*. Pyrrolnitrin, which is similarly generated by *Pseudomonads*, can prevent *Rhizoctonia solani* from infecting cotton plants during damping-off. Antibiotics, such as polymyxin, circulin, and colistin, generated by most of the *Bacillus* spp., are effective against many Gram-positive and Gram-negative microbes, as well as several pathogenic fungi. *Bacillus* spp. additionally produces zwittermicin A and kanosamine antibiotics, that help to inhibit pathogenic fungi and fungus-like bacteria in the soil environment (Beneduzi et al. 2012).

The antibiotics produced by PGPR are divided into two categories: volatile and non-volatile compounds. Alcohols, aldehydes, ketones, sulfides, and hydrogen cyanide are volatile antibiotics, while polyketides, cyclic lipopeptide amino polyols, phenylpyrrole, and heterocyclic nitrogenous substance are non-volatile antibiotics. This class of non-volatile antibiotics has wide antimicrobial activity against a variety of plant pathogens (Kenawy et al. 2019).

Several rhizobacteria produce hydrogen cyanide (HCN), a volatile compound that has a significant inhibitory impact upon several species. The cyanide ion suppresses metalloenzymes, particularly copper-containing mitochondrial cytochrome c oxidases, preventing transfer of electrons and disrupting the cell's energy supply, resulting in the death of organisms. HCN synthesis is mostly attributed with Gram-negative *Pseudomonas* spp., although several other bacteria species, such as *Alcaligenes, Aeromonas, Bacillus*, and *Rhizobium*, can also synthesize the same

(Abd El-Rahman et al. 2019). *P. putida* has been reported to inhibit *Fusarium solani* infection alfalfa plant by secreting HCN (Sarhan and Shehata 2014). Moreover, black root rot disease which is caused by the phytopathogen *Thielaviopsis basicola* (Kenneth et al. 2019) and *Macrophomina phaseolina* (Reetha et al. 2014) is also found to be prevented by *Pseudomonas fluorescens*. Both siderophores and HCN are recognized to have a role in a PGPR's antifungal properties. Production of hydrogen cyanide has been reported as an important antifungal trait of a good number of rhizobacteria to control root infecting fungi (Bakthavatchalu et al. 2012).

14.5 Endophytes Contribute to Plant Stress Adaptation

Global warming of recent years, climate change, and many other environmental adversaries cause hyper-salinities, water deficits, and extreme temperatures and have become prevalent abiotic stress factors that attribute decreased crop production globally. Studies have already revealed the microbiome of a plant to play an important role behind it's adaptability against many stress factors, directly or indirectly inciting plant's own physical and chemical mechanism to lead it toward the ultimate objective of survival and growth. As the environmental disturbances continue to increase the stresses, the microbiome and its ability to support plant adaptation become all the more significant, for survival and sustenance (Table 14.3).

Table 14.3 Plant microorganisms alleviating different stress conditions in plants

Microorganism	Associated plant	Stress response	References
Achromobacter piechaudii ARV8	Pepper and tomato	Increased drought tolerance by producing ACC deaminase	(Patel Priyanka et al. 2019)
	Solanum lycopersicum	Elevated acquisition of phosphorus and potassium under saline conditions	Mayak et al. (2004a)
Pseudomonas mendocina	Lettuce	Reduced oxidative stress under drought condition through antioxidant catalase activity	Kohler et al. (2008)
Glomus intraradices			
Pseudomonas fluorescens	Canola (*Brassica* spp.)	Enriched the plant with proteins related to energy metabolism and cell division to help tolerate salinity	Banaei-Asl et al. (2016)
Pseudomonas spp. AKM-P6	Sorghum	At being inoculated, improved physiological and metabolic performances under high heat stress	Ali et al. (2009)
Burkholderia phytofirmans PsJN	Grapevine	Root inoculation resulted in improved root, increased biomass, and enhanced physiological activity at low temperature (4 °C)	Ait Barka et al. (2006)

14.5.1 Hypersaline Habitat Adaptability

In saline conditions the excessive uptake of Na^+ and Cl^- ions disrupts the Na^+ homeostasis and nutritional balance, generating cytotoxicity, and fatally damaging the plant (Table 14.3). Some rhizobacteria as well as endophytic microbes evidently help maintain the Na^+ homeostasis in saline conditions as they enhance Na^+ exclusion and K^+, Ca^+ uptake thus improving the Na^+/K^+ ratio and sustaining the Na^+ homeostasis (Singh et al. 2021; Rojas-Tapias et al. 2012). Some plant growth promoting rhizobacteria (PGPR) may achieve this through the secretion of exopolysaccharides that bind cations and assist in biofilm production around root surfaces to restrict the entry of Na^+ (Ashraf and Harris 2004; Dodd and Pérez-Alfocea 2012). Some have shown the ability to downregulate the expression of high affinity K^+ transporters (HKT1) in the roots limiting Na^+ uptake, whereas inducing their expression in the shoots promoting root to shoot Na^+ recirculation and thus improving Na^+ homeostasis (Zhang et al. 2008). PGPR and endophytes are also reported to improve N, P, K, Ca, Fe, and other nutrient uptake through fixing, solubilizing, releasing siderophores, balancing nutrition in saline stress (Mayak et al. 2004b; Upadhyay et al. 2016). Reactive oxygen species (ROS) produced due to high salinity stress are also addressed by plant microbiome as they are seen to interfere with expression of related genes to detoxify ROS (Kim et al. 2014; Sukweenadhi et al. 2015). Even the volatile organic compounds (VOCs) produced by this microbiome promote salt tolerance in plants. Two such VOCs, hexanedioic acid and anoic acid, secreted by a *Betaproteobacteria Alcaligenes faecalis* JBCS1294 induce salt tolerance in plant through reprogrammed auxin and gibberellin pathways (Bhattacharyya et al. 2015). Colonized endophytes may provide salt tolerance through their signaling molecules, hormones, and even microRNAs, providing high amounts of osmolytes, decreasing water usage, and fighting production and the effects of ROS (Azad and Kaminskyj 2016; Woodward et al. 2012). Saline stress increases ethylene production in a plant that inhibits plant growth. PGPR exhibit the ability to produce the ACC (aminocyclopropane-1-carboxylate) deaminase which restricts the production of ethylene (Glick et al. 2007). A plant's response to saline condition may vary in accordance with its associated microbiome.

14.5.2 Alleviating Temperature Stress

Increased temperature directly affects plant growth by reducing membrane permeability, photosynthesis, respiration, and seed germination. These also enhance ROS production resulting in cell death. Plant microbiome may mitigate the effects of heat stress through Induced Systemic Tolerance (IST) (Table 14.3). Root endophytes were found to confer heat tolerance to its host plant by inducing constitutive chromatin modification (Shekhawat et al. 2021). The plant microbiome can alleviate the heat stress effects through symbiosis as observed in tropical panic grass

Dichanthelium lanuginosum and the fungus *Curvularia protuberate* (Márquez et al. 2007). Microbes also help plant growth and development in high temperature through inducing decreased production of abscisic acid (ABA), increased production of salicylic acid, increasing the generation of antioxidants, neglecting the effects of ROS, etc. (Khan et al. 2020). Not just high temperatures, microbiome supports plants in cold temperatures too. A psychrotolerant bacterial inoculation is reported to reduce freezing injuries, ice nucleating activities, and lipid peroxidation while lessening the effects of ROS (Tiryaki et al. 2019). Some microbes confer tolerance in both higher and lower temperatures, for example, *Burkholderia phytofirmans* strain PsJN provides heat tolerance in tomato but cold tolerance in grapevine (Issa et al. 2018; Miotto-Vilanova et al. 2016).

14.5.3 Drought Stress Reduction

Drought stress damages membranes, causes loss of photosynthesis, reduces leaf size, root proliferation and even shoot extension, lessens water use efficiency along with fatal accumulation of ROS. Plants showcase several mechanisms to counter these effects but the importance of plant microbiome is manifested by the way plants may have evolved to recruit drought specific microbes in prolonged drought conditions (Farooq et al. 2009; Naylor and Coleman-Derr 2018). In response to drought, PGPR have been reported to manipulate phytohormones such as Indole Acetic Acid (IAA), ABA, cytokinins, induce the production of ACC deaminase, and decrease that of ethylene (Poudel et al. 2021). Some root zone bacteria even produce osmotically active molecules and osmo-solutes such as proline to impart drought tolerance (Moreno-Galván et al. 2020). Arbuscular mycorrhizal fungi (AMF) enhance osmotic balance through increased uptake of K^+, Ca^+, and other cations (Ruiz-Lozano 2003). AMF also improve the water use efficiency and water status of a plant (Augé 2001). Plant microbial VOCs confer drought tolerance by monopolizing the expression of genes involved in maintaining the cell wall structure and affecting phytohormone signaling pathways (Asari et al. 2016; Poudel et al. 2021). Apart from that, a plant microbiome provides protection against the accumulation of ROS through enzymatic and non-enzymatic antioxidant system (Poudel et al. 2021).

Microbiome-associated plant adaptations against stress are adaptable themselves as they differ in different stress conditions. One microbe may have the ability to inaugurate different responses in different conditions and plants may recruit a specific species of microbe to withstand a specific stress condition. No matter what the case, in every stress condition prominence of plant–microbiome interactions is substantiated.

14.6 Applications of Microbiome Engineering

Microbial communities are so intricately connected to the host that there is a set of "core microbiome" specific for the host that is determined at Operational Taxonomic Unit (OTU) levels (Gopal et al. 2013). Any disruption in the composition or function of the core microbiome can prove to be catastrophic for the host. And as the hosts are interconnected in ecology so are the microbial communities. So, the effect does not only extend to a particular host but the whole ecosystem, as they may bring about physical, biochemical, and even genetic changes in the host and thus affecting the productivity and efficiency of the whole ecosystem. The extensive impact and potential of microbiome forces the focus of research around microbiome, with both urgency and interest. And as microorganisms are easier to handle, they become a convenient target for engineering to generate desired phenotypic and even genetic changes in the host and consequently other biotic elements in the ecosystem. The biotechnological advances regarding genome sequence, meta-omics tools, computational tools, genome wide functional genomics, etc. further encouraged microbiome engineering, efforts to "customize" the microbiota to sustain agriculture without damaging the environment by introducing biological ways to control, fertilize, and stimulate, for the welfare of humankind.

Microbiome transfer and host-mediated microbiome engineering (HMME) and synthetic microbiomes are widely used in microbiome engineering. Based on synthetic biology bottom-up or top-down approaches are notably used to engineer phytomicrobiomes (Ke et al. 2021). The bottom-up approach isolates the plant-associated microbes from environmental microbes, modifies genome for them to carry desired traits, and then reintroduces the synthetic microbial communities (SynCom) to the host plant for it to then express the outcome of the engineering (Toju et al. 2018; Vorholt et al. 2017). On the other hand, horizontal gene transfer (HZT) is used in top-down approach to inaugurate the desired traits in a broad range of hosts in situ. This can be achieved through incorporating mobile genetic elements, developing bacteriophage system, or by other necessary strategies to introduce the desired traits in the host (Ke et al. 2021).

Engineered microbiome may suppress different diseases of host as seen in the case of ACC deaminase engineered banana endospheric bacterial cell wall promoting suppression of *Fusarium* induced wilt disease (Kaul et al. 2021). Conferring stress tolerance is one of the most desired applications of plant microbiome engineering for sustainable food production. Microbes are specifically engineered to confront stress conditions including drought and salinity (Jochum et al. 2019; Mueller et al. 2021). Microbiome engineering can also be employed in biocontrol, which can be exemplified through plant pathogen *Burkholderia ambifaria* that produces a biocontrol component cepacin A. Various studies have proved that engineered *Burkholderia ambifaria* that does not have virulence can be used as an effective biocontrol agent (Coenye 2019; Mullins et al. 2019). Microbial system can be engineered to produce novel biocontrol compounds as in the case of *Serratia fonticola* and novel fungicidal phenazine (Greunke et al. 2018). Engineered microbiome exhibits excellent biofertilizer activity as demonstrated by Voigt and colleagues, where engineering of *nif* gene clusters in microbiome resulted in nitrogen fixation of a cereal crop (Ryu et al. 2020; Temme et al. 2012). Moreover, microbiome engineered to generate specific phytohormones can elevate host's ability to resist stress, work as a biostimulant and even suppress disease (Ke et al. 2021).

Biosafety, biosecurity, and biocontainments need to be evaluated before the field application of microbiome engineering, but once applied may open the door to exceptional advantages leading toward sustained agriculture and flourishing ecosystem.

14.7 Application of Endophytes as Potential Biofertilizer and Biocontrol

Chemical fertilizers serve a critical part in meeting the world population's ever-increasing food needs. Increased agricultural usage of chemical fertilizers, on the other hand, has negative consequences for ecosystems. These fertilizers are leached into water bodies due to poor absorption by plants, causing eutrophication. They can also have a variety of negative consequences on soils, such as depletion of water storage capacity and uneven soil fertility (Ongley et al. 2010). Another severe issue is the widespread use of pesticides, many of which are toxic to both people and animals, including pollinators (Sponsler et al. 2019). Pesticides may also change the dynamics of soil microbial communities associated with plants (Meena et al. 2020). As a result, finding ecologically acceptable alternatives is crucial.

For decades, microbiological tools such as biofertilizers and biocontrol agents, which are bacteria and fungi, capable of promoting plant growth and health, respectively, have been produced. Biofertilizers have the potential to increase agricultural yields while also being ecologically beneficial (Mahanty et al. 2017; Giri et al. 2019). Biological nitrogen fixation, initiated by the endophytic bacteria *Rhizobium* spp., colonizes and occupies the interior compartment of plant tissue with little or no injury to the host plant, is one of the most prevalent ways. This category of endophytic microbes can be used as a biofertilizer. Endophytic nitrogen-fixing bacteria isolated from various plants and plant components were found to supply around 48% of the nitrogen. Endophytes like *Gluconacetobacter diazotrophicus*, *Serratia marcescens*, and *Azoarcus* sp. are suitable bioformulations/biofertilizers and hence can be prepared to use in agricultural fields (Garima Gupta et al. 2012). Moreover, nitrogen-fixing alpha-proteobacterial genera of *Agrobacterium*, *Allorhizobium*, *Azorhizobium*, *Bradyrhizobium*, *Mesorhizobium*, *Rhizobium*, *Sinorhizobium*, *Methylobacterium*, *Ochrobactrum*, and *Phyllobacterium*, as well as beta-proteobacterial *Burkholderia* and *Cupriavidus*, have also been reported as potential biofertilizers (Franche et al. 2009).

Endophytic fungi have been rarely exploited as potential tools in agriculture, despite their various favorable effects on plants and promise in biocontrol and biofertilization. *Piriformospora indica*, which colonizes the roots of cereal plants like barley and maize, is one of the most promising and well-studied endophytes (Waller et al. 2005; Malla et al. 2004; Prasad et al. 2008, 2013; Mishra et al. 2014; Chadha et al. 2014, 2015; Gill et al. 2016). The fungus can promote phosphate and sulfur intake, increase biomass output, and stimulate early flowering and seed production (Oelmüller et al. 2009). Effects of *P. indica* have been widely investigated and evaluated in over 150 different plant species (Smriti Shrivastava and Varma 2014).

Biological control agents have shown the ability to colonize plants, or at the very least form a close bond with them. The majority of biological control endophyte (BCE) agents have been discovered in a variety of crops and have the ability to reduce major pathogens. Bacterial endophytes lessen or prevent deleterious effects of plant pathogens through antibiosis (antibiotic production), growth promotion, inducing host defenses (induced systemic resistance, ISR), parasitism, competition, and signal interference (quorum sensing) (Jorjani et al. 2012; Mansoori et al. 2013). A bacterial endophyte *Pseudomonas fluorescens* PICF7 was found to suppress *Verticillium* wilt, one of the most serious diseases affecting olive trees worldwide (Mercado-Blanco et al. 2004). In glasshouse and field trials, *Paenibacillus* K165 isolated from tomato root tips was demonstrated to minimize disease severity and symptoms caused by *Verticillium dahliae* in potato and eggplant (Tjamos et al. 2004). Bacterial strains Y30 and E36 identified as *Streptomyces virginiae* were isolated from tomato plants and found to suppress *Ralstonia solanacearum*, the causal agent of tomato bacterial wilt (Tan et al. 2011). In a recent work by Chowdhury and Bae (2018) discovered an endophyte *Burkholderia stabilis* EB159 (PG159) that has the strongest inhibitory efficacy against *B. cinerea*—caused leaf spot disease. *Bacillus cabrialesii* TE3T, a new *Bacillus* species identified as an endophytic bacteria associated with wheat, was also found as a biological control agent against a developing wheat phytopathogen (*Bipolaris sorokiniana*) in the Yaqui Valley, inducing spot blotch and lowering the infection frequencies (Villa-Rodriguez et al. 2019). Several bacterial endophytes were also found to inhibit pathogens during the post-harvest period. Aiello et al. (2019) recently evaluated the biocontrol properties of *Pseudomonas synxantha* DLS65, an isolated endophyte from kiwi fruit tissues, against *Monilinia fructicola* and *Monilinia fructigena*, which are causal agents of post-harvest brown rot of fruit with bone and found 100% of mycelial growth inhibition. Zaman et al. recently identified a novel bacterial endophyte *Burkholderia contaminans* NZ from jute plant that substantially suppress the growth of different fungal plant pathogens including *Macrophomina phaseolina* (Zaman et al. 2021).

Several fungal endophytes were also reported to have potential biocontrol properties. *Trichoderma* species have been found to antagonize *Sclerotinia sclerotiorum* and *Rhizoctonia solani* hyphae (Mukhopadhyay and Kumar 2020). *Purpureocillium lilacinum* is a classic biocontrol agent of plant parasitic nematodes and insects (Wang et al. 2016). In a study by Wei et al. (2019a) found that a novel *Fusarium* strain (*Fusarium solani* CEF559) could protect cotton plants against wilt in greenhouse tests with better than 60% control, similar to the control level (50%) reached by another *F. solani* strain (Bx215) in a greenhouse experiment. Non-toxigenic strains of *Aspergillus flavus* have been demonstrated to have the capacity to control mycotoxigenic *Aspergillus* in cereals in several investigations (Sarrocco and Vannacci 2018).

Despite the strong antagonistic nature and capacity of endophytes to benefit the host plant by inducing resistance and growth promotion, there are a number of obstacles to overcome when generating or using endophytes as biofertilizers and biocontrol agents. When administered to the roots, endophytes may encounter competition from the varied diversity of indigenous soil microflora, as well as the various degrees of their population densities, which are often influenced by environmental factors. Furthermore, despite the fact that some endophytes have excellent bioactivity in laboratory and greenhouse trials, they are not always used in the field, possibly due to poor viability during storage, poor colonization, toxicity to non-targeted organisms, or the fact that they are not suitable for large-scale applications due to low yield in cultures (Pirttilä et al. 2021; Kumar et al. 2017). Better understanding of the mechanisms involved in the growth promotion and antagonistic abilities of endophytes will lead to better crop production. Several approaches such as the use of genomics, in vivo expression technology, fluorescence experiments, and model plants can help achieve this objective.

14.8 Conclusion and Future Prospects

Plant microbe management is considered as an advanced approach in order to increase nutrient absorption and control disease. Therefore, most of the current researches are aimed at elucidating the processes of host–pathogen interaction in order to promote sustainable agriculture. Exploiting interactions between plant and soil microbial populations is a viable strategy for increasing food production for the world's rising population while minimizing ecological damage in the present climate change situation. The development of potential microbial inoculants or the modification of naturally occurring microbial populations are the two main techniques for regulating the plant-microbiome. The formulation, biosafety maintainence, and modalities of application of microbial inoculants are being given special attention. Challenges arise mostly from the demand of characterizing a broad variety of systems involving the complex microbial populations, where the majority of which are unculturable species. A growing number of culture-independent approaches are often used to unravel the unknown variety of microbes living in plant, and soil environments, as well as to characterize the molecular basis of plant–microbe interactions. Different forms of stressful conditions, such as salinity, drought, diseases, and pests, have negative impact on plants' functioning. So, plant–microbiome interactions have huge role in growth and survival of plants in stressful conditions by using a signaling network. Recognizing this network is critical for developing biotechnological solutions to increase plant adaption mechanisms and soil microorganisms' abilities to alleviate stress in crops. Several ways are now being investigated to see whether the plant microbiome may be manipulated to promote beneficial species while preventing disease presence. As a rational conclusion, numerous accomplishments have been made via the use of microbial biotechnology in agriculture, but many problems and possibilities need to be investigated in order to ensure future agricultural sustainability.

References

Abd El-Rahman A, Shaheen HA, Abd El-Aziz RM, Ibrahim DS (2019) Influence of hydrogen cyanide-producing rhizobacteria in controlling the crown gall and root-knot nematode, Meloidogyne incognita. Egypt J Biol Pest Control 29(1):41

Abdelfattah A, Li Destri Nicosia MG, Cacciola SO, Droby S, Schena L (2015) Metabarcoding analysis of fungal diversity in the phyllosphere and carposphere of olive (Olea europaea). PLoS One 10(7):e0131069

Ahemad M, Kibret M (2014) Mechanisms and applications of plant growth promoting rhizobacteria: current perspective. J King Saud Univ Sci 26(1):1–20

Aiello D, Restuccia C, Stefani E, Vitale A, Cirvilleri G (2019) Postharvest biocontrol ability of pseudomonas synxantha against Monilinia fructicola and Monilinia fructigena on stone fruit. Postharvest Biol Technol 149:83–89

Ait Barka E, Nowak J, Clément C (2006) Enhancement of chilling resistance of inoculated grapevine plantlets with a plant growth-promoting rhizobacterium, Burkholderia phytofirmans strain PsJN. Appl Environ Microbiol 72(11):7246–7252

Ali SZ, Sandhya V, Grover M, Kishore N, Rao LV, Venkateswarlu B (2009) Pseudomonas sp. strain AKM-P6 enhances tolerance of sorghum seedlings to elevated temperatures. Biol Fertil Soils 46(1):45–55

Amanullah K, Fahad S (2018) Nitrogen in agriculture: updates. BoD–Books on Demand, Norderstedt

Amend AS, Cobian GM, Laruson AJ, Remple K, Tucker SJ, Poff KE et al (2019) Phytobiomes are compositionally nested from the ground up. Peer J 7:e6609

Asari S et al (2016) Multiple effects of Bacillus amyloliquefaciens volatile compounds: plant growth promotion and growth inhibition of phytopathogens. FEMS Microbiol Ecol 92(6): fiw070

Ashraf M, Harris P (2004) Potential biochemical indicators of salinity tolerance in plants. Plant Sci 166(1):3–16

Atamna-Ismaeel N, Finkel O, Glaser F, von Mering C, Vorholt JA, Koblížek M et al (2012) Bacterial anoxygenic photosynthesis on plant leaf surfaces. Environ Microbiol Rep 4(2): 209–216

Augé RM (2001) Water relations, drought and vesicular-arbuscular mycorrhizal symbiosis. Mycorrhiza 11(1):3–42

Azad K, Kaminskyj S (2016) A fungal endophyte strategy for mitigating the effect of salt and drought stress on plant growth. Symbiosis 68(1):73–78

Badri DV, Vivanco JM (2009) Regulation and function of root exudates. Plant Cell Environ 32(6): 666–681

Bai Y, Müller DB, Srinivas G, Garrido-Oter R, Potthoff E, Rott M et al (2015) Functional overlap of the Arabidopsis leaf and root microbiota. Nature 528(7582):364–369

Bakthavatchalu S, Shivakumar S, Sullia SB (2012) Identification of multi-trait PGPR isolates and evaluation of their potential as biocontrol agents. Acta Biologica Indica 1(1):61–67

Balhara M, Chaudhary R, Ruhil S, Singh B, Dahiya N, Parmar VS et al (2016) Siderophores; iron scavengers: the novel & promising targets for pathogen specific antifungal therapy. Expert Opin Ther Targets 20(12):1477–1489

Banaei-Asl F, Farajzadeh D, Bandehagh A, Komatsu S (2016) Comprehensive proteomic analysis of canola leaf inoculated with a plant growth-promoting bacterium, Pseudomonas fluorescens, under salt stress. Biochimica et Biophysica Acta (BBA)-Proteins Proteomics 1864(9):1222–1236

Bashir I, War AF, Rafiq I, Reshi ZA, Rashid I, Shouche YS (2022) Phyllosphere microbiome: diversity and functions. Microbiol Res 254:126888

Beneduzi A, Ambrosini A, Passaglia LM (2012) Plant growth-promoting rhizobacteria (PGPR): their potential as antagonists and biocontrol agents. Genet Mol Biol 35(4):1044–1051

Berg G, Smalla K (2009) Plant species and soil type cooperatively shape the structure and function of microbial communities in the rhizosphere. FEMS Microbiol Ecol 68(1):1–13

Berg G, Rybakova D, Grube M, Köberl M (2016) The plant microbiome explored: implications for experimental botany. J Exp Bot 67(4):995–1002

Bhattacharyya D, Yu S-M, Lee YH (2015) Volatile compounds from Alcaligenes faecalis JBCS1294 confer salt tolerance in Arabidopsis thaliana through the auxin and gibberellin pathways and differential modulation of gene expression in root and shoot tissues. Plant Growth Regul 75(1):297–306

Bulgarelli D, Garrido-Oter R, Münch PC, Weiman A, Dröge J, Pan Y et al (2015) Structure and function of the bacterial root microbiota in wild and domesticated barley. Cell Host Microbe 17(3):392–403

Campbell R, Renwick A, Coe S (1986) Antagonism and siderophore production by biocontrol agents, plant growth promoting organisms and the general rhizosphere population. In: Iron, Siderophores, and plant diseases. Springer, New York, pp 179–188

Chadha N, Mishra M, Bandyoadhyay P, Agarwal A, Prasad R, Varma A (2014) Protocol for production of the beneficial fungus *Piriformospora indica* for field applications. J Endocytobiosis Cell Res 25:42–46

Chadha N, Prasad R, Varma A (2015) Plant promoting activities of fungal endophytes associated with tomato roots from central Himalaya, India and their interaction with *Piriformospora indica*. Int J Pharm BioSci 6(1):333–343

Charpentier M, Oldroyd G (2010) How close are we to nitrogen-fixing cereals? Curr Opin Plant Biol 13(5):556–564

Cheremisinoff NP (2010) Handbook of pollution prevention and cleaner production vol 3. In: Best practices in the agrochemical industry. William Andrew, Norwich

Chowdhury MEK, Bae H (2018) Bacterial endophytes isolated from mountain-cultivated ginseng (Panax ginseng Mayer) have biocontrol potential against ginseng pathogens. Biol Control 126: 97–108

Coenye T (2019) Uncoupling virulence and biocontrol. Nat Microbiol 4(6):908–909

Compant S, Mitter B, Colli-Mull JG, Gangl H, Sessitsch A (2011) Endophytes of grapevine flowers, berries, and seeds: identification of cultivable bacteria, comparison with other plant parts, and visualization of niches of colonization. Microb Ecol 62(1):188–197

Compant S, Samad A, Faist H, Sessitsch A (2019) A review on the plant microbiome: ecology, functions, and emerging trends in microbial application. J Adv Res 19:29–37

Cordovez V, Dini-Andreote F, Carrión VJ, Raaijmakers JM (2019) Ecology and evolution of plant microbiomes. Annu Rev Microbiol 73:69–88

Cornforth DM, Foster KR (2015) Antibiotics and the art of bacterial war. Proc Natl Acad Sci 112(35):10827–10828

Das A, Prasad R, Srivastava A, Giang PH, Bhatnagar K, Varma A (2007) Fungal siderophores: structure, functions and regulations. In: Varma A, Chincholkar SB (eds) Microbial siderophores. Springer-Verlag, Berlin, pp 1–42

Dastogeer KM, Tumpa FH, Sultana A, Akter MA, Chakraborty A (2020) Plant microbiome—an account of the factors that shape community composition and diversity. Curr Plant Biol 23: 100161

Deveau A, Bonito G, Uehling J, Paoletti M, Becker M, Bindschedler S et al (2018) Bacterial–fungal interactions: ecology, mechanisms and challenges. FEMS Microbiol Rev 42(3):335–352

Dickinson CH (2012) Microbiology of aerial plant surfaces. Elsevier, Amsterdam
Dodd IC, Pérez-Alfocea F (2012) Microbial amelioration of crop salinity stress. J Exp Bot 63 (9):3415–3428
Doornbos RF, van Loon LC, Bakker PA (2012) Impact of root exudates and plant defense signaling on bacterial communities in the rhizosphere. A review. Agron Sust Dev 32(1):227–243
Etesami H, Emami S, Alikhani HA (2017) Potassium solubilizing bacteria (KSB): mechanisms, promotion of plant growth, and future prospects. A review. J Soil Sci Plant Nutr 17(4):897–911
Farooq M et al (2009) Plant drought stress: effects, mechanisms and management. In: Sustainable agriculture. Springer, Dordrecht, pp 153–188
Ferreira CM, Vilas-Boas Â, Sousa CA, Soares HM, Soares EV (2019) Comparison of five bacterial strains producing siderophores with ability to chelate iron under alkaline conditions. AMB Express 9(1):78
Franche C, Lindström K, Elmerich C (2009) Nitrogen-fixing bacteria associated with leguminous and non-leguminous plants. Plant Soil 321(1):35–59
Fukami J, Cerezini P, Hungria M (2018a) Azospirillum: benefits that go far beyond biological nitrogen fixation. AMB Express 8:73
Fukami J, Cerezini P, Hungria M (2018b) Azospirillum: benefits that go far beyond biological nitrogen fixation. AMB Express 8(1):1–12
Fürnkranz M, Wanek W, Richter A, Abell G, Rasche F, Sessitsch A (2008) Nitrogen fixation by phyllosphere bacteria associated with higher plants and their colonizing epiphytes of a tropical lowland rainforest of Costa Rica. ISME J 2(5):561–570
Fürnkranz M, Lukesch B, Müller H, Huss H, Grube M, Berg G (2012) Microbial diversity inside pumpkins: microhabitat-specific communities display a high antagonistic potential against phytopathogens. Microb Ecol 63(2):418–428
Gans J, Wolinsky M, Dunbar J (2005) Computational improvements reveal great bacterial diversity and high metal toxicity in soil. Science 309(5739):1387–1390
Gill SS, Gill R, Trivedi DK, Anjum NA, Sharma KK, Ansari MW, Ansari AA, Johri AK, Prasad R, Pereira E, Varma A, Tuteja N (2016) *Piriformospora indica*: potential and significance in plant stress tolerance. Front Microbiol 7:332. https://doi.org/10.3389/fmicb.2016.00332
Giri B, Prasad R, Wu Q-S, Varma A (2019) Biofertilizers for sustainable agriculture and environment. Springer International Publishing. ISBN 978-3-030-18932-7. https://www.springer.com/gp/book/9783030189327
Glick BR, Cheng Z, Czarny J, Duan J (2007) Promotion of plant growth by ACC deaminase-producing soil bacteria. In: New perspectives and approaches in plant growth-promoting Rhizobacteria research. Springer, pp 329–339
Gong T, Xin XF (2021) Phyllosphere microbiota: community dynamics and its interaction with plant hosts. J Integr Plant Biol 63(2):297–304
Gopal M, Gupta A, Thomas GV (2013) Bespoke microbiome therapy to manage plant diseases. Front Microbiol 4:355
Goswami D, Thakker JN, Dhandhukia PC (2016) Portraying mechanics of plant growth promoting rhizobacteria (PGPR): a review. Cogent Food Agric 2(1):1127500
Greunke C et al (2018) Direct pathway cloning (DiPaC) to unlock natural product biosynthetic potential. Metab Eng 47:334–345
Gupta S, Pandey S (2019) ACC deaminase producing bacteria with multifarious plant growth promoting traits alleviates salinity stress in French bean (Phaseolus vulgaris) plants. Front Microbiol 1506
Gupta G, Panwar J, Akhtar MS, Jha PN (2012) Endophytic nitrogen-fixing bacteria as biofertilizer. In: Sustainable agriculture reviews. Springer, New York, pp 183–221
Hacquard S (2016) Disentangling the factors shaping microbiota composition across the plant holobiont. New Phytol 209(2):454–457
Haidar B, Ferdous M, Fatema B, Ferdous AS, Islam MR, Khan H (2018) Population diversity of bacterial endophytes from jute (Corchorus olitorius) and evaluation of their potential role as bioinoculants. Microbiol Res 208:43–53

Haney CH, Samuel BS, Bush J, Ausubel FM (2015) Associations with rhizosphere bacteria can confer an adaptive advantage to plants. Nat Plants 1(6):1–9

Hannula SE, Heinen R, Huberty M, Steinauer K, De Long JR, Jongen R et al (2021) Persistence of plant-mediated microbial soil legacy effects in soil and inside roots. Nat Commun 12(1):1–13

Hardoim PR, Van Overbeek LS, Berg G, Pirttilä AM, Compant S, Campisano A et al (2015) The hidden world within plants: ecological and evolutionary considerations for defining functioning of microbial endophytes. Microbiol Mol Biol Rev 79(3):293–320

Hartman K, van der Heijden MG, Wittwer RA, Banerjee S, Walser J-C, Schlaeppi K (2018) Cropping practices manipulate abundance patterns of root and soil microbiome members paving the way to smart farming. Microbiome 6(1):1–14

İnceoğlu Ö, Al-Soud WA, Salles JF, Semenov AV, van Elsas JD (2011) Comparative analysis of bacterial communities in a potato field as determined by pyrosequencing. PLoS One 6(8): e23321

Innerebner G, Knief C, Vorholt JA (2011) Protection of Arabidopsis thaliana against leaf-pathogenic Pseudomonas syringae by Sphingomonas strains in a controlled model system. Appl Environ Microbiol 77(10):3202–3210

Issa A et al (2018) Impacts of Paraburkholderia phytofirmans strain PsJN on tomato (Lycopersicon esculentum L.) under high temperature. Front Plant Sci 9:1397

Jacoby R, Peukert M, Succurro A, Koprivova A, Kopriva S (2017) The role of soil microorganisms in plant mineral nutrition—current knowledge and future directions. Front Plant Sci 8:1617

Jansson JK, Baker ES (2016) A multi-omic future for microbiome studies. Nat Microbiol 1(5):1–3

Ji SH, Gururani MA, Chun S-C (2014) Isolation and characterization of plant growth promoting endophytic diazotrophic bacteria from Korean rice cultivars. Microbiol Res 169(1):83–98

Jiao R, Munir S, He P, Yang H, Wu Y, Wang J et al (2020) Biocontrol potential of the endophytic Bacillus amyloliquefaciens YN201732 against tobacco powdery mildew and its growth promotion. Biol Control 143:104160

Jochum MD et al (2019) Host-mediated microbiome engineering (HMME) of drought tolerance in the wheat rhizosphere. PLoS One 14(12):e0225933

Jorjani M, Heydari A, Zamanizadeh HR, Rezaee S, Naraghi L, Zamzami P (2012) Controlling sugar beet mortality disease by application of new bioformulations. J Plant Protect Res 52:303–307

Kalayu G (2019) Phosphate solubilizing microorganisms: promising approach as biofertilizers. Int J Agron 2019:4917256

Kannahi M, Senbagam N (2014) Studies on siderophore production by microbial isolates obtained from rhizosphere soil and its antibacterial activity. J Chem Pharm Res 6(4):1142–1145

Kaul S et al (2021) Engineering host microbiome for crop improvement and sustainable agriculture. Front Microbiol 12:635917

Ke J, Wang B, Yoshikuni Y (2021) Microbiome engineering: synthetic biology of plant-associated microbiomes in sustainable agriculture. Trends Biotechnol 39(3):244–261

Kenawy A, Dailin DJ, Abo-Zaid GA, Abd Malek R, Ambehabati KK, Zakaria KHN et al (2019) Biosynthesis of antibiotics by PGPR and their roles in biocontrol of plant diseases. In: Plant growth promoting Rhizobacteria for sustainable stress management. Springer, New York, pp 1–35

Kenneth OC, Nwadibe EC, Kalu AU, Unah UV (2019) Plant growth promoting rhizobacteria (PGPR): a novel agent for sustainable food production. Am J Agric Biol Sci 14:35–54

Khan A et al (2020) Silicon and gibberellins: synergistic function in harnessing ABA signaling and heat stress tolerance in date palm (Phoenix dactylifera L.). Plan Theory 9(5):620

Kim K, Jang Y-J, Lee S-M, Oh B-T, Chae J-C, Lee K-J (2014) Alleviation of salt stress by Enterobacter sp. EJ01 in tomato and Arabidopsis is accompanied by up-regulation of conserved salinity responsive factors in plants. Mol Cells 37(2):109

Knief C, Delmotte N, Chaffron S, Stark M, Innerebner G, Wassmann R et al (2012) Metaproteogenomic analysis of microbial communities in the phyllosphere and rhizosphere of rice. ISME J 6(7):1378–1390

Kohler J, Hernández JA, Caravaca F, Roldán A (2008) Plant-growth-promoting rhizobacteria and arbuscular mycorrhizal fungi modify alleviation biochemical mechanisms in water-stressed plants. Funct Plant Biol 35(2):141–151

Kour D, Rana KL, Kaur T, Yadav N, Halder SK, Yadav AN et al (2020) Potassium solubilizing and mobilizing microbes: biodiversity, mechanisms of solubilization, and biotechnological implication for alleviations of abiotic stress. In: New and future developments in microbial biotechnology and bioengineering. Elsevier, Amsterdam, pp 177–202

Kuan KB, Othman R, Abdul Rahim K, Shamsuddin ZH (2016) Plant growth-promoting rhizobacteria inoculation to enhance vegetative growth, nitrogen fixation and nitrogen remobilisation of maize under greenhouse conditions. PLoS One 11(3):e0152478

Kumar M, Saxena R, Tomar RS (2017) Endophytic microorganisms: promising candidate as biofertilizer. In: Microorganisms for green revolution. Springer, New York, pp 77–85

Kumar P, Tapan Kumar T, Singh S, Tuteja N, Prasad R, Singh J (2020) Potassium: a key modulator for cell homeostasis. J Biotechnol 324:198–210. https://doi.org/10.1016/j.jbiotec.2020.10.018

Kumar A, Zhimo VY, Biasi A, Feygenberg O, Salim S, White JF et al (2021) Microbial volatiles: prospects for plant defense and disease management. In: Food security and plant disease management. Elsevier, pp 387–404

Ladha J, Barraquio W, Revilla L (1997) Isolation of endophytic diazotrophic bacteria from wetland rice. In: Opportunities for biological nitrogen fixation in rice and other non-legumes. Springer, New York, pp 15–24

Lareen A, Burton F, Schäfer P (2016) Plant root-microbe communication in shaping root microbiomes. Plant Mol Biol 90(6):575–587

Lauber CL, Hamady M, Knight R, Fierer N (2009) Soil pH as a predictor of soil bacterial community structure at the continental scale: a pyrosequencing-based assessment. Appl Environ Microbiol 75(15):5111–5120

Liu H, Carvalhais LC, Crawford M, Singh E, Dennis PG, Pieterse CM et al (2017) Inner plant values: diversity, colonization and benefits from endophytic bacteria. Front Microbiol 8:2552

Lucke M, Correa MG, Levy A (2020) The role of secretion systems, effectors, and secondary metabolites of beneficial rhizobacteria in interactions with plants and microbes. Front Plant Sci 11:1718

Lundberg DS, Lebeis SL, Paredes SH, Yourstone S, Gehring J, Malfatti S et al (2012) Defining the core Arabidopsis thaliana root microbiome. Nature 488(7409):86–90

Mahanty T, Bhattacharjee S, Goswami M, Bhattacharyya P, Das B, Ghosh A et al (2017) Biofertilizers: a potential approach for sustainable agriculture development. Environ Sci Pollut Res 24(4):3315–3335

Malla R, Prasad R, Kumari R, Giang PH, Pokharel U, Oelmueller R, Varma A (2004) Phosphorus solubilizing symbiotic fungus *Piriformospora indica*. Endocytobiosis Cell Res 15(2):579–600

Mansoori M, Heydari A, Hassanzadeh N, Rezaee S, Naraghi L (2013) Evaluation of Pseudomonas and Bacillus bacterial antagonists for biological control of cotton Verticillium wilt disease. J Plant Protect Res 53(2):154

Márquez LM et al (2007) A virus in a fungus in a plant: three-way symbiosis required for thermal tolerance. Science 315(5811):513–515

Mayak S, Tirosh T, Glick BR (2004a) Plant growth-promoting bacteria that confer resistance to water stress in tomatoes and peppers. Plant Sci 166(2):525–530

Mayak S, Tirosh T, Glick BR (2004b) Plant growth-promoting bacteria confer resistance in tomato plants to salt stress. Plant Physiol Biochem 42(6):565–572

Meena RS, Kumar S, Datta R, Lal R, Vijayakumar V, Brtnicky M et al (2020) Impact of agrochemicals on soil microbiota and management: a review. Land 9(2):34

Mendes R, Raaijmakers JM (2015) Cross-kingdom similarities in microbiome functions. ISME J 9(9):1905–1907

Mercado-Blanco J, Rodrıguez-Jurado D, Hervás A, Jiménez-Dıaz RM (2004) Suppression of Verticillium wilt in olive planting stocks by root-associated fluorescent Pseudomonas spp. Biol Control 30(2):474–486

Miotto-Vilanova L et al (2016) Burkholderia phytofirmans PsJN confers grapevine resistance against Botrytis cinerea via a direct antimicrobial effect combined with a better resource mobilization. Front Plant Sci 7:1236

Mishra M, Chadha N, Kumar V, Prasad R, Varma A (2014) Root endophytic fungi from two extreme geographical regions of India. J Endocytobiosis Cell Res 25:20–26

Mitra D, Anđelković S, Panneerselvam P, Senapati A, Vasić T, Ganeshamurthy A et al (2020) Phosphate-solubilizing microbes and biocontrol agent for plant nutrition and protection: current perspective. Commun Soil Sci Plant Anal 51(5):645–657

Mitter B, Petric A, Shin MW, Chain PS, Hauberg-Lotte L, Reinhold-Hurek B et al (2013) Comparative genome analysis of Burkholderia phytofirmans PsJN reveals a wide spectrum of endophytic lifestyles based on interaction strategies with host plants. Front Plant Sci 4:120

Montañez A, Abreu C, Gill PR, Hardarson G, Sicardi M (2009) Biological nitrogen fixation in maize (Zea mays L.) by 15N isotope-dilution and identification of associated culturable diazotrophs. Biol Fertil Soils 45(3):253–263

Moreno-Galván AE, Cortés-Patiño S, Romero-Perdomo F, Uribe-Vélez D, Bashan Y, Bonilla RR (2020) Proline accumulation and glutathione reductase activity induced by drought-tolerant rhizobacteria as potential mechanisms to alleviate drought stress in Guinea grass. Appl Soil Ecol 147:103367

Morris CE, Kinkel LL (2002) Fifty years of phyllosphere microbiology: significant contributions to research in related fields. Phyllosp Microbiol:365–375

Mueller UG et al (2021) Artificial selection on microbiomes to breed microbiomes that confer salt tolerance to plants. Msystems 6(6):e01125–e01121

Mukhopadhyay R, Kumar D (2020) Trichoderma: a beneficial antifungal agent and insights into its mechanism of biocontrol potential. Egypt J Biol Pest Control 30(1):1–8

Mullins AJ et al (2019) Genome mining identifies cepacin as a plant-protective metabolite of the biopesticidal bacterium Burkholderia ambifaria. Nat Microbiol 4(6):996–1005

Naylor D, Coleman-Derr D (2018) Drought stress and root-associated bacterial communities. Front Plant Sci 8:2223

Oelmüller R, Sherameti I, Tripathi S, Varma A (2009) Piriformospora indica, a cultivable root endophyte with multiple biotechnological applications. Symbiosis 49(1):1–17

Ongley ED, Xiaolan Z, Tao Y (2010) Current status of agricultural and rural non-point source pollution assessment in China. Environ Pollut 158(5):1159–1168

Pandya U, Saraf M (2014) In vitro evaluation of PGPR strains for their biocontrol potential against fungal pathogens. In: Microbial diversity and biotechnology in food security. Springer, New York, pp 293–305

Pang Z, Chen J, Wang T, Gao C, Li Z, Guo L et al (2021) Linking plant secondary metabolites and plant microbiomes: a review. Front Plant Sci 12:300

Pankievicz VC, do Amaral FP, Santos KF, Agtuca B, Xu Y, Schueller MJ et al (2015) Robust biological nitrogen fixation in a model grass–bacterial association. Plant J 81(6):907–919

Patel Priyanka J, Trivedi Goral R, Shah Rupal K, Saraf M (2019) Rhizospheric microflora: a natural alleviator of drought stress in agricultural crops. In: Plant growth promoting rhizobacteria for sustainable stress management. Springer, pp 103–115

Pena LC, Jungklaus GH, Savi DC, Ferreira-Maba L, Servienski A, Maia BH et al (2019) Muscodor brasiliensis sp. nov. produces volatile organic compounds with activity against Penicillium digitatum. Microbiol Res 221:28–35

Philippot L, Raaijmakers JM, Lemanceau P, Van Der Putten WH (2013) Going back to the roots: the microbial ecology of the rhizosphere. Nat Rev Microbiol 11(11):789–799

Phuakjaiphaeo C, Chang C, Ruangwong O, Kunasakdakul K (2016) Isolation and identification of an antifungal compound from endophytic Streptomyces sp. CEN 26 active against Alternaria brassicicola. Lett Appl Microbiol 63(1):38–44

Pirttilä AM, H MPT, Baruah N, Koskimäki JJ (2021) Biofertilizers and biocontrol agents for agriculture: how to identify and develop new potent microbial strains and traits. Microorganisms 9(4):817

Poudel M et al (2021) The role of plant-associated bacteria, fungi, and viruses in drought stress mitigation. Front Microbiol:3058

Prasad R, Bagde US, Pushpangdan P, Varma A (2008) *Bacopa monniera* L.: pharmacological aspects and case study involving *Piriformospora indica*. Int J Integr Biol 3:100–110

Prasad R, Kamal S, Sharma PK, Oelmueller R, Varma A (2013) Root endophyte *Piriformospora indica* DSM 11827 alters plant morphology, enhances biomass and antioxidant activity of medicinal plant *Bacopa monniera*. J Basic Microbiol 53(12):1016–1024

Puri A, Padda KP, Chanway CP (2018) Evidence of endophytic diazotrophic bacteria in lodgepole pine and hybrid white spruce trees growing in soils with different nutrient statuses in the west Chilcotin region of British Columbia, Canada. For Ecol Manag 430:558–565

Radzki W, Mañero FG, Algar E, García JL, García-Villaraco A, Solano BR (2013) Bacterial siderophores efficiently provide iron to iron-starved tomato plants in hydroponics culture. Antonie Van Leeuwenhoek 104(3):321–330

Raghuwanshi R, Prasad JK (2018) Perspectives of rhizobacteria with ACC deaminase activity in plant growth under abiotic stress. Root Biol:303–321

Raza A, Razzaq A, Mehmood SS, Zou X, Zhang X, Lv Y et al (2019) Impact of climate change on crops adaptation and strategies to tackle its outcome: a review. Plan Theory 8(2):34

Redford AJ, Bowers RM, Knight R, Linhart Y, Fierer N (2010) The ecology of the phyllosphere: geographic and phylogenetic variability in the distribution of bacteria on tree leaves. Environ Microbiol 12(11):2885–2893

Reetha AK, Pavani SL, Mohan S (2014) Hydrogen cyanide production ability by bacterial antagonist and their antibiotics inhibition potential on Macrophomina phaseolina (Tassi.) Goid. Int J Curr Microbiol App Sci 3(5):172–178

Rojas-Tapias D, Moreno-Galván A, Pardo-Díaz S, Obando M, Rivera D, Bonilla R (2012) Effect of inoculation with plant growth-promoting bacteria (PGPB) on amelioration of saline stress in maize (Zea mays). Appl Soil Ecol 61:264–272

Ruinen J (1965) The phyllosphere: III. Nitrogen fixation in the phyllosphere. Plant Soil:375–394

Ruiz-Lozano JM (2003) Arbuscular mycorrhizal symbiosis and alleviation of osmotic stress. New perspectives for molecular studies. Mycorrhiza 13(6):309–317

Ryu M-H et al (2020) Control of nitrogen fixation in bacteria that associate with cereals. Nat Microbiol 5(2):314–330

Saleem M, Arshad M, Hussain S, Bhatti AS (2007) Perspective of plant growth promoting rhizobacteria (PGPR) containing ACC deaminase in stress agriculture. J Ind Microbiol Biotechnol 34(10):635–648

Sandrini M, Nerva L, Sillo F, Balestrini R, Chitarra W, Zampieri E (2022) Abiotic stress and belowground microbiome: the potential of omics approaches. Int J Mol Sci 23(3):1091

Sang M, Kim KD (2012) The volatile-producing Flavobacterium johnsoniae strain GSE09 shows biocontrol activity against Phytophthora capsici in pepper. J Appl Microbiol 113(2):383–398

Sarhan EA, Shehata HS (2014) Potential plant growth-promoting activity of Pseudomonas spp. and Bacillus spp. as biocontrol agents against damping-off in alfalfa. Plant Pathol J 13(1):8–17

Sarrocco S, Vannacci G (2018) Preharvest application of beneficial fungi as a strategy to prevent postharvest mycotoxin contamination: a review. Crop Prot 110:160–170

Sayyed R, Chincholkar S, Reddy M, Gangurde N, Patel P (2013) Siderophore producing PGPR for crop nutrition and phytopathogen suppression. In: Bacteria in agrobiology: disease management. Springer, New York, pp 449–471

Scherer H (2005) Fertilizers and fertilization. In: Hillel D (ed) Encyclopedia of soils in the environment. Elsevier, Amsterdam

Schlaeppi K, Bulgarelli D (2015) The plant microbiome at work. Mol Plant-Microbe Interact 28(3): 212–217

Sergaki C, Lagunas B, Lidbury I, Gifford ML, Schäfer P (2018) Challenges and approaches in microbiome research: from fundamental to applied. Front Plant Sci 9:1205

Sessitsch A, Reiter B, Berg G (2004) Endophytic bacterial communities of field-grown potato plants and their plant-growth-promoting and antagonistic abilities. Can J Microbiol 50(4): 239–249

Shabanamol S, Divya K, George TK, Rishad K, Sreekumar T, Jisha M (2018) Characterization and in planta nitrogen fixation of plant growth promoting endophytic diazotrophic Lysinibacillus sphaericus isolated from rice (Oryza sativa). Physiol Mol Plant Pathol 102:46–54

Sharma SB, Sayyed RZ, Trivedi MH, Gobi TA (2013) Phosphate solubilizing microbes: sustainable approach for managing phosphorus deficiency in agricultural soils. Springerplus 2(1):1–14

Shekhawat K, Saad MM, Sheikh A, Mariappan K, Al-Mahmoudi H, Abdulhakim F et al (2021) Root endophyte induced plant thermotolerance by constitutive chromatin modification at heat stress memory gene loci. EMBO Rep 22(3):e51049

Shi J, Liu A, Li X, Feng S, Chen W (2010) Identification of endophytic bacterial strain MGP1 selected from papaya and its biocontrol effects on pathogens infecting harvested papaya fruit. J Sci Food Agric 90(2):227–232

Shrivastava S, Varma A (2014) From Piriformospora indica to rootonic: a review. Afr J Microbiol Res 8(32):2984–2992

Shrivastava M, Srivastava P, D'souza S (2018) Phosphate-solubilizing microbes: diversity and phosphates solubilization mechanism. In: Role of rhizospheric microbes in soil. Springer, New York, pp 137–165

Singh D, Raina TK, Kumar A, Singh J, Prasad R (2019) Plant microbiome: a reservoir of novel genes and metabolites. Plant Gene. https://doi.org/10.1016/j.plgene.2019.100177

Singh S, Singh UB, Trivdi M, Malviya D, Sahu PK, Roy M et al (2021) Restructuring the cellular responses: connecting microbial intervention with ecological fitness and adaptiveness to the maize (Zea mays L.) grown in saline–sodic soil. Front Microbiol 11:568325

Sponsler DB, Grozinger CM, Hitaj C, Rundlöf M, Botías C, Code A et al (2019) Pesticides and pollinators: a socioecological synthesis. Sci Total Environ 662:1012–1027

Sukweenadhi J, Kim Y-J, Choi E-S, Koh S-C, Lee S-W, Kim Y-J et al (2015) Paenibacillus yonginensis DCY84T induces changes in Arabidopsis thaliana gene expression against aluminum, drought, and salt stress. Microbiol Res 172:7–15

Tahir HAS, Gu Q, Wu H, Niu Y, Huo R, Gao X (2017) Bacillus volatiles adversely affect the physiology and ultrastructure of Ralstonia solanacearum and induce systemic resistance in tobacco against bacterial wilt. Sci Rep 7(1):1–15

Tan H, Zhou S, Deng Z, He M, Cao L (2011) Ribosomal-sequence-directed selection for endophytic streptomycete strains antagonistic to Ralstonia solanacearum to control tomato bacterial wilt. Biol Control 59(2):245–254

Temme K, Zhao D, Voigt CA (2012) Refactoring the nitrogen fixation gene cluster from Klebsiella oxytoca. Proc Natl Acad Sci 109(18):7085–7090

Teotia P, Kumar V, Kumar M, Shrivastava N, Varma A (2016) Rhizosphere microbes: potassium solubilization and crop productivity—present and future aspects. In: Potassium solubilizing microorganisms for sustainable agriculture. Springer, New York, pp 315–325

Tjamos S, Antoniou P, Kazantzidou A, Antonopoulos D, Papageorgiou I, Tjamos E (2004) Aspergillus Niger and Aspergillus carbonarius in Corinth raisin and wine-producing vineyards in Greece: population composition, ochratoxin A production and chemical control. J Phytopathol 152(4):250–255

Tiryaki D, Aydın İ, Atıcı Ö (2019) Psychrotolerant bacteria isolated from the leaf apoplast of cold-adapted wild plants improve the cold resistance of bean (Phaseolus vulgaris L.) under low temperature. Cryobiology 86:111–119

Tkacz A, Poole P (2021) The plant microbiome: the dark and dirty secrets of plant growth. Plants People Planet 3(2):124–129

Toju H et al (2018) Core microbiomes for sustainable agroecosystems. Nat Plants 4(5):247–257

Trivedi P, Leach JE, Tringe SG, Sa T, Singh BK (2020) Plant–microbiome interactions: from community assembly to plant health. Nat Rev Microbiol 18(11):607–621

Tsegay Z, Gizaw B, Tefera G (2016) Pharmacobiology and medical sciences. Int J Curr Trend Pharmacobiol Med Sci 1(2):37–44

Turner TR, James EK, Poole PS (2013) The plant microbiome. Genome Biol 14(6):1–10

Upadhyay S, Singh G, Singh D (2016) Mechanism and understanding of PGPR: an approach for sustainable agriculture under abiotic stresses. J Plant Sci 134(2):24–75

Uroz S, Buée M, Murat C, Frey-Klett P, Martin F (2010) Pyrosequencing reveals a contrasted bacterial diversity between oak rhizosphere and surrounding soil. Environ Microbiol Rep 2(2): 281–288

van der Heijden MG, Martin FM, Selosse MA, Sanders IR (2015) Mycorrhizal ecology and evolution: the past, the present, and the future. New Phytol 205(4):1406–1423

Velivelli SL, Sessitsch A, Prestwich BD (2014) The role of microbial inoculants in integrated crop management systems. Potato Res 57(3):291–309

Villa-Rodriguez E, Parra-Cota F, Castro-Longoria E, López-Cervantes J, de los Santos-Villalobos S (2019) Bacillus subtilis TE3: a promising biological control agent against Bipolaris sorokiniana, the causal agent of spot blotch in wheat (Triticum turgidum L. subsp. durum). Biol Control 132: 135–143

Vorholt JA (2012) Microbial life in the phyllosphere. Nat Rev Microbiol 10(12):828–840

Vorholt JA et al (2017) Establishing causality: opportunities of synthetic communities for plant microbiome research. Cell Host Microbe 22(2):142–155

Voříšková J, Baldrian P (2013) Fungal community on decomposing leaf litter undergoes rapid successional changes. ISME J 7(3):477–486

Wagner MR, Lundberg DS, Del Rio TG, Tringe SG, Dangl JL, Mitchell-Olds T (2016) Host genotype and age shape the leaf and root microbiomes of a wild perennial plant. Nat Commun 7(1):1–15

Waller F, Achatz B, Baltruschat H, Fodor J, Becker K, Fischer M et al (2005) The endophytic fungus Piriformospora indica reprograms barley to salt-stress tolerance, disease resistance, and higher yield. Proc Natl Acad Sci 102(38):13386–13391

Wang G, Liu Z, Lin R, Li E, Mao Z, Ling J et al (2016) Biosynthesis of antibiotic leucinostatins in bio-control fungus Purpureocillium lilacinum and their inhibition on Phytophthora revealed by genome mining. PLoS Pathog 12(7):e1005685

Watanabe K, Kohzu A, Suda W, Yamamura S, Takamatsu T, Takenaka A et al (2016) Microbial nitrification in throughfall of a Japanese cedar associated with archaea from the tree canopy. Springerplus 5(1):1–15

Wei F, Zhang Y, Shi Y, Feng H, Zhao L, Feng Z et al (2019a) Evaluation of the biocontrol potential of endophytic fungus fusarium solani CEF559 against Verticillium dahliae in cotton plant. Biomed Res Int 2019:1

Wei F, Zhao L, Xu X, Feng H, Shi Y, Deakin G et al (2019b) Cultivar-dependent variation of the cotton rhizosphere and endosphere microbiome under field conditions. Front Plant Sci 10:1659

Weselowski B, Nathoo N, Eastman AW, MacDonald J, Yuan Z-C (2016) Isolation, identification and characterization of Paenibacillus polymyxa CR1 with potentials for biopesticide, biofertilization, biomass degradation and biofuel production. BMC Microbiol 16(1):1–10

Wonglom P, Ito S-i, Sunpapao A (2020) Volatile organic compounds emitted from endophytic fungus Trichoderma asperellum T1 mediate antifungal activity, defense response and promote plant growth in lettuce (Lactuca sativa). Fungal Ecol 43:100867

Woodward C, Hansen L, Beckwith F, Redman RS, Rodriguez RJ (2012) Symbiogenics: an epigenetic approach to mitigating impacts of climate change on plants. HortScience 47 (6):699–703

Yin C, Casa Vargas JM, Schlatter DC, Hagerty CH, Hulbert SH, Paulitz TC (2021) Rhizosphere community selection reveals bacteria associated with reduced root disease. Microbiome 9(1): 1–18

Yu K, Pieterse CM, Bakker PA, Berendsen RL (2019) Beneficial microbes going underground of root immunity. Plant Cell Environ 42(10):2860–2870

Zaman NR, Chowdhury UF, Reza RN, Chowdhury FT, Sarker M, Hossain MM et al (2021) Plant growth promoting endophyte Burkholderia contaminans NZ antagonizes phytopathogen Macrophomina phaseolina through melanin synthesis and pyrrolnitrin inhibition. PLoS One 16(9):e0257863

Zhang H, Kim M-S, Sun Y, Dowd SE, Shi H, Paré PW (2008) Soil bacteria confer plant salt tolerance by tissuespecific regulation of the sodium transporter HKT1. Mol Plant-Microbe Interact 21(6):737–744

Zhao L-j, Yang X-n, Li X-y, Wei M, Feng L (2011) Antifungal, insecticidal and herbicidal properties of volatile components from Paenibacillus polymyxa strain BMP-11. Agric Sci China 10(5):728–736

Zheng B-X, Hao X-L, Ding K, Zhou G-W, Chen Q-L, Zhang J-B et al (2017) Long-term nitrogen fertilization decreased the abundance of inorganic phosphate solubilizing bacteria in an alkaline soil. Sci Rep 7(1):1–10

Chapter 15
Endophytic Phytohormone Production and Utilization of Functional Traits in Plant Growth Promotion

Aahuti Sharma, Pankaj Kumar, Vikas Pahal, Jitendra Kumar, and Shiv Shankar Pandey

Abstract Microorganisms such as bacteria, fungi, or actinomycetes play an important role in plant growth and development. They are ubiquitous in nature. Endophytes are beneficial microbes that live within host plant tissue without causing harmful effects to their host. They are present in all plants species and show symbiotic association with them. Population of the endophytes depends on region where the host plant grows and also affected by climatic condition. Endophytic microbes utilize various mechanisms to enhance plant growth and productivity and also play important role in plant defense mechanism against different environmental condition (biotic and abiotic stresses). Phytohormones are small molecule growth regulators which are synthesized during the plant metabolism. The most common phytohormones are auxins, cytokinin, gibberellin, abscisic acid, ethylene, brassinosteroids, salicylic acid, jasmonates, and strigolactones. Endophytes also produce phytohormones which promote plant growth and change the morphology and physiology of the plant. During stress responses, biosynthetic and signaling pathways of phytohormones play major role in coordinating the development of plant. We have focused on endophytic phytohormone production and their role in plant growth in this chapter.

A. Sharma · S. S. Pandey
Department of Biotechnology, CSIR-IHBT, Palampur, Himachal Pradesh, India

P. Kumar (✉)
Department of Microbiology, Dolphin (PG) Institute of Biomedical and Natural Science, Dehradun, Uttarakhand, India

V. Pahal
Department of Microbiology, Dolphin (PG) College of Science and Agriculture, Fatehgarh, Punjab, India

J. Kumar
Department of Biosciences, UIBT, Chandigarh University, Mohali, Punjab, India

S. Chhabra et al. (eds.), *Plant Microbiome for Plant Productivity and Sustainable Agriculture*, Microorganisms for Sustainability 37,
https://doi.org/10.1007/978-981-19-5029-2_15

Keywords Endophytes · Phytohormones · Auxin · Gibberellic acids · ACC · Jasmonic acid

Abbreviations

4-Cl-IAA	4-Chloroindole-3-Acetic Acid
ABA	Abscisic acid
ACC	1-Aminocyclopropane-1-Carboxylic Acid
CPP	Copalyl diphosphate
GA	Gibberellic Acid
IAA	Indole-3-Acetic Acid
IAld	Indole-3-Acetaldehyde
IAM	Indole-3-Acetamide
IAN	Indole-3-Acetonitrile
IAOX	Indole-3-Acetaldoxine
IBA	Indole-3-Butyric Acid
ICS	Isochorismate Synthase
IPA	Indole-3-Pyruvic Acid
IPP	Isopentenyl pyrophosphate
IPT	Isopentenyl Transferase
JA	Jasmonic Acid
MEP	Methyl Erythritol Phosphate
MVA	Mevalonic Acid
PAL	Phenylalanine Ammonia-Lyase
SA	Salicyclic Acid
TAM	Tryptamine
Trp	Tryptophan

15.1 Introduction

The term endophytes was first introduced by De Bary (1866), where endo means inside and phytes means plants. It is defined as any organisms that grow within plant tissue without harming their hosts (Yadav 2018). Endophytes exist in all plants and live asymptomatically in the plant cellular environment. During mutualistic interaction, endophytes perform specific function such as synthesis of secondary metabolite or signaling molecules that function as external and internal signals (Tidke et al. 2018). Endophytes show complex interaction with their host plant which involves mutualism and antagonism. Their association with plant can also be obligate and facultative. Microorganisms which do not reproduce outside the plant tissue, depends on plant metabolites for their survival are termed as obligate endophytes (Hardoim et al. 2008). Endophytes that live freely in soil at a definite stage of their life but they are usually entered to plants from the surrounding environment are defined as facultative endophytes (Abreu-Tarazi et al. 2010). Endophytes show symbiotic relationship with plants, which improve the physiological fitness and metabolite profile of plants while the plants provide shelter and food for endophytes (Chadha et al. 2015; Mishra et al. 2015; Kumar et al. 2017). Interaction of endophytes with plants depends on the capacity of microorganisms to use the exudates produced by the plant roots as their energy source. They are ubiquitous in nature and present in all species of plants. Endophytes population in plants species is highly variable and depends on different components such as host plant developmental stage, inoculums density, and climatic conditions (Dudeja and Giri 2014). Types of endophytes also vary from plant to plant and species to species. Endophytes are generally isolated from the internal tissue of plants after surface sterilization by using 1–2% sodium hypochlorite and 70% ethanol (Sahu et al. 2022). The role of endophytes is known to influence overall host plant growth, photosynthesis, nutrient absorption and physiological processes. Beneficial endophytes produce many compounds which are useful for protecting plants from environmental conditions, increase plant growth and sustainability, while living inside the host plant (Sarkar et al. 2021). Endophytic microbes also improve plant's ability to tolerate various stresses such as nutrient, temperature, salinity, drought and enhance the ability of resistance against plant insects and pests (Joseph and Priya 2011; Prasad et al. 2013, 2018). Due to abiotic stresses, the endogenous levels of phytohormones, such as auxins, gibberellins (GA), cytokinin, abscisic acid (ABA), jasmonic acid, and salicylic acid (SA), alter the plant growth pattern and play crucial role in plant-microbe interaction (Singh et al. 2021; Khan et al. 2014). Endophytic colonization can be local or systemic; they colonize in the stems, bark, roots, petiole, fruits, buds, seeds, and leaf segments with midribs (Specian et al. 2012; Stepniewska and Kuzniar 2013). They produce phytohormones and bioactive compounds which are useful in industries, agriculture, and medical science (Singh et al. 2021). Phytohormones are important growth regulators which have a prominent impact on plant metabolism and they are synthesized in defined organs of the plant (Kazan 2013). Endophytes secrete phytohormones and help in bidirectional nutrient transfer to improve nutritional level and plant health by protecting against phytopathogens (Andreozzi et al. 2019; Shen et al. 2019). Endophytes secrete various components such as secondary metabolites, serves as biocontrol agents, antimicrobial agents and they secrete antiviral compounds and also enhance natural antioxidants development, antibiotics, and insecticidal product (Gouda et al. 2016; Varma et al. 2017a, b). Endophytic microorganism helps in boosting plant growth through several mechanism. The general mechanisms employed by endophytic bacteria and fungi are indirect and direct mechanisms. In case of direct mechanisms microorganisms directly provide the nutrients and other resources that are necessary for plant growth or modulate plant hormonal levels. In indirect mechanisms, plant growth inhibition thwart by endophytic microbes by producing of siderophores, lytic enzymes, antibiotics, systemic resistance induction, alteration in ethylene levels, and direct competition with phytopathogens (Glick 2012; Chhabra and Dowling 2017).

15.2 Endophytic Phytohormones

15.2.1 Auxin

The first hormone discovered in plants was auxin. First time, Dutch botanist Fritz W. Went (1928) isolated auxin from the tip of oat coleoptiles in the gelatin block. Auxins are important plant hormones that play vital role in plant's daily growth and life (Grossmann 2010). It plays vital role in much of the early physiological processes such as plant cell expansion. Auxin enhances plant cell division, cell elongation, differentiation and extension, stimulates seed and tuber germination, and increases the rate of development of root and xylem (Fadiji and Babalola 2020). The most common natural auxin are indole-3-acetic acid (IAA), indole-3-butyric acid (IBA), and 4-Chloroindole-3-acetic acid (4-Cl-IAA) (Fig. 15.1). Indole -3-acetic acid (IAA) is the most common auxin, which is produced by plants and also by bacteria and fungi. Structurally, IAA is related to the amino acid tryptophan which is precursor for the biosynthesis of auxin (Duca et al. 2014). IAA production by endophytes improved apical dominance and also stimulates adventitious and lateral root development, also promotes plant root biomass, surface area, and mediate resistance stress condition (Chen et al. 2014; Ali et al. 2017). All the IAA producing bacteria are not beneficial to plants. Generally, at any particular time plants are very sensitive to the amount of IAA present in plant tissue. IAA-producing endophytes must provide appropriate concentration of IAA when combined with the amount of the hormones produced by the plant. Less concentration of IAA production by the endophytes can enhance plant growth, whereas the high amount of IAA prevents growth and development (Malik and Sindhu 2011). Mostly IAA are produced in young leaves and buds of plant. There are many ways by which IAA helps the plant growth. In young stem, IAA helps to increase cell wall extensibility.

IAA productions have been reported in both Gram-negative and Gram-positive bacteria. IAA-producing endophytic bacterial genera include *Acetobacter, Rhizobium, Azospirillum, Azotobacter, Klebsiella, Alcaligenes, Pantoea, Herbaspirillum, Burkholderia, Pseudomonas, Bacillus, Enterobacter, Methylobacterium, Citrobacter, Rhodococcus,* and *Streptomyces* (Apine and Jadhav 2011; Sun et al. 2013). Some fungi (e.g., *Colletotrichum acutatum, Fusarium proliferatum, F.*

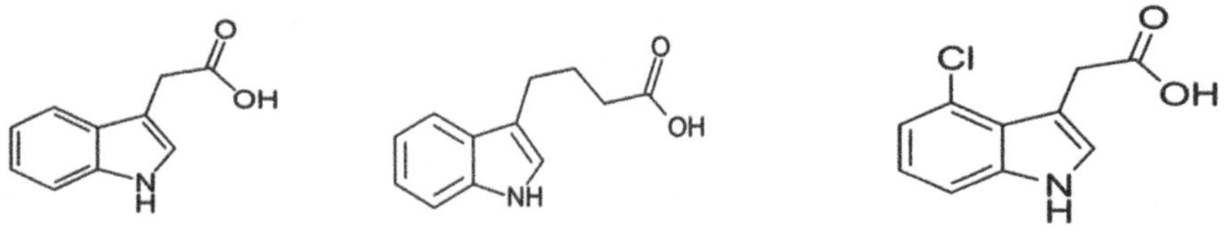

Indole-3-acetic acid (IAA)

Indole-3-butyric acid (IBA)

4-Chloroindole-3-acetic acid (4-Cl-IAA)

Fig. 15.1 Structure of three natural auxins

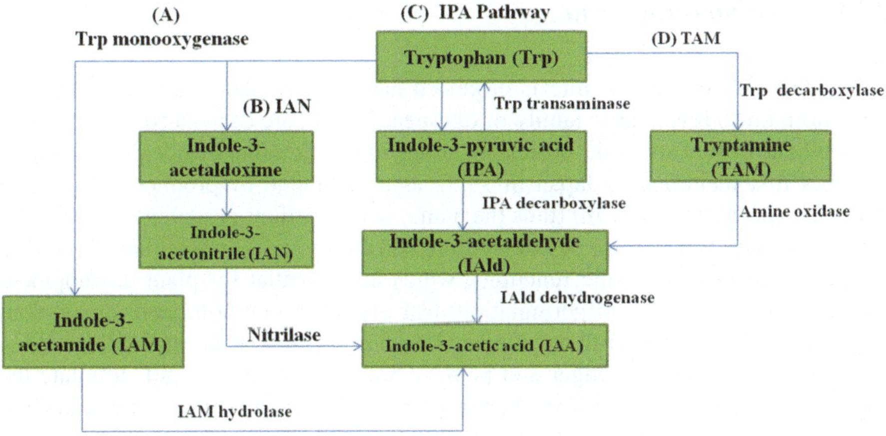

Fig. 15.2 Tryptophan-dependent pathways of IAA biosynthesis

fujikuroi, *F. oxysporum*, *Muscodor cinnamomi*, *Piriformospora indica*, *Ustilago esculenta*, and *U. maydis*) mainly use tryptophan-dependent pathways for IAA biosynthesis (Hilbert et al. 2012; Tsavkelova et al. 2012; Sun et al. 2014; Nutaratat et al. 2016). Moreover, amount of IAA in plants can also determine whether bacterial IAA promote or suppresses plant growth, as bacterial IAA production usually benefits those plants that have low endogenous level of IAA (Glick 2012). Endophytic bacteria produce IAA is the major effector molecule in pathogenesis, phytostimulation, and plant–microbe interaction (Gao and Tao 2012). IAA in plants is synthesized by different biosynthetic pathways. The main precursor for IAA synthesis is tryptophan, which helps to alter the level of biosynthesis (Kundan et al. 2015). IAA production is stimulated by tryptophan and it also regulates the IAA synthesis by inhibiting anthranilate because it reduces the IAA synthesis (Fig. 15.2). The types of pathway of IAA production by endophytes within plants depend on the nature of the resulting plant–microbe interactions. In many plant-beneficial bacteria, IAA production occurs via the indole-3-pyruvate pathway (IPA), whereas in many pathogenic bacteria indole-3-acetamide pathways (IAM) were used for the synthesis of IAA (Hardoim et al. 2008). Assumpção et al. (2009) reported several endophytes from soybean seeds as *Acinetobacter*, *Bacillus*, *Brevibacterium*, *Enterobacter*, *Methylobacterium*, *Paenibacillus*, *Pseudomonas*, *Ochrobactrum*, *Streptomyces*, and *Tsukamurella* produced IAA in vitro but only *Enterobacter* sp. increased the dry biomass of root of soybean. In another study, IAA-producing bacterial endophytes (*Bacillus cereus* and *B. subtilis*) and fungal endophytes (*Penicillium crustosum* and *Penicillium chrysogenum*) isolated from *Teucrium polium* stimulated the plant growth and increased biomass of maize in comparison to uninoculated plants (Hassan 2017). Chen et al. (2017) reported that endophytic IAA producing *Pseudomonas fluorescens* Sasm05 promoted the growth and photosynthesis of *Sedum alfredii*.

15.2.2 Gibberellic Acid

Gibberellins (GAs), are chemicals produced naturally by plants, are considered as phytohormones. It is a large family of diterpenoids that are derived from tetracyclic gibberellane carbon structure, which are arranged in either four or five ring structure. GA was first identified in Japan in 1926, as a metabolic by-product of the plant pathogen *Gibberella fujikuroi* (thus the name), which afflicts rice plants. *Gibberella fujikuroi* causes Bakane or Foolish seedling disease in rice. Gibberellins stimulate a number of plant's metabolic functions, which are essential for plant development and growth including seed germination, leaf growth promotion, stem elongation, stimulation of flowering, fruit development, and senescence (Khan et al. 2015). They also act as chemical messenger and help by breaking dormancy and stimulate the enzyme (Alpha amylase) and help in hydrolysis of starch present in many seeds into glucose to be used in cellular respiration (Kundan et al. 2015). There are many types of GAs but the main types of GAs include molecules with 20 carbon atom (C_{20}-GAs) and other molecules with 19 carbon (C_{19}-GAs), having variable fifth ring of a lactone (Fig. 15.3). The biologically active gibberellin is GA_1 which helps in controlling the stem growth and development (Bömke and Tudzynski 2009; Sponsel and Hedden 2010).

Gibberellin also plays an important role in mediating the effects of environmental changes on plant growth. Environmental factors including temperature and photoperiod can change the level of active gibberellins by affecting gene transcription for particular steps in biosynthetic pathway. The main precursor of gibberellins biosynthesis is isoprene unit (5C). The basic isoprenoid unit is isopentenyl diphosphate (IPP) which is generated by two ways: the mevalonic acid (MVA) and methyl erythritol phosphate (MEP) pathway. MVA pathway occurs in cytoplasm whereas the MEP pathway occurs in plastids (Hedden and Thomas 2012). Biosynthetic pathway of gibberellins can be divided into three stages according to their different cellular compartment and the enzyme involved (Fig. 15.4). First step is the production of terpenoid precursors (*ent*- Kaurene) occur in plastids. In second step, oxidation reaction occur in endoplasmic reticulum to form GA_{12} and GA_{53} and in last step formation of all other gibberellins from GA_{12} and GA_{53} occur in cytosol (Salazar-Cerezoa et al. 2018). *Bacillus siamensis* BE76 an endophytic bacteria

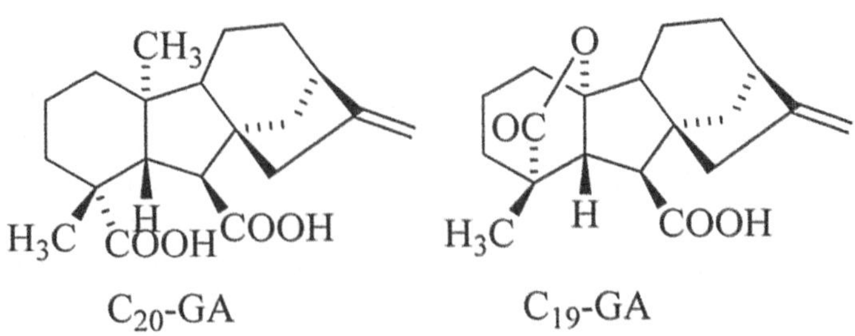

Fig. 15.3 Structure of C20-GAs and C19-GAs

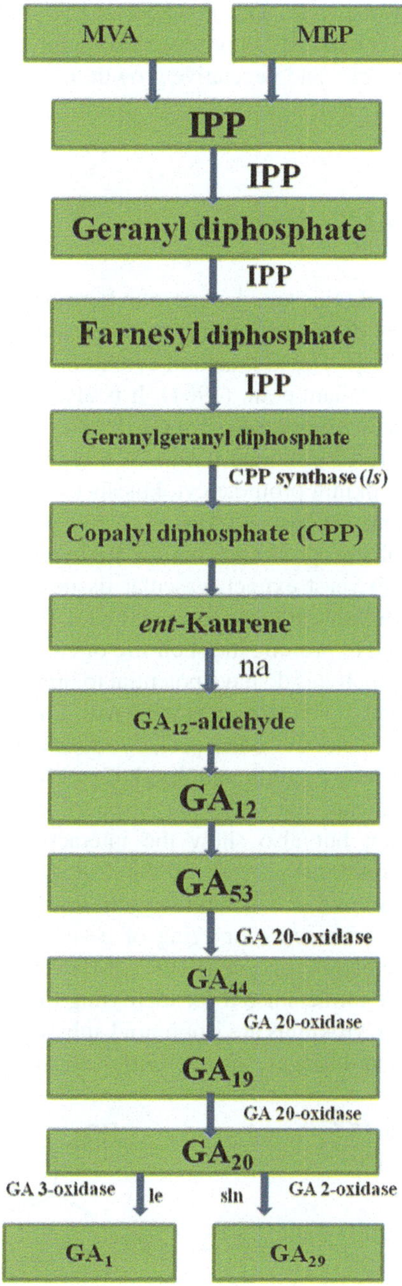

Fig. 15.4 Gibberellin biosynthetic pathway and the enzymes involved in the metabolic steps

isolated from stems of banana have the ability to produce a significant amount of gibberellic acid which provides beneficial effects on crop yield and prevent environmental pollution by avoiding excessive applications of chemical fertilizers

(Ambawade and Pathade 2015). In *Cicer arietinum* the endophytic bacteria *B. cereus* MEN8 produce maximum amount of GAs and increase seed germination capacity and vegetative growth, i.e., root/shoot length and fresh/dry weight (Baliyan et al. 2021).

15.2.3 Cytokinin

Cytokinins are plant growth hormones which are basic in nature. The first cytokinin was discovered from degraded autoclaved Herring sperm DNA by Miller et al. (1955). The first natural cytokinin was obtained from unripe maize grains or kernels by Letham et al. (1964). It is also present in coconut milk. Cytokinin is known as zeatin (6-hydroxy 3-methyl trans 2-butenyl amino-purine) (Fig. 15.5) or kinetin (6-furfuryl amino-purine). 18 types of Cytokinins have been discovered till date. Cytokinin promote cytokinesis (= cell division) either alone or in conjunction with auxin. Skoog and Tsui et al. (1948) found that callus from intermodal segments of Tobacco proliferate only when in addition to auxin, the nutrient medium is provided with yeast extract, vascular tissues, coconut milk or DNA because of presence of cytokinin in these sources. Cytokinin is also produced by endophytes which can be used to prolong the shelf life of cut flowers, fruits, and leafy vegetables. In addition, endophytes do have potential to produce many important pharmaceutically bioactive compounds (Bhore et al. 2010). Cytokinins like compounds are not only produced by plants but other prokaryotic and eukaryotic organisms also produce cytokinins such as bacteria and fungi. Cytokinins are important group of phytohormones that influence not only many aspects of plant physiology, growth, and development of plant but also show the interaction with microorganisms, including pathogens (Grosskinsky et al. 2016). Cytokinins help in plant to promote cell division of roots and shoots but the main functions of cytokinins are cell growth. It also delay the senescence or aging of tissue and thus effect the leaf growth, also inhibit the premature senescence of leaf (Hwang et al. 2012) and affect the apical dominancy. The farmers use it for increasing the overall yield and quality of crops.

Cytokinins are isoprenoid substituted adenine molecules. Under various environmental stages, the activity and type of cytokinin molecules differ remarkably

Fig. 15.5 Cis and trans -zeatin

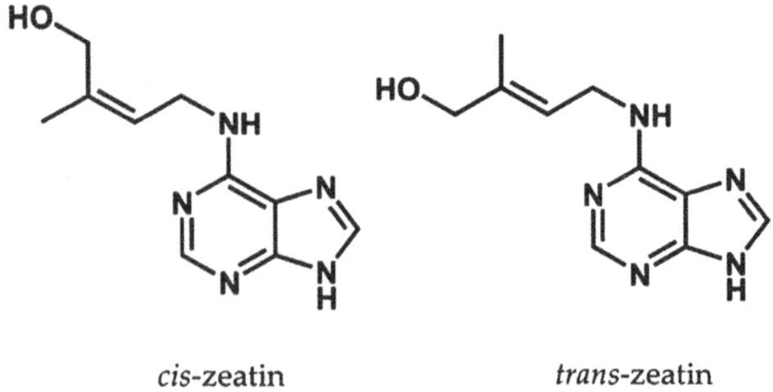

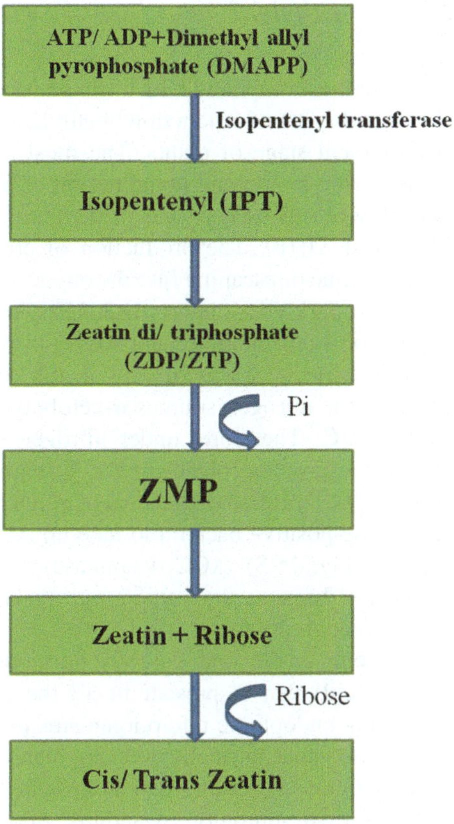

Fig. 15.6 Cytokinin synthesis pathway (Kieber and Schaller 2014)

between divergent plant species and tissues at different developmental stages. For the synthesis of cytokinins, the first enzyme is isopentenyl transferases (IPTs) which are involved in catalyzing isoprenoid to other various types of cytokinins including zeatin, N^6-isopentenyl-adenine (Fig. 15.6). The most common form of cytokinin in plants is Trans - zeatin. Cytokinins are metabolized and inactivated through conjugation to sugars or through degradation by cytokinin oxidases (Akhtar et al. 2019). Endophytic fungus *Phomopsis liquidambari* B3 can enhance cytokinin, auxin, ethylene level by establishing a mutualistic symbiotic relationship with *Oryza sativa* L. which enhanced nitrogen uptake, yield, and metabolism in rice (Li et al. 2018). Cytokinin-producing bacterial endophytes *Pseudomonas resinovorans* and *Paenibacillus polymyxa* were isolated from Sambung Nyawa (*Gynura procumbens* (Lour.) Merr.) and can also be used as plant growth-promoter (Bhore et al. 2010).

15.2.4 Ethylene

Ethylenes are compound which can be produced by almost all parts of higher plants, although the rate of production of ethylene depends on the type of tissue of plant and development stage of plants. The most active site for biosynthesis of ethylene is meristematic region and nodal region of plant. During leaf abscission, root nodulation, flower senescence, and fruit ripening, production of ethylene also increases (Sun et al. 2016). The production of ethylene depends on its rate of production versus its rate of escaping into the environment. It is produced more in dividing cells mostly in dark (Kundan et al. 2015). The main precursor of ethylene (Fig. 15.7) production is methionine, and ACC (1-aminocyclopropane-1-carboxylic acid) acts as an intermediate, which helps in the conversion of methionine to ethylene and provides the nitrogen sources α-ketobutyrate and ammonia from the ethylene precursor ACC. Therefore, under nitrogen-limitation conditions the bacterial endophytes can promote plant growth through the secretion of ACC deaminases (Sun et al. 2009). Decrease in the level of ethylene in plant relies on the ability of the ACC deaminase-positive bacteria to take up ACC before its oxidation by the plant ACC oxidase (Fig. 15.8). ACC deaminase-inducing endophytes are good plant-growth promoters, because they ameliorate plant stress by blocking ethylene production (Santoyo et al. 2016).

Ethylene is the first gaseous hormone formed by breakdown of methionine (amino acid) that is present in all the cells. The presence of ACC deaminase-containing endophytic microorganisms promotes tolerance against both abiotic and biotic stress that normally induces plant growth by inhibiting levels of ethylene. ACC deaminase-containing plant growth-promoting bacteria can effectively protect against growth inhibition by salinity, drought, flooding, cool, high levels of metals and organic contaminants, the presence of fungal and bacterial pathogens, as well as temperature stress (Gamalero et al. 2020). Under these stressed conditions, the endogenous production of ethylene is induced more to have the adverse effect on root growth and eventually on whole plant. Any type of wounds in plants can also induce ethylene biosynthesis. Biotic and abiotic stresses result in an increased ethylene production in plants that leads to inhibition of root elongation, formation of root hair, and development of lateral roots (Afzal et al. 2019). The ACC deaminase expressing endophyte *Pseudomonas* spp. enhances NaCl stress tolerance by reducing stress-related ethylene production, resulting in improved growth, photosynthetic performance, and ionic balance in tomato plants (Win et al. 2018). Khan et al. (2016) isolated edophytic bacteria from *Moringa peregrina* having ability to

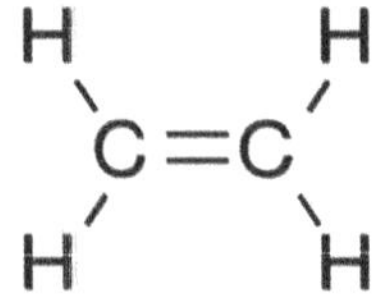

Fig. 15.7 Structure of ethylene

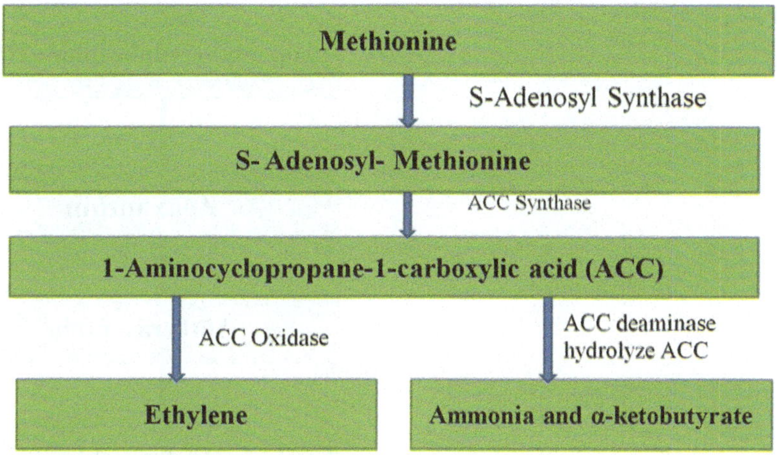

Fig. 15.8 Biosynthetic pathway of ethylene synthesis from methionine (Adams and Yang 1979)

Fig. 15.9 Structure of abscisic acid

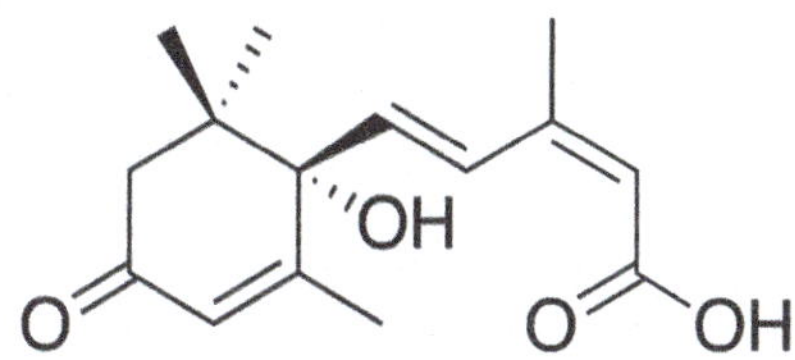

produce indole acetic acid and ACC deaminase enhanced the the growth of *Solanum lycopersicum*.

15.2.5 Abscisic Acid

Abscisic acid (ABA) is ubiquitous plant hormone present in vascular plants. Abscisic acid (Fig. 15.9) is a signaling molecule mediating bud growth, seed dormancy, and adaptation to various environmental stresses (Hauser et al. 2011). ABA also protects plant by inducing stomatal closure and thereby reduces water loss (Herrera-Medina et al. 2007). It is a derivative of sesquiterpenoid, which are group of phytohormones that helps in the regulation of plant growth.

Under drought stress condition, ABA shows different effects on the growth of root and shoot of the plant (Bano et al. 2012). ABA is required for the development of desiccation tolerance in the developing embryo, the synthesis of storage proteins, and the acquisition of dormancy (Nonogaki 2019). During abiotic stress the expression of stress responsive genes is regulated by ABA-induced and -mediated signaling, leading to better elicitation of tolerance responses (Sah et al. 2016).

ABA biosynthesis takes place in plastid and cytosol, which begins with isopentenyl diphosphate (as precursor of terpenoid) and leads to the synthesis of

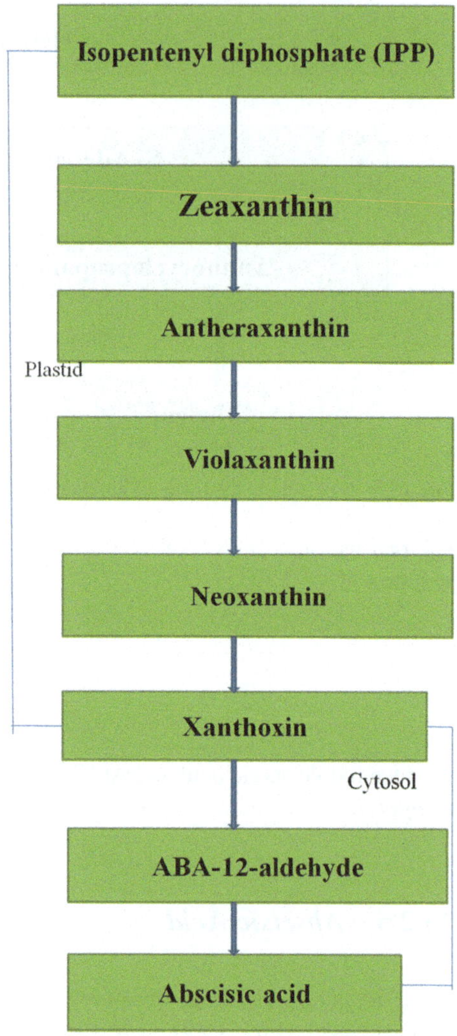

Fig. 15.10 Biosynthetic pathway of abscisic acid (Finkelstein 2013)

the C_{40}Xanthophyll (i.e., oxygenated carotenoid) Violaxanthin. Violaxanthin synthesis is catalyzed by Zeaxanthin epoxidase which convert Violaxanthin into C_{40} compound neoxanthin, which cleaved to form the C_{15} compound Xanthoxal, finally Xanthoxal is converted to ABA via oxidative steps involving the intermediates ABA aldehyde and the final step is catalyzed by aldehyde oxidases and synthesize ABA (Fig. 15.10). Cohen et al. (2009) suggested that both ABA and gas producing *Azospirillum* contribute to water-stress alleviation of plants. Seed born endophytic *B. amyloliquefaciens* RWL-1 produce ABA under saline conditions significantly increased the *Oryza sativa* growth (Shahzad et al. 2017).

15.2.6 Salicylic Acid

Salicylic acid (SA) is a phytohormone that plays a main role in plant defense under biotic and abiotic stimuli through the alteration of antioxidative enzyme activities (Ahmad et al. 2011; da Silva et al. 2017). SA is best known for endogenous signaling, mediating host responses against pathogen infection. Its biosynthesis in plants is not fully understood, but its role in plant defense activation has been well established. The IUPAC name of salicylic acid is 2-hydroxybenzoic acid (Fig. 15.11). SA is derived from two pathways, i.e., ICS and PAL pathway. Both the pathways start from chorismate (Lefevere et al. 2020). SA modulates several physiological processes involved in plant stress tolerance through stress-activated signal pathways and response mechanisms (Ahmad et al. 2011; Janda et al. 2012). SA treated plants showed better growth in terms of transcription, biomass accumulation, and promotion of cell differentiation and also showed antioxidant enzyme activity and higher photosynthetic rate of plant (da Silva et al. 2017).

Plants synthesize SA from cinnamate produced by the activity of phenylalanine ammonia lyase (PAL). Genetic studies, on the other hand, indicate that the bulk of SA is produced from isochorismate. In bacteria, SA is synthesized from chorismate through two reactions catalyzed by isochorismate synthase (ICS) and isochorismate pyruvate lyase (IPL). *Arabidopsis* contains two ICS genes but has no gene-encoding proteins similar to the bacterial IPL. Thus, how SA is synthesized in plants is not fully elucidated. Two recently identified *Arabidopsis* genes, PBS3 and EPS1, are important for pathogen-induced SA accumulation. PBS3 encodes a member of the acyl-adenylate/thioester-forming enzyme family and EPS1 encodes a member of the BAHD acyltransferase superfamily. PBS3 and EPS1 may be directly involved in the synthesis of an important precursor or regulatory molecule for SA biosynthesis (Chen et al. 2009). A study reported that the SA-producing *Pseudomonas tremae* EB-44 was most effective in suppressing tobacco wildfire disease and also has potential for use as an alternative, eco-friendly control measure for this disease in other crops (Islam et al. 2020).

15.2.7 Jasmonic Acid

Jasmonic acid (JA) is growth-regulating organic compound (Fig. 15.12). The molecule is a member of the jasmonate class of plant hormones. It is biosynthesized from

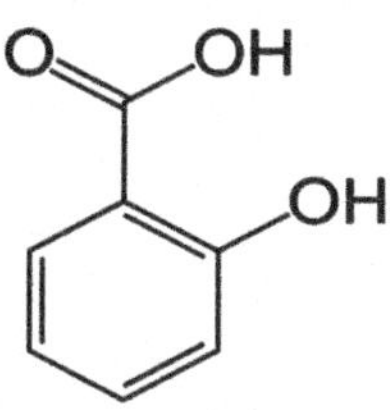

Fig. 15.11 Structure of salicylic acid

linolenic acid by the octadecanoid pathway. JA are lipid-derived signaling molecules which are essential for the adaptation of plants to adverse environment (Campos et al. 2014; Lefevere et al. 2020; Zhu et al. 2021). The exogenous application of JA also has a regulatory effect on plant growth. During stress condition, JA shows physiological responses including activation of antioxidant system and regulation of stomatal opening and closing (Wang et al. 2020).

15.3 Endophytic Fungi

Endophytic fungi are ubiquitous to plants, produce large number of chemical compounds which are essential for plant growth and development. First time, the plant hormone Gibberellins had been isolated from fungus *Gibberella fujikuroi.* These secondary metabolites are also present in different genera of fungi including *Aspergillus cadophora*, *Porosterum spadiceum*, *Penicillium*, *Neosartorya* sp., *Porostereum spadiceum* (Hamayun et al. 2017; You et al. 2013). Endophytic fungi include *Aspergillus fumigatus*, *Scolecobasidium tshawytschae*, *Aspergillus caespitosus, Porostereum spadiceum*, *Cadophora malorum* produces salicylic acid (SA), jasmonic acid (JA), and low concentration of abscisic acid (ABA) (Khan et al. 2011). The endophytic fungus *Paecilomyces formosus* LHL10 produced many physiologically active and inactive GAs and IAA, which helped the rice plants to grow well and significantly mitigated the negative impacts of salinity stress on cucumber plants (Khan et al. 2012). The co-inoculation of the endophytic fungus *Paecilomyces formosus* LHL10 and bacteria *Sphingomonas* sp. LK11 actively contributed to the tripartite mutualistic symbiosis in *Glycine max* under heavy metal stresses; this consortium could be used as an excellent strategy for sustainable agriculture in the heavy metal-contaminated fields (Bilal et al. 2018). We have summarized the various endophytic microorganisms (fungi and bacteria) having abilities to enhance the growth of plants in Table 15.1.

Fig. 15.12 Jasmonic acid

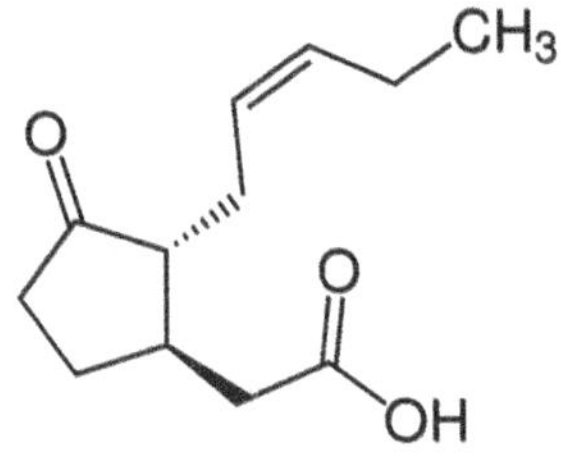

Table 15.1 Various plant growth hormones produced by endophytes

Plant hormones	Bacterial endophytes	References
Indole acetic acid (Auxin)	*Azospirillum, Azotobacter, Curtobacterium, Paenibacillus, Alcaligenes, Herbaspirillum, Enterobacter, Citrobacter, Pseudomonas, Klebsiella, Rhizobium, Burkholderia, Herbaspirillum, Enterobacter, Bacillus, Acinetobacter, Serratia*	Hauser et al. (2011), Rana et al. (2020), Singh et al. (2018), Zhao et al. (2018)
Gibberellins (GAs)	*Acetobacter diazotrophicus, Herbaspirillum seropedicae, Bacilus, Pseudomonas, Sinorhizobium, Rhizobium phaseoli, Sphingomonas* sp. *Leifsonia soli*	Ali et al. (2017), Shahzad et al. (2016), Sandhya (2017)
Cytokinin (CK)	*Paenibacillus polymyxa, Pseudomonas resinovorans, Bacillus subtilis, Pseudomonas, Sinorhizobium, Klebsiella*	Shahzad et al. (2017)
Abscisic acid (ABA)	*Bacillus amyloliquefaciens, Pseudomonas resinovorans*	Zhu and She (2018)
ACC deaminase	*Pseudomonas, Bacillus paralicheniformis, Streptomyces niveus, Paenibacillus polymyxa*	Islam et al. (2020), Borah and Thakur (2020)
Salicylic acid (SA)	*Pseudomonas tremae, P. fluorescens, P. aeruginosa* and *Serratia marcescens, Curtobacterium herbarum*	Chen et al. (2020), An and Mou (2011)
Jasmonic acid (JA)	*Azoarcus olearius, Azospirillum* spp.	Khan et al. (2014a)
Fungal endophytes		
Indole acetic acid	*Penicillium chrysogenum* and *Penicillium crustosum*	Hassan (2017)
Gibberellins	*Aspergillus, Porosterum spadiceum, Penicillium, Neosartorya* sp., *Porostereum spadiceum*	Hamayun et al. (2017)
Cytokinin and ethylene	*Phomopsis liquidambari*	Li et al. (2018)
Salicylic acid and Jasmonic acid	*Aspergillus, Cadophora, Porosterum*	Khan et al. (2011)

15.4 Conclusion

Endophytic microorganism supports plant development and growth by employing different mechanisms of action, i.e., indirect and direct mechanisms. Endophytes show mutualistic interaction with plants where the plants provide shelter and food to microbes and in return microbes provide benefits to plants by producing phytohormones and various secondary metabolites. Endophytes also have the ability to protect plants from various environmental stresses, including abiotic and biotic stress and resistance to pathogens by producing growth regulators, i.e., auxin, gibberellins, salicylic acid, cytokinin in host plant tissue. Endophytic microbes have a great potential to be used as biofertilizers, biopesticides, and usually used in agricultural field because they can synthesize and secrete chemical compounds which are useful for the development and growth of plant.

References

Abreu-Tarazi MF, Navarrete AA, Andreote FD, Almeida CV, Tsai SM, Almeida M (2010) Endophytic bacteria in long-term in vitro cultivated "axenic" pineapple microplants revealed by PCR–DGGE. World J Microbiol Biotechnol 26:555–560

Adams DO, Yang SF (1979) Ethylene biosynthesis: identification of 1-aminocyclopropane-1-carboxylic acid as an intermediate in the conversion of methionine to ethylene. Proc Natl Acad Sci U S A 76:170–174

Afzal I, Shinwari ZK, Sikandar S, Shahzad S (2019) Plant beneficial endophytic bacteria: mechanisms, diversity, host range and genetic determinants. Microbiol Res 221:36–49. https://doi.org/10.1016/j.micres.2019.02.001

Ahmad P, Nabi G, Ashraf M (2011) Cadmium-induced oxidative damage in mustard [*Brassica juncea* (L.) Czern.& Coss.]plants can be alleviated by salicylic acid. South Afr J Bot 77:36–44. https://doi.org/10.1016/j.sajb.2010.05.003

Akhtar S, Mekureyaw MF, Pandey C, Roitsch T (2019) Role of cytokinins for interactions of plants with microbial pathogens and pest insects. Front Plant Sci 10:1777. https://doi.org/10.3389/fpls.2019.01777

Ali S, Charles TC, Glick BR (2017) Endophytic phytohormones and their role in plant growth promotion. In: Doty SL (ed) Functional importance of the plant microbiome: implications for agriculture, forestry and bioenergy. Springer, Cham, pp 89–105

Ambawade MS, Pathade GR (2015) Production of gibberellic acid by *Bacillus siamensis* BE 76 isolated from Banana plant (*Musa* spp.). Int J Sci Res 4(7):394–398

An C, Mou Z (2011) Salicylic acid and its function in plant immunity. J Integr Plant Biol 53:412–428. https://doi.org/10.1111/j.1744-7909.2011.01043.x

Andreozzi A, Prieto P, Mercado-Blanco J, Monaco S, Zampieri E, Romano S (2019) Efficient colonization of the endophytes *Herbaspirillum huttiense* RCA24 and *Enterobacter cloacae* RCA25 influences the physiological parameters of *Oryza sativa* L. cv. Baldo rice. Environ Microbiol 21:3489–3504. https://doi.org/10.1111/1462-2920.14688

Apine OA, Jadhav JP (2011) Optimization of medium for indole-3-acetic acid production using *Pantoea agglomerans* strain PVM. J Appl Microbiol 110(5):1235–1244

Assumpção LDC, Lacava PT, Dias AC, de Azevedo JL, Menten JOM (2009) Diversity and biotechnological potential of endophytic bacterial community of soybean seeds. Pesqui Agropecu Bras 44(5):503–510

Baliyan N, Dhiman S, Dhiman S, Kumar S, Arora NK, Maheshwari DK (2021) Optimization of gibberellic acid production in endophytic *Bacillus cereus* using response surface methodology and its use as plant growth regulator in chickpea. J Plant Growth Regul. https://doi.org/10.1007/s00344-021-10492-2

Bano A, Ullah F, Nosheen A (2012) Role of abscisic acid and drought stress on the activities of antioxidant enzymes in wheat. Plant Soil Environ 58:181–185. https://doi.org/10.1371/journal.pone.0052439

Bhore SJ, Ravichantar N, Loh CY (2010) Screening of endophytic bacteria isolated from leaves of Sambung Nyawa [*Gynura procumbens* (Lour.) Merr.]for cytokinin-like compounds. Bioinformation 5(5):191–197. https://doi.org/10.6026/97320630005191

Bilal S, Shahzad R, Khan AL, Kang SM, Imran QM, Al-Harrasi A, Yun BW, Lee IJ (2018) Endophytic microbial consortia of phytohormones-producing fungus *Paecilomyces formosus* LH10 and bacteria *Sphingomonas* sp. LK11 to *Glycine max* L. regulates physio-hormonal changes to attenuate aluminum and zinc stresses. Front Plant Sci 9:1273. https://doi.org/10.3389/fpls.2018.01273

Bömke C, Tudzynski B (2009) Diversity, regulation, and evolution of the gibberellins biosynthetic pathway in fungi compared to plants and bacteria. Phytochemistry 70:1876–1893

Borah A, Thakur D (2020) Phylogenetic and functional characterization of culturable endophytic actinobacteria associated with *camellia* spp. for growth promotion in commercial tea cultivars. Front Microbiol 11:318

Campos ML, Kang JH, Howe GA (2014) Jasmonate-triggered plant immunity. J Chem Ecol 40:657–675

Chen Z, Zheng Z, Huang J, Lai Z, Fan B (2009) Biosynthesis of salicylic acid in plants. Plant Signal Behav 4(6):493–496. https://doi.org/10.4161/psb.4.6.8392

Chadha N, Prasad R, Varma A (2015) Plant promoting activities of fungal endophytes associated with tomato roots from central Himalaya, India and their interaction with *Piriformospora indica*. Int J Pharm BioSci 6(1):333–343

Chen B, Shen J, Zhang X, Pan F, Yang X, Feng Y (2014) The endophytic bacterium, *Sphingomonas* SaMR12, improves the potential for zinc phytoremediation by its host, Sedum alfredii. PLoS One 9:e106826. https://doi.org/10.1371/journal.pone.0106826

Chen X, Marszałkowska M, Reinhold-Hurek B (2020) Jasmonic acid, not salicyclic acid restricts endophytic root colonization of rice. Front Plant Sci 10:1758. https://doi.org/10.3389/fpls.2019.01758

Chen B, Luo S, Wu Y, Ye J, Wang Q, Xu X, Pan F, Khan KY, Feng Y, Yang X (2017) The effects of the endophytic bacterium Pseudomonas fluorescens Sasm05 and IAA on the plant growth and cadmium uptake of Sedum alfredii Hance. Front Microbiol 8:2538. https://doi.org/10.3389/fmicb.2017.02538

Chhabra S, Dowling DN (2017) Endophyte promoted nutrient acquisition: phosphorous and iron. In: Doty S (ed) Functional importance of the plant microbiome: implications for agriculture, forestry and bioenergy. Springer, Cham. https://doi.org/10.1007/978-3-319-65897

Cohen AC, Travaglia CN, Bottini R, Piccoli P (2009) Participation of abscisic acid and gibberellins produced by endophytic *Azospirillum* in the alleviation of drought effects in maize. Botany 87(5):455. https://doi.org/10.1139/B09-023

Da Silva AS, Suassuna JF, de Melo AS, Costa RR, de Andrade WL, da Silva DC (2017) Salicylic acid as attenuator of drought stress on germination and initial development of sesame. Rev Bras Eng Agríc Ambient 21:156–162. https://doi.org/10.1590/1807-1929/agriambi.v21n3p156-162

De Bary A (1866) Morphologieund Physiologieder Pilze, Flechtenund Myxomyceten, vol 2. Hofmeister's Handbook of Physiological Botany, Leipzig

Duca D, Lorv J, Patten CL, Rose D, Glick BR (2014) Indole-3-acetic acid in plant-microbe interactions. Antonie Van Leeuwenhoek 106(1):85–125

Dudeja SS, Giri R (2014) Beneficial properties, colonization, establishment and molecular diversity of endophytic bacteria in legume and non-legume. Afr J Microbiol Res 8:1562–1572. https://doi.org/10.5897/AJMR2013.6541

Fadiji AE, Babalola OO (2020) Elucidating mechanisms of endophytes used in plant protection and other bioactivities with multifunctional prospects. Front Bioeng Biotechnol 8:467. https://doi.org/10.3389/fbioe.2020.00467

Finkelstein R (2013) Abscisic acid synthesis and response. Arabidopsis Book 11:e0166. https://doi.org/10.1199/tab.0166

Gamalero E, Favale N, Bona E, Novello G, Cesaro P, Massa N, Glick BR, Orozco-Mosqueda MC, Berta G, Lingua G (2020) Screening of bacterial endophytes able to promote plant growth and increase salinity tolerance. Appl Sci 10:5767

Gao D, Tao Y (2012) Current molecular biologic techniques for characterizing environmental microbial community. Front Environ Sci Eng 6:82–97. https://doi.org/10.1007/s11783-011-0306-6

Glick BR (2012) Plant growth-promoting bacteria: mechanisms and applications. Scientifica 2012:963401

Gouda S, Das G, Sen SK, Shin H-S, Patra JK (2016) Endophytes: a treasure house of bioactive compounds of medicinal importance. Front Microbiol 7:1538. https://doi.org/10.3389/fmicb.2016.01538

Grosskinsky DK, Tafner R, Moreno MV, Stenglein SA, de Salamone IEG, Nelson LM, Novak O, Strnad M, van der Graaff E, Roitsch T (2016) Cytokinin production by *Pseudomonas fluorescens* G2018 determines biocontrol activity against *Pseudomonas syringae* in Arabidopsis. Sci Rep 6:23310. https://doi.org/10.1038/srep23310

Grossmann K (2010) Auxin herbicides: current status of mechanism and mode of action. Pest Manag Sci 66(2):113–120

Hamayun M, Hussain A, Khan SA, Kim HY, Khan Waqas M (2017) Gibberellins producing endophytic fungus *Porostereum spadiceum* AGH786 rescues growth of salt affected soybean. Front Microbiol 8:686

Hardoim PR, Van Overbeek LS, Elsas JD (2008) Properties of bacterial endophytes and their proposed role in plant growth. Trends Microbiol 16:463–471

Hassan SE-D (2017) Plant growth-promoting activities for bacterial and fungal endophytes isolated from medicinal plant of *Teucrium polium* L. J Adv Res 8(6):687–695. https://doi.org/10.1016/j.jare.2017.09.001

Hauser F, Waadt R, Schroeder JI (2011) Evolution of abscisic acid synthesis and signaling mechanisms. Curr Biol 21:R346–R355

Hedden P, Thomas SG (2012) Gibberellin biosynthesis and its regulation. Biochem J 444:11–25

Herrera-Medina MJ, Steinkellner S, Vieheilig H, Bote JAO, Garrido JMG (2007) Abscisic acid determines arbuscule development and functionality in the tomato arbuscular mycorrhiza. New Phytol 175:554–564

Hilbert M, Voll LM, Ding HJ, Sharma M, Zuccaro A (2012) Indole derivative production by the root endophyte *Piriformospora indica* is not required for growth promotion but for biotrophic colonization of barley roots. New Phytol 196:520–534. https://doi.org/10.1111/j.1469-8137.2012.04275.x

Hwang I, Sheen J, Muller B (2012) Cytokinin signaling networks. Annu Rev Plant Biol 63:353–380

Islam MN, Ali MS, Choi S-J, Park Y-I, Bae K-H (2020) Salicylic acid-producing endophytic bacteria increase nicotine accumulation and resistance against wildfire disease in tobacco plants. Microorganisms 8:31. https://doi.org/10.3390/microorganisms8010031

Janda K, Éva SGH, Kovác L, Janda T (2012) Salicylic acid may indirectly influence the photosynthetic electron transport. J Plant Physiol 169:971–978. https://doi.org/10.1016/j.jplph.2012.02.020

Joseph B, Priya RM (2011) Bioactive compounds from endophytes and their potential in pharmaceutical effect: a review. Am J Biochem Mol Biol 1:291–309. https://doi.org/10.3923/ajbmb.2011.291.309

Kazan K (2013) Auxin and the integration of environmental signals into plant root development. Ann Bot 112:1655–1665. https://doi.org/10.1093/aob/mct229

Khan AL, Hamayun M, Ahmad N, Hussain J, Kang SM, Kim YH (2011) Salinity stress resistance offered by endophytic fungal interaction between *Penicillium minioluteum* LHL09 and *Glycine max* L. J Microbiol Biotechnol 21:893–902

Khan AL, Hamayun M, Kang SM et al (2012) Endophytic fungal association via gibberellins and indole acetic acid can improve plant growth under abiotic stress: an example of *Paecilomyces formosus* LHL10. BMC Microbiol 12:3. https://doi.org/10.1186/1471-2180-12-3

Khan MIR, Asgher M, Khan NA (2014) Alleviation of salt-induced photosynthesis and growth inhibition by salicylic acid involves glycine betaine and ethylene in mungbean (*Vigna radiata* L.). Plant Physiol Biochem 80:67–74. https://doi.org/10.1016/j.plaphy.2014.03.026

Khan AL, Waqas M, Hussain J, Al-Harrasi A, Al-Rawahi A, Al-Hosni K (2014a) Endophytes *Aspergillus caespitosus* LK12 and *phoma* sp. LK13 of *Moringa peregrina* produce gibberellins and improve rice plant growth. J Plant Interact 9:731–737

Khan AL, Hussain J, Al-Harrasi A, Al-Rawahi A, Lee I (2015) Endophytic fungi: resource for gibberellins and crop abiotic stress resistance. Crit Rev Biotechnol 35(1):62–74

Khan AL, Halo BH, Elyassi A, Ali S, Al-Hosni K, Hussain J, Al-Harrasi A, Lee I (2016) Indole acetic acid and ACC deaminase from endophytic bacteria improves the growth of *Solanum lycopersicum*. Electron J Biotechnol 21:58–64

Kieber JJ, Schaller GE (2014) Cytokinins. Arabidopsis Book 12:e0168. https://doi.org/10.1199/tab.0168

Kumar V, Teotia P, Tuteja N, Prasad R, Goyal P, Varma A, Bisht S, Singh J, Kumar M (2017) Metabolomics-mediated characterization of endophytic species in recalcitrant tree species. In: Varma A, Sharma AK (eds) Modern tools and techniques to understand microbes. Springer, Cham, pp 251–257

Kundan R, Pant G, Jadon N, Agrawal PK (2015) Plant growth promoting Rhizobacteria: mechanism and current prospective. J Fertil Pestic 6:155. https://doi.org/10.4172/2471-2728.1000155

Lefevere H, Bauters L, Gheysen G (2020) Salicylic acid biosynthesis in plants. Front Plant Sci 11: 338. https://doi.org/10.3389/fpls.2020.00338

Letham DS, Shannon JS, McDonald IR (1964) Structure of zeatin factor inducing cell division. Proceedings of the Chemical Society of London, p 230. (JUL)

Li X, Zhou J, Xu RS et al (2018) Auxin, cytokinin, and ethylene involved in rice n availability improvement caused by endophyte *Phomopsis liquidambari*. J Plant Growth Regul 37:128–143. https://doi.org/10.1007/s00344-017-9712-8

Malik DK, Sindhu SS (2011) Production of indole acetic acid by *Pseudomonas* sp.: effect of coinoculation with *Mesorhizobium* sp. Cicer on nodulation and plant growth of chickpea (*Cicer arietinum*). Physiol Mol Biol Plants 17:25–32

Miller CO, Skoog F, Okumura FS, von Saltza MH, Strong FM (1955) Structure and synthesis of kinetin. J Am Chem Soc 77:2662–2663

Mishra M, Prasad R, Varma A (2015) Endophytic fungi: biodiversity and functions. Int J Pharm BioSci 6(1):18–46

Nonogaki H (2019) ABA responses during seed development and germination. Adv Bot Res 92: 171–217

Nutaratat P, Srisuk N, Arunrattiyakorn P, Limtong S (2016) Indole-3-acetic acid biosynthetic pathways in the basidiomycetous yeast *Rhodosporidium paludigenum*. Arch Microbiol 198: 429–437. https://doi.org/10.1007/s00203-016-1202-z

Prasad R, Kamal S, Sharma PK, Oelmueller R, Varma A (2013) Root endophyte *Piriformospora indica* DSM 11827 alters plant morphology, enhances biomass and antioxidant activity of medicinal plant *Bacopa monniera*. J Basic Microbiol 53(12):1016–1024

Prasad R, Gill SS, Tuteja N (2018) Crop improvement through microbial biotechnology. Elsevier. ISBN: 9780444639882

Rana KL, Kour D, Kaur T, Sheikh I, Yadav AN, Kumar V, Suman A, Dhaliwal HS (2020) Endophytic microbes from diverse wheat genotypes and their potential biotechnological applications in plant growth promotion and nutrient uptake. Proc Natl Acad Sci India Sect B Biol Sci 90:969–979

Sah SK, Reddy KR, Li J (2016) Abscisic acid and abiotic stress tolerance in crop plants. Front Plant Sci 7:571. https://doi.org/10.3389/fpls.2016.00571

Sahu PK, Tilgam J, Mishra S, Hamid S, Gupta A, Jayalakshmi K, Verma SK, Kharwar RN (2022) Surface sterilization for isolation of endophytes: ensuring what (not) to grow. J Basic Microbiol 62:647. https://doi.org/10.1002/jobm.202100462

Salazar-Cerezoa S, Martínez-Montiela N, García-Sáncheza J, Pérez-y-Terrónb R, Rebeca D, Contreras M (2018) Gibberellin biosynthesis and metabolism: a convergent route for plants, fungi and bacteria. Microbiol Res 208:85–98. https://doi.org/10.1016/j.micres.2018.01.010

Sandhya V (2017) Endophytes from maize with plant growth promotion and biocontrol activity under drought stress. Russ Agric Sci 43:22–34

Santoyo G, Moreno-Hagelsieb G, Orozco-Mosqueda M d C, Glick BR (2016) Plant growth-promoting bacterial endophytes. Microbiol Res 183:92–99. https://doi.org/10.1016/j.micres.2015.11.008

Sarkar S, Dey A, Kumar V, Batiha GE-S, El-Esawi MA, Tomczyk M, Ray P (2021) Fungal endophyte: an interactive endosymbiont with the capability of modulating host physiology in myriad ways. Front Plant Sci 12:701800. https://doi.org/10.3389/fpls.2021.701800

Shahzad R, Waqas M, Khan AL, Asaf S, Khan MA, Kang S-M, Yun B-W, Lee I-J (2016) Seed-borne endophytic *Bacillus amyloliquefaciens* RWL-1 produces gibberellins and regulates endogenous phytohormones of Oryza sativa. Plant Physiol Biochem 106:236–243

Shahzad R, Khan AL, Bilal S, Waqas M, Kang S-M, Lee I-J (2017) Inoculation of abscisic acid-producing endophytic bacteria enhances salinity stress tolerance in *Oryza sativa*. Environ Exper Bot 136:68–77. https://doi.org/10.1016/j.envexpbot.2017.01.010

Shen FT, Yen JH, Liao CS, Chen WC, Chao YT (2019) Screening of rice endophytic biofertilizers with fungicide tolerance and plant growth promoting characteristics. Sustainability 11:1133. https://doi.org/10.3390/su11041133

Singh D, Geat N, Rajawat MVS, Mahajan MM, Prasanna R, Singh S, Kaushik R, Singh RN, Kumar K, Saxena AK (2018) Deciphering the mechanisms of endophyte-mediated biofortification of Fe and Zn in wheat. J Plant Growth Regul 37:174–182

Singh S, Kumar V, Datta S, Ramamurthy PC, Singh J (2021) Molecular mechanism and signaling pathways interplay between plant hormones during plant-microbe crosstalk, microbial management of plant stresses, pp 93–105. https://doi.org/10.1016/B978-0-323-85193-0.00004-8

Skoog F, Tsui C (1948) Chemical control of growth and bud formation in tobacco stem segments and callus cultured in vitro. Am J Bot 35:782–787

Specian V, Sarragiotto MH, Pamphile JA, Clemente E (2012) Chemical characterization of bioactive compounds from the endophytic fungus *Diaporthe helianthi* isolated from *Luehea divaricata*. Braz J Microbiol 43:1174–1182. https://doi.org/10.1590/S1517-83822012000300045

Sponsel VM, Hedden P (2010) Gibberellin biosynthesis and inactivation. Plant hormones. Springer, Dordrecht, pp 63–94

Stepniewska Z, Kuzniar A (2013) Endophytic microorganisms-promising application s in bioremediation of green house gases. Appl Microbiol Biotechnol 97:9589–9596. https://doi.org/10.1007/s00253-013-5235-9

Sun Y, Cheng Z, Glick BR (2009) The presence of a 1-aminocyclopropane-1-carboxylate (ACC) deaminase deletion mutation alters the physiology of the endophytic plant growth-promoting bacterium *Burkholderia phytofirmans* PsJN. FEMS Microbiol Lett 296:131–136

Sun H, He Y, Xiao Q, Ye R, Tian Y (2013) Isolation, characterization and antimicrobial activity of endophytic bacteria from *Polygonum cuspidatum*. Afr J Microbiol Res 7:1496–1504. https://doi.org/10.5897/AJMR12.899

Sun P-F, Fang W-T, Shin L-Y, Wei J-Y, Fu S-F, Chou J-Y (2014) Indole-3-acetic acid-producing yeasts in the phyllosphere of the carnivorous plant *Drosera indica* L. PLoS One 9:e114196. https://doi.org/10.1371/journal.pone.0114196

Sun L, Wang X, Li Y (2016) Increased plant growth and copper uptake of host and nonhost plants by metal-resistant and plant growth-promoting endophytic bacteria. Int J Phytoremediation 18: 494–501

Tidke SA, Kiran S, Giridhar P, Gokare RA (2018) Current understanding and future perspectives of endophytic microbes vis-a-vis production of secondary metabolites. In: Jha S (ed) Endophytes and secondary metabolites. Reference series in Phytochemistry. Springer, Cham. https://doi.org/10.1007/978-3-319-76900-4_12-1

Tsavkelova E, Oeser B, Oren-young L, Israeli M, Sasson Y, Tudzynski B, Sharon A (2012) Identification and functional characterization of indole-3-acetamide-mediated IAA biosynthesis in plant-associated *Fusarium* species. Fungal Genet Biol 49:48–57. https://doi.org/10.1016/j.fgb.2011.10.005

Varma A, Prasad R, Tuteja N (2017a) Mycorrhiza: eco-physiology, secondary metabolites, nanomaterials. Springer International Publishing. ISBN: 978-3-319-57849-1. http://www.springer.com/us/book/9783319578484

Varma A, Prasad R, Tuteja N (2017b) Mycorrhiza: nutrient uptake, biocontrol, ecorestoration. Springer International Publishing. ISBN: 978-3-319-68867-1. http://www.springer.com/us/book/9783319688664

Wang J, Song L, Gong X, Xu J, Li M (2020) Functions of Jasmonic acid in plant regulation and response to abiotic stress. Int J Mol Sci 21:1446. https://doi.org/10.3390/ijms21041446

Went FW (1928) Wuchesstoff and Wachstum. Rec Trav Bota Neerl 25:1–116

Win KT, Tanaka F, Okazaki K, Ohwaki Y (2018) The ACC deaminase expressing endophyte *Pseudomonas* spp. enhances NaCl stress tolerance by reducing stress-related ethylene production, resulting in improved growth, photosynthetic performance, and ionic balance in tomato plants. Plant Physiol Biochem 127:599–607. https://doi.org/10.1016/j.plaphy.2018.04.038

Yadav AN (2018) Biodiversity and biotechnological applications of host specific endophytic fungi for sustainable agriculture and allied sectors. Acta Sci Microbiol 1:44

You YH, Yoon H, Kang SM, Woo JR, Choo YS, Lee IJ (2013) *Cadophora malorum* Cs-8-1 as a new fungal strain producing gibberellins isolated from *Calystegia soldanella*. J Basic Microbiol 53:630–634

Zhao L, Xu Y, Lai X (2018) Antagonistic endophytic bacteria associated with nodules of soybean (*Glycine max* L.) and plant growth-promoting properties. Braz J Microbiol 49:269–278

Zhu Y, She X (2018) Evaluation of the plant-growth-promoting abilities of endophytic bacteria from the psammophyte Ammodendron bifolium. Can J Microbiol 64:253–264

Zhu L, Guo J, Sun Y, Wang S, Zhou C (2021) Acetic acid-producing endophyte *Lysinibacillus fusiformis* orchestrates Jasmonic acid signaling and contributes to repression of cadmium uptake in tomato plants. Front Plant Sci 12:670216. https://doi.org/10.3389/fpls.2021.67021

Chapter 16
Role of Endophytic Microorganisms in Phosphate Solubilization and Phytoremediation of Degraded Soils

Dipita Ghosh, B. S. Manisha Singh, Manish Kumar, Subodh Kumar Maiti, and Nabin Kumar Dhal

Abstract Phosphorus (P) is considered as the second most important element in plant nutrient profile after nitrogen. It primarily exists incorporated in organic compounds or as mineral salts in soil. Despite these, phosphorus compounds are disbursed abundantly in agricultural soil, and the majority of them are of insoluble form. With the assistance of plant-associated bacteria, the inorganic phosphate solubilization is one of the significant mechanisms for plant growth promotion. The mechanism involves the solubilization of phosphate complexes into more available forms such as orthophosphate ions by organic acid secreted by microbes. The employment of plant growth promoting P bacterial inoculants as biofertilizers can provide favourable alternative to replace chemical fertilizer to some extent. Some examples of phosphate solubilizer are *Bacillus*, *Pseudomonas* and *Aspergillus,* while the phosphate absorber includes arbuscular mycorrhizal fungi (e.g. Glomus). Phytoremediation of heavy metals in association with phosphate-solubilizing bacteria are known to overcome metal stress on plants due to the contaminated substrate. In case of mine-degraded soils, endophyte assisted P-solubilization enhances the bioavailability of insoluble P to plants which in turn enhances the plant growth. Therefore, this chapter covers endophytes assisted

D. Ghosh · S. K. Maiti
Ecological Restoration Laboratory, Department of Environmental Science & Engineering, Indian Institute of Technology (ISM), Dhanbad, Jharkhand, India

B. S. M. Singh
Environment Sustainability Department, CSIR-Institute of Minerals and Materials Technology, Bhubaneswar, Odisha, India

Academy of Scientific and Innovative Research (AcSIR), CSIR-Human Resource Development Centre, (CSIR-HRDC) Campus, Ghaziabad, Uttar Pradesh, India

M. Kumar · N. K. Dhal (✉)
Environment Sustainability Department, CSIR-Institute of Minerals and Materials Technology, Bhubaneswar, Odisha, India
e-mail: nkdhal@immt.res.in

S. Chhabra et al. (eds.), *Plant Microbiome for Plant Productivity and Sustainable Agriculture*, Microorganisms for Sustainability 37,
https://doi.org/10.1007/978-981-19-5029-2_16

sustainable in-situ remediation of contaminated site which stimulates plant growth, defence against metal toxicity and soil fertility.

Keywords Biofertilizers · Soil · Heavy metals · Amendment · Phytoremediation · Microbes

16.1 Introduction

Mining activity causes complete loss of soil profile, vegetation and the biodiversity of a land. It also causes air and water pollution, disturbs drainage and permanently affects a landform (Ghosh and Maiti 2020; Mohapatra et al. 2020). Mine spoil is characterized by impoverished nutrient content, low organic content and cation exchange capacity and disturbed ambient soil physicochemical and biological properties (Basu et al. 2015; Ahirwal et al. 2021; Ghosh and Maiti 2021a). A degraded mine spoil is devoid of soil organic matter, microbial activity and the enzymatic activities associated with the soil fauna (Maiti 2013; Ghosh and Maiti 2021b). A mine spoil is devoid of essential soil nutrients and often the storehouse of potentially toxic elements (Ahirwal and Maiti 2017; Ghosh and Maiti 2021c). Phosphorous (P) is a crucial component for overall plant development and productivity (Rawat et al. 2021). Its properties constrain its free accessibility and make it a restraining nutrient for vegetation development (Mehta et al. 2017). Thus, an efficient amendment technique is required for mitigation of phosphate deficiency and heavy metal contaminations in mine spoils/tailings and technosol. Some common restoration practices for post-mining coal mine degraded land are forestry, agricultural practices, grass-legume seeding, fly ash amended plantation and biochar aided plantation (Šebelíková et al. 2019; Shukla and Lal 2005; Kumari et al. 2022; Świątek et al. 2019; Fellet et al. 2011; Ghosh et al. 2020).

In a natural soil ecosystem, plants interact with a number of symbiotic microorganisms (Domka et al. 2019). The plant–soil interaction includes synergy of plants with rhizobacteria and endophytic fungi (Maiti 2013; Domka et al. 2019; Varma et al. 2019a, b). Actinomycetes, bacterial and fungal endophytes perforate the plant through root zones along with flower, leaf, stem and cotyledon (Li et al. 2012). The microbiomes are such integral part of plants that they can be used as proxy to study the phenotypic variation of the plant genotype. The knowledge of plant–microbiome interactions can help improving the economic and environmental sustainability of mine spoil restoration through agriculture and forestry. A reduction in inputs, in terms of fertilizer, water, or chemical pesticides, would lead to significant cost savings (Prasad 2017, 2018; Prasad et al. 2021).

Endophytic microbes have the ability to grow throughout the host plant tissues and releases phytochemicals that provide resistance to disease and help in nutrient mineralization for host plant (Maiti 2013). Some endophytic fungi can also solubilize P and supply it to their non-mycorrhizal counterparts, encouraging its growth under nutrient environment (Mehta et al. 2017; Rawat et al. 2021). Thus, they help in improving the overall plant growth under stressed environmental conditions (Maiti 2013). *Curvularia geniculata* isolated from *Parthenium hysterophorus* roots is a dark septate root endophytic fungus which can improve plant growth by promoting P-solubilization and certain phytohormone secretion (Priyadharsini and Muthukumar 2017). Another important role played by endophytes includes resistance to heavy metals and assistance in phytoremediation of a metal contaminated site. Endophyte-assisted phytoremediation technology has been reported to be an efficient technique for in situ remediation of potentially toxic elements contaminated soils (Mastretta et al. 2009; Domka et al. 2019; Guerrero-Zúñiga et al. 2020). During the phytoremediation of polluted sites, heavy-metal contamination enduring endophytes can also improve plant growth, reduce metal phytotoxicity and influence translocation and accumulation of metal. Thus, this chapter focuses on the beneficial role of endophytes for phosphate solubilization and heavy metal remediation. In conclusion, this chapter provides an insight on how endophytes-assisted phytoremediation enhances soil properties.

16.2 Role of Endophytes for Mine Spoil Reclamation

16.2.1 Phytostimulation and Nutrient Cycling

Essential nutrients such as C, N, H, O and P are absolutely necessary for plant growth and development. These nutrients are in chemical form through atmosphere, soil, water and organic matter. Endophyte facilitates the uptake of nutrients by the roots of the plants (Nair and Padmavathy 2014). They have been reported to elicit different modes of actions for plant adaptation in P-deficient soil and facilitation of N uptake (Arachevaleta et al. 1989). Certain endophytic bacteria have been reported to produce phytohormones such as cytokinins, auxins and gibberellic acids which are essential plant growth regulators (Xin et al. 2009). Endophytes play vital role in biodegradation of the debris of its host flora (Mehta et al. 2017).

16.2.2 Enzyme Production, Antimicrobial Activity and Source of Bioactive

Soil micro-organisms are the source of a number of commercially important enzymes. This quest for alternative source of enzyme production has led to the discovery of certain endophytes which can produce vital enzymes. Endophytic fungi such as *Aspergillus japonicas*, *Cladosporium sphaerospermum*, *Nigrospora sphaerica*, *Penicillium aurantiogriseum*, *P. glandicola* and *Xylaria* sp. have been reported to produce enzymes such as pectinases, cellulases, xylanases and proteases (Nair and Padmavathy 2014). *Acremonium zeae*, isolated from maize, has also been reported to produce the enzyme hemicellulase (Bischoff et al. 2009). A number of isolated endophytes from plants have been reported to possess antimicrobial activity (He et al. 2020). Most endophytes show antimicrobial activity; however, the ones obtained from medicinal plants affects a broad spectrum of pathogenic microbes (Nair and Padmavathy 2014).

16.2.3 Bioremediation

Bioaccumulation, bio-stimulation, bio-deterioration, bio-leaching, bio-reduction and bio-sorption are some common bioremediation techniques used for heavy metal contamination. Endophytes possess the ability to breakdown complex compounds. Mastretta et al. (2009) reported that the inoculation of *Nicotiana tabacum* with endophytes resulted in improved plant growth under Cd toxicity and the phytoavailable Cd concentration was high in comparison with the one having no endophytic growth. According to Basu et al. (2015), a number of microorganisms catalyse the reduction of Cr (VI) to Cr (V) or Cr (III) in various environmental conditions. Cr (VI) reduction is shown to be metabolic in some species of bacteria but can also be dissimilatory/respiratory when exposed to anaerobic conditions. Although, most microbes are sensitive to Cr (VI), some microbes are highly resistant and can tolerate Cr (VI) toxicity in the soil. Metal reductase genes found on plasmids and chromosomes impart the resistance to these microbes for growth in Cr (VI) environment (Patra et al. 2017). Some common endophytes that have the potential for Cr remediation include *Acinetobacter*, *Arthrobacter*, *Bacillus* spp., *Cellulomonas* spp., *Escherichia coli*, *Enterobacter cloacae*, *Pseudomonas* and *Ochrobactrum* (Hossan et al. 2020). A review conducted by Pushkar et al. (2021) reported that major bacterial communities found at chromium contaminated sites are *Gammaproteobacteria*. Other bacteria reported to inhabit chromite contaminated sites includes *Serratia marcescens*, *Pseudomonas aeruginosa*, *Alcaligenes faecal* and *Klebsiella oxytoca*.

16.3 Role of Endophytes for Phosphate Solubilization

Phosphorus is an essential macronutrient for the proper metabolism, growth and plant development. Phosphorus is abundantly available in both inorganic and organic forms in soil; however, due to the complex formation with metal ions in soil, it is unavailable for plant uptake. Phosphate-solubilizing endophytes have the ability to solubilize the complex phosphates in the soil by various mechanisms. Some commonly used mechanisms used by these microbes include production of enzymes, organic acids and siderophores that have the ability to chelate the heavy metal ions and form complexes, making bioavailable phosphates for vegetation uptake (Rawat et al. 2021). These endophytes also produce certain phytohormones such as auxins, cytokinins and gibberellins which promote plant growth. 1-aminocyclopropane-1-carboxylic acid deaminase produced by endophytes has been reported to improve plant growth under stressful environment which improves its resistance to heavy metal toxicity (Fig. 16.1). A few examples of endophytes, their host plant and the role they play are given in Table 16.1.

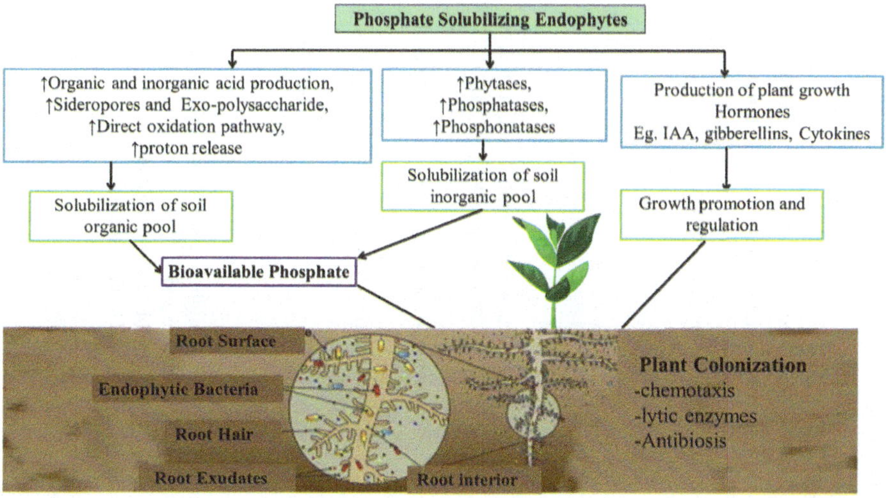

Fig. 16.1 Mechanism by which endophytes promotes plant growth and phosphate solubilization

16.4 Role of Endophytes for Phytoremediation

A number of endophytes have been reported to be heavy metals resistant. Endophyte-assisted phytoremediation is an effective technique for in situ remediation of contaminated soils (Prasad 2022). Microbes develop symbiotic relationships with their plant hosts and promote phytoremediation. Some common hyperaccumulating plants such as *Brassica juncea* (L.) Czern., *Pteris vittata*, *Sedum alfredii* and non-hyper-accumulators, such as *Arabidopsis thaliana, Brassica napus* and *Glycine max* have been reported to house a number of important endophytes (He et al. 2020). During pollutant phytoremediation association of heavy-metal-resistant endophytes can result in enhancement of plant development followed by decrease in metal phytotoxicity and affect translocation of metals in plants. They even produce certain enzymes which help in the degradation of contaminants that reduces the phytotoxicity of the potentially toxic elements. Application of endophytes for phytoremediation and their significance for the host plant growth has been given in Table 16.2.

16.5 Case Studies

16.5.1 Fungal Root Endophytes in Metal-Polluted Tailings

Flores-Torres et al. (2021) conducted a research identifying and assessing the plant and fungal root endophytes in bioremediation of polymetallic polluted tailings. The study revealed the significant role of native plants such as *Tagetes lunulata, Cerdia*

Table 16.1 Phosphate-solubilizing endophytes, their respective host, phosphate-solubilizing ability and their significance

Endophyte	Host	Isolation of endophytes and colonization	Phosphorous solubilized (μg mL^{-1})	Significance	References
Serratia plymuthica BMA1	*Vicia faba* L.	• Strain isolated from the extracted rock phosphate (RP) stockpiles (operated by the company of phosphate of Gafsa: CPG Tunisia). • Direct strain inoculation in plant at a cell concentration of approximately 3×10^6 CFU g^{-1} of soil	450	• Phosphorus uptake improved • Improved plant biomass and height	Borgi et al. (2020)
Pseudomonas sp.	*Triticum aestivum*	–	101	• Production of Siderophore • Improved biomass and dry matter	Liu et al. (2019)
Bacillus megaterium	*Brassica napus*	–	119	Phosphorus content and bio-mass production increased	Zheng et al. (2019)
Staphylococcus sciuri, Bacillus pumilus,	*Oryza sativa* L.	• From fresh soil (attached to roots of *O. sativa*) • Colonization of PSB in glucose based N-free medium for Azotobacter type free living N_2 fixers	192	Enhanced phosphorus uptake in shoot, root, grains	Rajapaksha and Senanayake (2011)
Enterobacter asburiae *Acinetobacter rhizosphaerae*	*Zea mays* L.	• Rhizosphere of *Hippophae rhamnoides* • Charcoal based bacterial inoculants (CFU of approx. 10^7/g)	750	• Plant height and root shoot length increased • Available ca and P content high in soil with endophytes.	Gulati et al. (2010)
Penicillium spp.	*Triticum aestivum*	• From disinfected wheat roots (3-7 mm root segments)	–	Plant fertility improved	Wakelin et al. (2004)

		• Cultured in semiselective medium for Penicillium spp. (Dichloran, rose bengal, chlor-amphenicol agar)			
Bacillus sp.	*Cicer arietinum*	• Pre isolated bacterial strain	753	• Improved phytohormones • Dry weight of nodules, root shoot biomass increment	Ditta et al. (2018)
Pseudomonas putida	*Pisum sativum*	N-fixer isolation from nodules of pea plants Isolated and colonized in yeast extract mannitol medium +2.5% Congo red dye	319	• Shoot biomass increased • High seed protein content and increased phosphorus in shoots	Ahmad et al. (2013)
Penicillium oxalicum	*Triticum aestivum, Zea mays* L.	• Rhizospheric soil from *Jatropha curcas* (rock phosphate landfills of Rajasthan state mines and minerals Ltd) • Fungi isolated in PVK agar +50 mg P_2O_5/100 ml medium	586	• Shoot height enhanced 1.5 times compared to control • 42% increase in yield of wheat compared to control • 82% increase in root shoot biomass compared to control	Singh and Reddy (2011)
Streptomyces laurentii	*Sorghum bicolour* L	Aerobic bacteria isolated from rhizospheric soil (Amaranthus, buckwheat, millets and maize soil attached to roots) by serial dilution method in growth medium	206	Siderophore production, enzymatic activity potassium solubilization	Kour et al. (2020)
Pantoea vagans, Pseudomonas psychrotolerans, Bacillus subtilis, Bacillus safensis and *Pantoea agglomerans*	*Festuca arundinacea; Solanum lycopersicum and Capsicum annuum*	• Endophytes isolated from plants grown in the foothills of Appalachian Mountains of USA (37.125372, 79.298415) unfertilized soil. • Sterilized random plants were sectioned in root leaf and	274–372	• Promoted *Festuca arundinacea* growth in vitro • Promoted pepper and tomato growth • The extracellular enzymes such as acid phosphatase and phytase secreted by the	Mei et al. (2021)

(continued)

Table 16.1 (continued)

Endophyte	Host	Isolation of endophytes and colonization	Phosphorous solubilized (μg mL^{-1})	Significance	References
		stem and they were individually grounded followed by centrifugation then 50 μL aliquot spreader on Luria agar		bacteria helped in decomposition and nutrient cycling	
Enterobacter sp. J49 *or Serratia* sp. *S119*	*Zea mays, Glycine max and Arachis hypogaea L.*	Isolated from cultivated peanut plants Two endophytic strains colonization was analysed by isolating bacterial cells from internal tissues of stem, leaves and roots of peanut, soybean and maize plants from the microcosm assay.	–	• Promoted the growth of maize and soybean plants and contributed significantly P to their tissues • Root exudates of the three produced changes in pectinase and cellulase activities • Potential sources for the formulation of biofertilizers	Lucero et al. (2021)

Table 16.2 Application of endophytes for phytoremediation and their significance for the host plant growth

Endophytes	Heavy metals (mg kg^{-1})	Host plant	Isolation of endophytes	Significance	References
Enterobacter sp. CBSB1 (OGM – gene)	Cd (45.2); Pb (876.7)	*Brassica juncea*	• Strain was isolated from the surface-sterilized *B. juncea* roots	Improved shoot length. A decrease in Cd and Pb concentration was observed in shoot biomass.	Qiu et al. 2014
Bacillus sp.	Cd (5.9), Zn (1236) Pb (153)	*Sedum plumbizincicola*	Isolated from the rhizospheric soil of *S. plumbizincicola* by serial dilution method	Plant biomass and chlorophyll content increased. Increased Cd, Pb and Zn soil bioavailability	Ma et al. 2015
Pseudomonas azotoformans	Cd (400) Ni (350) Zn (500)	*Trifolium arvense*	Strain isolated from the leaves of Ni hyperaccumulator plant *Alyssum serpyllifolium* (subsp. malacitanum)	Improved plant biomass and increased [Cd], [Ni] and [Zn] in shoots	Ma et al. 2017
Pseudomonas koreensis AGB-1	As (827) Cd (103) Cu (396) Pb (2431) Zn (3991)	*Miscanthus sinensis*	Isolated from fine roots of *M. Sinensis* growing in the mine site soil following Sun et al. (2010)	Improved shoot and root biomass, increased chlorophyll, protein content. Decrease in plant stress and increase in [As] and [Cd] in roots	Babu et al. 2015

congestiflora and *Lupinus campestris* as well as the exotic plant species *Asphodelus fistulosus*, and *Cortaderia selloana* in phytoextraction and/or phytostabilization of Zn, Pb and Cd. Molecular studies of fourteen endophytic fungi isolated from root inner zones of *Pennisetum villosum* and *T. lunulata* showed the prevalence of *Alternaria* and other *Pleosporales*. The dominance of endophytes in several plant root systems indicates the interaction and functioning of mycorrhiza in mine tailings. Exotic invasive plants *A. Fistulosus* and *P. villosum* showed more than 50% root colonization intensity by endophytes, which could ascertain its invasive capacity. The study reported that these endophytes could facilitate the advancement of *Ambrosia artemisiifolia* growing at polluted sites; therefore, mycorrhizal interactions can help in promoting local adaptation and/or reducing environmental stress. Thus, the study indicated that the employment of native endophytic fungi could emphasize the establishment of plants for reclamation of mine waste in semi-arid climate in biologically sustainable manner. Also, high efforts are needed to enhance the vegetation practice of mine wastes under study, which can efficiently reduce, in turn, their potential ecotoxicological impact on organisms, human populations and agricultural areas.

16.5.2 Root Colonizing Endophytes for Succession in a Mine Degraded Land

Kolaříková et al. (2017) studied the fungal community assembly during spontaneous primary succession in Sokolov brown-coal mining in Czech Republic. The fungal communities associated with the roots of *Betula pendula* and *Salix caprea* were studied in a mine spoil chronosequence (12–50 years old sites) site. The study showed that the fungal root endophytes, fungal plant pathogens and ectomycorrhizal fungi changed significantly along the age of reclamation. Ectomycorrhizal fungi and fungal plant pathogens communities have a direct impact on the development of the vegetation cover and the properties of the reclaimed mine spoil. Thus, the study concluded that plant community structure changed along the various stages of succession which was directly impacted by the endophyte and pathogen communities of the soil. The study provided a better understanding of community assembly of root-associated fungi and provided insight of fungal ecology in various stages of succession.

16.6 Conclusion

Phytoremediation with endophyte assistance can be a promising technique for the restoration of a degraded and contaminated soil. They are known to improve nutrient uptake, enhance growth, decrease phytotoxicity of heavy metals and effect their assimilation in plants body. These endophytes also solubilize the unavailable phosphorus in soil and restore the deficiency in soil. These endophytes also play a vital role in phytoremediation of heavy metal contaminated sites. Thus, endophytes as a mean to remediate contaminated sites should be explored for eco-friendly and effective remediation of heavy metal contaminated sites. Selection of potent endophytes with multifunctional role is essential for the commercialisation and reduction of coast of restoration of mine-degraded land. Thus, future researches should be done to develop and discover new strains from various ecological niches and for employing in degraded soil restoration.

References

Ahirwal J, Maiti SK (2017) Assessment of carbon sequestration potential of revegetated coal mine overburden dumps: a chronosequence study from dry tropical climate. J Environ Manag 201: 369–377. https://doi.org/10.1016/j.jenvman.2017.07.003

Ahirwal J, Kumari S, Singh AK, Kumar A, Maiti SK (2021) Changes in soil properties and carbon fluxes following afforestation and agriculture in tropical forest. Ecol Indic 123:107354. https://doi.org/10.1016/j.ecolind.2021.107354

Ahmad E, Khan M, Zaidi A (2013) ACC deaminase producing *Pseudomonas putida* strain PSE3 and *Rhizobium leguminosarum* strain RP2 in synergism improves growth, nodulation and yield of pea grown in alluvial soils. Symbiosis 61(2):93–104. https://doi.org/10.1007/s13199-013-0259-6

Arachevaleta M, Bacon CW, Hoveland CS, Radcliffe DE (1989) Effect of the tall fescue endophyte on plant response to environmental stress. Agron J 81(1):83–90, Dissertation, Univ. of Georgia, Athens. https://doi.org/10.2134/agronj1989.00021962008100010015x

Babu AG, Shea PJ, Sudhakar D et al (2015) Potential use of Pseudomonas koreensis AGB-1 in association with Miscanthus sinensis to remediate heavy metal (loid)-contaminated mining site soil. J Environ Manag 151:160–166. https://doi.org/10.1016/j.jenvman.2014.12.045

Basu A, Panda SS, Dhal NK (2015) Potential microbial diversity in chromium mining areas: *bulletin of environment*. Pharmacol Life Sci 4:158–169

Bischoff KM, Wicklow DT, Jordan DB et al (2009) Extracellular hemicellulolytic enzymes from the maize endophyte *Acremonium zeae*. Curr Microbiol 58(5):499–503. https://doi.org/10.1007/s00284-008-9353-z

Borgi MA, Saidi I, Moula A et al (2020) The attractive *Serratia plymuthica* BMA1 strain with high rock phosphate-solubilizing activity and its effect on the growth and phosphorus uptake by Vicia faba L. plants. Geomicrobiol J 37(5):437–445. https://doi.org/10.1080/01490451.2020.1716892

Ditta A, Imtiaz M, Mehmood S et al (2018) Rock phosphate-enriched organic fertilizer with phosphate-solubilizing microorganisms improves nodulation, growth, and yield of legumes. Commun Soil Sci Plant Anal 49(21):2715–2725. https://doi.org/10.1080/00103624.2018.1538374

Domka AM, Rozpaądek P, Turnau K (2019) Are fungal endophytes merely mycorrhizal copycats? The role of fungal endophytes in the adaptation of plants to metal toxicity. Front Microbiol 10: 371. https://doi.org/10.3389/fmicb.2019.00371

Fellet G, Marchiol L, Delle Vedove G, Peressotti A (2011) Application of biochar on mine tailings: effects and perspectives for land reclamation. Chemosphere 83(9):1262–1267. https://doi.org/10.1016/j.chemosphere.2011.03.053

Flores-Torres G, Solis-Hernández AP, Vela-Correa G et al (2021) Pioneer plant species and fungal root endophytes in metal-polluted tailings deposited near human populations and agricultural areas in Northern Mexico. Environ Sci Pollut Res 28(39):55072–55088. https://doi.org/10.1007/s11356-021-14716-6

Ghosh D, Maiti SK (2020) Can biochar reclaim coal mine spoil? J Environ Manag 272:111097. https://doi.org/10.1016/j.jenvman.2020.111097

Ghosh D, Maiti SK (2021a) Biochar assisted phytoremediation and biomass disposal in heavy metal contaminated mine soils: a review. Int J Phytoremediation 23(6):559–576. https://doi.org/10.1080/15226514.2020.1840510

Ghosh D, Maiti SK (2021b) Eco-restoration of coal mine spoil: biochar application and carbon sequestration for achieving UN sustainable development goals 13 and 15. Land 10(11):1112. https://doi.org/10.3390/land10111112

Ghosh D, Maiti SK (2021c) Effect of invasive weed biochar amendment on soil enzymatic activity and respiration of coal mine spoil: a laboratory experiment study. Biochar 3(4):519–533. https://doi.org/10.1007/s42773-021-00109-y

Ghosh D, Masto RE, Maiti SK (2020) Ameliorative effect of Lantana camara biochar on coal mine spoil and growth of maize (*Zea mays*). Soil Use Manag 36(4):726–739. https://doi.org/10.1111/sum.12626

Guerrero-Zúñiga AL, López-López E, Rodríguez-Tovar AV, Rodríguez-Dorantes A (2020) Functional diversity of plant endophytes and their role in assisted phytoremediation. In: Bioremediation of industrial waste for environmental safety, pp 237–255. https://doi.org/10.1007/978-981-13-3426-9_10

Gulati A, Sharma N, Vyas P et al (2010) Organic acid production and plant growth promotion as a function of phosphate solubilization by *Acinetobacter rhizosphaerae* strain BIHB 723 isolated from the cold deserts of the trans-Himalayas. Arch Microbiol 192(11):975–983. https://doi.org/10.1007/s00203-010-0615-3

He W, Megharaj M, Wu CY et al (2020) Endophyte-assisted phytoremediation: mechanisms and current application strategies for soil mixed pollutants. Crit Rev Biotechnol 40(1):31–45. https://doi.org/10.1080/07388551.2019.1675582

Hossan S, Hossain S, Islam MR et al (2020) Bioremediation of hexavalent chromium by chromium resistant bacteria reduces phytotoxicity. Int J Environ Res Public Health 17(17):6013. https://doi.org/10.3390/ijerph17176013

Kolaříková Z, Kohout P, Krüger C et al (2017) Root-associated fungal communities along a primary succession on a mine spoil: distinct ecological guilds assemble differently. Soil Biol Biochem 113:143–152. https://doi.org/10.1016/j.soilbio.2017.06.004

Kour D, Rana KL, Kaur T et al (2020) Microbe-mediated alleviation of drought stress and acquisition of phosphorus in great millet (*Sorghum bicolour* L.) by drought-adaptive and phosphorus-solubilizing microbes. Biocatal Agric Biotechnol 23:101501. https://doi.org/10.1016/j.bcab.2020.101501

Kumari S, Ahirwal J, Maiti SK (2022) Reclamation of industrial waste dump using grass-legume mixture: an experimental approach to combat land degradation. Ecol Eng 174:106443. https://doi.org/10.1016/j.ecoleng.2021.106443

Li HY, Wei DQ, Shen M, Zhou ZP (2012) Endophytes and their role in phytoremediation. Fungal Divers 54:11–18. https://doi.org/10.1007/s13225-012-0165-x

Liu C, Mou L, Yi J et al (2019) The endogene of *Burkholderia cenocepacia* strain 71-2 is involved in phosphate solubilization. Curr Microbiol 76:495–502. https://doi.org/10.1007/s00284-019-01642-7

Lucero CT, Lorda GS, Anzuay MS et al (2021) Peanut endophytic phosphate solubilizing bacteria increase growth and P content of soybean and maize plants. Curr Microbiol 78(5):1961–1972. https://doi.org/10.1007/s00284-021-02469-x

Ma Y, Oliveira RS, Wu L et al (2015) Inoculation with metal-mobilizing plant-growth-promoting rhizobacterium Bacillus sp. SC2b and its role in rhizoremediation. J Toxic Environ Health A 78(13–14):931–944. https://doi.org/10.1080/15287394.2015.1051205

Ma Y, Rajkumar M, Moreno A et al (2017) Serpentine endophytic bacterium *Pseudomonas azotoformans* ASS1 accelerates phytoremediation of soil metals under drought stress. Chemosphere 185:75–85. https://doi.org/10.1016/j.chemosphere.2017.06.135

Maiti SK (2013) Ecology and ecosystem in mine-degraded land. In: Ecorestoration of the coalmine degraded lands. Springer, India, pp 21–37. https://doi.org/10.1007/978-81-322-0851-8_2

Mastretta C, Taghavi S, Van Der Lelie D et al (2009) Endophytic bacteria from seeds of *Nicotiana tabacum* can reduce cadmium phytotoxicity. Int J Phytoremediation 11(3):251–267. https://doi.org/10.1080/15226510802432678

Mehta P, Sharma R, Putatunada C, Walia A (2017) Endophytic bacteria: role in phosphate solubilization. In: Singh BP (ed) Advances in endophytic fungal research, fungal biology, pp 61–93. https://doi.org/10.1007/978-3-319-66544-3_4

Mei C, Chretien RL, Amaradasa BS et al (2021) Characterization of phosphate solubilizing bacterial endophytes and plant growth promotion in vitro and in greenhouse. Microorganisms 9(9):1935. https://doi.org/10.3390/microorganisms9091935

Mohapatra S, Kumar M, Karim AA, Dhal NK (2020) Biochars evaluation for chromium pollution abatement in chromite mine wastewater and overburden of Sukinda, Odisha, India. Arab J Geosci 13(586). https://doi.org/10.1007/s12517-020-05532-2

Nair DN, Padmavathy S (2014) Impact of endophytic microorganisms on plants, environment and humans. Sci World J 2014:1–11. https://doi.org/10.1155/2014/250693

Patra JM, Panda SS, Dhal NK (2017) Biochar as a low-cost adsorbent for heavy metal removal : a review. Int J Res Biosci 6(1):1–7

Prasad R (2017) Mycoremediation and environmental sustainability, vol 1. Springer International Publishing. ISBN 978-3-319-68957-9. https://link.springer.com/book/10.1007/978-3-319-68957-9

Prasad R (2018) Mycoremediation and environmental sustainability, vol 2. Springer International Publishing. ISBN 978-3-319-77386-5. https://www.springer.com/us/book/9783319773858

Prasad R (2022) Phytoremediation for environmental sustainability. Springer, Singapore. ISBN 978-9811656200. https://doi.org/10.1007/978-981-16-5621-7

Prasad R, Nayak SC, Kharwar RN, Dubey NK (2021) Mycoremediation and environmental sustainability, vol 3. Springer International Publishing. ISBN: 978-3-030-54421-8. https://www.springer.com/gp/book/9783030544218

Priyadharsini P, Muthukumar T (2017) The root endophytic fungus *Curvularia geniculata* from *Parthenium hysterophorus* roots improves plant growth through phosphate solubilization and phytohormone production. Fungal Ecol 27:69–77. https://doi.org/10.1016/j.funeco.2017.02.007

Pushkar B, Sevak P, Parab S, Nilkanth N (2021) Chromium pollution and its bioremediation mechanisms in bacteria: a review. J Environ Manag 287(November 2020):112279. https://doi.org/10.1016/j.jenvman.2021.112279

Qiu Z, Tan H, Zhou S, Cao L (2014) Enhanced phytoremediation of toxic metals by inoculating endophytic Enterobacter sp. CBSB1 expressing bifunctional glutathione synthase. J Hazard Mater 267:17–20. https://doi.org/10.1016/j.jhazmat.2013.12.043

Rajapaksha CP, Senanayake AP (2011) Potential use of rock-phosphate- solubilizing bacteria associated with wild rice as inoculants for improved rice (*Oryza sativa*). Arch Agron Soil Sci 57(7):775–788. https://doi.org/10.1080/03650340.2010.493878

Rawat P, Das S, Shankhdhar D, Shankhdhar SC (2021) Phosphate-solubilizing microorganisms: mechanism and their role in phosphate solubilization and uptake. J Soil Sci Plant Nutr 21(1): 49–68. https://doi.org/10.1007/s42729-020-00342-7

Šebelíková L, Csicsek G, Kirmer A et al (2019) Spontaneous revegetation versus forestry reclamation—vegetation development in coal mining spoil heaps across Central Europe. Land Degrad Dev 30(3):348–356. https://doi.org/10.1002/ldr.3233

Shukla MK, Lal R (2005) Soil organic carbon stock for reclaimed mine- soils in northeastern Ohio. Land Degrad Dev 16(4):377–386. https://doi.org/10.1002/ldr.669

Singh H, Reddy MS (2011) Effect of inoculation with phosphate solubilizing fungus on growth and nutrient uptake of wheat and maize plants fertilized with rock phosphate in alkaline soils. Eur J Soil Ecol 47:30–34. https://doi.org/10.1016/j.ejsobi.2010.10.005

Sun LN, Zhang YF, He LY et al (2010) Genetic diversity and characterization of heavy metal-resistant-endophytic bacteria from two copper-tolerant plant species on copper mine wasteland. Bioresour Technol 101:501e509. https://doi.org/10.1016/j.biortech.2009.08.011

Świątek B, Woś B, Chodak M, Kumar S (2019) Geoderma fine root biomass and the associated C and nutrient pool under the alder (*Alnus* spp.) plantings on reclaimed technosols. Geoderma 337: 1021–1027. https://doi.org/10.1016/j.geoderma.2018.11.025

Varma A, Swati T, Prasad R (2019a) Plant biotic interactions: state of art. Springer International Publishing. ISBN 978-3-030-26657-8. https://www.springer.com/gp/book/9783030266561

Varma A, Swati T, Prasad R (2019b) Plant microbe interface. Springer International Publishing. ISBN 978-3-030-19831-2. https://www.springer.com/gp/book/9783030198305

Wakelin SA, Warren RA, Harvey PR et al (2004) Phosphate solubilization by *Penicillium* spp. closely associated with wheat roots. Biol Fertil Soil 40:36–43. https://doi.org/10.1007/s00374-004-0750-6

Xin G, Zhang G, Kang JW et al (2009) A diazotrophic, indole-3-acetic acid-producing endophyte from wild cottonwood. Biol Fertil Soils 45(6):669–674. https://doi.org/10.1007/s00374-009-0377-8

Zheng BX, Ding K, Yang XR (2019) Straw biochar increases the abundance of inorganic phosphate solubilizing bacterial community for better rape (*Brassica napus*) growth and phosphate uptake. Sci Total Environ 647:1113–1120. https://doi.org/10.1016/j.scitotenv.2018.07.454

Chapter 17
Techniques to Study Plant–Microbe Interactions that Lead to Efficient Sustainable Agriculture

Nishra Joshi, Janki A. Ruparelia, Meenu Saraf, and Chaitanya Kumar Jha

Abstract Plant and microbes interact with each other at many levels of their life cycle, which may be beneficial, harmful, or neutral, and these interactions can be seen on multiple interfaces of the plant like root, shoot, stem, or leaves. If we have a glance in the field of agriculture in this context, the demand of agriculture is rising unprecedentedly and the conventional agriculture practices cannot keep pace with it both economically and environmentally, and we as humankind need to execute new strategies that are both economical and environment friendly. A strategy that is gaining momentum these days is the use of plant growth-promoting rhizobacteria (PGPR) in the field to get an increased amount of crop production, to improve the nutrient profile of crops, to remove toxic compounds, or for elimination of pathogens. But to get ahead in this field, first we have to understand the complex relationships between plant and microbes. Also, we need to examine the interspecies relationship between microbes themselves as to how they are imparting beneficial or harmful effects on plants. Only then will we be able to open the door to sustainable agriculture to meet the rising demand of food, fuel, and fibers. It was a hefty task to study the microbiomes some decades ago but thanks to techniques developed in recent decades, such as polymerase chain reaction, enzyme-linked assays such as enzyme-linked immunosorbent assay (ELISA) and radioimmunoassay (RIA), nucleic acid sequencing (genome sequencing, 16S ribosomal RNA sequencing, clustered regularly interspaced short palindromic repeats [CRISPR]), and microscopic techniques (fluorescence microscopy, video microscopy), it is now easy to

N. Joshi · J. A. Ruparelia
Microbiology Department, Gujarat Arts and Science College, Ahmedabad, Gujarat, India

Department of Microbiology and Biotechnology, University School of Science, Gujarat University, Ahmedabad, Gujarat, India

M. Saraf
Department of Microbiology and Biotechnology, University School of Science, Gujarat University, Ahmedabad, Gujarat, India

C. K. Jha (✉)
Microbiology Department, Gujarat Arts and Science College, Ahmedabad, Gujarat, India

S. Chhabra et al. (eds.), *Plant Microbiome for Plant Productivity and Sustainable Agriculture*, Microorganisms for Sustainability 37,
https://doi.org/10.1007/978-981-19-5029-2_17

study the microbes and microbial interactions neatly. In this chapter, we will discuss the microbial interactions at the plant interface and various techniques to study microbiomes.

Keywords Plant microbiome · PGPR · Plant–microbe interface · Sustainable agriculture

17.1 Introduction

Every organism on the planet depends on interactions with its surroundings to stay alive. The term "interaction" merely brings up a mental image of two entities actively interacting in a certain situation. Interaction derives from the Latin inter, which means "between," and ago, which means "to do" or "to act." Since we human beings and other animals are heterotrophic in nature, we mainly depend on the plants for the food and fiber needs.

Crop production will be one of the primary challenges of the twenty-first century to provide enough food for the growing human population, renewable energy, and fundamental molecules in industrial processes. Farmers in the modern day can benefit from microbial solutions since they can help meet the demand for more sustainable farming approaches. Increased crop yields and a more sustainable industry impact profile are possible outcomes of such solutions, resulting in more food to feed a growing world and new chances to safeguard the environment (Jha et al. 2011; Prasad et al. 2018; Kaushal and Prasad 2021).

Naturally, plants are inhabited by a myriad of microorganisms that perform a range of associative functions with bacteria, fungi, algae, archaea, protozoa, and viruses. The natural interaction between plants and microorganisms has long been associated with higher plant growth, nutrition, and health (Berg and Smalla 2009). For application of microbes in crop cultivation, we have to first address the questions: "How the microbes are affecting the plant or the crop?" "Is it beneficial or detrimental to the plant?" "How microorganisms benefitted the plant?" "Whether it benefits the plant by eliminating the pathogens, or by making the plant more stress tolerant, or by aiding the plant in growth?" Consequently, we also have to address the same questions like: "What microorganisms are doing?" "How they are interacting with the plant?" "What is the effect of environmental stress on the interaction?" "If plant microbial interactions are beneficial, how can we utilize the beneficial interaction for sustainable agriculture?" "How we can help humankind to survive sustainably by all these research?"

Application of helpful microorganisms in plant cultivation is required to engineer a massive building of sustainable agriculture in order to increase soil fertility and productivity. The study of interactions at the plant–microbe (PM) interface is a critical component of environment-friendly agriculture (Varma et al. 2019a, b). This chapter discusses and describes both conventional and emerging tools for studying plant–microbe interactions.

17.2 Agroecosystems and the Importance of Plants and Microorganisms and Their Interactions

There is a growing desire for sustainable agricultural techniques that are capable of maintaining optimum soil fertility, plant growth, and, eventually, keeping the crop disease-free, which has driven growers to limit their use of chemical inputs. The use of microbiological solutions such as plant growth-promoting rhizobacteria (PGPR) has an alternative potential option to build eco-friendly techniques, which can be an important moving force toward this end. In recent decades, agrobiologists have concentrated their efforts on gaining a better understanding of the role of useful microorganisms and harnessing diverse beneficial plant–microbe interactions, which can be beneficial to agroecosystem's productivity (Jha and Saraf 2015; Prasad et al. 2015).

Plants and bacteria are in naturally co-evolved symbiotic partnerships, and the interfaces between host plants and microbes are one of the most important determinants of crop health and productivity (Garg et al. 2019). The rhizosphere is the region where plant and microbial interactions are most closely coordinated, and it has the ability to provide a wide range of system properties such as rapid root colonization, improved nutrient uptake, and resistance to biotic and abiotic influences. The plant–microbe interaction spectrum is vast, encompassing different microbial communities and organisms, and the health of a plant is determined by the establishment of plant–microbe interactions such as by positive and negative influences on the host plant (Sharma and Pathak 2014). Plant–microbe interactions have mostly been explored in terms of commensalism, mutualism, competition, and parasitism among all plant–microbe interactions (Wu et al. 2009). The positive typical contact via mutualism and commensalism is commonly known, in which both or one member gets benefit from the interaction, respectively. The PGPRs are potential candidates to contribute to sustainable crop production.

A wide range of symbiotic bacteria such as *Rhizobium* and *Bradyrhizobium* spp., or non-symbiotic bacteria such as *Azospirillum*, *Bacillus*, *Pseudomonas*, and *Klebsiella* spp., among others, are increasingly being employed to boost plant productivity around the world (Hayat et al. 2010). Mutual interactions between legumes and N-fixing rhizobia and arbuscular mycorrhizal fungi (AMF) species, which help host plants obtain nitrogen and phosphorus in nutrient-deficient situations, are well-known examples of beneficial plant–microbe interactions (Cao et al. 2017; Udvardi and Poole 2013). When the detrimental bacteria infect plants, negative interactions such as parasitism occur, causing collateral damage and obtaining resources from their hosts, whereas competition arises when plants and microbes have similar nutritional requirements (Shelake et al. 2019). An overview of the plant–microbe interactions is given in Table 17.1.

Table 17.1 Types of plant–microbe interaction

Type of plant–microbe interaction	Description	Example of microbes involved in interaction	Reference
Mutualism	Both interacting species get benefit from interaction; plants and symbiotic or free-living plant growth-promoting rhizobacteria (PGPR) both get benefit in this relationship	*Rhizobia* and *Frankia*, *Achromobacter*, *Corynebacterium*, *Derxia*, *Enterobacter*, *Klebsiella*, *Pseudomonas*, *Rhodospirillum*, *Rhodopseudomonas*	Saharan and Nehra (2011)
Commensalism	Only one can benefit from this interaction; usually, plant root exudates provide the necessary nutrients to the PGPR while plants remain unaffected in this interaction	*Helicobacter pylori*	Hirsch (2004)
Protocooperation	Protocooperation is another type of interaction where both the interacting species get benefitted, but the collaborating species do not rely on each other for existence	*Bacillus* and *Halobacillus*	Banik et al. (2018)
Parasitic	Parasites colonize the host plant, causing collateral harm and may compete for nutritional needs	*Stachybotrys elegans*	Chamoun et al. (2015)
Pathogenic	Pathogens can actively destroy the host plant for their own trophic advantage	*Pectobacterium atrosepticum*	Chamoun et al. (2015)

17.3 Methods to Study Plant–Microbe Interaction

Morphological studies and biochemical testing were the main methods to study the plant–microbe interactions a few decades ago. These methods are not only time-consuming and laborious but also inaccurate at some points. The development of advanced techniques such as enzyme-linked immunosorbent assay (ELISA), radio-immunoassay (RIA), various gene-sequencing including DNA sequencing and RNA sequencing, next-generation sequencing (NGS), and metagenomics gave way to have an insight into a deep world of microorganisms. A simplified evolution of study techniques of plant–microbe interactions is presented in Fig. 17.1.

17.3.1 Conventional Techniques

Conventionally, biochemical detection and microscopy were used to identify the microbes. Biochemical identification includes isolation of the microbes by the use of

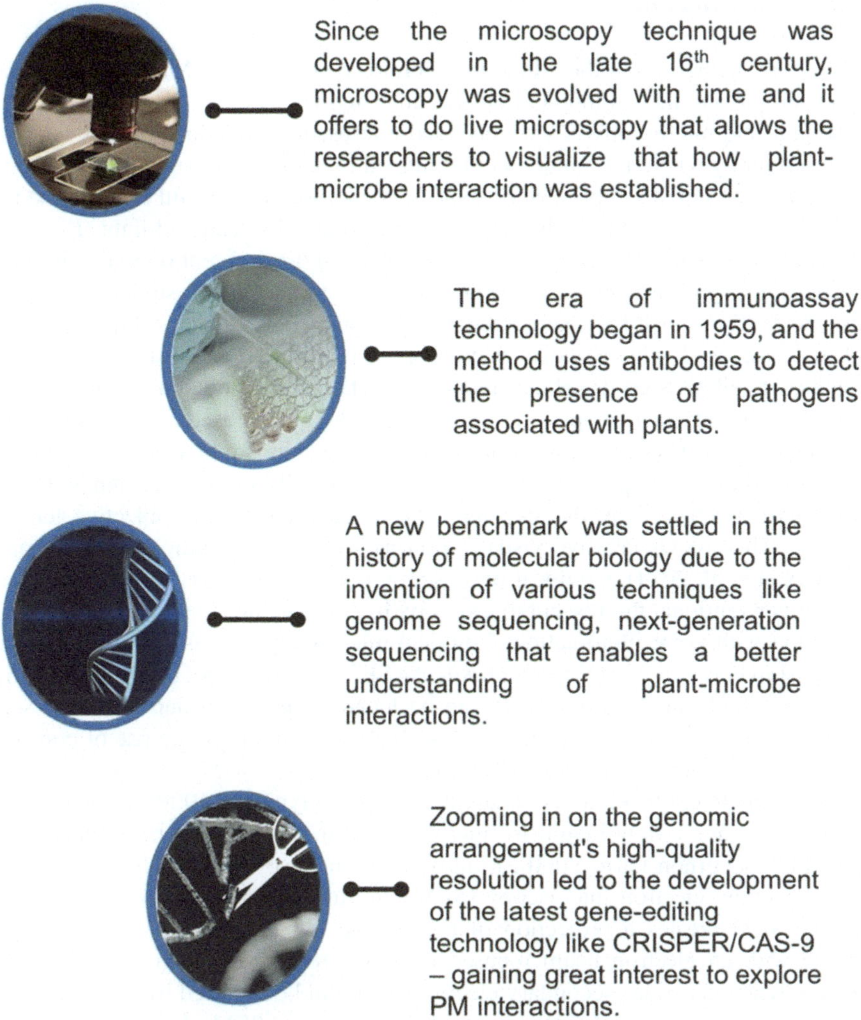

Fig. 17.1 Evolution of approaches to study interactions at plant–microbe (PM) interface (Dass and Mallick 2019; Prabhukarthikeyan et al. 2020)

agar media (routine, semi-synthetic, synthetic, semi-selective, selective, and specialized media), and studying colony morphology and biochemical characteristics. The main limitation of these culture-dependent techniques is that it only allows limited analysis of culturable microorganisms.

17.3.1.1 Microscopy

Microscopy is the reason the whole microbiology exists today. Microscopy started with a simple and then compound optical microscope.

Bright-field microscopy is a widely used microscopic technique mainly because of its simplicity where the sample is illuminated by white light, suitable for observing stained or colored samples, while dark-field microscopy is useful for unstained as well as living samples where the sample is illuminated by scattered light (Liu et al. 2014). Phase-contrast microscopy takes advantage of the different refractive indexes of the subject and provides a better projection of the unstained subject making it possible to distinguish between structures of similar transparencies. Phase-contrast optics when used with a video microscope can be used to observe time-lapse imaging of cell division or cellular movement (Hadjidemetriou et al. 2008).

In fluorescent microscopy, the shorter wavelength is used to illuminate the specimen either stained with fluorescent dyes or, in some cases, inductively expressed fluorescent proteins already present in cells or tissues, and a longer wavelength is emitted by the specimen that is captured by the objective lens or detector. This allows creative visualization as well as a resolution of the sample (Sanderson et al. 2014). Confocal microscopy is developed to bypass some limitations of conventional fluorescent microscopy by minimizing signal and maximizing resolution with point illumination. Confocal microscopy was used to illustrate the relationship between *Azospirillum brasilense* SP7 and wheat root by which bacteria, root tissue, and mucilaginous layer can be localized precisely and XY or Z scan image can be used to show them. It is also the first example of the use of confocal microscopy in microbial ecology (Cardinale 2014).

Maximum magnification in optical microscopes is typically 1000× while electron microscopy gave a new height to microscopy with a maximum magnification of 1,000,000× (Vernon-Parry 2000). Mainly, there are two types of electron microscopy: scanning electron microscopy (SEM) and transmission electron microscopy (TEM). SEM creates a projection of the sample by scanning the surface of the sample with an electron beam focused on it where the specimen excited by the electron beam generates secondary electrons that will be captured by the detector to create a three-dimensional (3D) image of the specimen. SEM also gives the depth effect to the sample, which allows us to study the surface characteristics and shapes of the microbes very easily (Vernon-Parry 2000; Elad et al. 1983). Hyphal interaction of *Trichoderma harzianum* with *Rhizoctonia solani* during the infection of prior on the latter was studied using SEM (Benhamou and Chet 1993). TEM, on the other hand, transmits the electron through ultra-thin specimens and these transmitted electrons are captured by the detector to create an image of the specimen. TEM can also create a 3D micrograph of the specimen and can obtain a magnification of up to 0.2 nm (Singh et al. 2021). TEM has been used to study the structure of the Cucumber Mosaic Virus (CMV). For the study of the structure of viruses, TEM is a better option (Khan et al. 2011).

Though microscopy is one of the primitive techniques, it remains as important in the modern world.

17.3.2 *Biochemical Techniques*

17.3.2.1 Immunoassays

Immunoassays can be categorized into two broad groups: (1) Labeled, and (2) Direct.

Labeled Immunoassays

This includes enzyme-linked immunosorbent assay (ELISA), radioimmunoassay (RIA), and immunofluorescence (Dass and Mallick 2019). Advance science many times uses biochemical and serological methods such as ELISA, which is mainly used for the detection and quantification of specific substances such as mycotoxins, viral proteins/particles, and other such compounds that are able to perform an antigen/antibody reaction (Zhang and Vrient 2020; Zheng et al. 2006). It is a specific, sensitive, relatively rapid, and cheap technique (Zhang and Vrient 2020). Mainly, three variations of ELISA exist: (1) direct, (2) indirect, and (3) sandwich. ELISA mainly includes the use of: antibodies specific to the antigen or secondary antibodies specific to primary antibodies; plant extract suspected to contain, or soil sample containing, the target antigen; and a substrate and an enzyme that changes the color of the substrate. When the enzyme is added, it changes the color of the substrate and produces a spectrometric signal and the target compound is detected by spectrophotometric determination. In direct ELISA, an antigen-containing buffer solution is added to the microtiter plate, and enzyme-conjugated antibodies specific to the target antigen are added subsequently, which creates an antigen–antibody complex. The substrate when added, changes color if the target antigen is present in the sample (Hema and Konakalla 2021; Fenner et al. 1987). The indirect ELISA, in addition to direct ELISA, includes the addition of primary antibody after the antigen has been added to microtiter wells. Enzyme-conjugated secondary antibodies will then be added, which is specific to primary antibodies and changes the color of the substrate added subsequently (Hema and Konakalla 2021; Varma and Singh 2020). In sandwich ELISA, the first step is to add primary antibodies in the wells. Then the suspected sample containing antibodies will be added, which follows enzyme conjugates and the substrate and the result are interpreted the same as above, spectrophotometrically (Varma and Singh 2020). It is also referred to as a triple-antibody sandwich (TAS) when the secondary antibodies are used to detect the primary antibody bound to the antigen. The sandwich ELISA technique can be further divided into double-antibody sandwich (DAS) ELISA, triple-antibody sandwich (TAS) ELISA, and Protein A sandwich (PAS) ELISA (Fenner et al. 1987;

Serçe and Ayyaz 2020). Many pathogens such as *Phytophthora cinnamomi*, the causative agent of turf disease, *Pythium* spp., *Rhizoctonia solani*, and *Sclerotinia homoeocarpa* are detected using ELISA (Dass and Mallick 2019).

Radioimmunoassay (RIA) is a technique similar to ELISA with the label of the radioactive compound, commonly ^{125}I but also ^{131}I or tritium (^{3}H), instead of enzymes in ELISA. Procedure and variations remain the same as ELISA involving direct and indirect methods except for detection, which is made under gamma counter as the radioactive label has been used (Kricka and Park 2014; Hema and Konakalla 2021). Some pathogens including *Botrytis cinerea* are detected using RIA (Dass and Mallick 2019).

Immunofluorescence is another much similar technique except that it uses fluorophores instead of enzymes as label and fluorescence microscopy or confocal microscopy is used to detect the labeled antibodies (Joshi and Yu 2017).

Direct Immunoassays

Direct immunoassays include immunoprecipitation, immunoagglutination, and immunodiffusion.

Immunoprecipitation is used to detect a specific protein present in the sample. Antibodies specific to the protein of interest are allowed to incubate with agarose beads so that agarose beads become coated with antibodies. Now, these antibody–agarose bead complexes are incubated with the sample so that the protein of interest can make the complex with the antibodies added. Co-immunoprecipitation is also used mainly to recognize the protein–protein interaction between the protein of interest we have isolated and other proteins detected by immunoblotting (Pazour 2009). This technique is useful when we are looking for specific proteins such as enzymes or hormones or toxins or antigens present in the pathogen and its interaction with other proteins (Corthell 2014).

Immunodiffusion is the other simpler technique that involves pouring the antigen in one and antibody in another well created on the agar media; they diffuse toward each other due to porous nature of agar media (Licata et al. 2016) and form a precipitation band in between the wells. This simple technique is used to check the presence of specific pathogens in a sample (Fenner et al. 1987).

Microscopic carrier particles (referred to as latex) are coated with antibodies in immune-agglutination protocol that are then incubated with the sample suspected for containing the antigen. If the antigen is present, agglutination reaction takes place here indicating the presence of the antigen. This method is called direct immune-agglutination. Indirect immune-agglutination is performed by the immobilization of antigen on latex particles and antibodies are detected based on the agglutination reaction (Molina bolívar and Galisteo gonzález 2005).

Immunoassays are used when there are possibilities of antigen–antibody reaction, mainly in plant pathology, which should indeed be an area of interest when we are talking about agriculture. For better growth of commercially important plants and better agricultural production, it is important to prevent infections in crops, and in this type of studies, immunoassays are very important for the detection of pathogens in crops.

17.3.3 *Molecular Techniques for Detection of Plant–Microbe Interaction*

The study of plant–microbe interactions is critical for the use of microbiomes, which play a critical role in sustainable agriculture. Knowledge of the mechanisms underpinning plant–microbe interactions in the rhizosphere was previously limited due to a lack of proper approaches; now, with advances in molecular biology, our understanding of the plant–microbe interaction is gradually becoming more transparent. In this regard, some molecular diagnostic approaches have been deciphered in this area.

17.3.3.1 Polymerase Chain Reaction (PCR) Technique

Polymerase chain reaction (PCR) is a scientific technique invented by Kary Mullis that revolutionized DNA research and led to numerous scientific breakthroughs such as genome sequencing, gene expression in recombinant systems, and the study of molecular genetic analysis (Singh et al. 2014). The polymerase chain reaction, also known as molecular photocopying, is a simple and low-cost process for amplifying or copying a specific piece of DNA (Singh et al. 2014).

The polymerase chain reaction (PCR) technology is extremely practical, and its main premise is to multiply certain DNA sequences based on their molecular structure. Denaturation, annealing, and the extension stage are the three simple processes that make up the overall process. The single-stranded DNA molecules are obtained in the denaturation step at 94 °C (Singh et al. 2014; Gachet et al. 1998). At lower temperatures such as 55 °C, short DNA molecules named primers bind to flanking regions of the target DNA during annealing. Finally, extension occurs at 72 °C, and new strands are synthesized with the assistance of DNA polymerase (Gachet et al. 1998; Singh et al. 2014). The increase in target DNA sequences results from a number of cycles carried out in a fully automated machine known as a thermocycler, which is designed to modify the temperature of each phase and may be finished in a couple of hours (Rajalakshmi 2017). Without PCR amplification, studying large amounts of specific isolated bits of DNA at the genetic and molecular levels in prokaryotic and eukaryotic organisms is nearly impossible.

Although procedures have altered with considerable increases in performance that have contributed to the emergence of multiple PCR variants, the core concepts of PCR have remained consistent over time. There have been several variations of PCR available over the years based on their applications, but the most prevalent varieties of PCR are reverse transcriptase-PCR (RT-PCR), real-time PCR, or quantitative PCR (qPCR), and RT-PCR/qPCR combined (Singh et al. 2014; Rajalakshmi 2017).

PCR variants such as reverse transcriptase-PCR (RT-PCR), real-time PCR, quantitative PCR (qPCR), and RT-PCR/qPCR combined have been created to achieve simultaneous identification of several bacteria, and differentiation and quantification of live bacterial cells. Quantitative PCR, often known as real-time PCR, is a technique for amplification and measurement of PCR products generated during each cycle of the process (Mo et al. 2012), while reverse transcriptase-PCR is a refined PCR technique in which RNA molecules are transformed into complementary DNA (cDNA) molecules before being amplified by PCR (Mo et al. 2012). The reverse transcriptase-quantitative PCR (RT-qPCR) technique represents a big step forward in the field of PCR. It was created with the goal of quantifying RNA expression using reverse-transcriptase PCR combined with qPCR (Singh et al. 2014).

17.3.3.2 16S rRNA Gene Sequencing for Bacterial Identification

The bacterial identification was harder to achieve in an evolutionary context than the macro-organisms. Primarily, to identify the organisms standard approaches such as changes in growth, shape, enzyme activity, and so on are typically used. However, due to tiny size of microbes, their phenotypic characteristics have a limited range of expression and typically fail because the phenotype is susceptible to interpretation biases. The restrictions, in ordinary strategies, were eliminated by the new molecular biology science procedures for bacterial identification. Notwithstanding, 16S ribosomal RNA (rRNA) genome sequencing has become a cornerstone for bacterial identification as they can recognize as well as propose the phylogenetic connection between microorganisms (Petti et al. 2005). Research of bacterial phylogeny and taxonomy using 16S rRNA gene sequences is a common practice because of a number of unusual properties like its presence in almost all bacteria, often as a multigene family, or operons; its sequence is sufficiently conserved to accurately evaluate random sequence changes in time (evolution) and contains areas of moderated, variable, and hypervariable sequence regions. The 16S rRNA quality (1500 bp) is sufficiently enormous to contain adequate data for identification and phylogenetic examination (Janda and Abbott 2007). Standard PCR clone sequencing, as well as Sanger sequencing, 454 pyrosequencing, and the PhyloChip, a specialized microarray for 16S surveys, are some of commonly used techniques for bacterial identification and taxonomic classification of bacteria.

Throughout the most recent couple of years, various examination by researchers have developed massive collection of genomically sequenced bacterial strains isolated from plant roots or rhizosphere soil (Jacoby and Kopriva 2019). The discovery of the plant microbiome allows building microbial networks to demonstrate stability and adaptability allowing them to fulfill specific capacities. Sequenced rhizosphere diversity might be explored to improve our functional knowledge of plant–microbe interactions, which could then be utilized in the field to increase plant productivity.

Due to a convergence of methodological advances, 16S is re-emerging as a standalone molecular tool.

17.3.3.3 Next-Generation Sequencing (NGS)

With the headway in molecular biology, next-generation sequencing technology has proven to be a revolutionary change in sequencing technology, allowing for greater output reads and the production of enormous sequence data at a reduced cost. Instruments utilized for NGS can be categorized as second-generation and third-generation sequencing technologies (Liu et al. 2012). Second-generation sequencing technology includes instruments from Roche 454, Illumina, and Life Technologies. SOLiD—Sequencing by Oligonucleotide Ligation and Detection—and Ion Torrent sequencers are the most common second-generation sequencing technologies. The PacBio RS is the sole commercially available third-generation sequencing system from Pacific Biosciences (Schadt et al. 2010; Kumari et al. 2017).

In the field of agrobiology, researchers can examine the microbiome associated with the plant and the interconnection among them employing NGS, which can also include utilizing transcriptomics, genomics, and metabolomics studies of microbes, which will eventually enable the identification of the framework for their subsistence and interconnections (Kumar and Sharma 2020). Many studies have underlined the importance of NGS; for example, *Gammaproteobacteria* and *Betaproteobacteria* were recognized in the soil to control *Rhizoctonia solani* infection, according to NGS technology used on sugar beet (Kumari et al. 2017; Mendes et al. 2011). Using NGS, taxonomic profiling investigations in the plant rhizosphere have attempted to obtain varied information about plant species, genotype, soil type, different growth stages, and various microbial community structures (Gupta et al. 2021).

17.3.3.4 CRISPR/Cas9

Understanding the basic concepts of the plant-specific microbiome seems to be a viable technique for its application in agriculture, as the plant-associated microbiota greatly influences the host's phenotype. In addition, future plant microbiome applications will require research into microbial and plant genes implicated in plant–microbe interactions (Shelake et al. 2019; Sarma et al. 2021).

Multiple intriguing genome-editing (GE) technologies (also known as genome engineering) have evolved in recent years as a strong tool for changing an organism's genome to explore gene function with precision and efficiency. The targeted genome editing is done by using programmed meganucleases to create double-stranded breaks (DSBs) in specific genes of interest. The formation of nuclease-induced double-stranded breaks (DSBs) is the starting point for targeted DNA modifications. This event activates the cellular DNA repair mechanism, which works on the double-stranded break to repair it by non-homologous end joining (NHEJ) or homology-directed repair (HDR). Insertions, deletions, substitutions, and DNA recombination may occur during the process (Zhang et al. 2017). Specific-target genetic modifications can be achieved in a variety of ways, but transcription activator-like effector nucleases (TALENs), zinc finger nucleases (ZFNs), and the clustered regularly interspaced short palindromic repeats (CRISPR)/CRISPR-associated (Cas) system have been the most widely used meganucleases (Shelake et al. 2019). These meganucleases, which include transcription activator-like effector nucleases (TALENs), zinc finger nucleases, and others, can be used to make genetic modifications in the early stages of genome-editing technology (Razzaq et al. 2019). The encroachment of genome-editing tools solves challenges such as time consumption and the need for long processes to achieve target specificity. In comparison to ZFNs and TALENs, CRISPR/Cas9 system is more adventurous in terms of simple design, versatility, cost-effectiveness, greater efficiency, multiplexing, and specificity (Razzaq et al. 2019).

CRISPR/Cas9 system has proven to be a game-changing genome-editing method with a wide range of potential applications in agriculture around the world. The most powerful genome-editing tool, CRISPR-associated 9 (Cas9) nuclease, is a newly found nuclease that was first recognized as an RNA-mediated adaptive immune mechanism in bacteria and archaea against viral invasion (El-Mounadi et al. 2020). CRISPR/Cas systems are commonly split into two classes, the first of which consists of multiprotein effector complexes further subdivided into types I, III, and IV, and the second of which consists of a single effector protein (types II, V, and VI). The Class 2 CRISPR/Cas9 system is one of the most well-known and widely employed for nucleic acid alteration (Li et al. 2020). The widely known CRISPR/Cas9 system, which relies on a single Cas protein that targets the specific DNA sequences and was developed from *Streptococcus pyogenes* for genome editing, is the most commonly used for the construction and usage of CRISPR-based tools due to its promising efficiency (Li et al. 2020). CRISPR/Cas9 is an appealing gene-editing technique because it targets specific DNA sequences. A single-stranded guide RNA (sgRNA) and a Cas9 endonuclease are the two key components of the CRISPR/Cas9 system. The sgRNA often contains a special 20-base-pair sequence that is meant to complement the target DNA site in a sequence-specific manner, and this must be followed by a short DNA sequence upstream known as the "protospacer adjacent motif" (PAM), which is essential for Cas9 protein compatibility (Li et al. 2020). Once sgRNA attaches to the target sequence via Watson–Crick base pairing, Cas9 accurately cleaves the DNA to cause a DSB. Following the DSB, DNA-DSB repair mechanisms commence genome repair. The CRISPR/Cas9 system can be used to make targeted genomic changes, such as small insertions and deletions, using the NHEJ or high-fidelity HDR pathways (Li et al. 2020).

Nowadays, CRISPR-based technologies have been used in a variety of plant–pathogen interaction investigations, including host responses to bacteria, fungi, oomycetes, viruses, and so on (Shelake et al. 2019). CRISPR-Cas9 has been used to improve: metabolic pathways; resistance to biotic stresses like fungal, bacterial, or viral diseases or abiotic stressors like cold, drought, and salt; and the nutritional content, yield, and crop quality (El-Mounadi et al. 2020). To understand better viral infection in plants, the bacterial CRISPR-Cas system can also be used to constrain viral genetic material employing Cas9's nuclease activity (Mahadevakumar and Sridhar 2020). CRISPR/Cas9-based disease management is proved to be a novel approach to plant disease control. The CRISPR/Cas9 technique was used to create mutations in the OsSWEET13 coding area, which resulted in increased resistance to Xoo infection caused by *Xanthomonas oryzae* pv. *oryzae* (Prabhukarthikeyan et al. 2020). The CRISPR-Cas9 technology has also been used to better understand the infection process of fungal infections through the creation of CRISPR-Cas9-assisted endogenous gene tagging (EGT) techniques (Wang and Coleman 2019). CRISPR-mediated editing of plant-microbe-associated genes involved in the production of secondary metabolite pathway is a novel and intriguing method for enhancing stable and bioactive secondary metabolite production. Today, many secondary metabolites are being used as nutrition, medications, repellents, perfumes, tastes, and coloring compounds since they are important in plant or microbial defense processes (Shelake et al. 2019; El-Mounadi et al. 2020).

Overall, CRISPR (clustered regularly interspaced short palindromic repeats)-based genome editing (GE) is a good platform for learning the foundations of PM interactions fast and enabling precise genetic alterations for increased crop productivity and disease resistance using genome-editing technology (Barrangou and Notebaart 2019; Singh and Ramakrishna 2021).

17.3.3.5 Other Approaches to Study the Plant–Microbe Interface

Other than the genome-based studies, studying RNA, protein, and metabolites has also nowadays evolved when it comes to identification as well as understanding the function and interaction of the species with another species or the environment, and the branches involved in the study is known as transcriptomics, proteomics, and metabolomics (Tan et al. 2009). An overview of various "-omics" approaches and their application for study of plant–microbe interface are given in Fig. 17.2.

The most popular among all the developing "-omics" approaches is metagenomics. Entire genomic DNA is prepared from samples, irrespective of its microbial composition, and is characterized by way of whole-genome sequencing in metagenomics, which is also known as gene-centric environmental genomics. Instead of relying on a single-genome-like traditional approaches, metagenomics looks for all genes and genomes from a microbial community at once (Martí 2019; Schloss and Handelsman 2005). The traditional genomic approach has been proven accurate in the identification and analysis of disease caused by a single genotype but when it comes to epidemiological studies, a single genomic approach might mask the population of highly similar but still distinguishable individual genotypes that may be responsible for the disease. Furthermore, a sample consisting of genotypes from different bacterial, fungal, archaeal, or eukaryotic species is to be analyzed; a single genomic approach would not be appropriate because not all the microbes are cultivable (Martí 2019). Other than that, the discovery of infections involving multiple microbes is increasing (Melcher et al. 2014). In both cases, the metagenomic study is useful to reveal the actual nature of plant–microbe or in

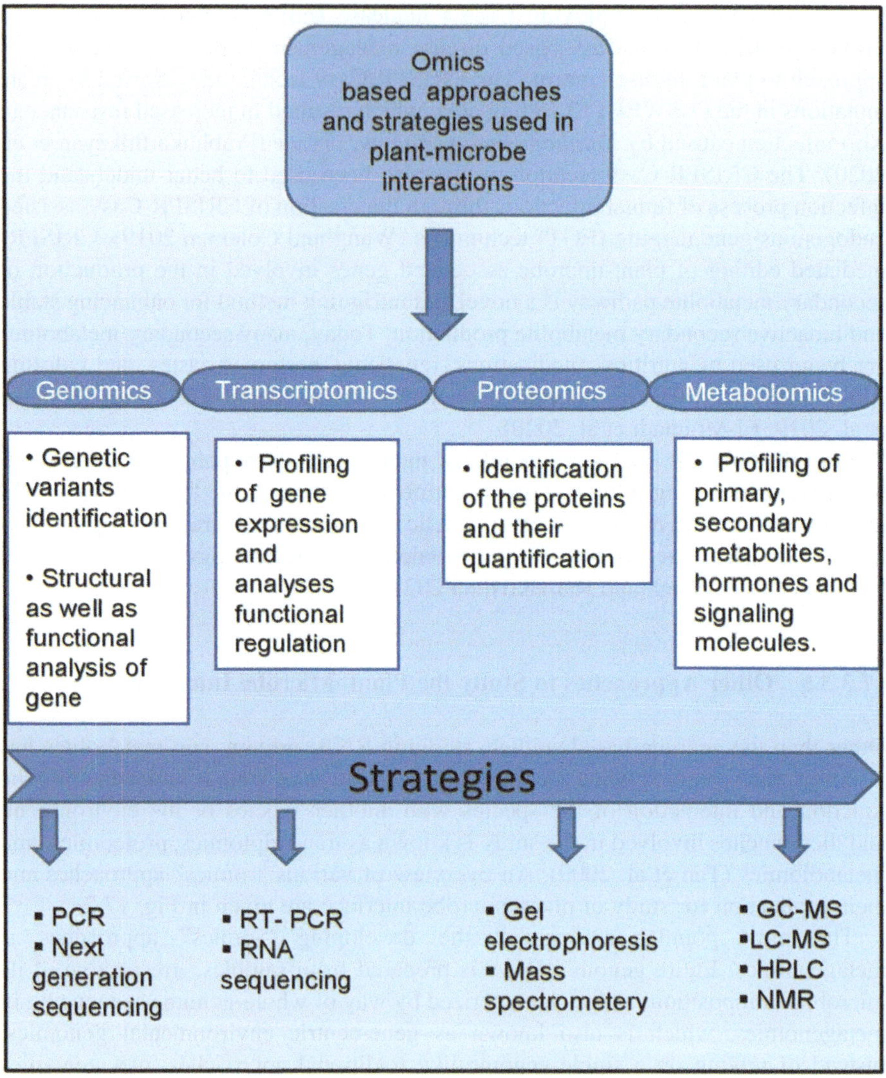

Fig. 17.2 Application of "-omics" approach to study PM interface

general host–microbe interactions (Martí 2019; Melcher et al. 2014). In spite of all the advantages of metagenomics, it should be noted that the field of metagenomics is still under development and next-generation sequencing techniques are changing the outlook of metagenomics and hence new types of bioinformatics analysis tools are expected in the future (Martí 2019). The diversity of endophytic bacteria present in maize (*Zea mays*) plants was studied using metagenomics. Fungal endophytic diversity of maize plants also has been revealed using metagenomics (Fadiji and Babalola 2020).

Apart from metagenomics, transcriptomics is also gaining momentum in the modern era. Analysis and study of the entire set of transcripts including messenger RNAs (mRNAs), microRNAs (miRNAs), and long non-coding RNAs (lncRNAs), also known as transcriptome, is called transcriptomics, which is also known as quantitative gene expression profiling (Tan et al. 2009). Initially, transcriptomic studies were carried out using the high-capacity microarray technique but using microarray, it is very difficult to study transcriptome as it uses defined gene or transcript probes. The next-generation sequencing technologies provide readings of transcription boundaries precisely and unmapped transcripts without any predefined gene or transcript. However, for transcriptome analysis, array-based technology is still widely in use as the sequencing technology has some limitations including cost, availability, complexity, and error vulnerability of sequence assembly (Metzker 2010). This approach can uncover gene expression, allow us to look into pathways triggered due to biotic or abiotic stress conditions, and it can be used for diagnostics and disease profiling (Garg et al. 2016; Wang et al. 2009; Tavassoly et al. 2018), while for the study of the microbial community, meta-transcriptomics is used, where transcriptome of the whole sample is analyzed in order to study the expression and regulation profiling of complex microbial communities at a given moment and under a specific condition (Aguiar-Pulido et al. 2016). Participation of small RNAs (sRNAs) in environmental processes such as carbon metabolism and nutrient acquisition was proved by the meta-transcriptomic study (Bharagava et al. 2019; Shi et al. 2009).

Like genomics, proteomics is a study of the entire protein content of a cell, tissue, or organism, which is also known as proteomes, resulting in insights full of information on protein expression and their modulation under a specific condition (Yu et al. 2010; Holger Husi 2014). Initially, two-dimensional (2-D) gel electrophoresis was used to study proteomes but nowadays mass spectrometry is used for the proteomic study. Typically, there are two types of approaches in the study of proteomes. The first is the bottom-up approach where protein is digested into peptides and is used to study complex samples, and the approach is also known as "shotgun proteomics." One major limitation of this approach is the loss of important information on post-translational modifications and limited protein sequence coverage by the identified peptides. Another approach is the top-down approach, where the proteome is studied directly as intact proteins. Now, this method is suitable for the simpler samples and it needs higher mass accuracy for analysis (Holger Husi 2014). Proteomics can be the reliable microscope to see enzymatic activity and how it affects the plant–microbe interaction; it also helps in studying the pathogenicity of a microbe and also the response of the host plant to the invading microbe. When the pathogen *Pseudomonas syringae* was implanted in the leaves and *Pseudomonas fluorescens* WCS417r-mediated induced systemic resistance (ISR) was inserted in the roots, significant differences in transcript levels in *Arabidopsis* shoots were observed (Sharma et al. 2020). Proteomics is an effective approach for studying plant–fungus pathogenic interactions. Proteomic study of tomato–*Cladosporium fulvum* interaction uncovers the first avirulence gene product (avr9) in fungi. Tomato–*Fusarium oxysporum* pathosystem is another example where proteomic study allowed identification of first effector of root invading fungi (Six1) (Schottens-Toma and de Wit 1988; Rep et al. 2004; Gonzalez-Fernandez and Jorrin-Novo 2012). Metaproteomics is an emerging side of proteomics. Like metagenomics, metaproteomics is an analysis of the protein samples collected from the microbial communities residing in a specific environment. Metaproteomics has the potential to unravel the genetic diversity and microbial activities in environmental communities and its impact on the ecosystem. However, uneven species distribution, large genetic heterogeneity within the microbial communities, or the broad range of protein expression levels within microorganisms are the main challenges for the metaproteomic analysis (Bharagava et al. 2019). Metaproteomic study of field-grown rice (*Oryza sativa*) showed that nitrogenase complexes containing bacteria were mainly from type II methanotrophic bacteria from the family Methylocystaceae (Bao et al. 2014).

Another reliable approach to study the plant–microbe interaction is metabolomics. Metabolomics is the study of metabolomes, the term used for all the metabolites present in an organism, tissue, or cell, which are the end products of the cellular processes (Burgess et al. 2014). In the beginning, gas chromatography-mass spectrometry (GC-MS) was used for metabolic profiling. Liquid chromatography coupled with mass spectrometry and capillary electrophoresis is used in metabolomic analysis (Burgess et al. 2014). Metabolomics holds useful applications in pathway analysis, drug discovery, and pharmacogenomics. When analyzing the environmental samples, metabolomics also includes analysis of metabolites secreted by the organisms in the immediate environment. In this way, metabolomics can demonstrate signaling processes between bacteria during communication in the community such as quorum sensing (Aguiar-Pulido et al. 2016; Bharagava et al. 2019).

17.4 Conclusion

There is a range of possibilities lying in the diversity of microbes at the plant–microbe interface and the interaction between plant and microbes as well as among microbes themselves. There are many traits like nitrogen fixation, phosphorus solubilization, pathogen resistance, or stress tolerance, which are either imparted or enhanced, directly or indirectly, by the microbes interacting with the plant, and ultimately resulting in the growth promotion of the plant. Thus, understanding those interactions can be a game changer as it may allow us to mimic the nature for imparting or enhancing particular characteristic of our interest. All the techniques described above are trails leading us to understanding and ultimately enhancing the growth promotion of plants, which can lead to economical as well as ecological benefits.

17.5 Future Perspective

From the era of agar media to the DNA writing era, humankind has not only progressed outside the Earth to the Moon and Mars but also inside the micro world, now that humankind has the capacity to edit the DNA. What seemed impossible to study some decades ago is not just possible but affordable today. Though we believe we have got way ahead in the field of science, what we have fathomed might be just a few meters of a mile deep sea. We are expecting the same in the interaction study at plant–microbe interface. We have so much data, research, and knowledge on the plant–microbe interactions, but we still have so much to explore. However, with the interaction studies opening the door of sustainable agriculture, it is fair to expect it as the answer to sustainable agriculture and also as the answer to sustainable development in the long run.

References

Aguiar-Pulido V, Huang W, Suarez-Ulloa V, Cickovski T, Mathee K, Narasimhan G (2016) Metagenomics, metatranscriptomics, and metabolomics approaches for microbiome analysis: supplementary issue: bioinformatics methods and applications for big metagenomics data. Evol Bioinforma 12:EBO.S36436

Banik A, Pandya P, Patel B, Rathod C, Dangar M (2018) Characterization of halotolerant, pigmented, plant growth promoting bacteria of groundnut rhizosphere and its in-vitro evaluation of plant-microbe protocooperation to withstand salinity and metal stress. Sci Total Environ 630: 231–242

Bao Z, Okubo T, Kubota K, Kasahara Y, Tsurumaru H, Anda M, Ikeda S, Minamisawa K (2014) Metaproteomic identification of diazotrophic methanotrophs and their localization in root tissues of field-grown rice plants. Appl Environ Microbiol 80(16):5043–5052

Barrangou R, Notebaart RA (2019) CRISPR-directed microbiome manipulation across the food supply chain. Trends Microbiol 27(6):489–496

Benhamou N, Chet I (1993) Hyphal interactions between *Trichoderma harzianum* and *Rhizoctonia solani*: ultrastructure and gold cytochemistry of the mycoparasitic process. Phytopathology 83: 1062–1062

Berg G, Smalla K (2009) Plant species and soil type cooperatively shape the structure and function of microbial communities in the rhizosphere. FEMS Microbiol Ecol 68(1):1–13

Bharagava RN, Purchase D, Saxena G, Mulla SI (2019) Applications of metagenomics in microbial bioremediation of pollutants: from genomics to environmental cleanup. In: Microbial diversity in the genomic era. Elsevier, San Diego, CA, pp 459–477

Burgess K, Rankin N, Weidt S (2014) Metabolomics. In: Handbook of pharmacogenomics and stratified medicine. Academic Press, London, pp 181–205

Cao Y, Halane MK, Gassmann W, Stacey G (2017) The role of plant innate immunity in the legume-rhizobium symbiosis. Annu Rev Plant Biol 68:535–561

Cardinale M (2014) Scanning a microhabitat: plant-microbe interactions revealed by confocal laser microscopy. Front Microbiol 5:94

Chamoun R, Aliferis KA, Jabaji S (2015) Identification of signatory secondary metabolites during mycoparasitism of *Rhizoctonia solani* by *Stachybotrys elegans*. Front Microbiol 6:353

Corthell JT (2014) Basic molecular protocols in neuroscience: tips, tricks, and pitfalls. Academic Press, London

Dass RS, Mallick R (2019) A concise compilation of the diverse detection methods to study plant-microbe interfaces at the cellular and molecular level: the past, present and future. In: Plant microbe interface. Springer, Cham, pp 291–308

Elad Y, Chet I, Boyle P, Henis Y (1983) Parasitism of Trichoderma spp. on *Rhizoctonia solani* and Sclerotium rolfsii-scanning electron microscopy and fluorescence microscopy. Phytopathology 73(1):85–88

El-Mounadi K, Morales-Floriano ML, Garcia-Ruiz H (2020) Principles, applications, and biosafety of plant genome editing using CRISPR-Cas9. Front Plant Sci 11:56

Fadiji AE, Babalola OO (2020) Metagenomics methods for the study of plant-associated microbial communities: a review. J Microbiol Methods 170:105860

Fenner F, Bachmann P, Gibbs EPJ, Murphy FA, Studdert MJ, White DO (1987) Laboratory diagnosis of viral diseases. Veterinary Virol 237–264

Gachet E, Martin G, Vigneau F, Meyer G (1998) Detection of genetically modified organisms (GMOs) by PCR: a brief review of methodologies available. Trends Food Sci Technol 9(11–12):380–388

Garg R, Shankar R, Thakkar B, Kudapa H, Krishnamurthy L, Mantri N, Varshney RK, Bhatia S, Jain M (2016) Transcriptome analyses reveal genotype-and developmental stage-specific molecular responses to drought and salinity stresses in chickpea. Sci Rep 6(1):1–15

Garg N, Bharti A, Sharma A, Bhalla S (2019) Plant-mycorrhizal and plant-rhizobial interfaces: underlying mechanisms and their roles in sustainable agroecosystems. In: Plant microbe interface. Springer, Cham, pp 27–67

Gonzalez-Fernandez R, Jorrin-Novo JV (2012) Contribution of proteomics to the study of plant pathogenic fungi. J Proteome Res 11(1):3–16

Gupta R, Anand G, Gaur R, Yadav D (2021) Plant–microbiome interactions for sustainable agriculture: a review. Physiol Mol Biol Plants 27(1):165–179

Hadjidemetriou S, Gabrielli B, Pike T, Stevens F, Mele K, Vallotton P (2008) Detection and tracking of cell divisions in phase contrast video microscopy. In: Proceeding of the third MICCAI workshop on microscopic image analysis with applications in biology

Hayat R, Ali S, Amara U, Khalid R, Ahmed I (2010) Soil beneficial bacteria and their role in plant growth promotion: a review. Ann Microbiol 60(4):579–598

Hema M, Konakalla NC (2021) Recent developments in detection and diagnosis of plant viruses. In: Recent developments in applied microbiology and biochemistry. Elsevier, London, pp 163–180

Hirsch AM (2004) Plant-microbe symbioses: a continuum from commensalism to parasitism. Symbiosis 37(1–3):345–363

Holger Husi AA (2014) Proteomics. In: Padmanabhan S (ed) Handbook of pharmacogenomics and stratified medicine. Academic Press, London, pp 147–179. https://doi.org/10.1016/B978-0-12-386882-4.00009-8

Jacoby RP, Kopriva S (2019) Metabolic niches in the rhizosphere microbiome: new tools and approaches to analyse metabolic mechanisms of plant–microbe nutrient exchange. J Exp Bot 70(4):1087–1094

Janda JM, Abbott SL (2007) 16S rRNA gene sequencing for bacterial identification in the diagnostic laboratory: pluses, perils, and pitfalls. J Clin Microbiol 45(9):2761–2764

Jha CK, Saraf M (2015) Plant growth promoting rhizobacteria (PGPR): a review. J Agric Res Dev 5(2):108–119

Jha CK, Aeron A, Patel BV, Maheshwari DK, Saraf M (2011) Enterobacter: role in plant growth promotion. In: Bacteria in agrobiology: plant growth responses. Springer, Berlin, pp 159–182

Joshi S, Yu D (2017) Immunofluorescence. In: Basic science methods for clinical researchers. Elsevier, San Diego, CA, pp 135–150

Kaushal M, Prasad R (2021) Microbial biotechnology in crop protection. Springer, Singapore. ISBN: 978-981-16-0049-4. https://www.springer.com/gp/book/9789811600487

Khan S, Jan AT, Aquil B, Haq QMR (2011) Coat protein gene based characterization of cucumber mosaic virus isolates infecting banana in India. J Phytology 3(2):94–101

Kricka L, Park J (2014) Assay principles in CLinical pathology. In: Pathobiology of human disease: a dynamic encyclopedia of disease mechanisms. Academic Press, Amsterdam, pp 3207–3221
Kumar P, Sharma PK (2020) Soil salinity and food security in India. Front Sustain Food Syst 4:174
Kumari A, Sumer S, Jalan B, Nongbri PL, Laskar MA (2017) Impact of next-generation sequencing technology in plant–microbe interaction study. In: Microbial applications, vol 1. Springer, Cham, pp 269–294
Li H, Yang Y, Hong W, Huang M, Wu M, Zhao X (2020) Applications of genome editing technology in the targeted therapy of human diseases: mechanisms, advances and prospects. Signal Transduct Target Ther 5(1):1–23
Licata NA, Mohari B, Fuqua C, Setayeshgar S (2016) Diffusion of bacterial cells in porous media. Biophys J 110(1):247–257
Liu L, Li Y, Li S, Hu N, He Y, Pong R, Lin D, Lu L, Law M (2012) Comparison of next-generation sequencing systems. J Biomed Biotechnol 2012:251364
Liu Z, Tian L, Liu S, Waller L (2014) Real-time brightfield, darkfield, and phase contrast imaging in a light-emitting diode array microscope. J Biomed Opt 19(10):106002
Mahadevakumar S, Sridhar KR (2020) Plant-microbe interaction: current developments and future challenges. In: Advances in plant microbiome and sustainable agriculture. Springer, Singapore, pp 1–38
Martí JM (2019) Recentrifuge: robust comparative analysis and contamination removal for metagenomics. PLoS Comput Biol 15(4):e1006967
Melcher U, Verma R, Schneider WL (2014) Metagenomic search strategies for interactions among plants and multiple microbes. Front Plant Sci 5:268
Mendes R, Kruijt M, De Bruijn I, Dekkers E, van der Voort M, Schneider JH, Piceno YM, DeSantis TZ, Andersen GL, Bakker PA (2011) Deciphering the rhizosphere microbiome for disease-suppressive bacteria. Science 332(6033):1097–1100
Metzker ML (2010) Sequencing technologies—the next generation. Nat Rev Genet 11(1):31–46
Mo Y, Wan R, Zhang Q (2012) Application of reverse transcription-PCR and real-time PCR in nanotoxicity research. Methods Mol Biol 926:99–112
Molina Bolívar J, Galisteo González F (2005) Latex immunoagglutination assays. J Macromol Sci Polym Rev 45(1):59–98
Pazour GJ (2009) Immunoprecipitation to examine protein complexes. Methods Cell Biol 91:135–142
Petti C, Polage CR, Schreckenberger P (2005) The role of 16S rRNA gene sequencing in identification of microorganisms misidentified by conventional methods. J Clin Microbiol 43(12):6123–6125
Prabhukarthikeyan SR, Parameswaran C, Keerthana U, Teli B, Jagannadham PT, Cayalvizhi B, Panneerselvam P, Senapati A, Nagendran K, Kumari S (2020) Understanding the plant-microbe interactions in CRISPR/Cas9 era: indeed a sprinting start in marathon. Curr Genomics 21(6): 429–443
Prasad R, Kumar M, Varma A (2015) Role of PGPR in soil fertility and plant health. In: Egamberdieva D, Shrivastava S, Varma A (eds) Plant growth-promoting rhizobacteria (PGPR) and medicinal plants. Springer International Publishing, Cham, pp 247–260
Prasad R, Gill SS, Tuteja N (2018) Crop improvement through microbial biotechnology. Elsevier. ISBN: 9780444639882
Rajalakshmi S (2017) Different types of PCR techniques and its applications. Int J Pharm Chem Biol Sci 7(3)
Razzaq A, Saleem F, Kanwal M, Mustafa G, Yousaf S, Imran Arshad HM, Hameed MK, Khan MS, Joyia FA (2019) Modern trends in plant genome editing: an inclusive review of the CRISPR/Cas9 toolbox. Int J Mol Sci 20(16):4045
Rep M, Van Der Does HC, Meijer M, Van Wijk R, Houterman PM, Dekker HL, De Koster CG, Cornelissen BJ (2004) A small, cysteine-rich protein secreted by fusarium oxysporum during colonization of xylem vessels is required for I-3-mediated resistance in tomato. Mol Microbiol 53(5):1373–1383

Saharan B, Nehra V (2011) Plant growth promoting rhizobacteria: a critical review. Life Sci Med Res 21(1):30
Sanderson MJ, Smith I, Parker I, Bootman MD (2014) Fluorescence microscopy. Cold Spring Harb Protoc 2014(10):pdb.top071795
Sarma H, Forid N, Prasad R, Prasad MNV, Ma LQ, Rinklebe J (2021) Enhancing phytoremediation of hazardous metal(loid)s using genome engineering CRISPR–Cas9 technology. J Hazard Mater. https://doi.org/10.1016/j.jhazmat.2021.125493
Schadt EE, Turner S, Kasarskis A (2010) A window into third-generation sequencing. Hum Mol Genet 19(R2):R227–R240
Schloss PD, Handelsman J (2005) Metagenomics for studying unculturable microorganisms: cutting the Gordian knot. Genome Biol 6(8):1–4
Schottens-Toma IM, de Wit PJ (1988) Purification and primary structure of a necrosis-inducing peptide from the apoplastic fluids of tomato infected with *Cladosporium fulvum* (syn. Fulvia fulva). Physiol Mol Plant Pathol 33(1):59–67
Serçe ÇU, Ayyaz M (2020) Diagnosis of the casual viruses of crop plants. In: Applied plant virology. Elsevier, pp 55–67
Sharma S, Pathak H (2014) Basic techniques of phytoremediation. Int J Sci Eng Res 5(4):584–604
Sharma M, Sudheer S, Usmani Z, Rani R, Gupta P (2020) Deciphering the omics of plant-microbe interaction: perspectives and new insights. Curr Genomics 21(5):343–362
Shelake RM, Pramanik D, Kim J-Y (2019) Exploration of plant-microbe interactions for sustainable agriculture in CRISPR era. Microorganisms 7(8):269
Shi Y, Tyson GW, DeLong EF (2009) Metatranscriptomics reveals unique microbial small RNAs in the ocean's water column. Nature 459(7244):266–269
Singh S, Ramakrishna W (2021) Application of CRISPR–Cas9 in plant–plant growth-promoting rhizobacteria interactions for next green revolution. 3 Biotech 11(12):1–25
Singh J, Birbian N, Sinha S, Goswami A (2014) A critical review on PCR, its types and applications. Int J Adv Res Biol Sci 1(7):65–80
Singh SS, Salem DR, Sani RK (2021) Spectroscopy, microscopy, and other techniques for characterization of bacterial nanocellulose and comparison with plant-derived nanocellulose. In: Microbial and natural macromolecules. Elsevier, pp 419–454
Tan KC, Ipcho SV, Trengove RD, Oliver RP, Solomon PS (2009) Assessing the impact of transcriptomics, proteomics and metabolomics on fungal phytopathology. Mol Plant Pathol 10(5):703–715
Tavassoly I, Goldfarb J, Iyengar R (2018) Systems biology primer: the basic methods and approaches. Essays Biochem 62(4):487–500
Udvardi M, Poole PS (2013) Transport and metabolism in legume-rhizobia symbioses. Annu Rev Plant Biol 64:781–805
Varma A, Swati T, Prasad R (2019a) Plant microbe interface. Springer International Publishing. ISBN 978-3-030-19831-2. https://www.springer.com/gp/book/9783030198305
Varma A, Swati T, Prasad R (2019b) Plant biotic interactions: state of art. Springer International Publishing. ISBN 978-3-030-26657-8. https://www.springer.com/gp/book/9783030266561
Varma A, Singh MK (2020) Diagnosis of plant virus diseases. In: Applied plant virology. Elsevier, pp 79–92
Vernon-Parry K (2000) Scanning electron microscopy: an introduction. III-Vs Rev 13(4):40–44
Wang Q, Coleman JJ (2019) CRISPR/Cas9-mediated endogenous gene tagging in Fusarium oxysporum. Fungal Genet Biol 126:17–24
Wang Z, Gerstein M, Snyder M (2009) RNA-Seq: a revolutionary tool for transcriptomics. Nat Rev Genet 10(1):57–63
Wu CH, Bernard SM, Andersen GL, Chen W (2009) Developing microbe–plant interactions for applications in plant-growth promotion and disease control, production of useful compounds, remediation and carbon sequestration. Microb Biotechnol 2(4):428–440

Yu L-R, Stewart NA, Veenstra TD (2010) Proteomics: the deciphering of the functional genome. In: Essentials of genomic and personalized medicine. Elsevier, pp 89–96
Zhang S, Vrient A (2020) Rapid detection of plant viruses and viroids. In: Applied plant virology. Elsevier, pp 101–109
Zhang K, Raboanatahiry N, Zhu B, Li M (2017) Progress in genome editing technology and its application in plants. Front Plant Sci 8:177
Zheng MZ, Richard JL, Binder J (2006) A review of rapid methods for the analysis of mycotoxins. Mycopathologia 161(5):261–273

Chapter 18
Plant Microbiome in Agroecosystems for Sustainable Agriculture and Environments

Songita Sonowal, Sagar Chhabra, Naga Raju Maddela, Narendra Tuteja, and Ram Prasad

Abstract Plants are occupied by an assembly of microorganisms, cooperatively called "microbiome." These microbiomes occupy various niches in plants as commensals, symbionts, or as pathogens. Microbes are referred to as phyllosphere microorganisms when they are found on a plant leaf, and rhizosphere microbes when they are present in the root system or at the soil-root interface. The microbes have been associated with various applications and are a useful green technology resource to attain Sustainable Development Goals (SDGs). Microbes are the main participant in biogeochemical cycles and help in enhancing agroecosystems' productivity by overcoming biotic and abiotic stresses. Due to rapid increase in global population, the microbiome engineering is considered useful for agriculture. This chapter is mainly prepared to focus on potentiality of plant microbiome for sustainability in agriculture and for sustainable developments.

Keywords Microbiome · Phyllosphere · Rhizosphere · Sustainability · Climate change

S. Sonowal · R. Prasad (✉)
Department of Botany, School of Life Sciences, Mahatma Gandhi Central University, Motihari, Bihar, India

S. Chhabra
Department of Biosciences, School of Basic and Applied Sciences, Galgotias University, Greater Noida, Uttar Pradesh, India

N. R. Maddela
Departamento de Ciencias Biológicas, Facultad de Ciencias de la Salud, Universidad Técnica de Manabí, Portoviejo, Ecuador

N. Tuteja
International Centre for Genetic Engineering and Biotechnology, Plant Molecular Biology Group, New Delhi, India

S. Chhabra et al. (eds.), *Plant Microbiome for Plant Productivity and Sustainable Agriculture*, Microorganisms for Sustainability 37,
https://doi.org/10.1007/978-981-19-5029-2_18

18.1 Introduction

The word microbiome is a derivative of the ancient Greek word where "micro" means small and "biome" means life (Berg et al. 2020; Santos and Olivares 2021). Microorganisms can occupy various niches in plant; for example, huge numbers of bacteria and fungi live on the surface of plants, which is known as phyllosphere, and when they occupy the soil–root interface or roots, it is known as rhizosphere or root microbiomes (Jansson and Hofmockel 2020). The region of rhizosphere is extended to about 2–80 mm from the root system; in this region, the rate of microbial growth and variation is very fast because of which it influences the root exudation of the plant. The rhizospheric microbes have various applications, like they can fix atmospheric nitrogen, help in solubilizing inorganic phosphate, help in improving iron and zinc availability, support in formation of root nodule, and promote plant growth. The microbes also help to resist against biotic and abiotic stresses and are associated to degradation of harmful environmental contaminants. A survey of different types of phyllosphere microbes has also recognized beneficial influences to plant because they protect the plant from toxic ozone rays and protect the environment from pollution (Liu et al. 2022). Microbes like *Euonymus japonicus*, *Curtobacterium*, *Marmoricola*, *Microbacterium*, *Cladosporium*, and *Alternaria* are mostly tolerant species to ozone exposure (Ding et al. 2019).

There are several latest technologies used to study plant microbiomes; for example, the use of informatics and robotics technology is currently introduced in agroecosystems to study microbiomes of plants to know the core of microbiomes and their application (Toju et al. 2018). So are the uses of agricultural techniques to solve the low agroecosystems' productivity problem. Now, people use the hydroponics system for the culture of agricultural crops apart from the conventional field production (Sharma et al. 2018). Construction of floating treatment wetland is also applicable in the agricultural practices. Though floating treatment wetland is mainly used for the remediation of wastewater, but they can also be used for agricultural purposes (Grosshans et al. 2019). To construct a floating treatment wetland, we first have to select the plant and bacteria that can grow in wastewater and form a plant–bacteria partnership because here microbes play main part in remediation of wastewater. As microbes use mechanisms like adsorption, absorption, and sedimentation, they can easily help in biofiltration of contaminants from water (Colares et al. 2020). The plant microbiome helps in improving the growth of the plant, and is useful along the plant breeding programs in maintaining agriculture sustainability (Gopal and Gupta 2016). There are several Sustainable Development Goals (SDGs) outlined by the United Nations (UN), to be achieved by the year 2030, where microbiome can be important to plants, agriculture sustainability and productivity, and environments. This chapter is mainly focused on potentiality of plant microbiome for sustainability in agriculture and for sustainable developments.

18.2 Microbiomes and Sustainability Concepts

Sustainability is a factor in economic development of a country without harming the environmental resources for future generation. The history of sustainability concept started from the time of Neolithic agricultural revolution (Dorin Paul 2008). Sustainability has three pillars, i.e., social, environmental, and economic, and they are dependent on each other (Purvis et al. 2019). The three pillars are depicted in Fig. 18.1. The primary goal of sustainability is to end poverty and hunger from the society and the main aim of sustainable development is to promote the kind of development that minimizes the environmental problems (Tilman et al. 2011). There are several Sustainable Development Goals (SDGs) outlined by the United Nations (UN), to be achieved by the year 2030, and microbiome's potential has been highlighted in several publications, for example in increasing the nutritious value of food and decreasing the vulnerability of crops to disease, resulting in increased yields, which is included under the SDG 2, i.e., "zero hunger." SDG 3, such as "healthy human consumption," includes the improvement of human microbiota to lower their risk of illness, including infectious and chronic diseases. SDG 6, "clean

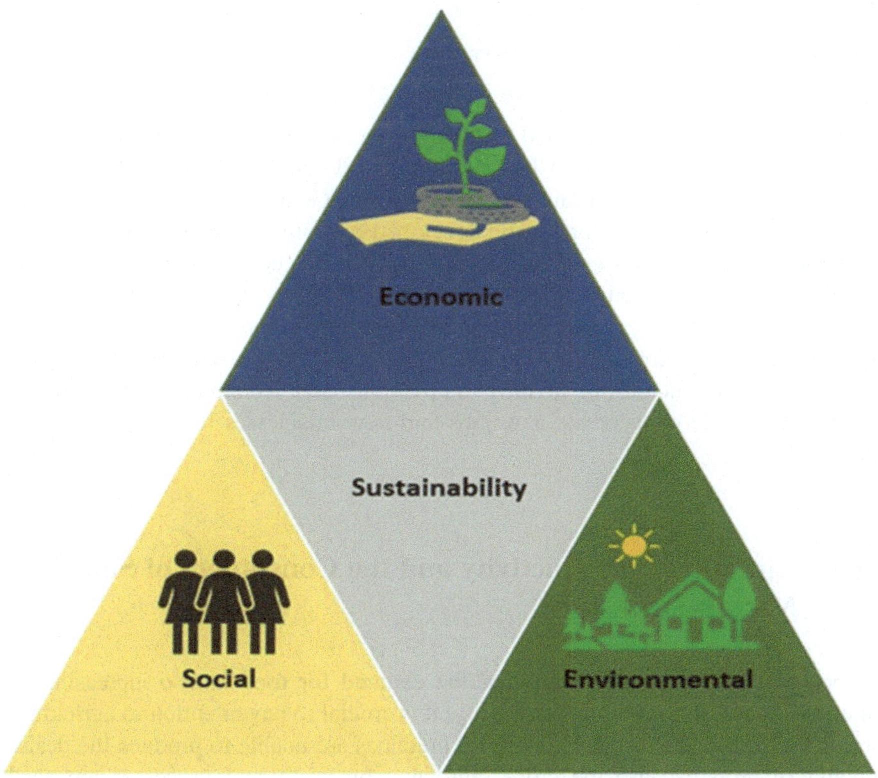

Fig. 18.1 Three pillars of sustainability development

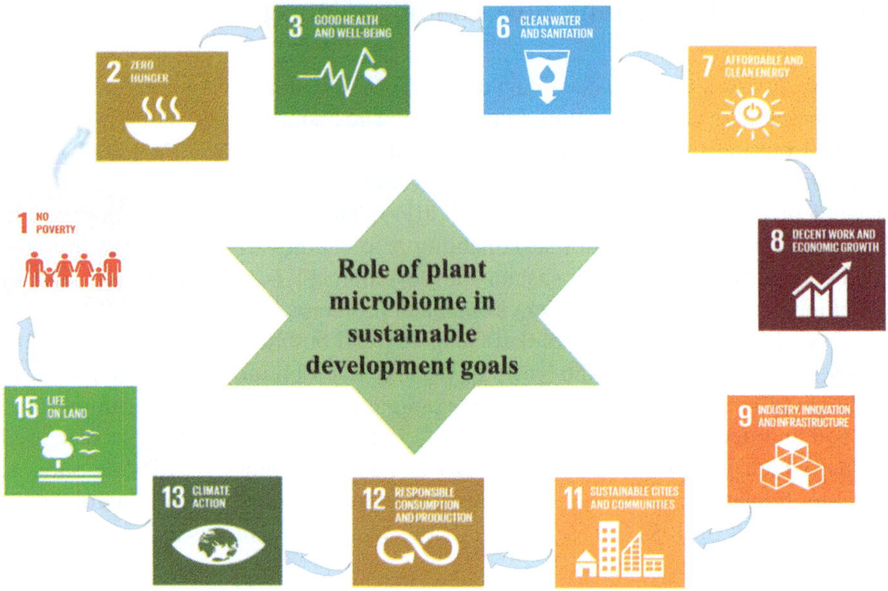

Fig. 18.2 Microbiome in sustainable development

water and sanitation," is fulfilled with the help of microbiome by purifying the contamination of water caused by the agricultural runoff. Microbiomes can play critical roles to achieve Sustainable Development Goals 7, 8, 9, 11, 12, 13, and 15 (Fig. 18.2). In both developing and developed economies, the purpose of SDGs or its implications is to remove poverty, and improve life, economic growth, industry innovation, sustainable cities, etc., and plant microbiomes can help achieve most of the SDGs in many ways. It is because the microbiomes are a crucial regulator of ecosystems and environments (Timmis et al. 2017). In present days, studies on microbiomes are increasing to have innovations and technologies on industrial microbiology side or to create new jobs and new businesses that can manage the unemployment.

18.3 Agriculture Productivity and the Constituent of Soil Microbiome

As population is growing so rapidly, the demand for food is also increasing. To attain the demand for food as a necessity, it is crucial to pay attention to agricultural productivity. But traditional agricultural practices are unable to produce the desired amount of crop because the crops cultivated by traditional processes are easily affected by the diseases and pests. Also, with the scarcity of land due to rapid growth of global population, it is not possible to produce the required amount of crop and food to reduce the hunger of people. Therefore, various modern agricultural techniques are applied to produce the crop. For example, techniques like genetically modified technique, precision farming, hydroponics, terrace gardening, etc. have evolved in recent era. In these techniques, the associations of microorganisms are must and needed. Plant microbiome in agroecosystems helps in supplying nutrients to plants, improving water uptake, and improving plant growth & productivity. The microbes applied externally to the agroecosystem can help the plants to survive in stress conditions (Jat et al. 2021).

18.3.1 Specialization of Microbes in Soil Fertility and Improving Agricultural Productivity

Microbes strengthen the health of soil by increasing the physical and chemical activities. The physical and chemical activities among the microbes and plants regulate the pH and fertility of soil (Kumar and Verma 2019). The plant microbiome of rhizospheric region, like mycorrhiza, helps in fixation of atmospheric nitrogen, enhances the growth of crop in agriculture, and reduces the environmental pollution. Plants need both macro and micro nutrients in a limited amount for the growth of plants, and microbes like bacteria, fungi, mycorrhizae, and actinomycetes provide those essential elements to the plant (Prasad et al. 2020). The rhizosphere region of plants is known as the hot spot of microorganisms because in this region the diversity of microorganism is very high. In this region microorganisms are united to form colony or biofilms to increase the biogeochemical process (Shrivastava et al. 2014; Prasad et al. 2015; Basu et al. 2021). Plants provide required nutrients like carbon rich food and in return microbes mine the soil for mineral to the plants from which both are equally benefited from each other in associations. Microbes like *Rhizobacteria*, *Cyanobacteria*, *Bacillus radiocola*, etc. can fix the atmospheric nitrogen in plants. As nitrogen is the base product of amino acid, protein, chlorophyll, hormones, and other vitamins, it is abundantly required by the plants. There are only a few prokaryotic microbes that can fix the atmospheric nitrogen to the plant (Pagano and Miransari 2016).

Like nitrogen, carbon is also another essential component for plant growth and without carbon the survival of a plant is unimaginable. In most soils, microorganisms help the soil to maintain their humus by decaying the soil organic matter like cellulose, lignin, hemicellulose, chitin, and lipids into carbon. Soil organic carbon is like a sink for the plant to supply nutrients (Khatoon et al. 2017) and microbes make it available in soil by recycling the waste organic product. Phosphorus is another essential element that is required by the plants in their metabolism and proper functioning. But in environment it is found in organic and inorganic forms and microbes can make it accessible such as by mineralization or solubilization process. There are different types of bacteria and fungi available in the soil, which can

Table 18.1 Microbes that help in soil fertility

Name of microbes	Role of microbes	References
Bacillus circulans, *Bacillus subtilis*, *Bacillus sircalmous*, *Enterobacter*, *Beggiatoa*, *Thiomargarita* sp., *Actinobacteria*, and *Micromonospora*	Help in phosphorus solubilization	Kumar et al. (2018)
Nitrosomonas europaea, *Nitrosospira briensis*, *Nitrosococcus oceanus*, *Nitrobacter winogradskyi*, *Thiobacillus denitrificans*, and *Micrococcus denitrificans*	Help in nitrogen fixation	Chen et al. (2003)
Actinomycetes, *Streptomyces*, *Nocardia*, *Trichoderma*, and *Verticillium*	Help in degradation of organic matter like cellulose, hemicellulose, lignin, lipid, etc.	Khatoon et al. (2017)
Thiobacillus sp., *Pseudomonas plecoglossicida*, *Pseudomonas aeruginosa*, *Pseudomonas fluorescens*, *Aspergillus*, *Penicillium*, *Rhizopus*, *Streptoverticillium*, and *Saccharomyces*	Help in removing heavy metal from the contaminated soil	Macaskie et al. (1992), Puranik and Paknikar (1997)

solubilize the inorganic phosphorus or organic form of phosphorus. Some of the microorganisms and their diversity are described in Table 18.1. The soil microbes are the hidden manager and participate in nutrient cycle such N, P, S, etc., and have the capacity to inhibit the soil-borne pathogens, which indirectly promotes the agricultural productivity (Sathya et al. 2016).

Due to rapid use of petroleum and mining of fossil fuel, the remediation of contaminated soil is necessary for environmental clean-up and healthy lifestyle or for sustainable environments and microbes play crucial role in bioremediation of contaminated environments (Masciandaro et al. 2013; Chhabra and Prasad 2020). Microbes can also help improve the water-holding capacity of soil by absorbing the moisture from the environment and thus help improve plant growth. Bacteria are one of the abundantly found microbes that can help to hold the water molecules in soil. It improves the soil quality by infiltration of contaminants and absorbing soil moisture (Doula and Sarris 2016).

18.3.2 Role of Microbes in Greenhouse Gas Reduction

Global warming is a serious threat in today's life because it negatively affects living organisms of earth. Rapid increases in temperature and climate change are also threats to the ecosystem, causing spread of disease, high mortality, and loss of natural habitat, which are also some of the outcomes of global warming or greenhouse effect (Kweku et al. 2018; Venkatramanan et al. 2020a, b). Greenhouse effect mainly increases due to burning of fossil fuel and biomass burning. When fossil fuels are used, it releases some toxic elements like chlorofluorocarbons, methane, and carbon dioxide. But soil microbes can convert the greenhouse gases into the usable form (Maximillian et al. 2019). The microorganisms release many enzymes that can easily convert greenhouse gases into usable form for the plant and soil. For example, microorganisms can convert CO_2 into methane with the help of enzyme methyl coenzyme M reductase; they can absorb the atmospheric methane and bring changes in the climate (Dutta and Dutta 2016).

18.3.3 Impact of Microbes in Biotic and Abiotic Stress Reduction in Plant

In agriculture, growth of plants is affected by the different types of stress like biotic and abiotic stresses. Under biotic stress, the plant is mostly affected by the different types of prokaryotic and eukaryotic microorganisms. Some pathogenic microbes like bacteria, fungi, virus, nematodes, and actinomycetes infect the plants and cause some lethal diseases that cause agricultural losses (Gull et al. 2019). For example, in a report of 2020 it was said that approximately 21.3% crop losses every year were due to the diseases caused by nematodes (Kumar et al. 2020). On the other side, abiotic stresses like drought, salinity, water logging, extreme temperature, and mineral toxicity also adversely impact on the plant and these cause losses in the production of crops. Some man-made activities, like farmers using different types of chemical fertilizers and pesticides to fertilize the soil and kill the pest, also negatively affect the normal growth of plants (Gull et al. 2019). The plant growth-promoting microbes help to lower the biotic and abiotic stresses in plant by stimulating different types of plant hormones and enzymes. For example, presence of rhizobacteria in xerophytic plant helps plants to tolerate the drought stress, stimulate the accumulation of proline, and reduce glutathione reductase activity. Another example is the bacteria associated with the foxtail millet that secrete 1-aminocyclopropane-1-carboxylic acid (ACC) deaminase and exopolysaccharide to survive in drought stress (Niu et al. 2018). Normally, plant cannot survive in high temperatures, especially above the 38 °C, but there are some microbes that can help plants survive at high temperatures, and they help the host plant by changing their physicochemical properties to adjust in adverse condition. Some of the examples of microbes that help in stress reduction of plant are given below Table 18.2, with their role in host plant. The stress tolerance potential of microbes are due to presence of stress proteins that help the microbes resist oxidative stress, high temperature, low pH, and hypoxia, but their role in plants is still unknown. A previous study revealed that common stress proteins YdaA and YnaF from *Salmonella typhimurium* help in the hydrolysis of adenosine triphosphate (ATP), cause cell membrane alteration, and cause chloride sensing of plant in stress conditions (Bangera et al. 2015).

Table 18.2 List of microbes that help in stress tolerance

Name of microbes	Host plant	Role of microbiome in biotic and abiotic stress	References
Pseudomonas aeruginosa	*Vigna radiata*	Helps plant to absorb water and nutrients	Riseh et al. (2021)
Bacillus sp.	*Megathyrsus maximus*	Helps in secretion of proline and glutathione reductase activity at the time of drought stress	Moreno-Galván et al. (2020)
Azotobacter sp.	*Triticum aestivum*	Helps in nitrogenous activity and in producing exopolysaccharide in drought stress	Saad et al. (2020)
Curvularia portuberata	*Dichanthelium lanuginosum*	Helps its host plant in heat tolerance	de Zelicourt et al. (2013)
Azospirillum brasilense	*Arabidopsis thaliana*	Reduces drought stress by inducing abscisic acid	Cohen et al. (2015)
Proteus penneri	*Zea mays*	Reduces drought stress by supplying water, protein, and sugar through the proline content to the host plant	Naseem and Bano (2014)
Pseudomonas aeruginosa			
Alcaligenes faecalis			
Klebsiella sp.	*Saccharum officinarum*	Helps in secretion of indole acetic acid (IAA) in extreme environmental conditions	Mishra et al. (2017)
Enterobacter sp.			
Bacillus sp.	*Momordica charantia*	Produces IAA to improve maize growth in Cd-contaminated soil	Ahmad et al. (2015)
Leifsonia sp.			
Enterobacter sp.			

18.4 Plant Microbial Associations for Sustainable Agroecosystems and Productivity

Investigators give importance in their study to microbiomes because they have potential applications like: they reduce antimicrobial growth, replace inorganic fertilizers and pesticides in agriculture, improve waste treatment, improve soil, help in carbon sequestration, help in biofuel production, resist and cure diet-related non-communicable diseases of human, and create sustainable business opportunities in market (Santos and Olivares 2021). Apart from this, microorganisms play a major role in ecosystems on the earth. They help to degrade the unwanted material from the earth and produce nutrients that are crucial for the growth of the plant (Hatzenpichler et al. 2020). Some of the specialized microbiomes in plant portions are described here.

18.4.1 Seed Microbiomes

Seed is the embryonic stage of a flowering plant. It is the source from where the microbiota is transferred from the one generation to the next generation (Shade et al. 2017). Different types of fungal, bacterial, and oomycete microbes are found in seed. Seed microbes are divided as endophytic (microbes growing on internal tissue) and epiphytic (microbes growing on surface of the seed). The endophytic microbes are transferred from one to the next generation (Brader et al. 2014; Nelson 2018). Presence of microbiota on seed can help adapt genotype in adverse conditions. In a previous study, it has been mentioned that the presence of microbes on seed provides plant enormous resistance capacity to fight with the pathogen and improves the metabolic activities in plant (Berg and Raaijmakers 2018). Microbes present in seed have also been recognized with antifungal and antibacterial properties, which keep the seed disease-free. For example, in a survey it was found that bacteria like *Sphingomonas* and *Methylobacterium* can assist in the growth and development of rice plant with disease-free adaptation (Eyre et al. 2019).

Among seed microbiome both pathogenic and beneficial microbes are present. The pathogenic microbes like *Pectobacterium carotovorum*, *Erwinia*, and *Pseudomonas viridiflava* cause some necrosis in plants, like in *Cucurbita pepo* and other plants. On the other hand, beneficial microbes like *Lysobacter* sp., *Paenibacillus* sp., etc. help in growth and high yield among crops (Adam et al. 2018). Apart from helping in growth and development of plant, seed microbiome can help in the management of weeds. Microorganisms present on seed help in germination and formation of the shape and structure of the plant. As the seed is the source of many microbes, it assists in the formation of immune system of a plant (Singh et al. 2020).

18.4.2 Rhizosphere Microbiomes

Many microbes exist in rhizosphere of plants. Plants secrete various kinds of exudates in the soil, which are appropriate for the development of various microorganisms. Plants provide required nutrients to the microbes and microbes provide resistance capability against pathogen and also enhance the growth and development of plants. For example, mycorrhizal organisms are an illustration of advantageous relationship with underlying foundations of higher plants. The ecto- and endomycorrhiza protect the plant root by absorbing the inorganic contaminants like heavy metals from soil and on the other hand plants provide required nutrients to mycorrhiza for the survival of the microbes (Galli et al. 1994). For the plants growing in heavy metal contaminated soil, the arbuscular mycorrhizal fungi are helpful in accumulation of nutrients and antioxidants from soil (Riaz et al. 2021). Plant roots secrete a variety of exudates like sugars, mucilage, secondary metabolites, etc. that influence the growth of different group of microbes in rhizosphere. Most commonly found rhizospheric microbes in bacteria are *Enterobacter*,

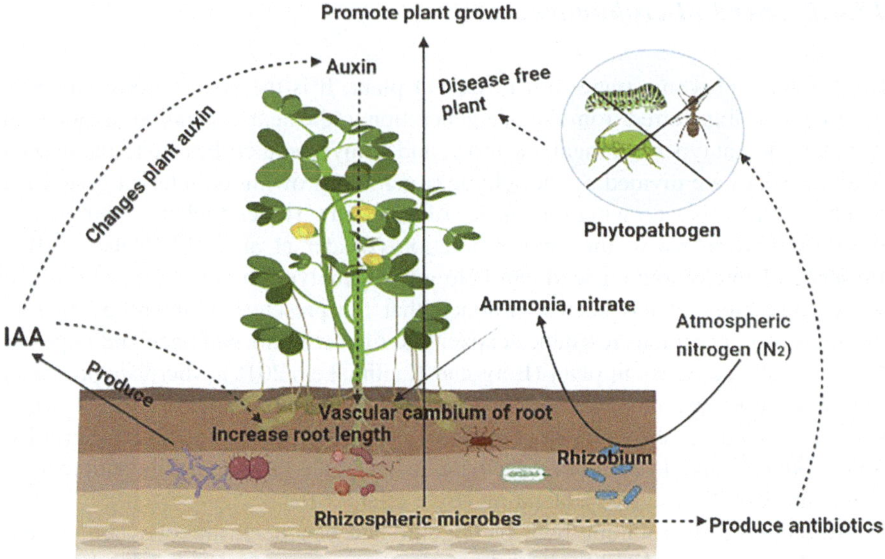

Fig. 18.3 Mechanism of rhizosphere microbes to promote plant growth and development

Rhizobium, *Pseudomonas*, *Azospirillum*, and *Burkholderia* (Babalola et al. 2021; Parray and Shameem 2020), but *Gluconacetobacter diazotrophicus*, *Bacillus amyloliquefaciens*, and *Pseudomonas putida* are also common (Parray and Shameem 2020). Among archaea, *Thermoprotei*, *Methanomicrobia*, *Halobacteria*, and *Methanobacteria* are most commonly found as rhizospheric microbes (Murthy and Naidu 2012). The rhizospheric microbes associated with plant root secrete indole-3-acetic acid that later changes the plant hormone auxin and it passes to the vascular tissue of root through the stem and increases the root length (Ali et al. 2010). These microbes are also associated with the production of antibiotics in plant, which keeps away the phytopathogen from the plant and also increases the disease resistance capacity (Fig. 18.3). Rhizosphere microbes are also responsible for the release of phytohormones such as cytokinin that help in the expansion of leaf, promotion of seed germination, and delay senescence of plant (Kaushal 2019).

Rhizospheric microbes assist in the phytoremediation of heavy metal contaminated soil by producing siderophores (Das et al. 2007; Rajkumar et al. 2010). The endophytic bacteria have also been described in literature that are well known for solubilization of minerals and release of root exudates, which help in phytoextraction of inorganic contaminants (Rajkumar et al. 2009). Rhizosphere is the hot spot that is responsible for different features and traits borne by the plants.

18.4.3 Phyllosphere Microbiomes

The microorganisms available in the aerial portion of the plant are known as the phyllosphere microbiome. The aerial surface of the plant is a habitat of different microbes and the microbes present in the aerial surface change the physiology and functioning of the plant (Bashir et al. 2022). The phyllosphere microbes protect the plants from the harmful ultraviolet radiation. For example, *Methylobacterium* sp. helps in absorbance of ultraviolet rays and it is also applied in production of pharmaceutical and cosmetic compounds (Kamo et al. 2018). Phyllosphere microbes are also responsible for the texture of leaf structure because microbes prepare the leaf surface according to their comfortability to accumulate nutrients. For example, *Pseudozyma antarctica* secretes some enzymes like esterase and xylanase that make the leaf of tomato plant thicker than the normal leaf surface and instigate the growth of plant and its management (Ueda et al. 2018). Some functions of phyllosphere microbes are given in Table 18.3. Figure 18.4 describes how phyllosphere microbes promote plant growth by reducing biotic and abiotic stresses.

18.5 The Current Approaches and Prospects of Microbiomes

Regular use of pesticides and chemical fertilizers in agricultural field causes adverse impact on the environment and hazardous influences on living organisms entering the food chain. The synthetic use of chemicals also increases soil erosion and leaching, and has been associated with barren land and agroecosystem productivity

Table 18.3 List of some phyllosphere microbes with their function

Name of microbes	Function of phyllosphere microbes	References
Microbacterium sp.	Improve growth and nutrient availability in plants by stimulating indole acetic acid and fixing atmospheric nitrogen	Abadi et al. (2020), Madhaiyan et al. (2015)
Stenotrophomonas sp.		
Methylobacterium sp.		
Pseudomonas sp.	Provide disease resistance capacity in *Arabidopsis* against *Botrytis cinerea*	Ritpitakphong et al. (2016)
Cladophora sp.	Help in N_2 fixation in *Hydrilla verticillata*	Bárta et al. (2021)
Pseudomonas syringae	Promote the growth of tomato plant	Ryffel et al. (2015)
Sphingomonas melonis	Promote plant growth	
Candida boidinii	Support methanol assimilation in plant	Kawaguchi et al. (2011)
Methylobacterium extorquens	Help in dinitrogen fixation in plant	Knief et al. (2011)

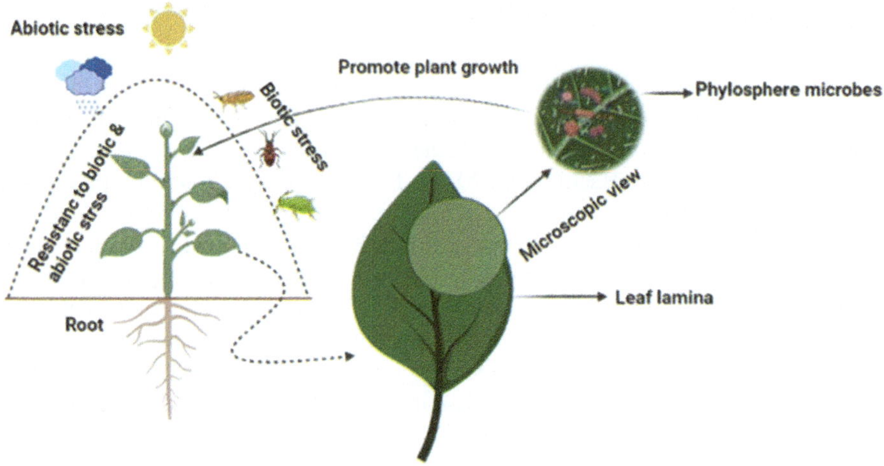

Fig. 18.4 Role of phyllosphere microbiome on the growth of plant and stress resistance

problems. Today, people are trying to lessen the utilization of chemical substances and pesticides and increase the use of green technology resources as an alternate source. Plant microbiomes are beneficial and are associated with many functions as a useful biological resource. Microbes produce different types of enzymes, biosurfactants, and siderophores, and they have the antagonistic effect/activities toward other pests and pathogenic microbes and against many abiotic stresses. Microbes are useful to attaining the sustainable development and for achieving the SDGs to overcome agroecosystems' productivity problems, and for environments, human health, and industrial production. The microbes are useful resource and shield agriculture, climate change, and help in ozone stress reduction; they are also useful for agriculture, industrial, and innovation purposes. The current approaches have unraveled range or assembly of microbes and can help improve agroecosystems' productivity and the environment for sustainable development.

References

Abadi VAJM, Sepehri M, Rahmani HA, Zarei M, Ronaghi A, Taghavi SM, Shamshiripour M (2020) Role of dominant phyllosphere bacteria with plant growth–promoting characteristics on growth and nutrition of maize (Zea mays L.). J Soil Sci Plant Nutr 20(4):2348–2363. https://doi.org/10.1007/S42729-020-00302-1

Adam E, Bernhart M, Müller H, Winkler J, Berg G (2018) The Cucurbita pepo seed microbiome: genotype-specific composition and implications for breeding. Plant Soil 422(1–2):35–49. https://doi.org/10.1007/S11104-016-3113-9/FIGURES/7

Ahmad I, Akhtar MJ, Asghar HN, Ghafoor U, Shahid M (2015) Differential effects of plant growth-promoting rhizobacteria on maize growth and cadmium uptake. J Plant Growth Regul 35(2): 303–315. https://doi.org/10.1007/S00344-015-9534-5

Ali B, Sabri AN, Hasnain S (2010) Rhizobacterial potential to alter auxin content and growth of *Vigna radiata* (L.). World J Microbiol Biotechnol 26(8):1379–1384. https://doi.org/10.1007/S11274-010-0310-1

Babalola OO, Emmanuel OC, Adeleke BS, Odelade KA, Nwachukwu BC, Ayiti OE, Adegboyega TT, Igiehon NO (2021) Rhizosphere microbiome cooperations: strategies for sustainable crop production. Curr Microbiol 78(4):1069–1085. https://doi.org/10.1007/S00284-021-02375-2

Bangera M, Panigrahi R, Sagurthi SR, Savithri HS, Murthy MRN (2015) Structural and functional analysis of two universal stress proteins YdaA and YnaF from *Salmonella typhimurium*: possible roles in microbial stress tolerance. J Struct Biol 189(3):238–250. https://doi.org/10.1016/J.JSB.2015.01.003

Bárta J, Pires De Paula CC, Rejmánková E, Lin Q, Kohoutová I, Sirová D (2021) Complex phyllosphere microbiome aids in the establishment of the invasive macrophyte *Hydrilla verticillata* (L.) under conditions of nitrogen scarcity. BioRxiv. https://doi.org/10.1101/2021.01.11.426196

Bashir I, War AF, Rafiq I, Reshi ZA, Rashid I, Shouche YS (2022) Phyllosphere microbiome: diversity and functions. Microbiol Res 254:126888. https://doi.org/10.1016/J.MICRES.2021.126888

Basu S, Kumar G, Chhabra S, Prasad R (2021) Role of soil microbes in biogeochemical cycle for enhancing soil fertility. In: Verma JP, Macdonald C, Gupta VK, Podile AR (eds) New and future developments in microbial biotechnology and bioengineering: phytomicrobiome for sustainable agriculture. Elsevier, Amsterdam, pp 149–157

Berg G, Raaijmakers JM (2018) Saving seed microbiomes. ISME J 12(5):1167–1170. https://doi.org/10.1038/s41396-017-0028-2

Berg G, Rybakova D, Fischer D, Cernava T, Vergès MCC, Charles T, Chen X, Cocolin L, Eversole K, Corral GH, Kazou M, Kinkel L, Lange L, Lima N, Loy A, Macklin JA, Maguin E, Mauchline T, McClure R, Schloter M (2020) Microbiome definition re-visited: old concepts and new challenges. Microbiome 8(1):1–22. https://doi.org/10.1186/S40168-020-00875-0/FIGURES/7

Brader G, Compan S, Mitter B, Trognitz F, Sessitsch A (2014) Metabolic potential of endophytic bacteria. Curr Opin Biotechnol 27:30–37. https://doi.org/10.1016/J.COPBIO.2013.09.012

Chhabra S, Prasad R (2020) Book review: sustainable green technologies for environmental management. J Clean Prod. https://doi.org/10.1016/j.jclepro.2020.124754

Chen G, Zhu H, Zhang Y (2003) Soil microbial activities and carbon and nitrogen fixation. Res Microbiol 154(6):393–398. https://doi.org/10.1016/S0923-2508(03)00082-2

Cohen AC, Bottini R, Pontin M, Berli FJ, Moreno D, Boccanlandro H, Travaglia CN, Piccoli PN (2015) *Azospirillum brasilense* ameliorates the response of *Arabidopsis thaliana* to drought mainly via enhancement of ABA levels. Physiol Plant 153(1):79–90. https://doi.org/10.1111/PPL.12221

Colares GS, Dell'Osbel N, Wiesel PG, Oliveira GA, Lemos PHZ, da Silva FP, Lutterbeck CA, Kist LT, Machado ÊL (2020) Floating treatment wetlands: a review and bibliometric analysis. Sci Total Environ 714:136776. https://doi.org/10.1016/J.SCITOTENV.2020.136776

Das A, Prasad R, Srivastava A, Giang PH, Bhatnagar K, Varma A (2007) Fungal siderophores: structure, functions and regulations. In: Varma A, Chincholkar SB (eds) Microbial siderophores. Springer-Verlag, Berlin, pp 1–42

de Zelicourt A, Al-Yousif M, Hirt H (2013) Rhizosphere microbes as essential partners for plant stress tolerance. Mol Plant 6(2):242–245. https://doi.org/10.1093/mp/sst028

Ding W, Jin W, Cao S, Zhou X, Wang C, Jiang Q, Huang H, Tu R, Han SF, Wang Q (2019) Ozone disinfection of chlorine-resistant bacteria in drinking water. Water Res 160:339–349. https://doi.org/10.1016/J.WATRES.2019.05.014

Dorin Paul B (2008) A history of the concept of sustainable development: literature review. Econ Sci Ser 17(2):576–580

Doula MK, Sarris A (2016) Soil environment. Environ Dev 213–286. https://doi.org/10.1016/B978-0-444-62733-9.00004-6

Dutta H, Dutta A (2016) The microbial aspect of climate change. Energ Ecol Environ 1:209–232. https://doi.org/10.1007/s40974-016-0034-7

Eyre AW, Wang M, Oh Y, Dean RA (2019) Identification and characterization of the core rice seed microbiome. Phytobiomes J 3(2):148–157. https://doi.org/10.1094/PBIOMES-01-19-0009-R/ASSET/IMAGES/LARGE/PBIOMES-01-19-0009-RT4.JPEG

Galli U, Schüepp H, Brunold C (1994) Heavy metal binding by mycorrhizal fungi. Physiol Plant 92(2):364–368. https://doi.org/10.1111/J.1399-3054.1994.TB05349.X

Gopal M, Gupta A (2016) Microbiome selection could spur next-generation plant breeding strategies. Front Microbiol 7(DEC):1971. https://doi.org/10.3389/FMICB.2016.01971/BIBTEX

Grosshans R, Lewtas K, Gunn G, Stanley M (2019) Floating treatment wetlands and plant bioremediation: nutrient treatment in eutrophic freshwater lakes AN IISD-ELA AND PELICAN LAKE, MANITOBA CASE STUDY. www.iisd.org/ela

Gull A, Lone AA, Wani NUI (2019) Biotic and abiotic stresses in plants. In: Bosco de Oliveira A (ed) Abiotic and biotic stress in plants. IntechOpen. https://doi.org/10.5772/INTECHOPEN.85832

Hatzenpichler R, Krukenberg V, Spietz RL, Jay ZJ (2020) Next-generation physiology approaches to study microbiome function at the single cell level. Nat Rev Microbiol 18(4):241. https://doi.org/10.1038/S41579-020-0323-1

Jansson JK, Hofmockel KS (2020) Soil microbiomes and climate change. Nat Rev Microbiol 18(1):35–46. https://doi.org/10.1038/S41579-019-0265-7

Jat SL, Suby SB, Parihar CM, Gambhir G, Kumar N, Rakshit S (2021) Microbiome for sustainable agriculture: a review with special reference to the corn production system. Arch Microbiol 203(6):2771–2793. https://doi.org/10.1007/S00203-021-02320-8

Kamo T, Hiradate S, Suzuki K, Fujita I, Yamaki S, Yoneda T, Koitabashi M, Yoshida S (2018) Methylobamine, a UVA-absorbing compound from the plant-associated bacteria methylobacterium sp. Nat Prod Commun 13(2):141–143. https://doi.org/10.1177/1934578X1801300208

Kaushal M (2019) Climatic resilient agriculture for root, tuber, and banana crops using plant growth-promoting microbes. In: Climate change and agricultural ecosystems: current challenges and adaptation. Woodhead Publishing, Cambridge, MA, pp 307–329. https://doi.org/10.1016/B978-0-12-816483-9.00012-8

Kawaguchi K, Yurimoto H, Oku M, Sakai Y (2011) Yeast methylotrophy and autophagy in a methanol-oscillating environment on growing *Arabidopsis thaliana* leaves. PLoS One 6(9):e25257. https://doi.org/10.1371/JOURNAL.PONE.0025257

Khatoon BHG, Khatoon H, Solanki P, Narayan M, Tewari L, Rai J, Hina Khatoon C (2017) Role of microbes in organic carbon decomposition and maintenance of soil ecosystem Bio-remediation of chlorpyrifos by bio-slurry reactor and its optimization by Data Enveloping and Response Surface Methodology View project Enzyme and Microbial Technology View project Role of microbes in organic carbon decomposition and maintenance of soil ecosystem. Int J Chem Stud 5(6):1648–1656. https://www.researchgate.net/publication/327051620

Knief C, Delmotte N, Chaffron S, Stark M, Innerebner G, Wassmann R, von Mering C, Vorholt JA (2011) Metaproteogenomic analysis of microbial communities in the phyllosphere and rhizosphere of rice. ISME J 6(7):1378–1390. https://doi.org/10.1038/ismej.2011.192

Kumar A, Verma JP (2019) The role of microbes to improve crop productivity and soil health. Springer, Singapore, pp 249–265. https://doi.org/10.1007/978-981-13-0149-0_14

Kumar A, Kumar A, Patel H (2018) Role of microbes in phosphorus availability and acquisition by plants. Int J Curr Microbiol Appl Sci 7:1344–1347. https://doi.org/10.20546/ijcmas.2018.705.161

Kumar V, Khan MR, Walia RK (2020) Crop loss estimations due to plant-parasitic nematodes in major crops in India. Natl Acad Sci Lett 43(5):409–412. https://doi.org/10.1007/S40009-020-00895-2

Kweku D, Bismark O, Maxwell A, Desmond K, Danso K, Oti-Mensah E, Quachie A, Adormaa B (2018) Greenhouse effect: greenhouse gases and their impact on global warming. J Sci Res Reports 17(6):1–9. https://doi.org/10.9734/jsrr/2017/39630

Liu J, Song M, Wei X, Zhang H, Bai Z, Zhuang X (2022) Responses of Phyllosphere microbiome to ozone stress: abundance, community compositions and functions. Microorganisms 10(4):680. https://doi.org/10.3390/MICROORGANISMS10040680

Macaskie LE, Empson RM, Cheetham AK, Grey CP, Skarnulis AJ (1992) Uranium bioaccumulation by a Citrobacter sp. as a result of enzymically mediated growth of polycrystalline HUO2PO4. Science 257(5071):782–784. https://doi.org/10.1126/SCIENCE.1496397

Masciandaro G, Macci C, Peruzzi E et al (2013) Organic matter–microorganism–plant in soil bioremediation: a synergic approach. Rev Environ Sci Biotechnol 12:399–419. https://doi.org/10.1007/s11157-013-9313-3

Madhaiyan M, Alex THH, te Ngoh S, Prithiviraj B, Ji L (2015) Leaf-residing Methylobacterium species fix nitrogen and promote biomass and seed production in Jatropha curcas. Biotechnol Biofuels 8(1):1–14. https://doi.org/10.1186/S13068-015-0404-Y/TABLES/4

Maximillian J, Brusseau ML, Glenn EP, Matthias AD (2019) Pollution and environmental perturbations in the global system. In: Environmental and pollution science. Academic Press, London, pp 457–476. https://doi.org/10.1016/B978-0-12-814719-1.00025-2

Mishra SK, Khan MH, Misra S, Dixit VK, Khare P, Srivastava S, Chauhan PS (2017) Characterisation of Pseudomonas spp. and Ochrobactrum sp. isolated from volcanic soil. Antonie van Leeuwenhoek 110(2):253–270. https://doi.org/10.1007/S10482-016-0796-0

Moreno-Galván AE, Cortés-Patiño S, Romero-Perdomo F, Uribe-Vélez D, Bashan Y, Bonilla RR (2020) Proline accumulation and glutathione reductase activity induced by drought-tolerant rhizobacteria as potential mechanisms to alleviate drought stress in Guinea grass. Appl Soil Ecol 147:103367. https://doi.org/10.1016/J.APSOIL.2019.103367

Murthy PS, Naidu MM (2012) Sustainable management of coffee industry by-products and value addition – A review. Resour Conserv Recycl 66:45–58. https://doi.org/10.1016/j.resconrec.2012.06.005

Naseem H, Bano A (2014) Role of plant growth-promoting rhizobacteria and their exopolysaccharide in drought tolerance of maize. J Plant Interact 9(1):689–701. https://doi.org/10.1080/17429145.2014.902125

Nelson EB (2018) The seed microbiome: origins, interactions, and impacts. Plant Soil 422(1–2):7–34. https://doi.org/10.1007/s11104-017-3289-7

Niu X, Song L, Xiao Y, Ge W (2018) Drought-tolerant plant growth-promoting rhizobacteria associated with foxtail millet in a semi-arid agroecosystem and their potential in alleviating drought stress. Front Microbiol 8(JAN):2580. https://doi.org/10.3389/FMICB.2017.02580

Pagano MC, Miransari M (2016) The importance of soybean production worldwide. In: Abiotic and biotic stresses in soybean production, vol 1. Academic Press, Amsterdam, pp 1–26. https://doi.org/10.1016/B978-0-12-801536-0.00001-3

Parray JA, Shameem N (2020) Molecular mechanism of plant-microbe interactions. Sustain Agric 85–136. https://doi.org/10.1016/B978-0-12-817109-7.00003-1

Prasad R, Kumar M, Varma A (2015) Role of PGPR in soil fertility and plant health. In: Egamberdieva D, Shrivastava S, Varma A (eds) Plant growth-promoting rhizobacteria (PGPR) and medicinal plants. Springer, Cham, pp 247–260

Prasad R, Chhabra S, Gill SS, Singh PK, Tuteja N (2020) The microbial symbionts: potential for thecrop improvement in changing environments. In: Tuteja N, Tuteja R, Passricha N, Saifi SK (eds) Advancement in crop improvement techniques. Elsevier, Amsterdam, pp 233–240

Puranik PR, Paknikar KM (1997) Biosorption of lead and zinc from solutions using Streptoverticillium cinnamoneum waste biomass. J Biotechnol 55(2):113–124. https://doi.org/10.1016/S0168-1656(97)00067-9

Purvis B, Mao Y, Robinson D (2019) Three pillars of sustainability: in search of conceptual origins. Sustain Sci 14(3):681–695. https://doi.org/10.1007/S11625-018-0627-5/FIGURES/1

Rajkumar M, Ae N, Freitas H (2009) Endophytic bacteria and their potential to enhance heavy metal phytoextraction. Chemosphere 77(2):153–160. https://doi.org/10.1016/J.CHEMOSPHERE.2009.06.047

Rajkumar M, Ae N, Prasad MNV, Freitas H (2010) Potential of siderophore-producing bacteria for improving heavy metal phytoextraction. Trends Biotechnol 28(3):142–149. https://doi.org/10.1016/J.TIBTECH.2009.12.002

Riaz M, Kamran M, Fang Y, Wang Q, Cao H, Yang G, Deng L, Wang Y, Zhou Y, Anastopoulos I, Wang X (2021) Arbuscular mycorrhizal fungi-induced mitigation of heavy metal phytotoxicity in metal contaminated soils: a critical review. J Hazard Mater 402:123919. https://doi.org/10.1016/J.JHAZMAT.2020.123919

Riseh RS, Ebrahimi-Zarandi M, Gholizadeh Vazvani M, Skorik YA (2021) Reducing drought stress in plants by encapsulating plant growth-promoting bacteria with polysaccharides. Int J Mol Sci 22(23):12979. https://doi.org/10.3390/ijms222312979

Ritpitakphong U, Falquet L, Vimoltust A, Berger A, Métraux JP, L'Haridon F (2016) The microbiome of the leaf surface of Arabidopsis protects against a fungal pathogen. New Phytol 210(3):1033–1043. https://doi.org/10.1111/NPH.13808

Ryffel F, Helfrich EJN, Kiefer P, Peyriga L, Portais JC, Piel J, Vorholt JA (2015) Metabolic footprint of epiphytic bacteria on *Arabidopsis thaliana* leaves. ISME J 10(3):632–643. https://doi.org/10.1038/ismej.2015.141

Saad MM, Eida AA, Hirt H, Doerner P (2020) Tailoring plant-associated microbial inoculants in agriculture: a roadmap for successful application. J Exp Bot 71(13):3878–3901. https://doi.org/10.1093/JXB/ERAA111

Santos LF, Olivares FL (2021) Plant microbiome structure and benefits for sustainable agriculture. Curr Plant Biol 26:100198. https://doi.org/10.1016/J.CPB.2021.100198

Sathya A, Vijayabharathi R, Gopalakrishnan S (2016) Soil microbes: the invisible managers of soil fertility. In: Microbial inoculants in sustainable agricultural productivity, Functional applications, vol 2. Springer, New Delhi, pp 1–16. https://doi.org/10.1007/978-81-322-2644-4_1

Shade A, Jacques MA, Barret M (2017) Ecological patterns of seed microbiome diversity, transmission, and assembly. Curr Opin Microbiol 37:15–22. https://doi.org/10.1016/J.MIB.2017.03.010

Sharma N, Acharya S, Kumar K, Singh N, Chaurasia OP (2018) Hydroponics as an advanced technique for vegetable production: an overview. J Soil Water Conserv *17*(4):364. https://doi.org/10.5958/2455-7145.2018.00056.5

Shrivastava S, Prasad R, Varma A (2014) Anatomy of root from eyes of a microbiologist. In: Morte A, Varma A (eds) Root Engineering, vol 40. Springer-Verlag, Berlin, pp 3–22

Singh BK, Trivedi P, Egidi E, Macdonald CA, Delgado-Baquerizo M (2020) Crop microbiome and sustainable agriculture. Nat Rev Microbiol 18(11):601–602. https://doi.org/10.1038/S41579-020-00446-Y

Tilman D, Balzer C, Hill J, Befort BL (2011) Global food demand and the sustainable intensification of agriculture. Proc Natl Acad Sci U S A 108(50):20260–20264. https://doi.org/10.1073/PNAS.1116437108

Timmis K, de Vos WM, Ramos JL, Vlaeminck SE, Prieto A, Danchin A, Verstraete W, de Lorenzo V, Lee SY, Brüssow H, Timmis JK, Singh BK (2017) The contribution of microbial biotechnology to sustainable development goals. Microb Biotechnol 10(5):984–987. https://doi.org/10.1111/1751-7915.12818

Toju H, Peay KG, Yamamichi M, Narisawa K, Hiruma K, Naito K, Fukuda S, Ushio M, Nakaoka S, Onoda Y, Yoshida K, Schlaeppi K, Bai Y, Sugiura R, Ichihashi Y, Minamisawa K, Kiers ET (2018) Core microbiomes for sustainable agroecosystems. Nat Plants 4(5):247–257. https://doi.org/10.1038/s41477-018-0139-4

Ueda H, Kurose D, Kugimiya S, Mitsuhara I, Yoshida S, Tabata J, Suzuki K, Kitamoto H (2018) Disease severity enhancement by an esterase from non-phytopathogenic yeast *Pseudozyma antarctica* and its potential as adjuvant for biocontrol agents. Sci Reports 8(1):1–12. https://doi.org/10.1038/s41598-018-34705-z

Venkatramanan V, Shah S, Prasad R (2020a) Global climate change and environmental policy: resilient and smart agriculture. Springer, Singapore. ISBN: 978-981-329-855-2. https://www.springer.com/gp/book/9789813298552

Venkatramanan V, Shah S, Prasad R (2020b) Global climate change and environmental policy: agriculture perspectives. Springer, Singapore. ISBN: 978-981-13-9569-7. https://doi.org/10.1007/978-981-13-9570-3

The manufacturer's authorised representative in the EU is Springer Nature Customer Service Centre GmbH, Europaplatz 3, 69115 Heidelberg, Germany. If you have any concerns regarding our products, please contact ProductSafety@springernature.com

Printed and bound by CPI Group (UK) Ltd, Croydon, CR0 4YY
30/09/2026
02225245-0013